C0-AVJ-358

Meteor Burst Communications

WILEY SERIES IN TELECOMMUNICATIONS

Worldwide Telecommunications Guide for the Business Manager
Walter L. Vignault

Expert System Applications to Telecommunications
Jay Liebowitz

Business Earth Stations for Telecommunications
Walter L. Morgan and Denis Rouffet

Introduction to Communications Engineering, 2nd Edition
Robert M. Gagliardi

Satellite Communications: The First Quarter Century of Science
David W. E. Rees

Synchronization in Digital Communications, Volume 1
Heinrich Meyr et al.

Telecommunication System Engineering, 2nd Edition
Roger L. Freeman

Computational Methods of Signal Recovery and Recognition
Richard J. Mammone et al.

Digital Telephony, 2nd Edition
John Bellamy

Elements of Information Theory
Thomas M. Cover and Joy A. Thomas

Telecommunication Transmission Handbook, 3rd Edition
Roger L. Freeman

Telecommunication Circuit Design
Patrick D. van der Puije

Meteor Burst Communications: Theory and Practice
Donald L. Schilling

Meteor Burst Communications
Theory and Practice

Edited by

Donald L. Schilling

A Wiley-Interscience Publication
JOHN WILEY & SONS, INC.
New York • Chichester • Brisbane • Toronto • Singapore

This text is printed on acid-free paper.

Library of Congress Cataloging in Publication Data:

Schilling, Donald L.
Meteor burst communications: theory and practice/Donald L.
Schilling.
p. cm. -- (Wiley series in telecommunications)
"A Wiley-Interscience publication."
Includes index.
ISBN 0-471-52212-0
1. Meteor burst communications. 2. Earth stations (Satellite
telecommunication) I. Title. II. Series.
TK6562.S5S35 1992
621.382'38-dc20 91-42445

Printed in the United States of America

10 9 8 7 6 5 4 3 2 1

CONTRIBUTORS

Tuvia Apelewicz, *SCS Telecom, Inc., 85 Old Shore Road, Suite 200, Port Washington, NY 11050*

Sheldon S. L. Chang, *Department of Electrical Engineering, SUNY at Stony Brook, Stony Brook, NY 11794*

Sorin Davidovici, *18 Riverbend Drive, North Brunswick, NJ 08902*

Les Dennis, *Alascom, Inc., 210 East Bluff Road, P. O. Box 196607, Anchorage, AK 99519*

Robert Desourdis, *SAIC, 255 Hudson Road, Stow, MA 01775*

Goran Djuknic, *Department of Electrical Engineering and Computer Science, Stevens Institute of Technology, Hoboken, NJ 07030*

Emmanuel G. Kanterakis, *SCS Telecom, Inc., 85 Old Shore Road, Suite 200, Port Washington, NY 11050*

Gary R. Lomp, *SCS Telecom, Inc., 85 Old Shore Road, Suite 200, Port Washington, NY 11050*

Lark Lundberg, Margie Dyer, *Martin Marietta Energy Systems, Inc., P. O. Box 2003, Oak Ridge, TN 37831*

Scott L. Miller, Laurence B. Milstein, *Department of Electrical Engineering and Computer Science, Mail Code R007, University of California at San Diego, Room 4805, La Jolla, CA 92093*

Jay Weitzen, *Department of Electrical Engineering, University of Lowell, 1 University Avenue, Lowell, MA 01804*

CONTENTS

PREFACE

This book is an outgrowth of research performed by the editor and the authors in the area of meteor burst communications. I was first introduced to the subject about ten years ago when I was asked to consider the meteor burst channel as an alternative means of communications to the HF channel. Most recently I was involved in the development of a meteor burst system as an alternative means of communication to a satellite system. It seems that many people either are not aware of the meteor burst channel or give it second-rate status as an "alternative." On the other hand, I think of it as the "poor person's satellite channel," since the channel is free!

The meteor burst channel offers something to everyone: To the researcher, characterizing and optimizing its performance result in numerous papers and PhD dissertations. To the military, the system is difficult to detect and/or intercept. In the commercial world, there are a myriad of applications, including truck monitoring and low-cost data communications and telemetry transfer, to mention but a few.

The chapters in this book, each written by an expert in the field with years of practical experience, touch on all of these matters. Our introductory chapter entitled "Buck Rogers Technology" introduces a technology that amazes many. Chapter Two, "Meteor Scatter Communication," presents an easily read description of this channel. Chapter Three describes the theory and modeling techniques used to simulate the meteor burst channel. Chapter Four describes the use of variable data rate transmission to efficiently communicate, and Chapter Five describes a method of implementing such a system using FM techniques. Experimental results are presented and compared with theory. Chapter Six presents some additional theoretical results obtained for the variable data rate techniques. Chapters Seven and Eight discuss the improvement obtained through the use of forward error correction coding.

I would like to thank the authors for their time and the effort they spent writing these chapters. I also thank Victoria Benzinger for her patience with me and the authors in the preparation of the manuscript. The authors and I express our gratitude for the assistance of our colleagues and friends for their helpful comments and suggestions.

DONALD L. SCHILLING

Port Washington, New York
February 1993

Meteor Burst Communications

1

BUCK ROGERS TECHNOLOGY—A STEP IN THE RIGHT DIRECTION

Les Dennnis, Lt. Col. USAF (Ret.)
Alascom, Inc., Anchorage, Alaska

When Russia put the first Sputnik into space in 1957, shock waves reverberated throughout the free world and the mad dash to the moon began. Fortunately, as a result of the space race, communications technology has taken a quantum leap forward. Because of space and weight considerations, engineering spawned the microelectronics era. The proliferation of semiconductors, chips, microchips, and superconductors made new communication technology possible with the most unlikely and unusual media. When comic strips told of space travel, rocket ships, ray runs, and other fantasies, the knowledgeable snickered and passes these ideas off as rubbish. The truth of the matter is, the cartoonists who wrote the Buck Rogers and Flash Gordon comics were just a few years ahead of their time.

Today, we know that to be true because man has walked on the moon, space travel has become a common event, high-energy lasers are used in the medical field and other areas, communication satellites carry voice and data circuits to all points of the globe, and there is meteor burst technology. Meteor burst technology? What's this? Are we bursting meteors? No, just another communications medium that until the early 1970s had been virtually ignored. Discovered in 1935 by a gentleman named Skellet, the technology was considered antediluvian in face of the then current communication systems. Basically, Skellet found that when a meteor entered the earth's atmosphere, the denser air caused the meteor to heat up and eventually burn, creating an ionized trail. He discovered that the ionized trail could be used to bounce a radio signal back to earth.

Scientists have known for decades that the earth is continually bombarded by meteors; that there are literally billions of meteors entering the earth's atmosphere daily. One hundred billion meteors have been estimated to enter or pass through the earth's atmosphere in a 24-hour period. From the 1950s through the 1970s, meteor burst technology was studied and actual tests were conducted to determine the feasibility of using the meteor trails to an advantage. The result of that research produced some interesting information.

It was found that some meteors occur in showers, but most are sporadic. Generally, meteor distribution was found to be random and roughly distributed in numbers proportional to the reciprocal of their mass. When one visualizes the earth moving through a stationary cloud of space debris, it is easy to acquire a simple picture of meteor trail distribution. Further, it was found that useful ionized trails occur in an altitude of about 80 to 120 km above the earth's surface. Trails with useful electron densities for reflecting radio signals in the range of 40 to 50 MHz were found to be plentiful enough to provide communications over a range of roughly 2000 km. The minimum range limitation was found to be 400 km, as determined by the scattering geometry and electron density. Ionized trails were found to have a lifetime of only a few tenths of a second, creating the need for rapid exchange of communication. The transmission rate had to be very fast (a burst of data if you will) to take advantage of the ionized trail. Hence the term "meteor burst" was coined.

Unfortunately, until the availability of integrated solid-state microcomputers, meteor burst as a communication medium was not considered practical except for slow-speed data systems. The rapid transmission rate of satellite systems was considered the state-of-the art, and the waiting time between meteor trails seemed too long for modern use. But interest in meteor burst did not completely die. A usable system became operational when Canada installed the JANET system between Toronto and Port Arthur in the 1950s. Another one-way link was installed between Bozeman, Montana, and Stanford, California, but the JANET system was the first complete and practical hardware implementation of meteor burst technology.

As mentioned previously, other tests and experiments were conducted in the 1960s and 1970s, but the technology was slow evolving. In the late 1970s, the Alaska SNOTEL system was installed to provide meteorological information from remote locations throughout Alaska. Because of terrain considerations, the meteor burst system was believed to be conceptually simple and relatively inexpensive in comparison to other systems evaluated.

Alaska is not heavily populated and there is no terrestrial communication grid system such as is found in populated areas. Communications to remote locations, referred to as the "bush," is via commercial satellite, which replaced the old White Alice tropospheric scatter network, originally installed and operated by the government. Road and rail networks are limited and provide connectivity to only a few locations. Most towns and villages are accessible only by aircraft or boat, and many locations can be reached only by air.

Even the Alaska state capital of Juneau is accessible only by air or ship.

There is no road system linking the capital with the rest of the state. However, the two most populated cities, Anchorage and Fairbanks, are linked by rail and highway. A microwave system was built in the early 1960s along the highway between these two cities at great expense. The microwave system was extended into the Kenai Peninsula to provide telephone service to the towns of Seward, Soldotna, Homer, and Kenai. Again, this was made possible because a road system had been built to these areas over the years. However, it must be stressed that getting from point 'A' to point 'B' in the interior can be a real problem.

Long, harsh winters exacerbate an already difficult operations and maintenance (O&M) effort. Any logistical task in the Alaskan environment can at times be overwhelming. In the 'bush', simple things like power, water systems, sewer treatment systems, and modern conveniences are nonexistent. In a word, Alaska can best be described as undeveloped. The map of Alaska, when superimposed over that of the United States, covers one-third of the entire country. The sheer size of the state explains the difficulty in providing a network like the SNOTEL net, and why meteor burst became a viable option.

The SNOTEL system consists of small, self-contained, battery-powered transceivers, placed by helicopter in remote and difficult to access locations. These are linked into the master base station located in Anchorage. Operationally, the system provides information such as temperature, humidity, and wind velocity to meteorologists. It is a very reliable system and provides a necessary service at low cost. In terms of data rate, however, it is considered very slow.

In 1980, the Alaska Air Command was gearing up to completely restyle the way they were doing business. The plan was to consolidate into the Joint Surveillance System (JSS), implement the Air Sovereignty mission into a new Regional Operations Control Center (ROCC) at Elmendorf AFB, and phase down military presence at the Long Range Radars throughout the state. Collectively, these actions were designed to save millions of dollars in the military budget (which has been proven to be the case). When the Commanding General realized his Command and Control Communications (C3) was single thread and vulnerable to outside influences, he expressed the need for a system that would provide reliable backup.

Satellite systems, although they are excellent communication mediums, are costly and fragile, and in the face of hostile actions, can easily be jammed. More importantly, the wide terrestrial footprint of the satellite enables surreptitious or clandestine intelligence operations. Obviously, a backup system would have to address these issues. In response to the commander's request, an investigation was immediately undertaken to narrow down the options that best fit the need.

It may come as no surprise to the reader that the medium ultimately selected was meteor burst (MB). Cost comparisons showed the MB system to be considerably less expensive than a microwave system. Another satellite network was not considered suitable because of the inherent deficiencies mentioned above and, of course, the cost. Experience in Alaska with high-frequency radio (HF) made that medium less than desirable. MB seemed to fit the bill. It is very difficult to intrude into a MB system because of the narrow footprint of the transmitted signal. Due to the operating frequencies (40–50 MHz), recovery in a hostile environment such as nuclear fallout is rapid: MB will begin passing traffic long before other systems under this scenario. Further, it is not easy to jam MB, and the operating distances in Alaska were well within the MB optimum capability. The decision was made to conduct tests and see just what a high-powered MB system was capable of in the Alaska theater.

Tests were conducted in 1983 with a base station located in Anchorage and a remove transceiver installed at Tin City AFS on the western coast of Alaska. Using a 5-kW base unit in Anchorage, a path was established between the two locations. The designed transmission rate was eight (8) kilobits per second (kbs) in order to complete enough messages, once a meteor trail occurred, to obtain meaningful data. With the 8-kb bit transmission rate, six messages were passed in 180 ms. Average waiting time between meteor trails was found to be 0.03 s. An interesting finding was that with a 500-W transmit power, waiting time between meteor trails increased to an average of 6 s. It became obvious that a large powered transmitter was needed to take advantage of smaller meteor trails. Based on the data collected, the decision was made to install a MB system at all of the Long Range Radar sites.

A half-duplex system was engineered and installed in 1985. Transmit power was programmed to be 6–8 kW. Because of the short construction season in Alaska, the effort to meet the winter deadline defied reason. Sightings and surveys had to be made, concrete pads constructed, utilities made available, land acquisition completed, and a myriad of support details had to be addressed.

While hardware design engineering was in progress, a team went to work on the software. Conceptualizing is one thing; putting the program into a software package is another. Many hard hours were spent putting together the necessary bits and pieces that would allow radar data to be passed over the system and integrated into the existing Regional Operations Control Center computer display system. As the software took shape, the various items of electronic controllers were being built. A special high-power amplifier was designed from the ground up. Specially designed antennas were needed that could handle not only the high-power output, but also the weather extremes that Alaska offered. A transfer switch, along with the necessary circuitry, was designed so that the antennas were "shunted" when the transmitter was activated. This kept the high-powered signal from feeding back into the equipment. The stated operational requirement made the system unique in terms of design considerations.

As the hardware became available from the manufacturer and other support items such as utilities were put in place, the task of sighting the distant terminals began. Although it would seem easy to take a topographical map and work the sightings from behind a desk, the truth of the matter was this particular phase of the program became the big problem area. It was almost a show stopper. What worked theoretically on paper was sometimes found to be just a theory. The unusual aspect of this particular effort was that what should work did not and vice versa. Path problems seemed to be directly related to distance and environmental factors. An example was that when a four-wheel ATV was left running near the antennas (to keep the engine warm), ignition noise destroyed the signal path.

Another unusual aspect of Alaska is there are many areas where electromagnetic fields make anything electronic behave in a peculiar fashion. An engineer making a sighting had to take this information into account when deciding where to locate a terminal. Those locations closer than 600 nautical miles to the master station presented more of a problem than those located at the optimum distance. In one instance, when the field terminal would not "talk" ot the master station, it was relocated a few feet to one side and the signal became strong. Subsequent investigation revealed that the signal path was actually a groundwave as opposed to a meteor trail. In this instance, the system was able to pass almost unlimited data; a definite plus.

Antenna orientation was a hit and miss proposition initially. As installers became smarter with each installation, things went much faster. As mentioned before, this particular application with meteor burst had never been tried before and the 'trial and error' method slowed the learning curve somewhat. Eventually, the system was installed and made operational just before winter set in. Data was then interfaced into the Control Center and target information displayed on the operator's scopes. Because the system is designed in a half-duplex, it is not considered to be as capable as the satellite system in terms of data throughput. However, it is better than nothing and MB has proven to be an inexpensive backup system with the ability to provide mission required communications when an interruption to the satellite system occurs.

Once the operators had radar data, the need for contacting the aircraft during a mission when the satellite was not functioning became apparent. Interfacing with the UHF radios at the radar sites, MB enabled the Elmendorf controller to key the radios remotely. A voice synthesizer interface was designed so that preprogrammed messages could be triggered via MB through the UHF radios and then on to the aircraft. To the pilot, the commands should like those heard from a modern-day robot. Yes, just like Darth Vadar! The system worked, and has been in use ever sine becoming operational in 1985. In that year, the MB system was considered state of the art. Five years later, advancements in electronic technology now make the system seem archaic. It would not surprise this writer to see a few changes made to that system to enhance throughput and reliability.

The Alaskan Air Command was the first to install a high-powered, 8 kb voice-synthesized MB system as an operational communication medium. Uniquely designed, it supports the operational requirements for Air Sovereignty in Alaska. Other uses may come to light that will parallel those of the Alaska Air Command. However, there will always be a need for inexpensive data systems within the military and in industry. Currently, NORAD is testing a C3 meteor burst network that will connect the Continental United States, Alaska and Canada.

The Alaska National Guard recently installed a MB system that ties the headquarters to remote locations throughout the state. Again, the cost of acquiring a MB system is considerably less than that of other systems, especially in the Alaskan environment. Other countries are now looking into the benefits of a MB system for specific applications, applications where great distances are involved and civil engineering support is too costly for other remote systems. A MB system has been installed between Sondrestrom AB and Thule AB, Greenland. The north–south link operates between 45 and 104 MHz. The system is a test bed to investigate performance during polar cap absorption (PCA) events.

The potential of MB in industry is just beginning to surface. An article recently appeared in *Popular Mechanics* that described a MB system for truckers to keep in touch with their home office while on the move. As a commercial offering, that kind of system could be cost effective under certain conditions. In the tactical arena, small portable transceivers could be used by the military in a variety of applications such as resupply nets, status of forces nets, intelligence nets, and others. The Navy could make use of a MB unit that was released under water and brought to the surface by buoy. This application would enable communications while minimizing detection. An oil platform out in the middle of a sea is the ideal application for MB. Red Cross and Civil Defence organizations could use portable MB systems to provide logistic support during natural disasters such as earthquake, flood, and fire. Alarm systems, emergency or logistical networks, and similar systems could be provided in remote locations at a fraction of the cost of comparable systems. Anywhere there is a need for data, MB is an inexpensive alternative that warrants consideration. The applications are limitless and await only the imagination of the far-sighted for identification.

In our fast moving technological world, we seem to be caught up in the misconception that communications must be instantaneous. The fast-paced environment we live and work in lends credibility to that thought. The cars we drive are capable of speeds faster than the safe limit; commercial airliners fly at the speed of sound; typewriters are now rated by the baud rate and not words per minute; the home computers we use are now called "turbos" and processing speeds have increased enormously. Data modems have increased their baud rate to previously unheard of speeds. The thought of having to wait .03 s for data update is considered unacceptable.

It is true that faster is better in terms of system processing times. More information is pushed through the pipe than ever before in the same amount of time. However, there are applications where "instant" is not part of the vocabulary. The Alaska SNOTEL net is a good example. Meteorological information is required on a regular basis; however, it is collected over a 24-h period. A slow data base update is acceptable and is tailored to the data collection requirement. The prerequisite for that system is that the data must get through in a timely manner. The requirement is not critical enough to demand instant throughput.

One only need analyze the basic information requirement, that is, life-cycle and time-to-live needs, to make decisions in choosing a baseline communications medium. Immediate update is not always the best answer to an equation. However, in face of the high-speed modems now coming on line, it becomes more difficult to argue that point. In spite of the popular demand for more speed, systems like meteor burst still fill special application needs. Certainly, MB systems can be designed to perform as quickly as other systems through the use of new products now available in today's market. In terms of cost, survivability, and intrusion resistance, meteor burst is hard to beat. Furthermore, it has that special Buck Rogers mystique that keeps one wondering how else MB can be used.

In recent months, meteor burst has become a very popular subject. One hears about more research in the field and more commercial companys coming on line with a product or software for MB application. More articles than ever before are appearing in trade journals and similar publications. One would think that MB was a completely new field recently discovered. It is exciting though, because the medium has largely been ignored. To some, the evolving technology of MB presents a new area of interest that has a lot of potential. The truth of the matter is, one could read the Sunday comics and get an idea of what is on the horizon. Thank goodness for Buck Rogers!

REFERENCES

Science Times 22 Aug 1989.

Communications Magazine Nov 1986.

Geophysics Laboratory Pub. GL-TR-89-0123, May 1989.

2

METEOR SCATTER COMMUNICATION: A NEW UNDERSTANDING

Jay A. Weitzen
University of Massachusetts Lowell, Center for Atmospheric Research
Lowell, Massachusetts

2.1 INTRODUCTION

The fact that ionized trails of meteors entering the earth's atmosphere will reflect radio waves has been known since the early 1930s. Only recently, however, has our improved understanding of the media coupled with improvements in technology allowed the potential of this media to be realized. What is this meteor scatter communication phenomena and how can it fit into the mix of modern telecommunications? Meteor scatter communication, operating in the low VHF band (30–100 MHz) uses reflection or scatter from the ionized trails of meteors entering the atmosphere to communicate short bursts of data. Due to the nature of the propagation mechanism, it has several advantages over conventional techniques such as high frequency (HF) and even satellites for low data rate beyond line of sight communication.

Meteor scatter communication, though intermittent, has the advantage that it can operate on a single frequency because it does not depend on the dynamic ionosphere for propagation. This can simplify antenna, rf hardware and network design. In addition, due to the higher frequency, relatively high gain antennas can be built at a fraction of the size of HF antennas. Meteor scatter communication has an inherent degree of privacy due to the relatively small ground illumination footprint of a meteor trail.

For high-latitude communication, meteor scatter is especially useful because it operates at higher frequencies than conventional HF propagation (2–30 MHz), which suffers from frequent outages due to absorption events and other high-latitude propagation phenomenon. During many polar cap absorption events, HF can be totally blacked out for a week or more, whereas meteor scatter will generally remain active.

Meteor scatter communication equipment can operate at ranges from line of sight out a single hop maximum of approximately 1000–1500 km. With multiple hop relays, the range is unlimited. The cost of meteor communication relative to other media makes it very attractive in the low data rate role. For these reasons, meteor scatter has received a great deal of attention.

Because the scattering is from a small finite length tube of ionization as opposed to a large layer as in HF propagation, the region on the ground illuminated is much smaller than for ionospheric propagation. This provides meteor scatter communication with an inherent degree of privacy and tends to make network design simpler than for ionospheric scattering systems.

Before discussing meteor scatter communication one must have a somewhat sophisticated understanding of the meteor scatter propagation phenomena. This chapter summarizes recent advances in the understanding of the media as a communication channel. The objective is to supplement rather than repeat the classic survey works by Sugar [114], McKinley [84], and Lovell [70].

2.1.1 Brief History of Meteor Scatter Communication

In the early 1930s when radio communication was in its infancy, Pickard [106] noticed that bursts of long-distance, high-frequency propagation occurred at times of major meteor showers. Skellet [112] is given credit with postulating that the propagation mechanism was reflection or scattering from electrons in meteor trails. During World War II radio engineers observed meteor trail echos, which were sometimes confused with incoming missiles. It was not until after the war when radio technology had extended into the VHF and UHF bands that radio engineers became interested in the meteor scatter phenomena.

After World War II, research proceeded in earnest toward understanding the potential of this new media, and a number of experiments were conducted in the United States, Canada, Great Britain, and the Soviet Union. These experiments established the fundamental nature of the scattering mechanism. One of the interesting properties associated with meteor scatter was the very small ground illumination footprint of individual trails, which provided meteor scatter communication with an inherent degree of privacy. After this discovery, much of the work in the area was classified and remained classified until the middle 1950s when the December 1957 issue of the *IRE Transactions* was devoted to the subject of meteor scatter.

Pineo of the National Bureau of Standards is credited by Montgomery and Sugar [92] with proposing in 1951 that scatter from the ionized trails of meteors entering the atmosphere could be used for beyond line of sight communications. The next few years saw an explosion of research since no organization wanted to be left behind in the exploitation of this media.

One of the first experimental communication links was the JANET system established by the Radio Physics Laboratory of the Canadian Defense Research Board in 1953, which provided simple point to point Teletype communication between Ottawa and Port Arthur [23, 45, 119]. A patent was later granted to the Canadian group for the use of meteor trails as a communication media [43]. The bi-directional system transmitted on a nominal frequency of 50 MHz with a 1-MHz spacing between the two transmitter frequencies. Each station transmitted a cw probe signal with the modulation being switched on when a usable channel was detected. Each station had a 500-W rf amplifier and used antennas with approximately 12 dBi gain. As with all systems of the era, data were stored on magnetic tape or in magnetic core memory.

In early 1953 the National Bureau of Standards (NBS) established links throughout North America to measure VHF propagation, including meteor scatter and ionosphereic scatter propagation [7, 18, 54, 92, 113, 115, 136]. The first link operated between Cedar Rapids, Iowa, and Sterling, Virginia. In 1955 a second link between Erie, Colorado, and Long Branch, Illinois, was installed. Other links in Alaska and Canada were established to measure characteristics of the meteor and other VHF propagation mechanisms. NBS was the first organization to measure the multipath spread of the channel using a high-resolution (300-kHz) pulsed waveform [19]. It was shown that the channel could support wideband communication, although technology to support high-speed digital communication via meteor scatter was 20 years in the future.

Another major research link in the United States was operated by Stanford Research Institute (SRI) under contract from the U.S. Air Force. SRI operated a test link from Palo Alto, California, to Bozeman, Montana, beginning in the early 1950s. This link was used primarily for propagation research and led to some of the fundamental discoveries on the nature of the meteor propagation mechanism [33–39, 72–77, 117, 118]. Much of the theoretical and experimental physics was performed by the Harvard/Smithsonian Astrophysical Observatory [47–50, 137, 138].

As the state of technology continued to improve in the late 1950s to include the newly invented transistor, a number of applications for meteor scatter were postulated, including facsimile communication [11], remote sensing and simple point-to-point data communication. Hughes Aircraft, under contract to the U.S. Air Force, investigated the use of meteor scatter to extend the range of conventional aircraft communication [64].

The launch of *Sputnik* was the high-water mark of meteor communication research. After that, interest in satellite communication caused interest in meteor scatter communication to wane in the early 1960s. By the late 1960s interest in meteor scatter was limited primarily to scientific applications related to using meteor echoes to investigate winds in the lower ionosphere. The 1960s did see the first operational military meteor scatter communication system deployed in Europe, the COMET system [8, 56, 96] operated by NATO SHAPE Technical Center. The system provided communication from The Hague, Netherlands, to southern France. This link was the first to introduce more advanced signaling protocols such as ARQ (automatic request), in which a data packet is transmitted and then either acknowledged or a request for retransmission is made. The system was used to transmit conventional Teletype messages.

In the 1970s, rapid advances in digital technology, including the development of the first microprocessors, made small, inexpensive meteor scatter communication equipment possible. Meteor scatter due to a number of inherent properties became the natural choice for communication systems used to relay small amounts of remote sensing data from sites that could not be connected to conventional telephone communications (mobile satellite terminals had not yet been invented).

The first major application of the new meteor scatter technology for relay of small amounts of meteorological and snow depth data from hundreds of solar-powered stations located in the western United States. The system, SNOTEL, developed and fielded by the fledgling Meteor Communication Corporation and operated by the Department of Agriculture, consisted of approximately 500 microprocessor-based remote stations and two master stations organized into a star network configuration [58, 66].

By the early 1980s the perceived vulnerability of satellite and terrestrial land line communication for military applications [99] stimulated a revival of meteor scatter communication as a backup communication media. During the 1980s new research into meteor scatter as a communication channel coupled with further advances in technology led to fielding of more advanced equipment. Networks with multiple interconnected master stations capable of intercontinental range were fielded. Advanced coding techniques were incorporated into terminals to improve their performance on weak scatter channels [89, 91, 93, 109].

Theoretical work [120, 129] showed that by using techniques that match information transfer rate to the dynamic capacity of the channel, large increases in throughput could be achieved. The theoretical work was verified with the fielding of the first adaptive data rate or protocol system [65]. Improvements in antenna technology allowed meteor scatter to compete with more conventional techniques such as HF.

Due to the relatively low cost of the equipment relative to that of satellite terminals, a number of new potential applications are developing. The sections that follow describe advances in the past 20 years of our fundamental understanding of the meteor scatter communication channel and how these improvements coupled with modern technology have found new applications.

2.2 THE VHF BEYOND-LINE-OF-SIGHT COMMUNICATION CHANNEL

A radio communication channel is described by statistics of the signal amplitude in terms of duty cycle above a threshold and probability distribution, by the multipath and fading (Doppler) profiles, and by the inherent privacy and spatial multiplexing (diversity).

This section concentrates on the characteristics of beyond line-of-sight propagation mechanisms at VHF, including meteor scatter, tropospheric scatter, auroral scatter, and sporadic-E propagation. The focus is on recent advances in our understanding of the properties of the meteor scatter communication channel; background information is summarized in the 1964 paper by Sugar [114] and the 1961 book by McKinley [84].

2.2.1 The Meteor Scatter Communication Channel

2.2.1.1 Meteors and Meteor Trails

2.2.1.1.1 Origin and Flux of Meteoric Particles. The description of the meteor communication channel begins with the meteors that form the trails that form the channels. Billions of tiny meteoric particles enter the earth's atmosphere each day, ranging in size from milligram micrometeors smaller than a grain of sand to large stones that can leave craters on the surface of the earth. Physical properties of meteors and meteor orbits have been determined from visual, photographic and radar observations of the trails formed by meteors as they enter the atmosphere [2, 3, 5, 6, 20, 28, 40, 57, 59–61, 70, 80–83, 85, 86, 100, 101, 116]. The research is summarized in books by Lovell [70] and McKinley [84].

Current theory states that the majority of the meteor particles have their origin in the solar system. Before being trapped by the gravitational field of the earth, these particles orbit around the sun. Meteors are believed to be of cometary or asteroidal origin, but a substantial fraction are believed to be loosely bound agglomerates sometimes called "dustballs." The distribution of meteors is often divided into two general classes: shower meteors and sporadic meteors. Shower meteors are collections of particles all moving at the same velocity in well-defined orbits or streams around the sun. Shower meteors are believed to be the remnants of comets or asteroids. The orbits of the streams intersect the earth at the same time each year and at these times "meteor showers" are observed. The most well-known shower in the Northern Hemisphere is the Persiads shower, which occurs in August of every year. If the distribution of particles in the orbit is fairly uniform, then the size of the shower each year will be relatively constant; if the distribution of particles is not uniform, the size of the shower can vary greatly from year to year. Shower meteors account for only a small fraction of the total meteor flux; however, during the summer months in the Northern

Hemisphere, shower meteors are a significant contributor to the increase in the observed middle latitude meteor flux.

The nonshower or sporadic meteors comprise the majority of meteors of interest for radio propagation. Sporadic meteors do not have well-defined streams; instead they have a random orbital distribution. A number of researchers have used radar and optical methods [9, 24, 25, 28–31, 39, 47, 53, 61, 62, 67, 81, 85, 87, 108, 110, 111] to characterize the distribution of meteor orbits. A majority of orbits tend to group toward the ecliptic plane (the path of earth in its orbit around the sun). Higher latitude orbits are also important contributors to the overall meteor flux. Meteors tend to move in the same direction around the sun as the earth and the planets.

The radiant (the point in the sky from which the meteor appears to originate) distribution of meteors can be computed from the orbital distribution of meteors [24, 39, 47, 53, 87, 108, 110]. These distributions have been used as the basis for a number of models used to predict the arrival rate of meteor trails useful for communication [12, 27, 53, 65, 87, 110, 122]. The topic of modeling is discussed in greater detail in Chapter 3.

The meteor flux rate shows a strong diurnal variation peaking at approximately 0600 local time when the motion of the earth tends to sweep in meteors and with a distinct minimum at approximately 1800 local time when meteors (which tend to group in the same orbital direction) must overtake the earth. This well-known phenomenon is illustrated in Fig. 2.1. The ratio of the maximum flux to minimum flux is a function of latitude. The higher the latitude, the less is the variation; the lower the latitude, the greater is the diurnal variation.

A general theory of sporadic meteors is that there are approximately equal total masses of each size of particle. For example, at 10^{-3} g there are approximately 10 times as many particles as at 10^{-2} g. The velocities of particles range from 11.3 km/s (the escape velocity of a particle from the earth) to 72 km/s (a combination of 30 km/s orbital velocity of the earth and the 42 km/s escape velocity of the solar system, assuming that all meteors have their origin in the solar system).

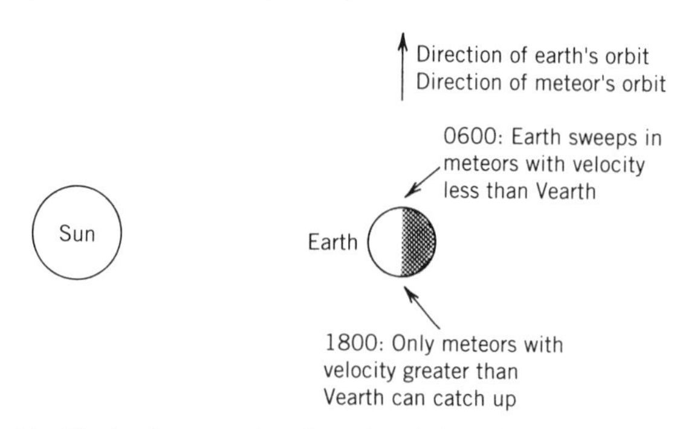

Figure 2.1. Mechanism causing diurnal variation of observed meteor arrivals.

Meteor arrivals are assumed to be independent of previous arrivals and to satisfy Poisson statistics. The probability of k trails in a time T is given by

$$p(k \text{ trails in time } T) = \frac{(\Omega(t)T)^k}{k!} e^{-\Omega(t)T} \tag{2.1}$$

where the time-varying parameter $\Omega(t)$ is the average arrival rate of meteor trails exceeding some threshold. Poisson processes have the interesting property that the variance equals the mean. Confidence intervals on estimation of the mean decrease much more slowly with increasing number of events than do standard normal processes.

The waiting time $w(t)$ between arrivals is given by the Poisson equation as

$$p_w(w) = \Omega(t)e^{-\Omega(t)w} \tag{2.2}$$

2.2.1.1.2 Formation of Meteor Trails. When a meteor enters the atmosphere, friction between the particle and the increasingly dense atmosphere beginning at approximately 120 km above the earth causes meteor particles to boil off. The particles collide with the air molecules and ionize, forming positive ions and free electrons. Radio wave scattering is primarily due to the effects of the electrons, since the positive ions are too massive to vibrate under the influence of an electric field. Immediately after formation, the trail begins to experience ambipolar diffusion, which is the primary mechanism leading to the decay of meteor trail echoes. Meteor trails are generally characterized by their electron line density, q, which in a three-dimensional cylindrical coordinate system is given by

$$q(z) = \int_{\theta=0}^{2\pi} \int_{r=0}^{\infty} N(r, \theta, z)\, dr\, d\theta \tag{2.3}$$

where N is the electron density (electrons/m^3) in a cylindrical coordinate system. The electron line density q (electrons/m) remains constant as the trail radially diffuses, whereas the electron density decreases as a function of time. Electron density N is commonly assumed to be Gaussian in the radial direction and the ionization may or may not be uniform in the axial direction. The total length of the ionization region of a trail is approximately 20–40 km from start to finish [72].

For a specular scatter communication channel to be established, two conditions must be satisfied. First, the line defining the meteor trail axis must be tangent to an ellipsoid of revolution with focii at the receiver and the transmitter. It can be easily shown that for any line in space and a pair of points on the ground, a point of tangency can be found. This brings up a second and critical requirement for coherent scatter that leads to the inherent privacy in meteor scatter communication: At the point of tangency,

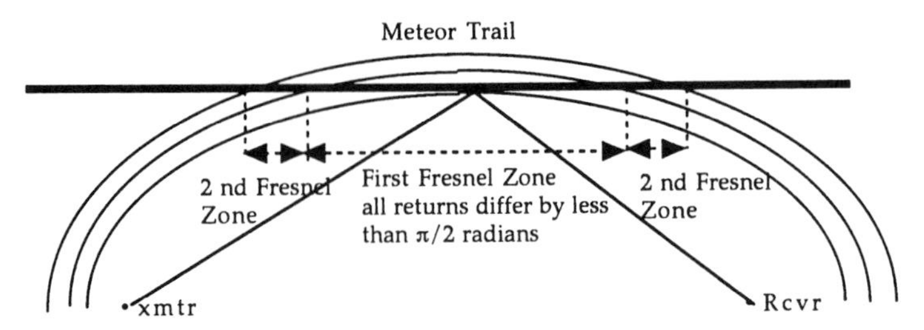

Figure 2.2. Geometry of scatter from meteor trail showing first and second Fresnel zones.

there must be significant ionization. Figure 2.2 shows a meteor trail oriented to support specular reflection. Superimposed on the figure are the Fresnel zones, or zones of constant phase. Each zone represents a region for which the phase differs by less than $\pm\pi/2$ radians.

All scatters within $\pm\pi/2$ radians of phase distance from the phase at the tangent point will contribute constructively to the echo strength. All scatters within the next region will contribute destructively, the next region constructively, and so on. Note that the length of the first region is greater than all the other regions. This region is referred to as the primary, principal, or first Fresnel zone. Scatterers within this region are the primary contributors to the echo strength.

The length of the nth Fresnel region is a function of geometry of the link, orientation of the trail, and wavelength and is given by Sugar as

$$F_n = [n^{0.5} - (n-1)^{0.5}]\left[\frac{\lambda R_R R_T}{(R_R + R_T)(1 - \sin^2\phi \cos^2\beta)}\right]^{0.5} \tag{2.4}$$

where the angles β and ϕ are described in equation 2.5. R_R is the distance from the electron column (trail) to the receiver and R_T is the distance from the electron column to the transmitter.

The length of the primary Fresnel zone ranges from 1 km for backscatter to tens of kilometers for the case of a trail oriented along the great circle path from receiver to transmitter. The greater the length of the principal Fresnel zone, the more electrons contribute to the total power and the greater is the received amplitude. The length of the Fresnel zone is dependent on frequency and is one of the reasons why meteor scatter has a strong frequency dependence. As the frequency increases, the Fresnel zone decreases in length, with correspondingly fewer scatterers contributing to the signal.

2.2.1.2 Basic Models for Received Signal Level. The meteor scatter propagation phenomena, while often assumed to be relatively simple due to the use of several closed form approximations, is actually very complicated due to the interaction of electrons and ions in the meteor trail, which is a plasma cloud with all the inherent complexity. Consider the simplest case of a trail that is modeled as a semi-infinite linear cylinder of electrons with constant electron density in the axial direction and Gaussian electron distribution in the radial direction. Assume that the electron density is low enough so that each electron can act as a Hertzian dipole vibrating in the direction of the incident electric field with no interaction between individual electrons. This model, though relatively simplistic, applies to a broad class of meteor trails and is called the classic underdense meteor propagation model. These trails are formed by micrometeors with masses from about 10^{-5} to about 10^{-3} g.

Because the model is relatively simplistic, closed form solutions for the rise and decay of these trails have been developed. Eshleman [34, 37] integrated the contribution from individual electrons over the assumed electron density profile to develop a closed form model for the received signal power as a function of time given by

$$\mathrm{RSL} = \frac{P_T G_T G_R \lambda^3 q^3 r_e^2 S}{16\pi^2 R_R R_T (R_T + R_R)(1 - \cos^2\beta \sin^2\phi)} \exp - \left(\frac{8\pi^2 r_0^2}{\lambda^2 \sec^2\phi} + \frac{32\pi^2 Dt}{\lambda^2 \sec^2\phi} \right) \tag{2.5}$$

where P_T = transmitter power
G_T = transmitter antenna gain at the point x, y, z
G_R = receiver antenna gain at the point x, y, z
R_T = distance from the transmitter to the trail (m)
R_R = distance from the receiver to the trail (m)
λ = wavelength (m)
q = electron line density (electrons/meter)
r_e = classic radius of an electron (2.818×10^{-15} m)
D = diffusion constant, which is a function of trail height
S = polarization coupling factor (see Weitzen [127])
ϕ = propagation angle formed by vectors R_T, R_R
β = angle of the trail relative to the plane formed by R_T and R_R

Underdense trails are characterized by their rapid rise and exponential decay with time. Figs. 2.3 and 2.4 show examples of meteor trails observed at 45 MHz on the 1260-km USAF high-latitude meteor scatter test bed operated by Phillips Laboratory [103, 133]. The three upper trails in Fig. 2.3 show examples of underdense trails with different amplitudes and decay rates. Also note the transient behavior at the beginning of the trail, which corresponds to the time when the meteor particle transits the first Fresnel zone, providing the large rise; then the second Fresnel zone contributes destructively, the third contributes constructively and so on. The term S in equation 2.5 represents the polarization coupling factor and has been modified by Weitzen [127] from the original work of Eshleman to incorporate arbitrary polarizations other than horizontal including generalized complex (elliptical) polarized waves.

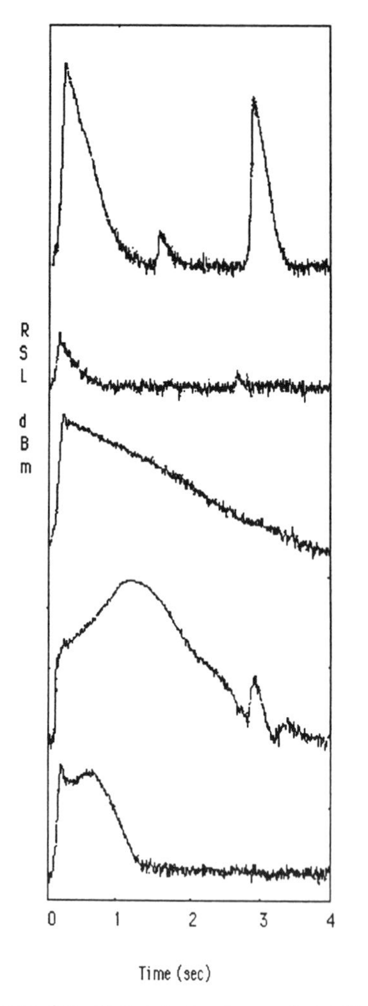

Figure 2.3. Meteor trails observed at 45 MHz on USAF High Latitude Meteor Scatter Test bed. First three waveforms: underdense trails with different amplitudes and durations. Next two waveforms: overdense meteor trails [133].

Figure 2.4. More waveforms from USAF High Latitude Meteor Scatter Test bed. First waveform: overdense meteor trail with fading: second waveform: trail with underdense and overdense characteristics; third waveform: underdense trail and overdense trail in which underdense trail has greater amplitude; fourth waveform: nonspecular overdense trail [133].

Meteors occur at random heights from 85 to 120 km above the earth. Due to the height dependence of the diffusion constant D in Eq. 2.5 the decay time is a function of the height at which the trail occurs, and recent data indicate that rate of the exponential decay of underdense trails

$$\tau = \frac{32\pi^2 Dt}{\lambda^2 \sec^2 \phi} \tag{2.6}$$

can be modeled as a Rayleigh or Log normal distributed random variable [133] with the average determined at the midlink using the technique of Brown and Williams [13].

As the electron density increases, the trail appears less like a tubular cloud of independently scattering electrons and more like a tubular plasma cloud. The assumption that each electron scatters the incident wave independently of all the other electrons no longer holds and the incident wave does not pass through the trail. As the electron density increases toward the critical density at which the effective dielectric constant becomes negative, a polarization resonance effect begins to occur which is not modeled in the classic equation of Eshleman. Enhancement of the coefficient of reflection for electric field vectors transverse to the trail axis has been predicted and observed [10, 21, 22, 68, 95]. Phase shifts are also believed to be associated with resonance.

As the electron density increases, the interactions between electrons become more and more complicated. The basic shape of the received echo changes from the simple exponential described earlier to a complicated waveform for which closed form solutions for the received signal power as a function of time do not exist. Trails for which the underdense scattering model does not hold are broadly referred to as overdense. Several numerical techniques have been developed to model overdense trails, but even these require a number of assumptions that limit their range of validity [4, 14, 44, 50, 52, 63, 75, 85, 90].

The most commonly used approximation (it should be noted that this is a very approximate approximation), developed by Hines and Forsythe, assumes that the trails are so dense that there is perfect reflection (the incident wave does not penetrate beyond the critical radius) from the surface of an expanding tubular metallic cylinder with a radius given by the critical radius at which the effective dielectric constant becomes negative. The received power due to reflection from an overdense trail approximated by Hines and Forsythe [52] is given by

$$\mathrm{RSL}(t) = \frac{P_{\mathrm{T}} G_{\mathrm{T}} G_{\mathrm{R}} \lambda^2 S}{32\pi^2 R_{\mathrm{R}} R_{\mathrm{T}} (R_{\mathrm{T}} + R_{\mathrm{R}})(1 - \cos^2\beta \sin^2\phi)} \times \left[\frac{4Dt + r_0}{\sec^2\phi} \ln\left(\frac{r_e q \lambda^2 \sec^2\phi}{\pi^2 (4Dt + r_0^2)} \right) \right]^{0.5} \tag{2.7}$$

Overdense trails are characterized by an initial rapid rise as the meteor transits the first Fresnel zone, continued slow rise as the metallic cylinder expands, followed by a slow decay as the cylinder diffuses. As the trail continues to diffuse, the density is reduced to the point where the trail begins to behave like an underdense trail. Overdense trails generally have a longer duration than underdense trails but do not necessarily have greater amplitude. The lower two trails in Fig. 2.3 show typical overdense trails at

45 MHz. The second trail in Fig. 2.4 has the slow risetime characteristic of an overdense trail and the exponential decay characteristic of an underdense trail.

The transition from underdense to overdense trails is not sharp; rather, it is very gradual. Trails in the transition region have characteristics of both classic underdense and overdense models. A commonly used approximation for the transition between overdense and underdense trails is the electron line density q at which the peak received signal power is equal for both the classic underdense and classic overdense approximations. Trails with q greater than the transition are considered overdense, while trails with q less than the threshold are underdense. At frequencies in the 50-MHz range, $q = 10^{14}$ e/m is often taken as a fixed transition [114].

One common misconception is that the received signal from an overdense trail is always much larger than that from an underdense trail. From Eqs. 2.5 and 2.7 it is observed that the received signal power is a function of the electron line density q, the trail orientation β, and the polarization coupling factor S. Depending on the orientation and the polarization coupling, the return from an underdense trail with one set of parameters could be greater than that from an overdense trail with a different set of parameters. The third trace in Fig. 2.4 shows returns from both an underdense and overdense trail within a 4-s period. The underdense trail return is larger than that from the overdense trail.

While many trails satisfy the rather stringent requirements required for the classic solutions, large numbers of trails do not, and more sophisticated models are required. One of the basic assumptions is that the meteor particle is moving fast enough so that the time required for the particle to transit the first or principle Fresnel zone is short relative to the decay constant of the trails. This assumption insures that the electron line density of the trail is uniform in the axial direction. When this assumption does not hold, the classic trail models of Eqs. 2.5 and 2.7 are not valid. Eshleman [37], who did much of the pioneering work in the development of trail models, describes approximations for the decay of low- and high-density trails under the assumption that the region of ionization is small enough to be considered as a parabola of revolution. Hawkins and Winters [50] develop similar models assuming that the ionization region is Gaussian instead of parabolic.

The basic overdense model of Eq. 2.7 assumes that a trail is so dense that it can be modeled as a metallic cylinder with radius given by the critical radius. The model does not consider contributions due to electrons outside the critical radius, (the sheath effect) nor does it consider the cases of trails that are dense enough that the underdense assumption is not valid but not so dense that the overdense assumption can be used. None of the classic models address resonance effects in the transition region between underdense and overdense trails.

2.2.1.3 Multipath and Fading Profile of the VHF Channels Formed by Meteor Trails. The meteor channel has been shown for the most part to be a power limited channel as opposed to multipath limited channels such as troposcatter; however, as burst data rates increase, it is increasingly important to understand and model multipath and fading on the meteor channel. Several multipath mechanisms have been identified on meteor burst channels. Manning [74] showed that high-altitude wind shears could warp a trail so that portions of the trail initially in the first Fresnel zone, contributing constructively to the signal, would later contribute destructively, causing fading. The multipath spread resulting from this mechanism is limited to several Fresnel zones or several hundred nanoseconds at 50 MHz. Since the multipath spread of this type of mechanism is so small, for most applications, the channel appears to be a flat fading channel.

Warping-induced flat fading tends to occur most frequently on long duration overdense meteor trails. Experiments using television signals [1, 46, 131], narrow radar pulses [19, 115], and a pseudo random sequence [32] showed that for the most part data rates in excess of 500 kbps could be supported without severe multipath interference problems.

Manning also showed that on occasion winds could warp the trail so that secondary areas satisfy the geometric conditions for specular scatter. This mechanism (glinting) can result in fading and delay spreads as high as several microseconds for long trails, which has been observed in experiments [1, 19, 32, 46, 115, 131].

The simultaneous occurrence of more than one meteor on a link can cause multipath spread in excess of 1–2 ms, depending on the geometry. Fortunately, for communication links with 500 to 1000-W transmitters, this phenomenon is observed in less than 1 to 2% of trails [115].

Because the scattering mechanism is for the most part specular, large Doppler spreads are generally not observed on meteor channels, except during the initial formation of the trail. Reflections from the head of the trail as it forms can cause peak Doppler shifts of 50–75 Hz during the short time that the meteor is transiting the principal Fresnel zone [78]. Trail drift due to high-altitude winds can account for Doppler shifts of typically 2–10 Hz [78, 120].

Ostergaard et al. [102] show that on some trails frequency diversity can be used to improve throughput. Frequency diversity (operating at multiple frequencies) is beneficial when the frequency separation is greater than the inverse of the multipath spread. For trails in which the multipath is due to multiple trail interference, small separations are required, however when the multipath is the result of glinting [74] several hundred Kilohertz to 1–2 MHz is required. Cannon et al. [16] showed that space diversity (multiple antennas spaced up to several hundred meters apart) could also provide some degree of improvement, especially toward the end of trails when wind-induced warping effects were greatest.

Recent studies by Phillips Laboratory (formerly Air Force Geophysics Laboratory) showed that as the duration of a trail increase the probability of a fade deep enough to cause data errors grows approximately linearly with trail duration [132]. For trails of less than 100-ms duration, the probability of a fade is negligible; however, as the duration approaches 400–600 ms, the probability of a fade increases to the order of 30%.

2.2.1.4 Distribution of Meteor Trail Amplitudes. This section analyzes statistics of the peak amplitudes of meteor trail echoes. In many meteor scatter system engineering applications parametric analysis is used to assess the effects on performance of changing various link and system parameters. One measure of performance commonly considered is the arrival rate of meteor trails exceeding a specified minimum signal threshold. The model in general use for the distribution of trails as a function of peak received signal power is described by Sugar [114]. The primary assumption is that all trails are of the more commonly occurring underdense variety and that the received signal power satisfies the classic equation of Eshleman [34] given in Eq. 2.5.

Sugar assumed that all parameters in Eq. 2.5 were fixed at mid-link values, with the exception of electron line density q which varies from trail to trail. For underdense trails, q ranges from 10^{12} to 10^{14} electrons/meter with the probability density function (pdf) approximated by

$$f_q(q) = \frac{C}{q^2}\,[u(q - q_{\min}) - u(q - q_{\max})] \tag{2.8}$$

$$\text{where} \quad C = \frac{q_{\min} q_{\max}}{q_{\max} - q_{\min}}$$

and $u(x)$ is the unit step function. The density function in eq. 2.8 is based upon assumptions that the electron line density of a trail is proportional to the mass of the meteor forming the trail and that the number of meteors at a given mass is inversely proportional to the mass.

To develop a distribution for the number of trails versus received signal, Sugar rewrote Eshleman's equation in the general form

$$\text{RSL} = P_1 q^2 \tag{2.9}$$

where P_1 incorporates all link-related constant terms. Sugar solved Eq. 2.9 for q and substituted the result into Eq. (2.8) to yield a distribution for the number of trails as a function of the peak received signal power given by

$$P\{\text{RSL}_0 > \text{RSL}\} = P\left\{q_0 > \left(\frac{\text{RSL}}{p_1}\right)^{0.5}\right\} = C\left(\frac{P_1}{\text{RSL}}\right)^{0.5} \tag{2.10}$$

for $RSL_{min} \leq RSL \leq RSL_{max}$ where RSL_{min} and RSL_{max} are calculated from Eq. 2.9 at q_{min} and q_{max}, respectively. To make the mathematics easier, Sugar assumed that all trails are underdense so that q_{max} is infinite.

The relationship described by Eq. 2.10 is appealing because of its mathematical simplicity and is commonly used as the basis for parametric trade-off analysis. Equation 2.10 implies that a 3-dB increase in transmitter power P_T will increase the number of meteor trails observed by a factor of $2^{0.5}$. The increase predicted by Eq. 2.9 is independent of signal threshold.

Analysis of data from the High Latitude Meteor Scatter Program sponsored by the US Air Force [125] showed that the classic distributions of peak underdense trail amplitudes (Eq. 2.10) are not supported by the data (Fig. 2.5). In particular, it was observed that the number of underdense meteor trails decreases faster with increasing signal level than predicted and the general shape of Eq. 2.10 does not correspond to the data. See Weitzen and Tolman [135] for an explanation of how underdense and overdense meteor trails are classified for use in the analysis.

Consider the peak received signal power due to an underdense trail given by evaluating Eq. 2.5 at $t = 0$ as

$$RSL_{peak} = \frac{P_T G_T G_R \lambda^3 q^2 r_e^2 \cos^2\alpha}{16\pi^2 R_R R_T (R_T + R_R)(1 - \cos^2\beta \sin^2\phi)} \tag{2.11}$$

Meteor trails occur throughout the receiver–transmitter antenna pattern with different orientations and line densities. Instead of a single random variable in the power equation assumed by Sugar, let α, q, G_T, G_R, R_R, R_T, β, and ϕ vary from trail to trail. Let electron line density q have distribution given by Eq. 2.8. Let the trail orientation factor β be uniformly distributed over the interval $(0, 0.5\pi)$. Let the polarization coupling factor α be uniformly distributed over the interval $(\varepsilon, \pi/2)$ so that $RSL > 0$, which is

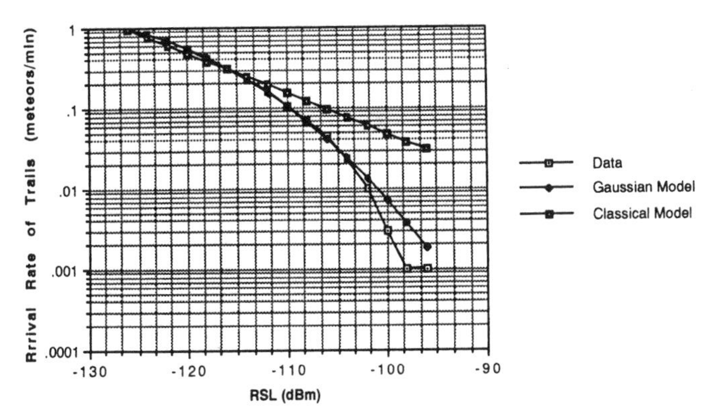

Figure 2.5. Number of meteors vs. received signal level, March 1989: data at 65 MHz, Gaussian model, and classical model.

required for the analysis. Let G_T, G_R, R_R, R_T, and ϕ be uniformly distributed over relatively narrow intervals about the mid-line averages.

Calculation of the distribution function of underdense meteor trail peak signal levels as a function of received signal level assuming Eq. 2.11 represents a probability transformation function of 8 random variables. Using the method of auxiliary variables, it is necessary to compute an 8-dimensional Jacobian matrix and then integrate over the 7 auxiliary variables to yield a solution for the density function as a function of received signal level. This process does not yield closed form solutions for the distribution of trail amplitudes.

Consider instead the received signal level decibels related to 1 W (dBW) rather than absolute power:

$$\mathrm{RSL}_{\mathrm{dBW}} = 10 \log_{10}(\mathrm{RSL}) \tag{2.12}$$

Equation 2.11 is written using Eq. 2.12 as

$$\begin{aligned}\mathrm{RSL}_{\mathrm{dBW}} = 10\{&\log P + \log G_T + \log G_R - \log[R_T R_R (R_T + R_T)] \\ &- \log(1 - \cos^2\beta \sin^2\phi) + 2 \log q + \log \cos^2\alpha\}\end{aligned} \tag{2.13}$$

where all Logarithms are base 10 and $P = \dfrac{P_T \lambda^3 r_e^2}{16\pi^2}$

incorporates link constant terms. The probability density function of the sum of independent random variables is calculated from the convolution of the density functions of the independent variables.

The central limit theorem shows that under certain conditions, the density function of the sum of an increasing number of independent random variables tends towards a Gaussian. Lindberg's theorem, described in [42], shows that an arbitrary sequence of random variables will, in the limit, obey the central limit theorem if the individual random variables are uniformly bounded (absolute value of all of the variables is finite). This condition is satisfied by the variable definitions described previously.

Using central limit theorem arguments, the density function of underdense meteor trail amplitudes (in dBW) is approximated by a Gaussian as

$$f(\mathrm{RSL}) = \frac{\exp\{-[(\mathrm{RSL} - \overline{\mathrm{RSL}}_{\mathrm{dB}})]^2/2\sigma^2_{\mathrm{RSL}}\}}{\sqrt{2\pi}\sigma_{\mathrm{RSL}}} \tag{2.14}$$

Where $\overline{\mathrm{RSL}}_{\mathrm{dB}}$ is the average received signal level in dBW computed from the sum of the average values, of the individual variables and the standard deviation σ_{RSL} is computed from the sum of the individual variances. Weitzen et al. [130] calculates distribution and density functions assuming that α, β, and q are random and shows that with only three random variables, the probability distribution closely approximates a Gaussian distribution.

Comparing a model to data is complicated by the fact that the entire range of data are not available because only trails with peak amplitude greater than the noise threshold are receivable. All trails with amplitudes below the noise cannot be observed. Define the complementary cumulative distribution function $F_c(\mathrm{RSL})$ as the probability that trail amplitude exceeds a specified RSL:

$$F_c(\mathrm{RSL}) = \frac{1}{\sqrt{2\pi\sigma^2_{\mathrm{RSL}}}} \int_{\mathrm{RSL}}^{\infty} \exp\left(-\frac{(\mathrm{RSL}_0 - \overline{\mathrm{RSL}}_{\mathrm{dB}})^2}{2\sigma^2_{\mathrm{RSL}}}\right) d\mathrm{RSL}_0$$

$$= Q\left(\frac{\mathrm{RSL} - \overline{\mathrm{RSL}}}{\sigma_{\mathrm{RSL}}}\ \mathrm{dB}\right) \tag{2.15}$$

While Eq. 2.15 does not exist in closed form, tables and properties of the function are well known. To be useful for communication system engineering and to evaluate the goodness of fit to a Gaussian, Eq. 2.15 must be normalized relative to a data set, and the mean and variance of the process must be estimated. The three parameters (normalization constant, mean, and variance) were determined by Weitzen et al. [130] using the multidimensional downhill simplex method described in Press et al. [107] so as to minimize the absolute value of the difference between Eq. 2.15 and the data over all data points that exceed a minimum threshold 10 dB above the noise.

Consider the correspondence between data and predictions of Eq. 2.10 and 2.15. Figure 2.5 plots a cumulative distribution of the number of underdense meteors per minute as a function of peak signal level (dBm) using data from [130]. Superimposed on the figure is the best fit to a Gaussian distribution determined by the described method. Also plotted is the classic distribution function from Eq. 2.10. The figure shows the trend for the number of underdense meteor trails to decrease faster with increasing signal level than predicted by the classic equation. It also shows that Eq. 2.15 provides a good fit to the data.

Goodness of fit was tested for 20 cases at 45, 65, 85, and 104 MHz for 5 months during 1989. The Gaussian hypothesis was accepted for 90% of the cases. In those cases for which it was not accepted, there is a high likelihood that sporadic-E propagation contaminated the data set.

Equations 2.14 and 2.15 have a number of potentially important implications for the design of meteor scatter communication systems, especially for the cost versus benefit trade-off for changes in the link power budget. One important finding from the analysis embodied in Eq. 2.14 is that the improvement or degradation in performance due to changes in the link power budget will vary as a function of the operating point of the system. As an example, consider the effect of changing the transmitter power on the number of meteor trails observed per unit time (performance). Let P_{Ti} and

P_{Tn} be the initial and new transmitter powers, respectively. The classic model described by Eq. 2.10 implies that the increase (or decrease) in the number of trails with changes in the power factor is independent of the signal level and is given by

$$\text{Improvement factor} = \left(\frac{P_{\text{Tn}}}{P_{\text{Ti}}}\right)^{0.5} \tag{2.16}$$

Using the Log-normal distribution (Eq. 2.14) the effect of changing the transmitter power is dependent upon where on the signal threshold versus arrival rate curve a system is operating. The improvement is given by

$$\text{Improvement factor} = \frac{Q[(\text{RSL}_{\text{T}} - \overline{\text{RSL}}_{\text{n}})/\sigma_{\text{RSL}}]}{Q[(\text{RSL}_{\text{T}} - \overline{\text{RSL}}_{\text{i}})/\sigma_{\text{RSL}}]} \tag{2.17}$$

assuming that a 3-dB increase in the transmitter power corresponds to a 3-dB increase in $\overline{\text{RSL}}$. Figure 2.6 plots the increase in the number of trails observed per unit time, assuming a change from 500 to 1000 W, as a function of operating threshold (RSL_{T}) based on data at 65 MHz [130]. Over typical operating ranges for RSL_{T}, the expected performance improvement is greater than that predicted by classic theory. Depending on where on the curve a system is operating, the improvement due to increasing transmitter power may be much greater than implied in the classic model. The cost versus benefit of additional transmitter power may have to be reevaluated in light of this new curve.

As another example of the implications of the log normal approximation, consider the effect of different coding techniques for meteor scatter communication. Coding gain can be viewed as effectively reducing the minimum required signal threshold RSL_{T} for detection at a fixed bit error rate.

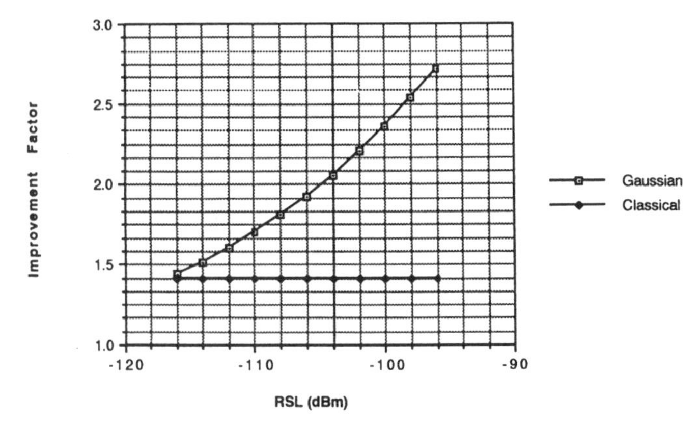

Figure 2.6. Improvement factor for a 3-dB increase in transmitter power, Gaussian and classical models (based on 65-MHz data).

Consider a code providing 3-dB effective coding gain. For the classic model of Eq. 2.10, the improvement factor would be 1.414. Using the model of Eq. 2.15 the improvement factor would be

$$\text{Improvement factor} = \frac{Q[(\text{RSL}_\text{T} - 3.0 - \overline{\text{RSL}})/\sigma_{\text{RSL}}]}{Q[(\text{RSL}_\text{T} - \overline{\text{RSL}})/\sigma_{\text{RSL}}]} \tag{2.18}$$

and Fig. 2.6 can be used to calculate the improvement.

Since it has been shown that the classic model for the effect of changes in performance as a function of link power budget is not accurate, the question arises as to whether there is a computationally simple relation that more accurately reflects trends in the data. As a first-order approximation to Eq. 2.14 providing an improved performance trend prediction over Eq. 2.10, the distribution of trail peak amplitudes can be approximated over a limited range of signal levels by

$$F_y(\text{RSL}) = \frac{C}{\text{RSL}^{\psi}} \tag{2.19}$$

where ψ is the channel improvement exponent determined from the best fit to a set of data over a range of potential signal thresholds (for the classic model $\psi = 0.5$). Although Eq. 2.19 provides a useful basis for link performance calculations that is potentially more accurate than the classic model with $\psi = 0.5$, the problem with the approximation is that ψ is a function of link range (a strong determinant), frequency, and received signal threshold [133].

Over the range of signal levels considered in most links, ψ decreases from 0.83 at long ranges to approach the classic case $\psi = 0.5$ at short ranges. While the approximation of Eq. 2.19 does not appear to be as accurate as that of Eq. 2.15, especially at the higher signal levels, it is appealing because of its simplicity. Hibshoosh et al. [51] use this approach with $\psi = 0.83$ based on data from Weitzen [125] to estimate channel capacity.

Figure 2.7 plots ψ as a function of signal level for several frequencies

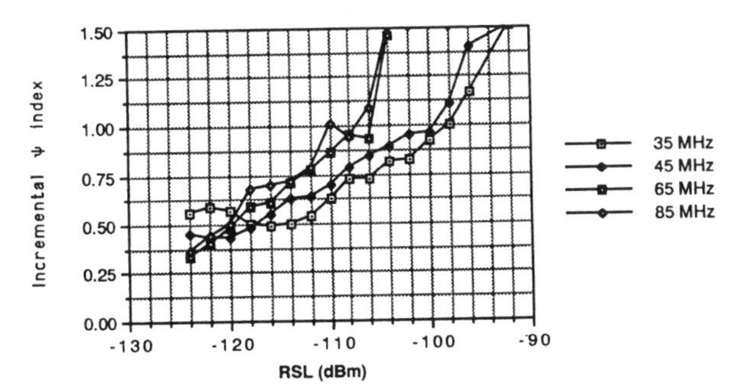

Figure 2.7. Index ψ as a function of signal threshold at 35, 45, 65, and 85 MHz for March 1989.

based on data from March 1989 and illustrates how the factor increases as the signal level increases [130]. At very low levels, the advantage of additional signal power is less than at higher signal thresholds.

2.2.2 Other VHF Beyond-Line-of-Sight Propagation Mechanisms

No survey of meteor burst communication would be complete or accurate without a discussion of propagation mechanisms other than meteor trails that are commonly observed on meteor communication links. These mechanisms can change the characteristics of the meteor communication channel from a weak-signal, intermittent environment with only one link connected at a time, to a continuous, high signal level, high multipath, and fast fading channel, with every node in a network connected. Propagation mechanisms, in addition to meteor scatter, observed on meteor links operating in the low VHF band include sporadic-E propagation, auroral scatter, ionospheric (E and F layer) scatter, knife-edge diffraction, line-of-sight propagation, and tropospheric scatter.

2.2.2.1 Sporadic-E Propagation and Ionospheric Scatter. Varieties of E and F region scatter and sporadic-E propagation occur frequently at all latitudes. E and F region scatter is generally a weak, above critical frequency scatter mechanism which occurs due to irregularities in the ionosphere. It is observed on systems with high-gain antennas and high-power transmitters. It also occurs at times of high solar activity when layer critical frequencies are elevated. The characteristics of this type of propagation are similar to E-layer HF propagation [7]. Scattering due to elevated critical frequencies can result in large portions of a network being connected and network protocols must be capable of recognizing and reacting to a change in the propagation conditions. Simultaneous scattering from both the E and F regions can result in severe multipath conditions. Fortunately, this does not occur frequently except in auroral regions.

Sporadic-E propagation occurs when patches of enhanced ionization are created and drift into the mid-link illumination region. While many mechanisms can create sporadic-E patches, the effect is similar: high signal level, slowly fading continuous propagation. The primary multipath mechanism on the sporadic-E channel is, ironically, the occurrence of meteor trails. Typical Doppler spreads are less than 10 Hz [19, 131, 134]. At 40–100 MHz, sporadic-E is unpredictable; however, at middle latitudes, it occurs more frequently during the summer months close to local noon. At high latitudes a number of different mechanisms are observed [17]. At equatorial latitudes it can be near continuous.

Figure 2.8 due to [121] plots a number of multipath profile snapshots from a 650-km link in Alaska. The snapshots, taken every 5 seconds, have a resolution of 125 μs. Multipath was determined using a 127-bit pseudorandom code continuously repeated. The slow fading of the channel is evident in the 5 min of data. Also evident are secondary reflections with

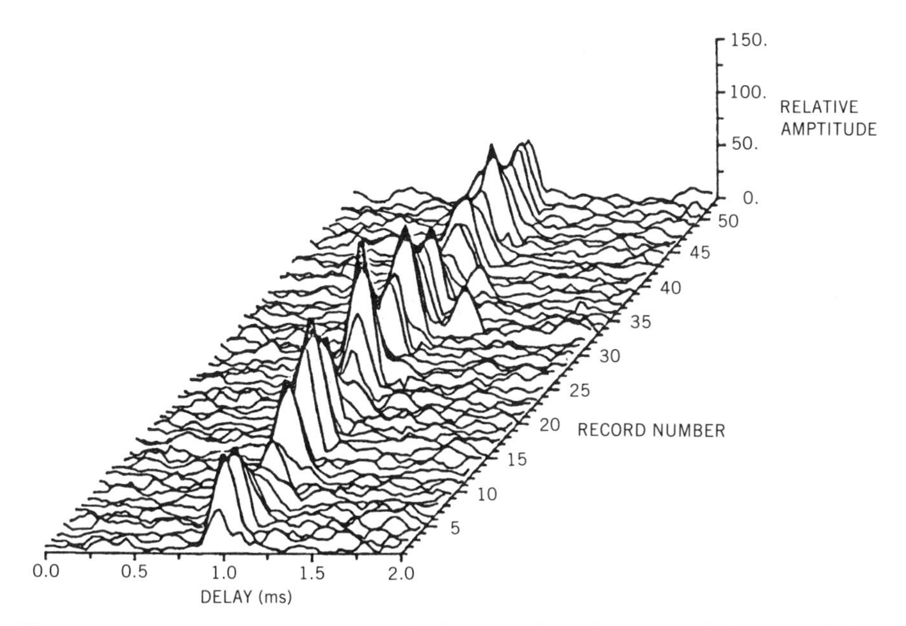

Figure 2.8. Multipath snapshots of sporadic-E channel from Weitzen and Tolman [135]. Note secondary echoes, most likely due to meteor trails in several records.

different path distances most likely due to meteor trails occurring in the common volume.

Sporadic-E propagation lasts from several minutes to hours depending on conditions. When it occurs, it can greatly increase the capacity in the channel. It can also greatly increase the complexity of protocols in a large network. Instead of a low duty cycle channel in which individual nodes are connected to the master via meteor trails, the entire network can be simultaneously connected and remain connected for hours. Protocols must be capable of determining that sporadic-E exists and of reacting to the change in the network connectivity.

2.2.2.2 Auroral Scatter Propagation. It is becoming increasingly apparent that at high latitudes auroral scatter can play an important role in determining the performance of meteor burst communication systems. Auroral scatter can transform the relatively benign, though intermittent, meteor channel into a relatively high-signal level multipath-limited channel. In a recent experiment to characterize the multipath, fading, and signal profiles of nonmeteoric propagation mechanisms on a 650-km link from Anchorage to Bethel in Alaska, rms multipath spreads as high as 1 ms with Doppler spreads in excess of 200 Hz were observed when auroral scatter was the dominant propagation mechanism [121, 126].

Communication in this environment requires changes in the modulation, data rate, and signaling protocols. Figures 2.9–2.12 show the multipath and

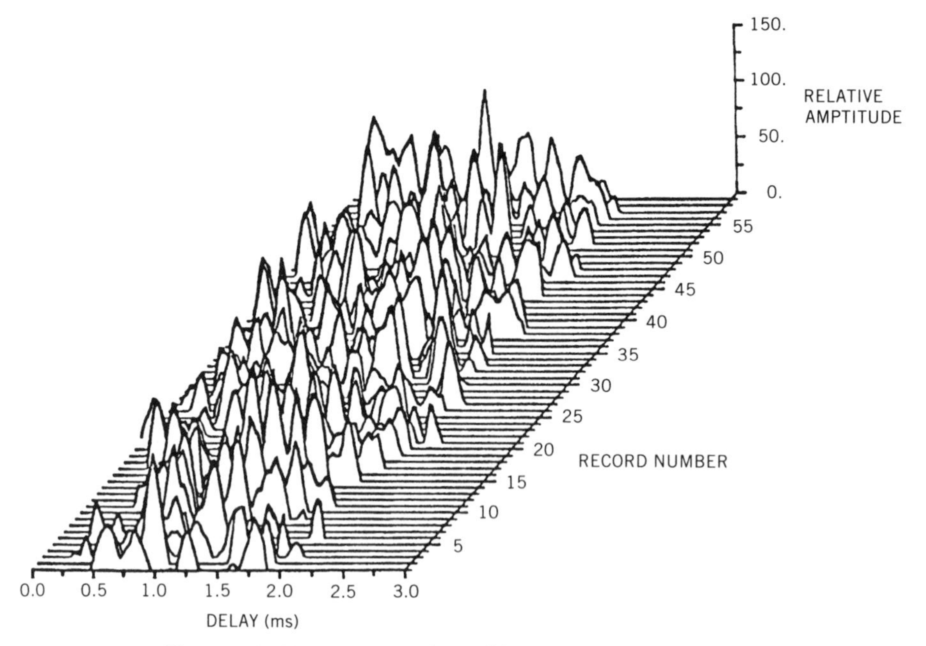

Figure 2.9. Multipath snapshots of large auroral scatter event.

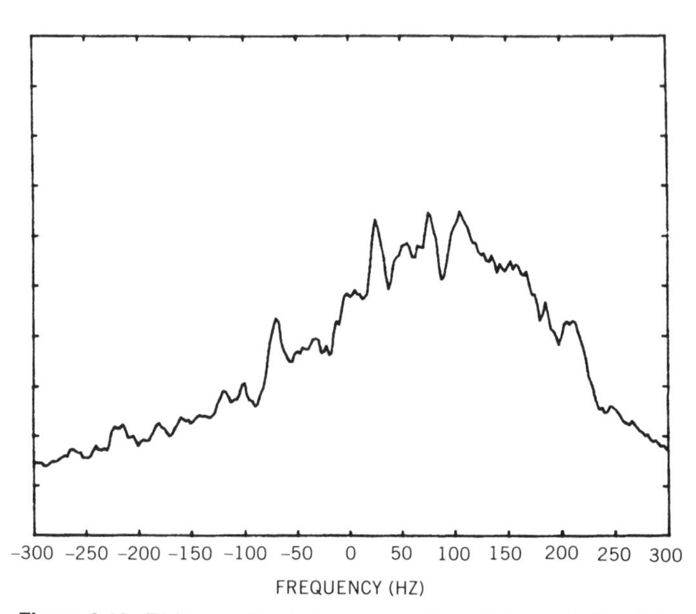

Figure 2.10. Fading profile of strong auroral event shown in Fig. 2.9.

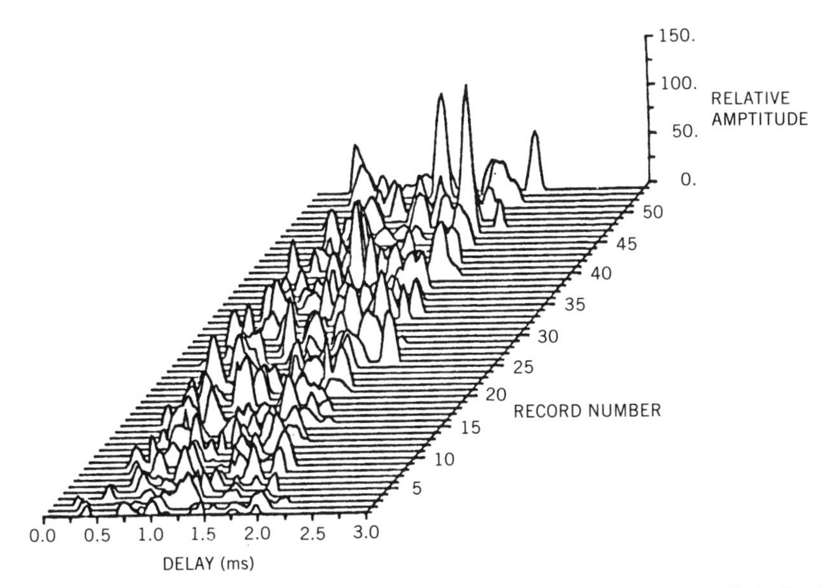

Figure 2.11. Multipath snapshots of weak auroral scatter event. Large peaks are likely due to strong meteor echoes.

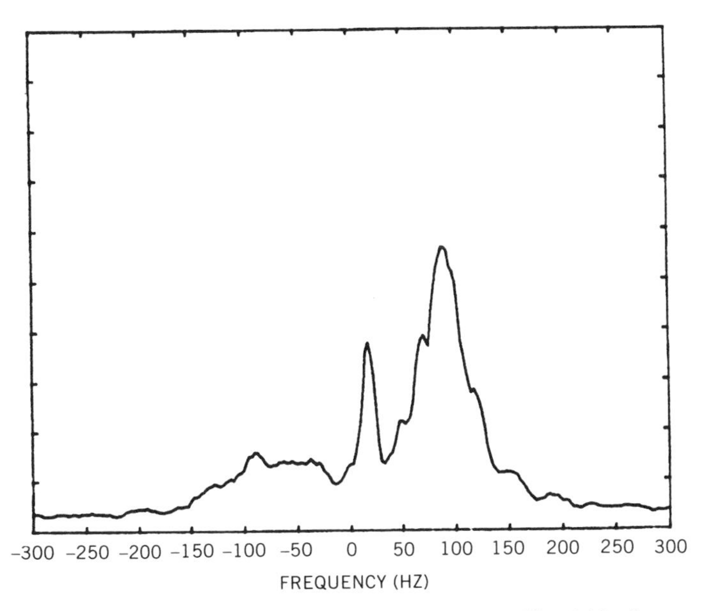

Figure 2.12. Fading profile of weak auroral event shown in Fig. 2.11. Component at 0 frequency shift is due to meteor trails, while component centered at 80 Hz is due to auroral scatter.

fading profiles of a VHF 42-MHz channel during several large auroral events in July 1985.

Figures 2.10 and 2.12 plot the Doppler fading profiles for two events on the 650-km link in Alaska described in [121]. Figure 2.12 shows two components. The component at zero Doppler shift represents meteor trail propagation and the component centered at 80 Hz is due to auroral scatter. Figure 2.11 shows the multipath profile for this event. Figure 2.10 shows the fading profile of a very large auroral event in which the rms spread is in excess of 100 Hz. Coherent communication in that type of environment would be difficult at best. Figure 2.9 shows the multipath profile for the same event. Multipath spreads (rms) in excess of 1 ms are observed.

Auroral scatter propagation, as opposed to other auroral effects such as enhanced absorption or auroral sporadic-E propagation, occurs only when a complicated set of requirements are satisfied. Auroral irregularities tend to align themselves along the earth's magnetic field lines [71]. Additionally, the link must be oriented relative to the aurora so that reflection from the curtain rising from the E through the F region can occur. For example, auroral scatter could occur when the visible aurora is north of an east–west link and the antenna pattern is broad enough that energy off the great circle path can reflect from the aurora. When the aurora is above the great circle path of a link, sporadic-E propagation or absorption tend to occur. For a given auroral event, some links in the auroral region will experience enhanced E-layer propagation, some will experience auroral absorption, and some may experience auroral scatter [71].

2.2.2.3 Diffraction, Troposcatter, and Line-of-Sight. At ranges less than several hundred kilometers, meteor scatter may not be the dominant propagation mechanism. There is an increasing appreciation of the roles of diffraction, tropospheric scatter, and line-of-light propagation on meteor scatter communications. At short ranges, diffraction and line-of-sight effects tend to dominate over all other propagation mechanisms. As the range approaches 100 km, a different propagation mechanism, tropospheric scatter, may occur.

Melton and Darnell [88] describe an experiment to characterize tropospheric scatter at VHF. A receiver was placed at ranges out to approximately 300 km from an existing master station. It was determined by observing received waveforms that at ranges up to about this range, meteor scatter was not the dominant propagation mechanism. Since the range was too short for sporadic-E propagation, in which the angle of incidence ϕ must be as large as possible, and too long for diffraction effects, as analyzed using the model of Longley-Rice, the conclusion was that a troposcatter propagation mechanism was supported. The existence of this propagation mechanism provides additional capacity into meteor scatter systems operating at short ranges in addition to that provided by conventional end-path illumination [65, 123].

2.2.2.4 Absorption Effects in Meteor Scatter Systems.

Meteor scatter propagation, while more resistant than HF to ionospheric absorption events, is nevertheless not immune. Two different high-latitude phenomena, polar cap absorption and auroral absorption, can result in enhanced ionization in the D region with the resultant HF blackout.

2.2.2.4.1 Polar Cap Absorption.

Polar cap absorption (PCA) is caused by high-energy protons emanating from the sun during large flare events. The particles are funneled into the polar cap ionosphere along the magnetic field lines of the earth. Collisions between the protons and ionospheric particles in the D region 45–90 km above the earth lead to ionization. Since the collision frequency at these altitudes is high, collision between the electrons vibrating under the influence of the incident radio wave and the neutral atmospheric particles transfers radio energy into kinetic energy, leading to absorption of radio waves passing through this region. The greater the ionization, the more the wave is absorbed. HF communication may be severely degraded by even relatively small absorption events. Maynard [79] postulated that due to the much higher operating frequency of VHF meteor scatter, the effect of a given absorption event should be much smaller.

To assess the effects of polar cap absorption on high-latitude meteor scatter communication, USAF Phillips Laboratory (formerly Air Force Geophysics Laboratory) established a high-latitude test bed from Sondrestrom to Thule in Greenland. This link lies entirely within the polar cap region. The link operates at 5 frequencies (35, 45, 65, 85, and 104 MHz) to measure the absorption at different frequencies. The system cycles between the various frequencies every 2 h. Details of the link are described elsewhere [103, 104, 105, 124, 135].

Absorption (in dB) is estimated from the classic formula given in Ostergaard et al. [105]:

$$A_{f1} = A_{f0}\left(\frac{f0}{f1}\right)^n \sec z \quad \text{(dB)} \tag{2.20}$$

where $f0$ is a reference frequency generally less than $f1$, A_{f0} is the zenith absorption, and z is the angle between the incident ray and the zenith. Generally, $f0$ is 30 MHz and A_{f0} is the difference between the normal galactic noise and the noise measured during the absorption event at 30-MHz frequency. Due to the multiplying effect of the secant term, zenith absorption of 6 dB (1 way) could translate into an oblique path absorption of 25 dB or more as the secant effect increases. A signal propagating between two meteor stations is attenuated twice, once from the transmitter to the trail and once from the trail to the receiver. Galactic noise that passes through the ionosphere only once is attenuated only once.

The frequency variation factor of PCAs, n in 2.20, ranges from $n = 0$ to $n = 2$ or slightly greater. As an example, consider the effect of a major PCA which occurred during August 1989, described in Ostergaard et al. [104, 105]. The 30-MHz vertical absorption reached a peak of approximately 13 dB. The entire enhanced absorption event lasted for approximately 1 week during which HF was blacked out most of the time.

During the peak of the absorption, 35- and 45-MHz meteor scatter arrival rates were reduced by a factor of approximately 10 at 45 MHz and 20 at 35 MHz. The 65-MHz arrival rate was reduced by a factor of about 5 and at 104 MHz the reduction was negligible. Under normal conditions, lower frequency meteor scatter outperforms high-frequency systems; however, during periods of severe absorption, there is an advantage to using higher frequencies.

2.2.2.4.2 Auroral Absorption. Absorption due to the aurora occurs in the sub-polar cap auroral region when the radio wave passes through the enhanced ionization of the aurora in the D region. Auroral absorption regions generally do not coincide with the regions of the visible aurora. Auroral absorption events are generally of short duration (less than 30 min), as opposed to several days for polar cap absorption events. Auroral absorption is concentrated in a small region, as opposed to PCA which is generalized throughout the sunlit polar cap region. Li [69] studied auroral absorption on a USAF Phillips Laboratory high-latitude meteor scatter link from Sondrestrom to Narssarssuaq in southern Greenland and observed absorption events on the order of 0.75 to 1 dB and on occasion 2-dB, 30-MHz vertical absorption.

Events with greater than 0.9 dB vertical absorption could cause noticible reduction in the meteor arrival rate. Because the events are of short duration, long-term outages due to auroral absorption are not observed, although a 15-min reduction might be observed. Auroral absorption effects at frequencies above 65 MHz are minimal.

2.2.5 Space Diversity, Privacy, and Spatial Multiplexing in Meteor Scatter Communication

In meteor scatter communication, radio wave reflection is from a small cylinder of electrons rather than from a large layer. As such, the region on the ground that is illuminated by a meteor trail is smaller than that of HF communication. This phenomenon provides meteor scatter with an inherent degree of spatial multiplexing relative to other scatter communication techniques. The relative small region illuminated on the ground also has a number of ramifications on the design of network protocols.

The first step in the analysis is to define some key terms, especially the general term *footprint* since it has several meanings, depending on the context. The *instantaneous footprint* is defined as the region on the ground that is illuminated at a given instant of time by a given meteor trail. Due to the finite velocity of meteor particles as they enter the atmosphere the instantaneous footprint moves along the ground as regions of the trail decay and new regions of ionization near the head of the trail are formed. The region on the ground that is illuminated at any time during the lifetime of a trail is referred to as the *trail footprint*. In the past, the trail footprint has been referred to as the footprint. Section 2.2.5.1 describes a new technique for computing trail and instantaneous footprints.

Some researchers are interested in the probability that a second node can receive a piece of a message, given that it was received by a node at the center of a grid. This is sometimes called the *composite footprint*, but is actually the one-dimensional conditional probability of reception. This quantity is more accurately an *intercept probability* than a footprint. It is calculated as a function of location relative to the center of a grid. A number of studies have been undertaken to characterize the intercept probability [26, 55, 56, 94, 97, 98, 128], and these are summarized in Section 2.2.5.2.

Other researchers are interested in the number of independent (degrees of diversity) in a network of meteor scatter receivers. This occurs since several nodes may receive a probe and attempt to respond, but only one can be serviced. The number of independent arrival factors in a network is the diversity factor of a network and is discussed in Section 2.2.5.4. The probability of k nodes responding to a given probing signal is discussed in Section 2.2.5.3.

The small ground illumination region attributable to meteor scatter occurs as the result of two fundamental properties of the meteor scatter propagation mechanism. The first condition is the classic tangency theorem, which requires that the line forming the trail be tangent to an ellipsoid of revolution with foci at the receiver and the transmitter. It has been shown that for a given line and a receiver and transmitter, a point of tangency exists along the line [128]. For a set of receiver nodes with a common transmitter, the points of tangency for each link will lie at different points along the line forming the trail. For a scatter channel to exist to the given node, there must be significant ionization at the point of tangency.

Figure 2.13 illustrates how the combination of the two conditions give rise to a dynamic ground illumination footprint. The confusion between the instantaneous footprint and trail footprint exists because the velocity of the meteor forming the trail is relatively slow (20–40 km/s) relative to the total ionized length of a trail (20–40 km). The region of ionization begins at the head of the moving meteor. Immediately after formation, the trail begins to expand due to diffusion and the received signal decays. Beyond some distance behind the meteor head, the electron density has decayed to a point at which communication is no longer supported by that region. Nodes for which the point of tangency lie within the region of ionization at any time are said to be the instantaneous footprint of the trail.

As the meteor continues along its path, new nodes in a network may become connected to the transmitter as the region of significant ionization

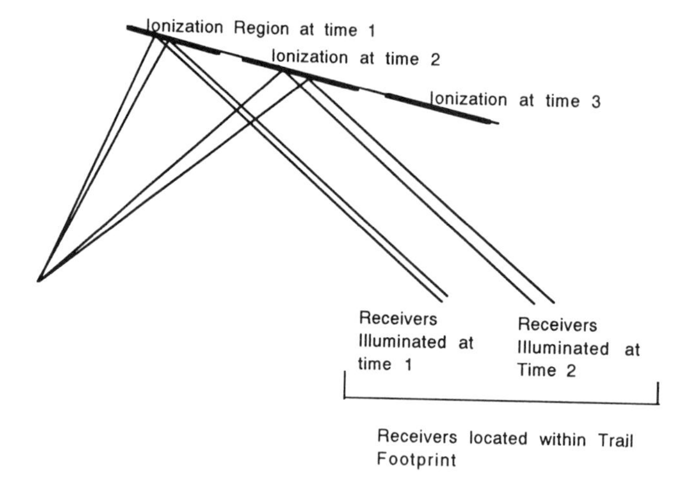

Figure 2.13. Motion of the ground illumination region due to trail formation and decay.

coincides with their points of tangency. Other nodes become disconnected from the transmitter when the region of ionization moves beyond their points of tangency. If the meteor velocity were semi-infinite, there would be no difference between the trail footprint and the instantaneous footprint. Because of the relatively slow meteor velocity, the instantaneous footprint region is a subset of the total trail footprint region. The illuminated region on the ground is a function of the length of the ionization region and the orientation of the trail relative to the geometry of the link. It is also a function of meteor velocity, and frequency of operation.

2.2.5.1 Ground Illumination Footprint of Individual Meteor Trails. Consider the Cartesian coordinate system shown in Fig. 2.14 with the origin at the midpoint of the chord from the transmitter to the receiver. Let the

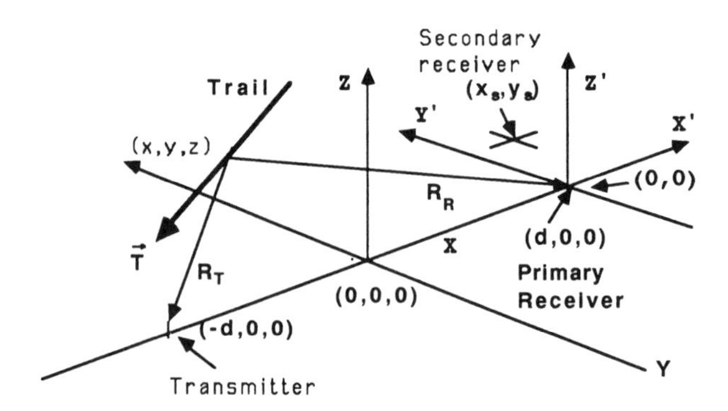

Figure 2.14. Primary (sky-based) and secondary (ground) Cartesian coordinate systems.

direction from the origin to the receiver represent the positive X axis. The Z axis represents the direction from the center of the earth to the origin and the Y axis is defined by the requirements for a right-hand coordinate system. Assume that a transmitter is located at $(-d, 0, 0)$ and a primary receiver is located at $(d, 0, 0)$ relative to this coordinate system.

Define a secondary, primed, Cartesian coordinate system parallel to the primary coordinate system with origin at the location of the primary receiver $(d, 0, 0)$, and with negative X' axis in the direction of the transmitter. In the analysis we consider only the $X'Y'$ plane and for clarity points in the primed coordinate system are denoted by two coordinates (x', y') and points in the primary coordinate system are denoted by three coordinates (x, y, z). In the primed coordinate system, the primary receiver is located at $(0, 0)$. Let a secondary receiver have coordinates (x_s, y_s) in this system.

For coherent scatter from a meteor trail, the line defining the axis of a cylindrical trail must be tangent to an ellipsoid of revolution with foci at receiver and the transmitter. Additionally, for the trail to be observable, there must be significant ionization in the region of the point of tangency. Consider a meteor trail tangent at (x, y, z) to an ellipsoid of revolution with foci at receiver and transmitter. Let the ionization region of the trail extend to $\pm L/2$ on each side of the tangent point, so that it is observable at the primary receiver. Assume that the ionization is uniform over this region. Lengths of typical meteor trails range from 10 to 40 km, and the ionization tapers at the ends of the trails.

A meteor trail is described in terms of its tangent point (x, y, z), elevation angle β relative to the XY plane, azimuth angle α relative to the XZ plane, and electron line density q. Given azimuth angle α and trail center (x, y, z), the trail elevation angle β for an observable trail is determined to satisfy the tangency conditions. Let $\vec{\mathbf{T}}$ describe the direction vector of the trail in the Cartesian coordinate system.

The classical approach to calculating the ground illumination region of an individual trail is described in Niedenfuhr [98] based on Huygen–Fresnel ray theory. The pattern of reradiated energy from a uniform linear trail can be approximated by drawing cones at each end of the trail with the cone axis colinear to the trail and half angles given by the propagation angles from transmitter to the ends of the trail as shown in Figure 2.15. The region formed by the intersection of the two cones and the planar earth represents the area illuminated by the trail.

The disadvantage of the classic approach is that it provides little physical insight into the mechanisms that result in the inherent privacy in the channel. Meteor trails have lengths ranging from 10 to 40 km; however, the region of constructive interference (first Fresnel zone) which contributes to the received signal is only on the order of 3 km in length. The classic technique does not indicate the location on trail of the reflection region for a given station, nor is it capable of considering effects of nonuniformities in the ionization. The technique also assumes that trails form instantaneously,

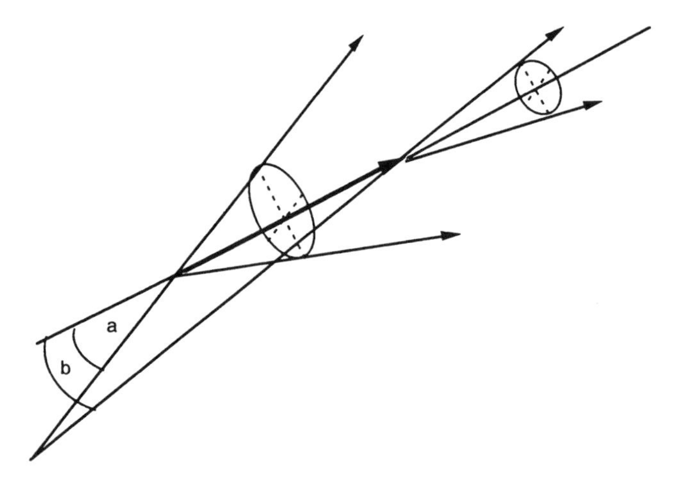

Figure 2.15. Ground illumination pattern using the Huygen–Fresnel theory.

when in practice meteor particles enter the atmosphere with velocities from 11 to 72 km/s. Differences in the time of formation between ends of a trail can affect the size and shape of the ground illumination region.

To provide additional insight into the privacy mechanisms, consider a meteor trail with mid point at (x, y, z) and with direction vector $\vec{\mathbf{T}}$ which is observed at the primary receiver at $(0, 0)$ in the ground coordinate system. Moving on the ground away from the primary receiver, the point on the trail at which the tangency conditions are satisfied for the new receiver location moves. If the new tangent point moves far enough from the original receiver so that the new point is no longer in the ionized region of the trail, the new receiver location will not be illuminated by the trail.

Let (x_t, y_t, z_t) be the point of tangency for a secondary receiver at (x_s, y_s), assuming a trail with midpoint (x, y, z), length L, and direction vector $\vec{\mathbf{T}}$. The mid point (x, y, z) represents the tangent point when the receiver is located at $(0, 0)$ in the ground coordinate system so that the trail $\vec{\mathbf{T}}$ creates a channel from the transmitter to the primary receiver.

The tangency condition for specular scatter at any ground location can be expressed as

$$\vec{\mathbf{N}} \cdot \vec{\mathbf{T}} = 0 \tag{2.21}$$

where $\vec{\mathbf{N}}$ is the normal vector to the ellipsoid at the point (x_t, y_t, z_t). The requirement that the new tangent point lies within the region of ionization is expressed as

$$|(x, y, z) - (x_t, y_t, z_t)| \leq \frac{L}{2} \tag{2.22}$$

Express the trail direction vector in Cartesian coordinates as

$$\vec{\mathbf{T}} = a\hat{a}_x + b\hat{a}_y + c\hat{a}_z \tag{2.23}$$

A point on the trail (x_t, y_t, z_t) satisfies the parametric equations for a straight line with direction vector $\vec{\mathbf{T}}$ given by

$$\frac{x - x_t}{a} = \frac{y - y_t}{b} = \frac{z - z_t}{c} \tag{2.24}$$

Let

$$\vec{\mathbf{R}}_T = (x_t + d)\hat{a}_x + y_t\hat{a}_y + z_t\hat{a}_z \tag{2.25}$$

and

$$\vec{\mathbf{R}}_R = (x_t - d - x_s)\hat{a}_x + (y_t - y_s)\hat{a}_y + z_t\hat{a}_z$$

represent vectors from the tangent point (x_t, y_t, z_t) to the transmitter $(-d, 0, 0)$ and a secondary receiver at $(d + x_s, y_s, z_s)$ respectively. Rudie [110] showed that the normal to an ellipsoid with foci at receiver and transmitter is given by

$$\vec{\mathbf{N}} = 0.5\left[\frac{\vec{\mathbf{R}}_T}{|\vec{\mathbf{R}}_T|} + \frac{\vec{\mathbf{R}}_R}{|\vec{\mathbf{R}}_R|}\right] \tag{2.26}$$

Substituting Eqs. 2.23, 2.25, and 2.26 into Eq. 2.21 yields

$$|\vec{\mathbf{R}}_T|^2[a(x_t - d - x_s) + b(y_t - y_s) + cz_t] = -|\vec{\mathbf{R}}_R|^2[a(x_t + d) + b(y_t) + cz_t] \tag{2.27}$$

Equations 2.24 and 2.27 are combined to solve for the tangent point (x_t, y_t, z_t) using numerical methods. If Eq. 2.22 is satisfied, then a secondary receiver at (x_s, y_s) can receive a signal received at $(0, 0)$.

Let $C_f(x_s, y_s, x, y, z, \alpha, \beta)$ represent the ground illumination region of a trail with midpoint at (x, y, z) and orientation (α, β) as a function of location on the ground. For mathematical expediency, define $C_f(x_s, y_s, x, y, z, \alpha, \beta)$ as a binary function: 1 if the point (x_s, y_s) is illuminated and 0 if it is not illuminated by the given trail. The size and shape of the ground illumination footprint is a function of the orientation of the trail and its distance from the receiver.

To demonstrate how the size and shape of the footprint varies with location and orientation of a trail relative to the receiver and transmitter, and to show how the receiver/transmitter antenna pattern can affect the ground illumination region of meteor scatter communication, consider a

660-km link with primary receiver at approximately (330, 0, 0) and transmitter at approximately (−330, 0, 0) in the primary coordinate system of Fig. 2.14. Ground illumination footprints relative to the primed coordinate system are computed for a series of trails in Figs. 2.17–2.20. Figure 2.16 shows the locations and orientations of trails considered in the figures. Trails have been selected for demonstration purposes and satisfy the tangency conditions for observability at the primary receiver. All trails are assumed to have a length of 20 km.

Figures 2.17 and 2.18 show ground illumination footprints, C_f for trails oriented at 0 and 90° relative to the great circle path at the link midpoint at an altitude of 95 km. Trails oriented along the great circle path produce ground illumination patterns that approximate hyperbolas, while trails oriented normal to the path produce ground illumination patterns that are long and narrow.

As the location of a trail moves from close to the transmitter to close to the receiver, the extent of the ground illumination region is reduced. Figures 2.19 and 2.20 consider the ground illumination footprint of trails close to the receiver and transmitter, respectively. Trails that are located close to the receiver have smaller ground illumination regions than trails that are located far from the receiver.

The implication of the analysis of individual meteor trail footprints is that links that have antenna patterns illuminating the region above the transmitter will have larger ground illumination regions than links in which the

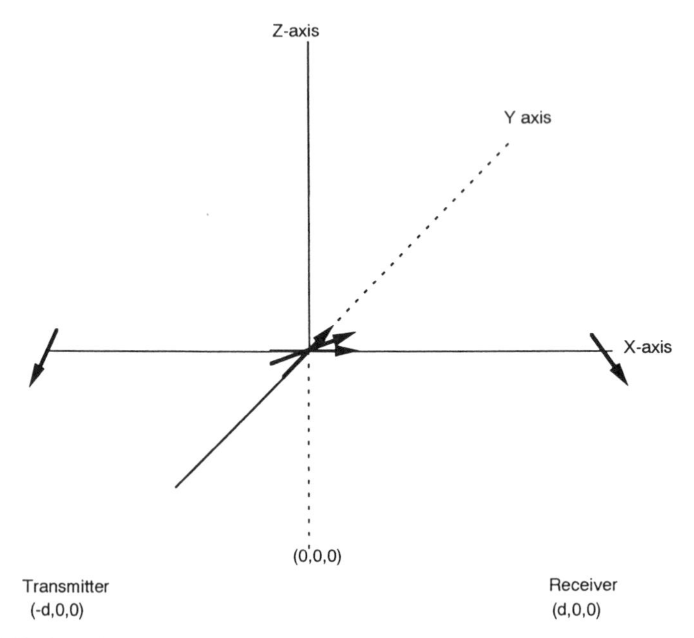

Figure 2.16. Locations in the primary coordinate system for which footprints are calculated in Figs. 2.17–2.20.

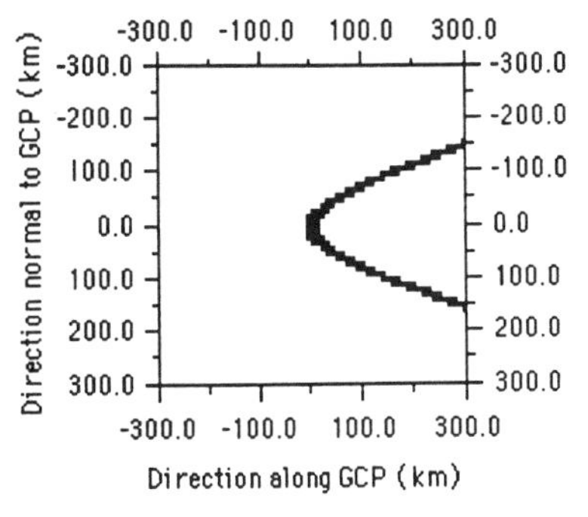

Figure 2.17. Ground illumination footprint due to trail oriented 0° relative to great circle path for a 660-km link.

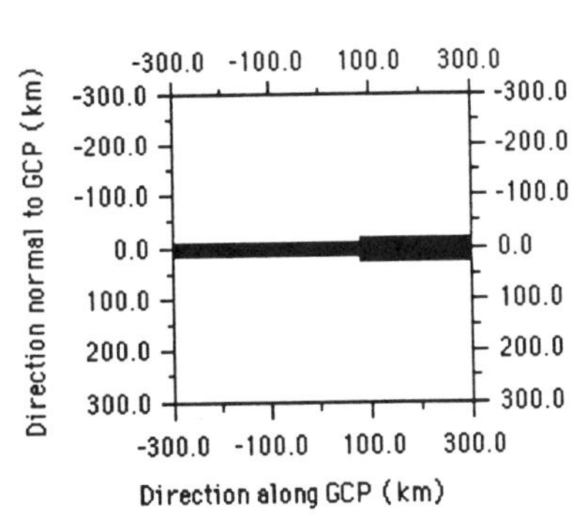

Figure 2.18. Ground illumination footprint due to trail oriented 90° relative to great circle path for a 660-km link.

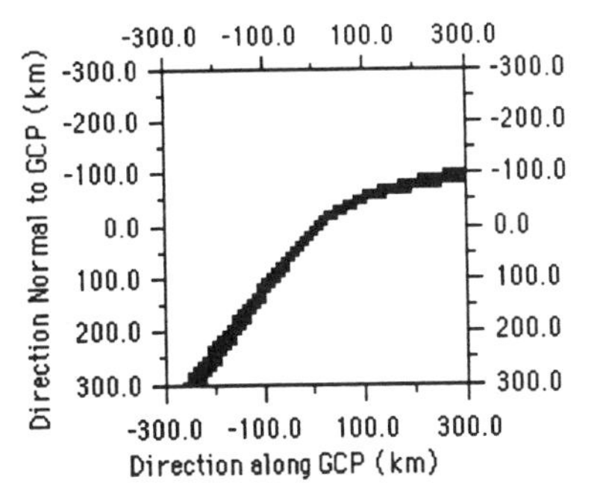

Figure 2.19. Ground illumination footprint due to trail located above receiver for a 660-km link.

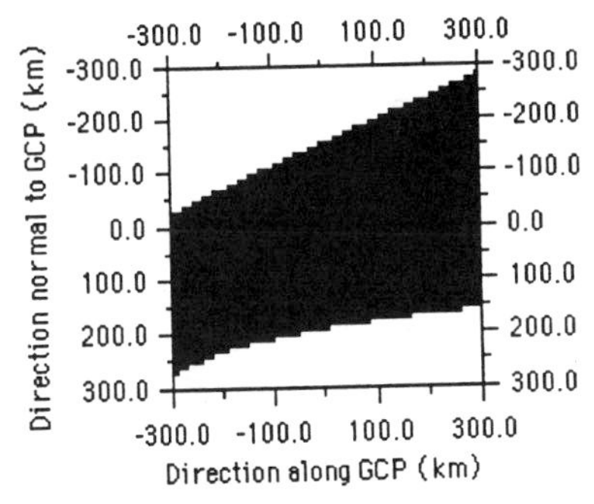

Figure 2.20. Ground illumination footprint due to trail located above transmitter for a 660-km link.

antenna pattern illuminates regions close to the intended receiver. This phenomenon is illustrated in Fig. 2.21.

C_f represents the region on the ground that is illuminated by a given trail. The region of instantaneous reception at any given time is a subset of C_f and moves during the lifetime of the trail. Up to this point it has been assumed that the trail forms instantaneously. Meteors enter the atmosphere with

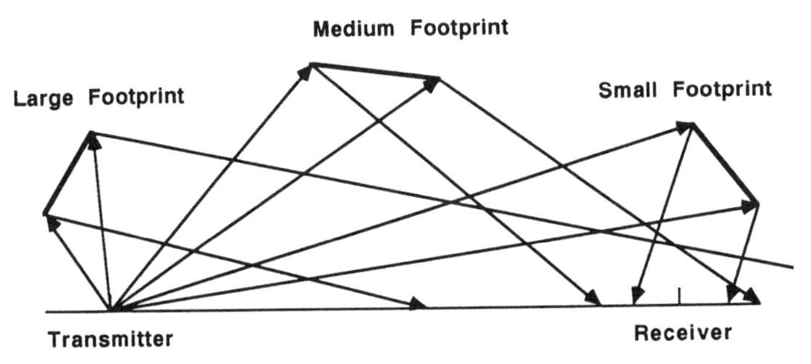

Figure 2.21. Effect of trail location on size of footprint.

velocities ranging from 11 to 72 km/s, forming trails that vary in length from 10 to 40 km. A 40-km trail formed by a meteor with a velocity of 40 km/s would require 1 s to form. Regions on the ground that are the result of tangent points close to the beginning of the trail will be illuminated and decay away before the meteor reaches the end of the trail line.

As an example, consider a trail 30 km in length formed by a meteor with an entry velocity of 60 km/s. The trail requires 500 ms to form completely. Let a station receive a message from a region at the center of the trail. Consider secondary receivers with tangent points at the beginning and trailing ends of the meteor trail. The trail at the first station forms 250 ms prior to the primary receiver and will decay faster because of the higher altitude [13]. If the trail at the first secondary receiver has decayed below the receive threshold prior to the beginning of the message, then there will be no adjacent station reception. Consider the station connected to the region at the end of the trail. Assuming a 100-ms message, the signal at this receiver will not begin until 150 ms after the message has been transmitted. As a result of this phenomenon the effective length of trails is reduced and inherent privacy in the channel is increased.

2.2.5.2 *First-Order Conditional Probability of Reception.* Define the probability of adjacent station reception, $P_i\{(x_s, y_s)|(0, 0)\}$ as the conditional probability that any portion of trail observed at a receiver located at $(0, 0)$ is observable at a secondary receiver located at (x_s, y_s). This is equivalent to a probability of intercept as a function of location relative to the primary receiver.

There are several assumptions built in to this definition. Assume that the noise level at the secondary receiver is less than or equal to that at the primary receiver. Also assume that the antenna gain pattern of the secondary receiver illuminates approximately the same region of the sky as the primary receiver. It is further assumed that long messages are exchanged, so that effects of trail formation time can be ignored for the present. These assumptions insure that any receiver located within the ground illumination footprint receives at least a portion of a transmission destined for the primary receiver.

Footprints resulting from trails centered at (x, y, z) with direction (α, β) are weighted by the number of trails at each orientation and location to form a composite ground illumination pattern. The probability of receiving the transmission at a point on the ground (x_s, y_s), given that it is received at the primary receiver located at $(0, 0)$, is given by

$$P_i(x_s, y_s) = \frac{\iiiint A(x, y, z, \alpha, \beta) C_f(x_s, y_s, x, y, z, \alpha, \beta)\, d\alpha\, dx\, dy\, dz}{\iiiint A(x, y, z, \alpha, \beta)\, d\alpha\, dx\, dy\, dz} \tag{2.28}$$

where $A(x, y, z, \alpha, \beta)$ is the number of trails per unit time at the point (x, y, z) with orientation (α, β) observed at the primary receiver. The model described in [122] provides the weighting function A. Figure 2.22 predicts probability of adjacent station reception versus location on the ground using equation (2.28) and the spatial arrival model of [122] for a 660-km north–south link during July at 0900 local time. Receiver and transmitter use 5-element Yagi-Uda antennas elevated 30 ft above ground with a 3° radio horizon.

The conditional reception pattern $P_i\{(x_s, y_s)|(0,0)\}$ is somewhat elliptic in shape, relatively narrow in the transverse direction, and extends to the rear of the primary receiver.

The inherent privacy of meteor scatter communication is sensitive to the region of space illuminated by the receiver/transmitter antenna pattern. Individual trails forming to the rear of the transmitter, farthest from the receiver, have large ground illumination footprints, while trails close to the receiver have relatively small ground illumination footprints. The extent to which trails to the rear of the transmitter are illuminated influences the size of the ground illumination region. Factors that affect common volume illumination are radio horizon and antenna patterns. Variation of the radio horizon by one degree changes the regions of the sky that are illuminated and, therefore, predictions of the size of the ground illumination contours.

The technique described in Weitzen [128], which weights each footprint by the relative number of arrivals at the location of the trail, provides more detailed analysis of the effect of antenna patterns, radio horizon, link distance, time of day, and season on the spatial arrival pattern of trails. The model predicts a diurnal variation in the ground illumination pattern not predicted by the uniform arrival model. Meteor trails to the rear of the transmitter and receiver are largely vertical due to the tangency requirements, while trails in the midlink region are largely horizontal. During the

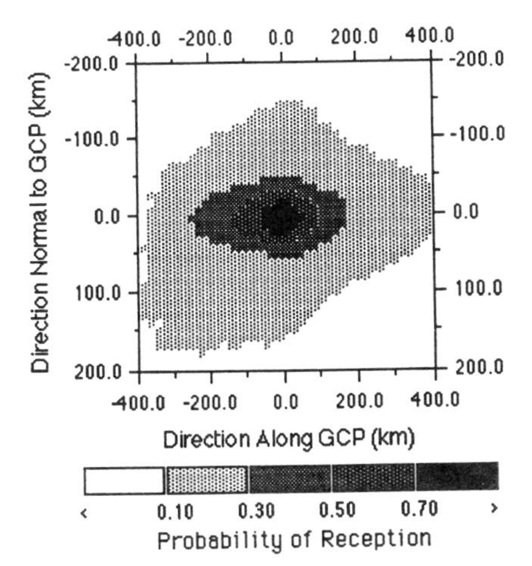

Figure 2.22. Conditional probability of simultaneous reception for a 660-km link.

early morning (0600 local time) the arrival rate of predominantly vertical meteors to the rear of the transmitter and receiver increases due to the orientation of the earth sweeping in meteors. Depending on how the antenna patterns illuminate these regions, the size of the ground illumination footprint may increase or decrease.

Probability of adjacent station reception for meteor scatter links is not reciprocal. Consider a link with stations A and B. In the first case, station B is transmitting to station A. If the antenna patterns illuminate the region of the sky to the rear of station A, the size of the region of adjacent station reception will be low. Now consider the reverse case in which station A is transmitting to station B. The antenna patterns illuminate the region of the sky to the rear of the transmitter so that the individual meteor trail footprints are larger and the region of adjacent station reception will be larger.

The region of adjacent station reception should be smaller than predicted by classical methods due to trail formation time. Consider a station receiving a short 100-ms message. A second station will receive that message if it is within the general footprint of the trail as described in the classic analysis and the point of tangency lies within a region of the trail that has formed and not decayed within the message transmission period.

2.2.5.3 Large-Scale Space Diversity in Meteor Scatter Communication Systems. While the ground illumination footprint of meteor propagation is small relative to that of conventional ionospheric propagation, it is large enough that it cannot automatically be assumed that only one station can receive a message on the trail. Due to the footprint region, more than one meteor scatter receiver may observe the same meteor trail. Since only one node can communicate with the master station at any one time, when estimating network performance, it is not "fair" to count one trail received at two remote nodes as two openings to the network when it is actually one opening.

The degree of diversity in a network is a function of the antenna patterns and the deployment of the receivers. Antennas that have their illumination close to the master station will provide less inherent diversity than those farther away since the ground illumination footprints will be larger for the latter case. Receiver deployments that have a large component transverse to the axis of the secondary coordinate system have more inherent diversity than those that are deployed laterally. A key factor when assessing the performance of a distributed network is how many independent meteor openings per unit time exist into the network. The network diversity factor is defined as the ratio of the arrival rate of trails into the network divided by the average arrival rate of an individual node:

$$\text{Network Diversity Factor} = \frac{\text{Arrival rate into network}}{\text{Average arrival rate to individual node}}$$

The network diversity factor (ndf) ranges from 1 to N, where N is the number of units comprising the network. The greater the diversity factor the more independent paths there are to the network and the less likelihood there is that more than one system will attempt to respond on the same trail to a network probe.

Consider a simple example. Two identical units, closely spaced (less than a few miles apart), will see for the most part the same set of meteor trails. The arrival rate into this simple network is the arrival rate to one of the stations and the space diversity factor is therefore close to 1. Now consider two units that are spaced far enough apart that they would see different sets of meteors. The arrival rate into the network would be twice that of an individual unit and the diversity factor would be close to 2. Expanding on this analogy, N closely spaced receivers would see a similar set of meteors and the network diversity factor would remain close to 1. N widely spaced nodes would each see different meteors and the diversity factor would be close to N.

Consider a network consisting of N units with an arbitrary deployment described in the secondary coordinate system of Fig. 2.14. The inherent diversity of the network is calculated using the following technique.

The ground illumination footprint due to trails at each position and with different orientations is different. To calculate the net effect of the ensemble of different meteor footprints we must loop over all locations within the common volume of the network and all possible trail orientations. Positions and orientations of trails are relative to the primary coordinate system. For each trail orientation and location the binary illumination function $C_f(x_{sn}, y_{sn}, x, y, z, \alpha, \beta)$ (0 for not illuminated, 1 for illuminated) is calculated for each station in the network. Each trail will illuminate a different combination of stations within the network.

In the next calculation, arrival rate of trails with that orientation and location is calculated for each receiver in the network that is illuminated by the given trail. This calculation considers differences in noise at each receiver, distance from the transmitter, antenna gain, and orientation differences that affect trail duration and amplitude. A modified version of the model described in Weitzen [122] provides this result, which is added to a running sum (looping over all orientations and positions) for each station. Distance from the transmitter tends to affect trail duration and in networks where the width of the network is on the order of the distance to the center of the network, differences in trail duration between stations at the front and rear of the network can be considerable.

Since a trail observed at one or more stations in the network counts as a trail observed in the network, the contribution to the network arrival rate due to trails with a given orientation and location is taken as the maximum arrival rate over all stations observing trails from that orientation and position. To provide insight into how the number of stations and their deployment affects diversity in a network, a running sum of the diversity as a function of the number of receivers is also computed.

In conjunction with the efforts to develop models for the inherent space diversity in a widely spaced network, an experiment was developed and performed in the western United States [94] during the winter of 1988/89. The receiver array consisted of 33 stations deployed on concentric rings of 5, 15, 30, and 60 miles radius with an additional four receivers located at a distance of 120-mile radius. Experiments were performed at ranges of 175, 400, and 600 miles (from the center of the array) on north–south and east–west paths. The deployment of receivers is shown in Fig. 2.23.

At each range, diversity was measured using both vertical and horizontal polarization. The horizontal configuration consists of a Yagi at the transmitter and a crossed dipole antenna at the receiver. For vertical polarization a Yagi antenna was used at the transmitter and a vertical whip was used at the receiver. In an additional test at 175-mile range, crossed horizontal dipoles and vertical whip antennas were substituted for the directional antennas at the transmitter. Individual messages were tagged with a code so calculations of footprint of individual trails and space diversity could be made. Data were collected and analyzed independently of the prediction effort.

Figures 2.24 and 2.25 compare predicted and measured diversity as a function of range in the 30- and 60-mile rings for trails of 120-ms duration. In the 60-mile ring a network of 29 receivers provides approximately 8–10 orders of diversity. On the average, 3 stations observe each meteor trail.

Consider two applications of space diversity in the design of a meteor scatter communication networks. First consider a broadcast scenario in which contacting any station in a network constitutes communicating with the network. In this scenario, the waiting time to contact the network is reduced by the diversity factor d. The effect of space diversity can be the

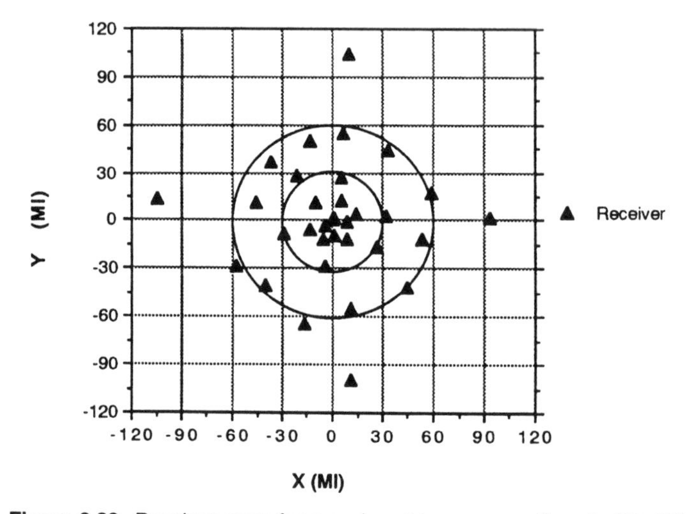

Figure 2.23. Receiver array for experiment to measure diversity [10, 95].

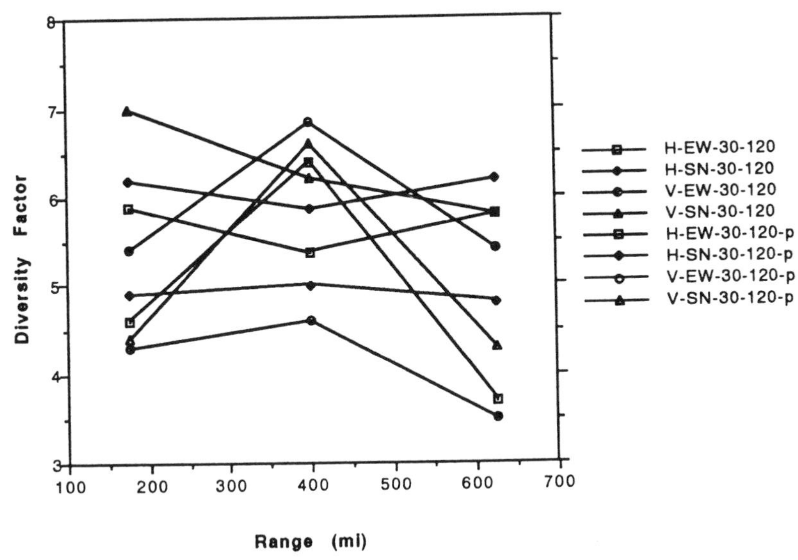

Figure 2.24. Diversity vs. range in a 30-mile ring. In the figures, "p" in the legend denotes prediction while those without "p" indicate measured data, "sn" denotes south to north links, "ew" denotes east west links, "h" denotes horizontal polarization, and "v" denotes vertical polarization.

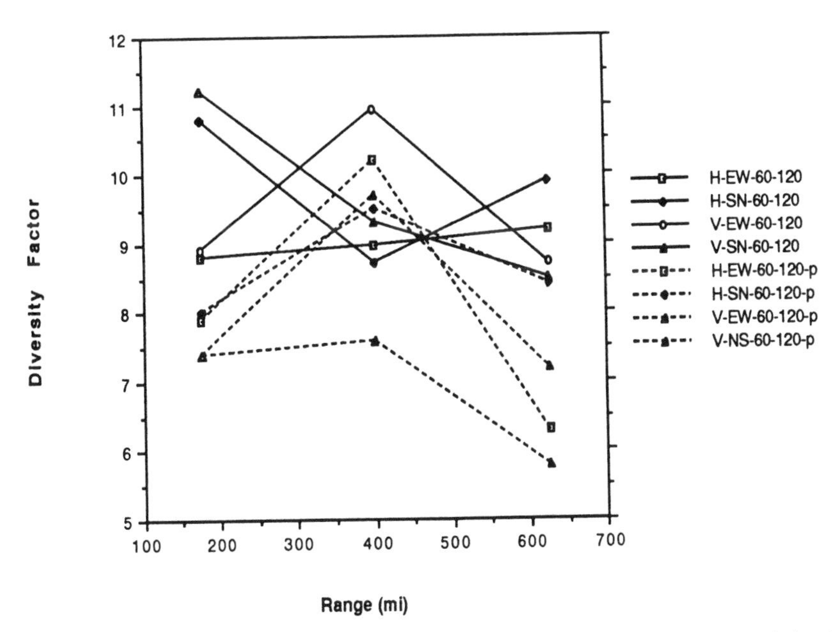

Figure 2.25. Diversity vs. range in a 60-mile ring. In the figures, "p" in the legend denotes prediction while those without "p" indicate measured data, "sn" denotes south to north links, "ew" denotes east west links, "h" denotes horizontal polarization, and "v" denotes vertical polarization.

difference between meeting acceptable message delivery times and not meeting required times.

Now consider a large network with 2000 systems in which the space diversity factor relative to a single unit is 750. This implies that there is a degree of correlation between arrivals at different units. On average, 2.66 stations will receive each message and could potentially respond. Degradation due to multiple responses would result unless additional spatial multiplexity was provided by the protocol structure. The network controller must be able to handle traffic at a rate 750 times that of a single station. The performance of the network will be based on 750 times the single trail arrival rate rather than 2000 as assumed in most analysis.

It has been demonstrated that by using the technique a reasonable degree of accuracy in the prediction of space diversity can be achieved. In future work this technique will be applied to the problem of predicting the performance of large meteor burst communication networks.

2.2.5.4 Probability of k Stations Receiving a Given Message Segment. The analysis of Section 2.2.5.2 described a technique to compute the conditional probability of a second station receiving a piece of a message, given that a first station had received the message. A second measure of interest for the design of networks with a large number of remote terminals is the probability that k nodes in a network will respond to a probing signal. This differs from the previous analysis in that probabilities are not relative to a centrally located node, but are computed for the entire network.

Consider the coordinate system used in the analysis of the previous sections. To compute the probability of k nodes being illuminated simultaneously, the following set of calculations are performed. First, loop over all points in the meteor region of sky coordinate system (x, y, z). Next, at each point in the sky, loop over a number of meteor trail orientations in azimuth and elevation (α, β). Trails of 40 km total length, centered at the point (x, y, z) in the primary coordinate system are considered in the analysis. For each node in the network, compute the point of tangency on the trail using the technique described in the previous section.

Assume that all meteors enter the atmosphere with an average velocity of 35 km/s and begin forming an ion trail at time $t = 0$. Consider messages of duration M. The time the meteor forming the trail requires to traverse the 40-km trail length is divided into intervals of duration M. The time interval at which each node is first illuminated by the trail is computed.

Looping over all the M intervals, count how many nodes are illuminated at this time or within 2 underdense time constants of this time. Two underdense time constants corresponds to approximately 10-dB signal decay, assuming the underdense decay model. This is done for each 40-ms interval and the distribution of $P(k)$ is computed for the individual trail orientation.

Next, the $P(k)$ distribution for a given trail orientation is weighted by the relative occurrence of trails at this orientation and location in the common volume using standard techniques. After looping over all locations within the common volume, normalize to determine the relative probability of k receptions.

Because of the computational complexity of the model, several inherent assumptions are made to make the computation time reasonable. The main assumption is that all nodes are identical in terms of noise conditions and antenna patterns. This is a reasonable assumption for galactic noise limited systems in which all nodes are identical. The estimate described becomes a worst-case upper bound for systems in which the antenna patterns of the nodes or the noise environment are different.

A second inherent assumption is the 2-time constant region for the individual trail footprint. This is based on signals at one point being 10 dB stronger than the weakest signal, which is based on the underdense model assumption. The footprint for overdense trails will be slightly larger than the underdense case.

One case not explicitly discussed is the case of two co-located identical receivers. A conventional model would predict that $P(k)$ would be 1 for this case. A number of experimenters have observed that $P(k)$ will be slightly less than 1 for this case. This can be attributed to the fact that on trails for which the signal-to-noise ratio is not exceedingly high (noise at the two receivers is at least partially independent), there is a nonzero probability that one receiver will make an error, while the other receiver does not make an error. There is also the small-scale space diversity effect discussed by Cannon [16]. Thus, some top-side error is inherent in the model and for the case of a small closely spaced data set tends toward an upper bound on $P(k)$.

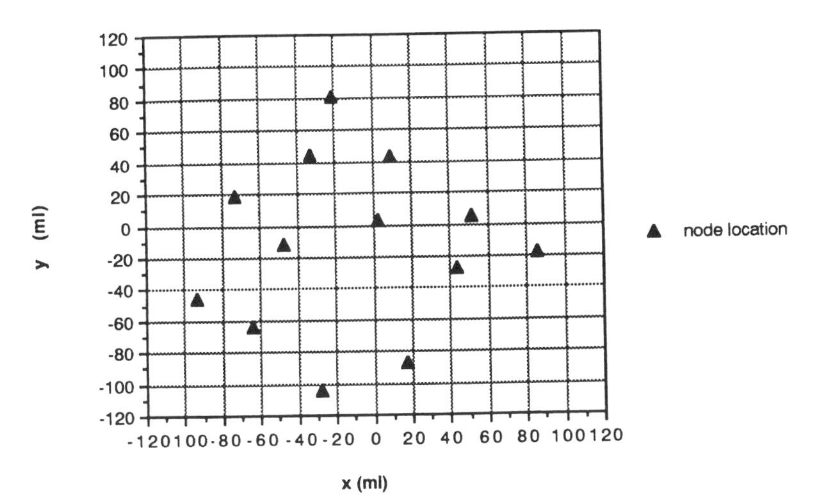

Figure 2.26. Array of nodes for sample cases.

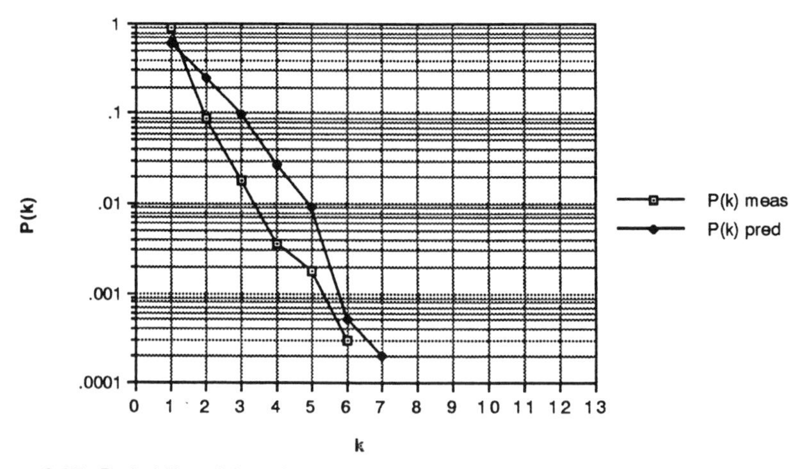

Figure 2.27. Probability of k nodes receiving a 40-ms message. Center of network is 175 miles from transmitter.

As an example of the technique, consider the network with remote nodes shown in Fig. 2.26 which is a subset of figure 2.23. The center of the array is located 175 miles from the transmitter. Message segments of 40-ms duration are used in the analysis.

Figures 2.27–2.29 predict $P(k)$ as a function of k for links of 175, 400, and 625 miles from the center of the array shown in Fig. 2.23. Predictions are compared to data from an experiment described in Mui [94]. In future work, this technique will be extended to networks with thousands of nodes and be used to develop protocols to control such a network.

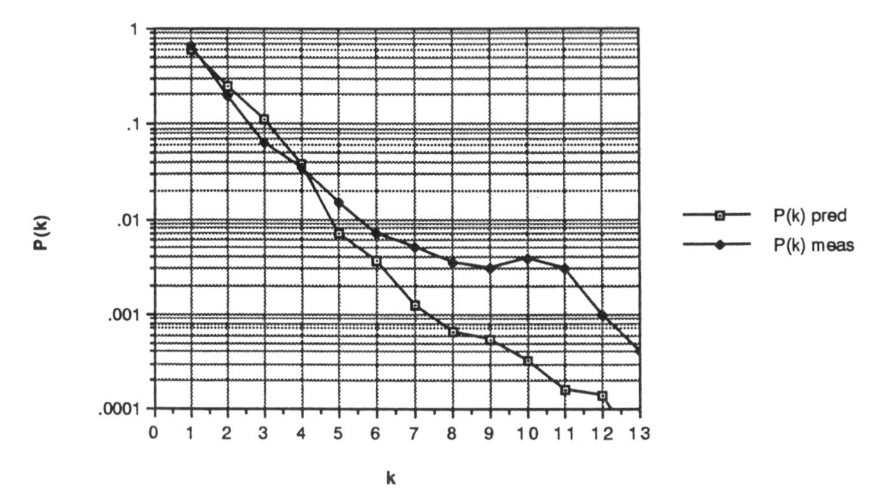

Figure 2.28. Probability of k nodes receiving a 40-ms message. Center of network is 400 miles from transmitter.

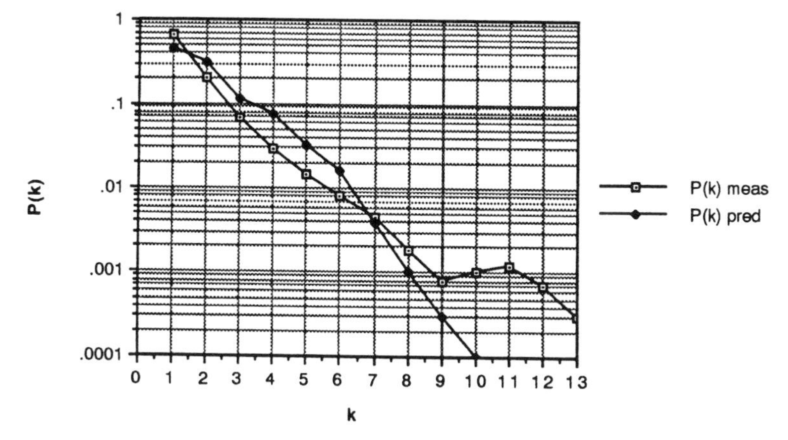

Figure 2.29. Probability of *k* nodes receiving a 40-ms message. Center of network is 625 miles from transmitter.

REFERENCES

1. F. Akram, A. Sheikh, A. Javed, and G. M. D. Grossi, Impulse response of a meteor burst communications channel determined by ray-tracing techniques. *IEEE Trans. Commun.* 467–471 (Apr 1977).
2. N. S. Andrianov, R. A. Kurganov, A. M. Nasirov, and V. V. Sidorov, Oblique-scattering method for measuring individual radiants and meteor velocities. In *Physics and Dynamics of Meteors*, Kresak and Millman (eds.), pp. 14–26. New York: Springer, 1967.
3. W. J. Baggaley, The determination of the initial radii of meteor trains. *Mon. Not. R. Astron. Soc.* 147: 231–243 (1970).
4. W. J. Baggaley, The deionization of dense meteor trains. *Planet Space Sci.* 26: 979–981 (1978).
5. W. J. Baggaley, Meteoroid structure ionization heights. *Mon. Not. R. Astron. Soc.* 101: 9–12 (1981).
6. W. J. Baggely and T. H. Webb, Measurements of the ionization height of sporadic radio meteors. *Mon. Not. R. Astron. Soc.* 191: 829–839 (1980).
7. D. K. Bailey, R. Bateman, and R. C. Kirby, Radio transmission at VHF by scattering and other processes in the lower ionosphere. *IRE Proc.* 43: 1181–1231 (1955).
8. P. J. Bartholeme and I. M. Vogt, COMET: a new meteor burst system incorporating ARQ and diversity reception. *IEEE Trans. Commun.* COM-16: 268–278 (1968).
9. O. I. Belkovic and J. A. Pupysev, The variation of sporadic meteor radiant density and the mass law exponent over the celestial sphere. In *Physics and Dynamics of Meteors*, Kresak and Millman (eds.), pp. 373–381. New York: Springer, 1967.

10. E. R. Billam and I. C. Browne, Characteristics of radio echoes from meteor trails, IV: Polarization effects. *Proc. Phys. Soc.* 69: 98–113 (1956).
11. W. H. Bliss, R. J. Wagner, and G. S. Wickizer, Wide-band facsimile transmission over a 900-mile path using meteor ionization. *IRE Trans. Commun.* CS-7: 252–256 (1959).
12. D. W. Brown, A physical meteor burst propagation model and some significant results for communication system design. *J. Selec. Areas Commun.* SAC-3: 745–755 (1985).
13. D. W. Brown and H. P. Williams, The performance of meteor burst communications at different frequencies. *Proc. AGARD Symposium, Aspects of Electromagnetic Scattering in Radio Communications*, Cambridge, MA, 3–7 Oct 1977.
14. H. Brysk, Electromagnetic scattering by high density meteor trails. *IRE PGAP* 330–336 (Dec 1989).
15. P. S. Cannon, Polarization rotation in meteor burst communication systems. *Radio Sci.* 21: 501–509 (1986).
16. P. S. Cannon, A. K. Shukla, J. N. Tyler, and A. H. Dickenson, Space diversity reception in meteor burst communications systems: Is it an advantage? *Proceedings of the SHAPE Meteor Burst Communications Symposium*, The Hague, Netherlands, 4–5 Nov 1987.
17. P. S. Cannon, J. A. Weitzen, J. C. Ostergaard, and A. D. Bailey, Seasonal variations in meteoric and non-meteoric duty cycle in the polar cap and auroral regions. *Proc. IEE Conference on HF Communications*, Scotland, 1991.
18. R. J. Carpenter and G. R. Ochs, The NBS meteor burst communication system. *IRE Trans. Commun.* CS-7: 263–271 (1959).
19. R. J. Carpenter and G. R. Ochs, High resolution pulse measurements of meteor-burst propagation at 41 Mc/s over a 1295-km path. *NBS J. Res.* 66D: 249–263 (1962).
20. Z. Ceplecha, Atmospheric corrections to meteor velocities and the atmospheric density gradiant. In *Meteors*, Kaiser (ed.), pp. 81–85. New York: Pergamon, 1955.
21. J. A. Clegg and R. L. Closs, Plasma oscillations in meteor trails. *Proc. Phys. Soc. B* 64: 718–179 (1951).
22. R. L. Closs, J. A. Clegg, and T. R. Kaiser, An experimental study of radio reflections from meteor trails. *Philos. Mag.* 44: 313–324 (1953).
23. J. H. Crysdale, Analysis of the performance of the Edmonton–Yellowknife Janet circuit. *IRE PGCS* CS-8: 33–40 (1960).
24. J. G. Davies, Radio measurements of individual meteor orbits. In *Meteors*, Kaiser (ed.), pp. 157–161. New York: Pergamon, 1955.
25. J. G. Davies, Radio observation of meteors. *Advances in Electronics and Electron Physics*, Marton (ed.), pp. 95–127. New York: Academic, 1957.
26. R. I. Desourdis and S. C. Merrill, Meteor scatter signal footprint variation with range, power margin, and time of day. *Proc. IEEE Military Communications Conference*, Boston, MA, 1989.
27. R. I. Desourdis, S. C. Merrill, J. J. Wojtsazek, and K. Hernandez, Meteor burst link performance sensitivity to antenna pattern, power, and range. *Proc. MILCOM Communications Conference*, San Diego, CA, 23–26 Oct, 1988.

28. W. G. Elford, Meteor shower mass distribution from radar echo counts. In *Physics and Dynamics of Meteors*, Kresak and Millman (eds.), pp. 353–361. New York: Springer, 1967.
29. C. D. Ellyett, Solar influence on meteor rates and atmospheric density variations at meteor heights. *J. Geophys. Res.* 82: 1455–1462 (1977).
30. C. Ellyett and C. S. L. Keay, Southern Hemisphere meteor rates. *Mon. Not. R. Astron. Soc.* 125: 326–346 (1963).
31. C. D. Ellyett and J. A. Kennewell, Radar meteor rates and atmospheric density changes. *Nature* 287: 521–522 (1980).
32. G. Eriksson, An investigation of the meteor channel for adaptive communication systems. *Proc. IEEE Military Communications Conference*, Monterey, CA, 1990.
33. V. R. Eshleman, Meteors and radio propagation, part A: meteor ionization trails, their formation and radio echoing properties. *Technical Report 44*, Radio Propagation Laboratory, Stanford University, Palo Alto, CA, 1955.
34. V.R. Eshleman, Theory of reflections from electron-ion clouds. *IRE PGAP* 32–39 (Jan 1955).
35. V.R. Eshleman, On the wavelength dependence of the information capacity of meteor burst propagation. *Proc. IRE* 45: 12 (1957).
36. V. R. Eshleman, The theoretical length distribution of ionized meteor trails. *J. Atmos. Terr. Phys.* 10: 57–72 (1957).
37. V. R. Eshleman, Meteor scatter. In *Radio Noise Spectrum*, pp. 49–79. Cambridge, MA: Cambridge U P, 1960.
38. V. R. Eshleman and L. A. Manning, Radio communication by scattering from meteoric ionization. *IRE Proc.* 42: 530–535 (1954).
39. V. R. Eshleman and R. F. Mlodnosky, Directional characteristics of meteor propagation derived from radar measurements. *IRE Proc.* 45: 1715–1723 (1957).
40. S. Evans, Atmospheric pressures and scale heights from radio echo observations of meteors. In *Meteors*, Kaiser (ed.), pp. 86–91. New York: Pergamon, 1955.
41. S. Evans and J. E. Hall, Meteor ionizing and luminous efficiencies. In *Meteors*, Kaiser (ed.), pp. 18–23. New York: Pergamon, 1955.
42. W. Feller, *An Introduction To Probability*. New York: Wiley, 1968.
43. Forsythe, P. A., Beyond the horizon communication system utilizing signal strength controlled propagation U.S. patent 3,054,895, Sept. 1962.
44. P. A. Forsythe, The forward-scattered radio signal from an overdense meteor trail. *Can. J. Phys.* 36: 1112–1114 (1958).
45. P. A. Forsythe, E. L. Vogan, D. R. Hansen, and C. O. Hines, The principles of JANET: a meteor burst communication system. *IRE Proc.* 45: 1642–1457 (1957).
46. M. D. Grossi and A. Javed, Time and frequency spread in meteor burst propagation paths. *Proc. AGARD-Electromagnetic Wave Propagation Panel, 23rd Symposium on Aspects of EM Wave Scattering in Radio Communications*, Cambrdige, MA, 1977.

47. G. S. Hawkins, A radio survey of sporadic meteor radiants. *Mon. Not. R. Astron. Soc.* 116: 92–104 (1956).
48. G. S. Hawkins, The Harvard radio meteor project. *Proceedings of the Symposium on the Astronomy and Physics of Meteors*, Cambridge, MA, 28 Aug–1 Sep 1961.
49. G. S. Hawkins and R. B. Southworth, The statistics of meteors in the earth's atmosphere. *Smithsonian Contrib. Astrophys.* 2: 359–364 (1958).
50. G. S. Hawkins and D. F. Winter, Radar echoes from overdense meteor trails under conditions of severe diffusion. *Proc. IRE* 45: 1290–1291 (1957).
51. E. Hibshoosh, D. L. Schilling, and J. A. Weitzen, Optimum bit rate predictions for meteor communication. *Proc. IEEE MILCOM Communication Conference*, San Diego, CA, 23–26 Oct 1988.
52. C. O. Hines and P. A. Forsythe, The forward scattering of radio waves from overdense meteor trails. *Can. J. Phys.* 35: 1033–1041 (1957).
53. C. O. Hines and R. E. Pugh, The spatial distribution of signal sources in meteoric forward scattering. *Can. J. Phys.* 34: 1005–1015 (1956).
54. D. E. Hornback, L. D. Breyfogle, and G. R. Sugar, The NBS meteor-burst propagation project: a progress report. *NBS Technical Note 86*, 1960.
55. A. N. Ince, Interception of signals transmitted via meteor trails. *Proc. AGARD Symposium Aspects of Electromagnetic Scattering in Radio Communications*, Cambridge, MA, 3–7 Oct 1977.
56. A. N. Ince, Spatial properties of meteor-burst propagation, *IEEE Trans. Commun.* COM-28: 841–849 (1980).
57. L. G. Jacchia, Fragmentation as the cause of the faint meteor anomaly. In *Meteors*, Kaiser (ed.), pp. 36–42. New York: Pergamon, 1955.
58. D. E. Johnson, Ten years experience with the SNOTEL meteor burst data acquisition system. *Proc. SHAPE Meteor Burst Communications Symposium*, The Hague, Netherlands, 4–5 Nov 1987.
59. T. R. Kaiser, The incident flux of meteors and the total meteoric ionization. *Meteors*, Kaiser (ed.)., pp. 119–130. New York: Pergamon, 1955.
60. T. R. Kaiser, The interpretation of radio echoes from meteor trails. In *Meteors*, Kaiser (ed.), pp. 44–64. New York: Pergamon.
61. T. R. Kaiser, The interplanetary dust cloud. In *Physics and Dynamics of Meteors*, Kresak and Millman (eds.). New York: Springer, 1967.
62. T. R. Kaiser and R. L. Closs, Theory of radio reflections from meteor trails, I. *Philos. Mag.* 43: 1–32 (1952).
63. G. H. Keitel, Certain mode solutions of forward scattering by meteor trails. *Proc. IRE* 43: 1481–1487 (1955).
64. T. G. Knight, Applicability of multipath protection to meteor burst communications. *IRE PGCS* 209–210 (1959).
65. J. D. Larsen, S. W. Melville, and R. S. Mawrey, Adaptive data rate capacity of meteor-burst communications. *Proc. IEEE Military Communications Conference*, Monterey, CA, 1990.
66. R. E. Leader and J. B. Jolly, Remote sensing via meteor trails in remote sensing applications in marine science and technology. *Proc. Adv. Study Initiative*, Dundee Scotland, 1983.

67. V. N. Lebedenec, Radar meteor orbits. In *Physics and Dynamics of Meteors*, Kresak and Millman (eds.) pp. 241–264. New York: Springer, 1967.
68. V. N. Lebedenec and A. K. Sosonova, Radio relfection from meteor trails. *Physics and Dynamics of Meteors*, Kresak and Millman (eds.), pp. 27–44. New York: Springer, 1967.
69. S. W. Li, Auroral Absorption Effect on Meteor Scatter Communication. 1992. Ph. D. dissertation, Univ. of Mass Lowell, Lowellma manuscript in preparation.
70. P. C. B. Lovell, *Meteor Astronomy*. New York: Oxford U P, 1954.
71. A. Malaga, Theory of VHF scattering by field aligned irregularities in the ionosphere. Interim Report, Rome Air Development Center, *RADC-TR-86-116*, 1986.
72. L. A. Manning, Meteoric radio echoes. *IRE PGAP* 4: 82–90 (1954).
73. L. A. Manning, The initial radius of meteoric ionization. *Astron. J.* 63: 283–291 (1958).
74. L. A. Manning, Air motions and the fading, diversity, and aspect sensitivity of meteoric echoes. *JGR* 64: 1415–1425 (1959).
75. L. A. Manning, Oblique echoes from overdense meteor trails. *J. Atmos. Terr. Phys* 14: 82–93 (1959).
76. L. A. Manning and V. R. Eshleman, Meteors in the ionosphere. *Proc. IRE* 47: 2 (1959).
77. L. A. Manning, O. G. Villard, and A. M. Peterson, Double Doppler study of meteoric echoes. *J. Geophys. Res.* 57: 3 (1952).
78. D. N. March, The phase stability of a VHF meteor trail forward scatter channel and an application: a time synchronization system. 1966. Ph. D. dissertation, Montana State Univeristy, Bozeman Montana.
79. L. A. Maynard, Meteor burst communications in the arctic. *Proc. NATO Institute on Ionospheric Radio Communications in the Arctic*, Finse, Norway, 1968.
80. R. E. McCrosky, Orbits of photographic meteors. In *Physics and Dynamics of Meteors*, Kresak and Millman (eds.), pp. 265–279. New York: Springer, 1967.
81. B. A. McIntosh, Meteor mass distribution from radar observations. In *Physics and Dynamics of Meteors*, Kresak and Millman (eds.), pp. 343–351. New York: Springer, 1967.
82. B. A. McIntosh, The effect of wind shear on the decay constant of meteor echoes. *Can. J. Phys.* 47: 1337–1341 (1969).
83. D. W. R. McKinley, The meteoric head echo. *Meteors*, Kaiser (ed.), pp. 65–72. New York: Pergamon, 1955.
84. D. W. R. McKinley, *Meteor Science and Engineering*. New York: McGraw Hill, 1961.
85. D. W. R. McKinley and A. G. McNamara, Meteoric echoes observed simultaneously by back scatter and forward scatter. *Can. J. Phys.* 34: 625–637 (1956).
86. D. W. R. McKinley and P. M. Millman, Determination of the elements of meteor paths from radar observations. *Can. J. Phys. A* 27: 53–66 (1949).
87. M. L. Meeks and J. C. James, On the influence of meteor-radiant distributions in meteor scatter communication. *Proc. IRE* 45: 1724–1732 (1957).

88. D. Melton and M. Darnell, A troposcatter system in the Low VHF band proc 3rd Bangor Symposium on Comm, 29–30 May 1991, Bangor, Wales UK.

89. J. J. Metzner, Improved coding strategies for meteor burst communication. *Trans. Commun.* COM-38: 133–136 (1990).

90. G. H. Millman, HF scatter from overdense meteor trails. *Proc. AGARD Symposium, Aspects of Electromagnetic Scattering in Radio Communications*, Cambridge, MA, 3–7 Oct 1977.

91. L. B. Milstein, D. L. Schilling, R. L. Pickholtz, J. Sellman, S. Davidovici, A. Pavelchek, A. Schneider and G. Eichmann, Performance of meteor burst communication channels. *JSAC* SAC-5: 146–154 (1987).

92. G. F. Montgomery and G. R. Sugar, The utility of meteor bursts for intermitent radio communication. *Proc. IRE* 45: 12 (1957).

93. S. Y. Mui, Coding for meteor burst communication. *Proceedings of International Conference on Communications*, Boston, MA, June 1989.

94. S. Y. Mui, Meteor trail footprint statistics, *Proc. IEEE Military Communications Conference*, Monterey, CA, 1990.

95. E. K. Nemirova, Some results of preliminary observations of resonance scattering of radio waves by meteor trails. *Sov. Astron.* 3: 371–373 (1959).

96. H. Ness, Propagation characteristics of meteor burst communication systems. *Technical Mem STC TM-710*, SHAPE Technical Center, 1983.

97. H. Ness, Dimensioning technique for meteor-burst communication systems. *IEE Proc. F* 132: 505–510 (1985).

98. F. E. Niedenfuhr, An analytical model of meteor burst communication reception zones. *Proceedings of the SHAPE Meteor Burst Communications Symposium*, The Hague, Netherlands 4–5 Nov 1987.

99. J. D. Oetting, An analytis of meteor burst communications for military applications. *IEEE Trans. Commun.* COM-28: 1591–1601 (1980).

100. E. J. Opik, The masses and stucture of meteors. In *Meteors*, Kaiser (ed.), pp. 33–35. New York: Pergamon, 1955.

101. E. J. Opik, Meteor excitation, ionization, and atomic luminous efficiency. In *Meteors*, Kaiser (ed.) pp. 29–32. New York: Pergamon, 1955.

102. J. C. Ostergaard, S. W. Li, and A. D. Bailey, Investigation of frequency diversity effects on meteor scatter links in Greenland. *Proc. IEEE MILCOM Conference*, Washington, DC, 1991.

103. J. C. Ostergaard, J. E. Rasmussen, M. J. Sowa, J. M. Quinn, and P. A. Kossey, Characteristics of high latitude meteor scatter propagation parameters over the 45–104 MHz band. *AGARD Conf. Proc. CP-382*, 1985.

104. J. C. Ostergaard, J. A. Weitzen, P. M. Bench, P. A. Kossey, A. D. Bailey, S. W. Li, J. R. Katan, A. J. Coriaty, and J. E. Rasmussen, Effects of absorption on high-latitude meteor scatter communication systems. *Proc. 1990 IEEE MILCOM Military Communication Conference*, Monterey, CA, 1990.

105. J. C. Ostergaard, J. A. Weitzen, P. M. Bench, P. A. Kossey, A. D. Bailey, S. W. Li, J. R. Katan, A. J. Coriaty, and J. E. Rasmussen, Effects of high-latitude absorption on meteor scatter communication. *Radio Science* Vol 26, No 4, July/Aug 1991 pp 931–943.

106. G. W. Pickard, A note on the relation of meteor showers and radio reception. *Proc. IRE* 19: 1166–1170 (1931).
107. W. H. Press, B. P. Flannery, S. A. Teukolsky, and W. T. Vetterling, *Numerical Recipes*. Cambridge, UK: Cambridge U P, 1986.
108. R. E. Pugh, The number density of meteor trails observable by the forwasrd scattering of radio waves. *Can. J. Phys.* 34: 997–1004 (1956).
109. M. B. Pursley and S. D. Sandberg, Variable-rate coding for meteor-burst communications. *Trans. Commun.* COM-37: 1105–1112 (1989).
110. N. J. Rudie, The relative distribution of observable meteor trails in forward-scatter meteor communications. 1967. Ph. D dissertation, Montana State University, Bozeman, Montana.
111. M. Simek and B. A. McIntosh, Meteor mass distribution from underdense-meteor trail echoes. In *Physics and Dynamics of Meteors*, Kresak and Millman (eds.), pp. 362–371. New York: Springer, 1967.
112. A. M. Skellet, The ionization effects of meteors. *Proc. IRE* 23: 6 (1935).
113. G. R. Sugar, Loss in channel capacity resulting from staring delay in meteor-burst communication. *NBS J. Res.* 64D: 493–495 (1960).
114. G. R. Sugar, Radio propagation by reflection from meteor trails. *Proc. IEEE* 52: 116–136 (1964).
115. G. R. Sugar, R. J. Carpenter and G. R. Ochs, Elementary considerations of the effects of multipath propagation in meteor-burst communication. *J. Res. NBS* 64D: 495–500 (1960).
116. R. N. Thomas, Heat transfer and the ablation process in meteors. In *Meteors*, Kaiser (ed.). New York: Pergamon, 1955.
117. O. G. Villard et al., Some properties of oblique radio reflection in meteor ionization. *J. Geophys. Res.* (June 1956).
118. O. G. Villard, V. R. Eschleman, L. A. Manning, and A. M. Peterson, The role of meteors in extended-range VHF propagation, *Proc. IRE* 43: 1473–1480 (1955).
119. E. L. Vogan and P. A. Forsythe, Privacy in system "JANET". *Project Report*, Defence Research Tellecommunications Establishment, 12-3-4, 1953.
120. J. A. Weitzen, The Feasibility of High Speed Digital Communication Using the Meteor Scatter Channel. 1983 Ph. D dissertation, Univeristy of Wisconsin, Madison, WI.
121. J. A. Weitzen, Characterizing the multipath and Doppler profiles of the high-latitude meteor scatter channel. Rome Air Development Center, *Technical Report RADC-TR-86-165*, 1986.
122. J. A. Weitzen, Predicting the arrival of meteors useful for meteor burst communication. *Radio Sci.* 21: 1009–1020 (1986).
123. J. A. Weitzen, Communicating via meteor burst at short ranges. *IEEE Trans. Commun.* COM-35: 1217–1222 (1987).
124. J. A. Weitzen, A data base approach to analysis of meteor burst data. *Radio Sci.* 22: 133–140 (1987).
125. J. A. Weitzen, Communicating via meteor scatter at high latitudes. Final Report, Rome Air Development Center, *RADC-TR-89-172*, 1989.

126. J. A. Weitzen, Effects of auroral multipath interference on meteor communication. *Proc. IEEE Communication Theory Workshop*, Hawks Cay, FL, 9–12 Apr 1989.

127. J. A. Weitzen, Effects of polarization coupling loss mechanism on design of meteor scatter antennas for short and long range communication. *Radio Sci.* 24: 549–558 (1989).

128. J. A. Weitzen, A study of ground illumination pattern of meteor scatter communication. *IEEE Trans. Commun.* COM-38: 426–431 (1990).

129. J. A. Weitzen, W. P. Birkemeier, and M. D. Grossi, A high speed digital modem for the meteor scatter channel. *Proceedings of Seventeenth Annual Conferene on Information Science and Systems*, Johns Hopkins University, Baltimore, MD, 23–25 Mar 1983.

130. J. A. Weitzen, S. Bourque, P. M. Bench, A. D. Bailey, and J. C. Ostergaard, Distributions of meteor trail amplitudes and its application to meteor scatter communication system design. *Radio Sci.* 26: 451–458 (1991).

131. J. A. Weitzen, M. D. Grossi, and W. P. Birkemeier, High-resolution multipath measurements of the meteor scatter channel, *Radio Sci.* 19: 375–381 (1984).

132. J. A. Weitzen and J. C. Ostergaard, Statistical characterization of fading on meteor scatter communication channels. Geophysics Laboratory, Air Force Systems Command, *Technical Report GL-TR-90-0362*, 1990.

133. J. A. Weitzen and W. T. Ralston, Meteor scatter: an overview. *IEEE Trans. Ant. Prop.* AP-36: 1813–1819 (1988).

134. J. A. Weitzen, M. J. Sowa, J. Quinn, and R. A. Scofidio, Characterizing the multipath and Doppler profiles of the meteor burst communication channel. *IEEE Trans. Commun.* 35: 1050–1058 (1987).

135. J. A. Weitzen and S. Tolman, A technique for automatic classification of meteor trails and other propagation mechanisms for the Air Force high latitude meteor burst test bed. Rome Air Development Center, *Technical Report RADC-TR-86-117*, 1986.

136. A. D. Wheelen, Amplitude distribution for radio signals reflected by meteor trails, II. *J. Res. NBS* 66D: 241–247 (1962).

137. F. L. Whipple and R. F. Hughes, On the velocities and orbits of meteors, fireballs, and meteorites. In *Meteors*, Kaiser (ed.), pp. 149–156. New York: Pergamon, 1955.

138. F. W. Whipple, On meteor luminosity and ionization. In *Meteors*, Kaiser (ed.), pp. 16–17. New York: Pergamon, 1955.

3

MODELING AND ANALYSIS OF METEOR BURST COMMUNICATIONS

Robert I. Desourdis, Jr.
Science Applications International Corporation
Marlborough, Massachusetts

3.1 INTRODUCTION

3.1.1 Background

3.1.1.1 Meteor Burst Communications. Meteor burst (MB) communication systems have been employed for government nonmilitary and commercial applications for low data rate telemetry from remote sensors. Examples of such systems include the Department of Agriculture's SNOTEL (for SNOpack TELemetry) system and the Northern Natural Gas Company's network for pipeline monitoring [1]. In these applications, message transmission delays of many minutes to hours may be acceptable, with average throughputs on the order of 10 bits per second (bps). Military applications, such as transmission of warning, command, and status messages, may tolerate maximum delays of at most seconds to minutes. Consideration of MB to provide time-urgent communications must be tempered by an understanding of the inherent channel variability as well as the sensitivity of the medium to antenna pattern, link power margin, and range.

Under ideal conditions, state-of-the-art MB terminals can provide daily average 1000-character message waiting times of less than one minute [2] for average throughputs in excess of 100 bps. Typically, these terminals employ

1 kW transmit power, operate between 40 and 50 MHz, and use high gain, horizontally polarized Yagi antennas located at "quiet" sites. Fixed rate transmission below 10 kbps, coherent modulation, and error detection are employed, while variable data rate, forward error correction, and impulsive noise mitigation schemes are also available [3]. Protocols support both half- and full-duplex operation with automatic repeat request (ARQ) for link communications and routing tables for network communications. Attempting to operate these terminals below the design link power margin can result in long transmission delays and negligible throughput.

Poor link performance typically results from the use of low gain antennas at long range or highly directional antennas at short range, coupled with inadequate power margin for operation in the local noise environment. This result is essentially independent of the modulation and coding techniques employed because a negligible number of meteoric signals are detected, and if detected produce too little duration to complete the link protocol and support communications. In this case, MB link power budget sensitivity combines with the significant temporal variation in meteor rates to yield orders of magnitude variability in link performance.

At the other extreme, the use of multikiloWatt transmit powers and high gain antennas (e.g., 20 dBi peak gain) at appropriate link ranges can produce a continuous low-level received signal, composed primarily of E- and D-region scatter enhanced by frequent meteor bursts, known as ionoscatter [4]. The use of variable data (or code) rate to optimally communicate during both the continuous low level ionoscatter signal and the high-level, albeit intermittent, MB signal was proposed as early as 1957 [5]. Variable rate operation employs real-time channel quality measurements to enable trans-burst modem selection of the appropriate information transmission rate. For example, transmission rates as low as 500 bps would be supported by the continuous low-level ionoscatter signal while rates exceeding 100 kbps would be possible during sufficiently intense meteor bursts or sporadic E propagation events. The resulting link performance would then be characterized by negligible transmission delays and average information throughputs exceeding 1000 bps.

As communication link performance requirements vary from message delivery delays of hours to seconds, or alternatively, from average throughputs of 10 bps to 1000 bps, the requirements for antenna pattern and link power margin also vary significantly. In other words, good rf link design must be achieved before consideration of advanced modulation or coding techniques and high-speed transmission to meet communication requirements. In this context, computer modeling and analysis of rf link design are necessary for the proper evaluation of MB communications for specific applications.

3.1.1.2 Meteor Scatter Phenomena.

The earth is immersed in a complex system of meteoroids orbiting the sun with a diversity of orbital periods and inclinations, as shown in Fig. 3.1. These meteoroids are divided into two

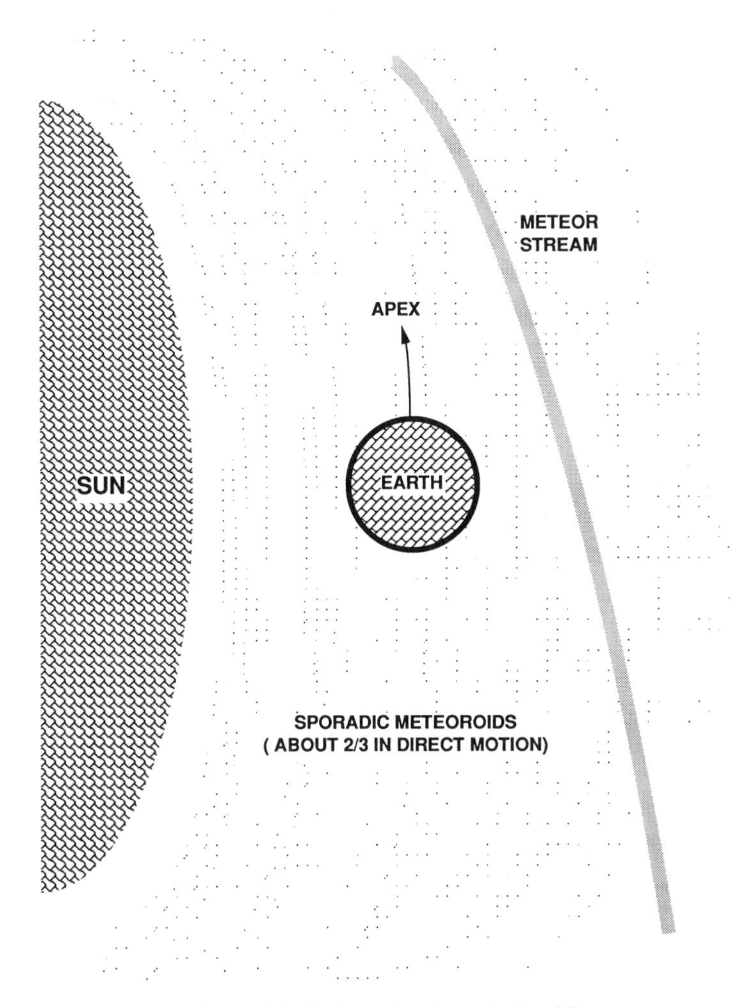

Figure 3.1. Earth and meteoroids in orbit.

categories: stream and sporadic. Stream meteoroids travel in confined solar orbits intersecting the earth's orbit at predictable times each year. Stream meteoroids become shower meteors within the earth's atmosphere and are located by their radiant, the point in the sky from which the shower appears to originate. The orbits of hundreds of major and minor streams have been catalogued [6] as well as their respective shower radiants, meteor arrival rates, meteor speeds, and observed durations. Major showers that produce visually observed rates of 20 to 50 meteors per hour [7] and recur for brief periods (hours to days) each year have been dubbed the "permanent" showers. Examples of permanent showers are the Perseids in August, achieving a peak rate about 12 Aug and the Geminids in December, peaking on 14 Dec. When they occur, shower meteors can produce significant performance enhancement to MB communication links. Their contribution is short-lived and, depending on link performance, may be negligible compared to the sporadic meteor contribution.

Sporadic meteoroids travel in dispersed orbits, usually characterized by a distribution of orbital parameters [8] or, equivalently, a sporadic meteor radiant density (SMRD) map. A SMRD map, has been determined covering most of the celestial sphere for each month from combined radar measurements performed at the Engelhardt Observatory in the former Soviet Union and Mogadishu, Somalia, in Africa [9]. Sporadic meteoroids produce the majority of meteors present in the atmosphere during all hours of each day in the year. In fact, about 10^{12} meteoroids have been estimated to enter the earth's atmosphere each day, amounting to a daily rate of more than ten tons of meteoric material [10]. The vast majority of these meteors disintegrate before contacting the earth's surface, but they nonetheless serve as the dominant source of meteor trails available to support MB communications. Since meteor arrival rates increase exponentially with decreasing meteoroid (pre-burn meteor) mass, the smallest meteor whose trail supports MB link communications determines the rate of usable meteor bursts. For a well-designed link, meteor masses below 10^{-3} g may be expected to produce trails useful for MB communications.

The earth's orbital motion, combined with the distribution of sporadic meteoroid velocities (speed and direction), causes the earth to "sweep up" meteroids as it moves through space. As a result, greater meteor arrival rates and speeds are observed on the front side of the earth, which corresponds to the morning hours (sunrise) at mid-latitudes. At this time, an observer (a MB link, for example) is positioned closest to the front, or apex, of the earth's way (see Fig. 3.2*a*) in its solar orbit. Since fewer meteoroids overtake the earth from the back side (antapex direction) to become meteors in the evening hours (sunset) than are swept up in the morning hours, a diurnal variation in the rate of usable meteor bursts occurs at mid-latitude. This diurnal variation is minimized near the earth's poles, where the MB link is exposed to a less variable meteor flux throughout the day.

The density of meteoroids along the earth's orbit achieves a maximum in June to September and a minimum in January to March, as depicted in Fig. 3.2b. This maximum is partially due to a significant increase in stream activity during the latter half of the year coupled with an increase in the sporadic background. Monthly one-day measurements [11] in the Northern Hemisphere were used to suggest that sporadic meteor density was uniformly distributed about the earth's orbit [12]. Nevertheless, more extensive radar measurements of sporadic meteor arrival rates [13, 14] at equatorial latitudes showed a minimum density encountered during the first several months of the year, while achieving a significantly greater density during the latter months of the year. The earth's orbital plane (ecliptic plane) tilt relative to its axis, called the obliquity of the ecliptic, causes latitude-dependent distortion of the seasonal variation in the observed meteor arrival rate. A MB link therefore experiences both diurnal and monthly variations

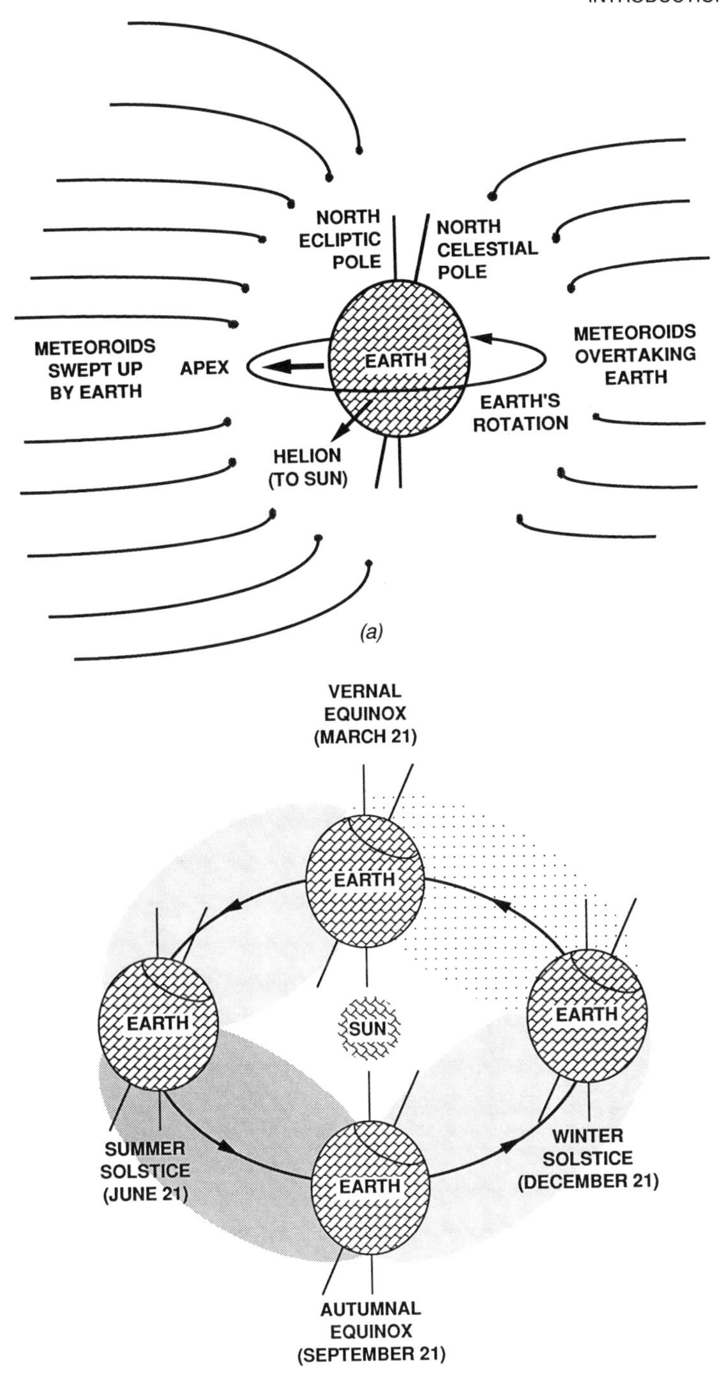

NOTE: CHANGES IN METEOROID ORBITAL DENSITY
ARE MORE GRADUAL THAN SHOWN

(b)

Figure 3.2. (*a*) Diurnal variation in meteor rate; (*b*) seasonal variation in meteor rate

in the hourly average rate of usable meteor bursts. In fact, a maximum to minimum average rate ratio of 25 to 1, or greater, may be experienced over all hours of the year on mid-latitude MB links.

A MB link therefore exhibits both diurnal and monthly variations in the hourly average usable meteor rate, with the instantaneous meteor rate exhibiting a Poisson-like distribution about these average values. A day-to-day variation in the average meteor rate also occurs for corresponding hour intervals on subsequent days. This variation is due to changes in the dominant meteor radiants through the month, the non-uniform distribution of meteoroids in space, and short-term variation in the earth's upper atmosphere. Moreover, hourly average values in a given month will change from year-to-year. This yearly variation may be due in part to solar effects on the earth's atmosphere, which are probably responsible for the observed solar cycle variation in usable meteor rates.

The *observed* meteor velocity in the earth's atmosphere, $\mathbf{V}_o$ is determined by the meteoroid's orbital (sun-relative or heliocentric) velocity, $\mathbf{V}_H$, the earth's heliocentric velocity, $\mathbf{V}_E$, the earth's gravitational attraction, atmospheric deceleration, and the earth's rotation (see Figs. 3.3*a*–3.3*d*, respectively). The vector difference between $\mathbf{V}_H$ and $\mathbf{V}_E$ resolves the *geocentric*, or earth-relative velocity, $\mathbf{V}_G = \mathbf{V}_H - \mathbf{V}_E$, of the meteoroid. Next, gravity accelerates the meteoroid and perturbs its trajectory. As a result, the meteoroid enters the atmosphere as a meteor with the *apparent* velocity, $\mathbf{V}_a$. This

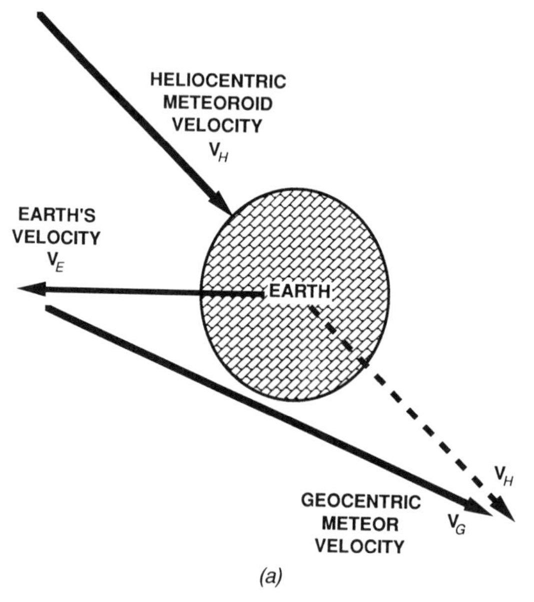

Figure 3.3. (*a*) Geocentric velocity; (*b*) Gravitation effects; (*c*) atmospheric deceleration; (*d*) diurnal aberration

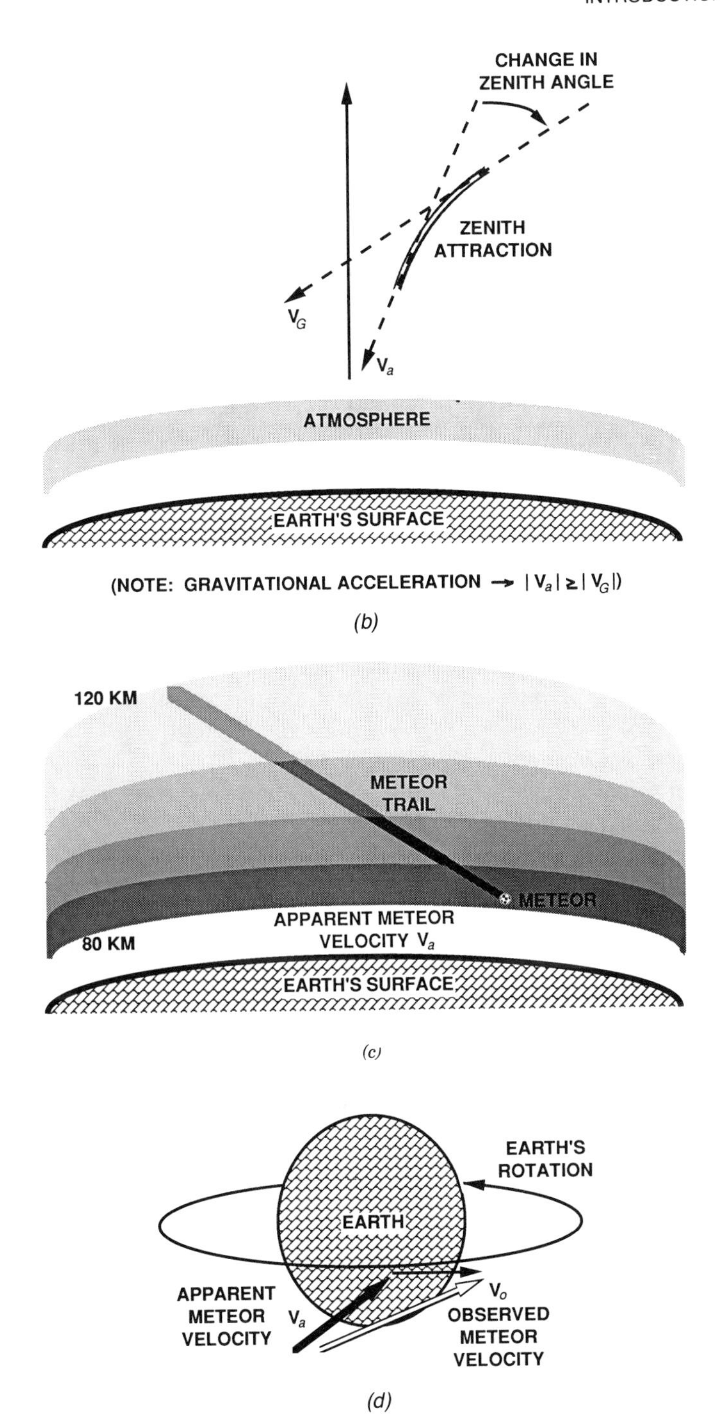

(b)

(c)

(d)

Figure 3.3. (*Continued*)

effect, called *zenith attraction* [15], has the greatest impact on meteoroids that have the lowest geocentric velocities [16]. The apparent meteor speed, $|\mathbf{V}_a|$, varies between about 11 km/s for meteoroids with zero geocentric velocity approaching the earth from the antapex direction (behind) to about 72 km/s for meteoroids approaching from the apex direction (head on).

A further modification of the apparent velocity, $\mathbf{V}_a$, is caused by the earth's rotation and the concomitant movement of an earth-bound observer. This effect, called *diurnal aberration* [17], is greatest for observers located at the celestial equator. It produces less than ± 0.5 km/s change in the apparent meteor speed for any observer's latitude as well as a small direction change. Diurnal aberration is important for precise astronomical measurements but may be ignored in meteor burst link modeling. The meteor velocity adjusted for the diurnal aberration is the so-called *no-atmosphere* velocity and is designated $\mathbf{V}_\infty$.

The meteor collides with atmospheric particles and decelerates as it descends into the atmosphere. Deceleration is negligible for most of the meteor's flight in the upper atmosphere where it may form usable trails for meteoric trail-scatter. The resulting observed velocity, $\mathbf{V}_o$, differs from the no-atmosphere velocity, $\mathbf{V}_\infty$, by less than 1 km/s. The phenomena of near constant meteor velocity through the upper atmosphere is called *velocity persistence* [18]. Ultimately, the exponentially increasing atmospheric particle density is sufficient to rapidly slow the meteor, if it has not disintegrated, to its terminal velocity.

The meteor's contact with atmospheric particles converts the kinetic energy of motion to heat, which evaporates atoms from the meteor's surface. Collision of these atoms with atmospheric particles produces heat, light, and ionization, which forms a trail (or train) behind the meteor. The mass, size, shape, and velocity of the meteor combined with the atmospheric density at the meteor's altitude determine the trail's electron volume density at that altitude. Faster meteors produce more efficient ionization, creating greater electron densities. In addition, faster meteors produce trails with greater initial diameters than slower meteors. The resulting electron volume density, combined with trail geometry relative to the transmit and receive antennas, determines the initial strength and lifetime of the trail-scattered signal above a specified received signal level (RSL).

The free electrons in the meteor trail serve as re-radiators of an incident rf signal. Maximum incident signal is coherently "scattered" from these electrons when the trail is oriented tangent to an ellipsoidal surface of revolution whose foci are located at the transmit and receive antennas of a radio link. Figure 3.4 shows typical trail-scatter geometry, in which maximum signal is received once the trail has completely traversed the central Fresnel zone. This zone determines that portion of the meteor trail providing the greatest contribution to the RSL. The portion of the trail within the central Fresnel zone is defined such that ray path lengths from the transmit antenna to the trail and then from the trail to the receive antenna differ by

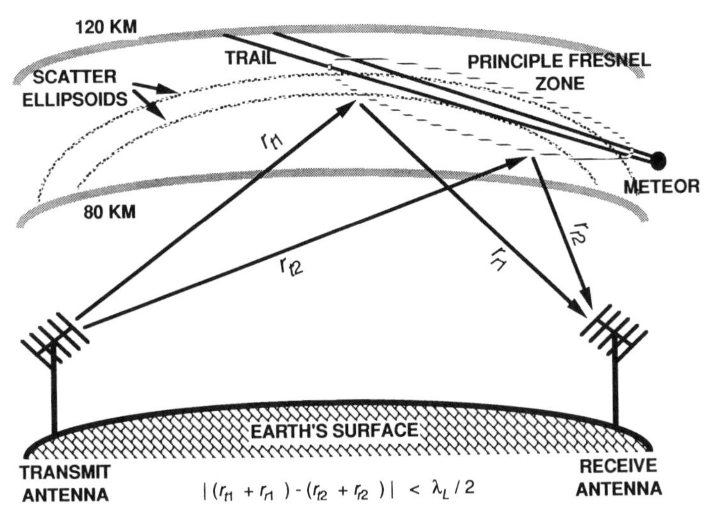

Figure 3.4. Link geometry for meteor trail-scatter.

less than half a signal wavelength. Rays with total path lengths that add constructively at the receiver meet the criterion for coherent scatter. As the meteor passes through subsequent Fresnel zones, alternating constructive and destructive interference results, yielding a velocity-dependent sinusoidal variation in the received signal amplitude as shown in Fig. 3.5. Atmospheric meteor speed may be accurately determined from these amplitude measurements [19].

The asymmetrical distribution of meteor radiants over the sky is equivalent to a directional sensitivity in the meteor arrival rate. This sensitivity, combined with diurnal variation in meteor rates, geometrical scatter requirements, and antenna patterns, determine those regions of the sky contributing the greatest number of meteor trail-scattered signals. These sky regions, called hot spots, are illustrated in Figs. 3.6*a–d* on a mid-latitude north-south MB link at 6 hrs local time (LT), 12, 18, and 24 (or 0) LT, respectively. This figure shows the number of suitably oriented meteor trails as line segments, with segment length indicating the relative meteor arrival rate.

Figure 3.6 is most easily interpreted by considering that most meteors, due to their earth-relative velocities, appear to approach the earth from the apex direction. At 6 LT (Fig. 3.6*a*), usable trail hot spots are formed to both east and west sides of the link Great Circle path (GCP). At noon (12 LT, Fig. 3.6*b*) usable orientations should form on the east side of the GCP as suitable meteor trajectories are descending from west to east. At 18 LT (Fig. 3.6*c*) trails should again be usable to either side of the GCP, albeit at significantly reduced arrival rates (i.e., shorter line segment lengths in the figure). Finally, the noon behavior is reversed at midnight (Fig. 3.6*d*), for which most trajectories descend from east to west forming a hot spot to the west of the GCP.

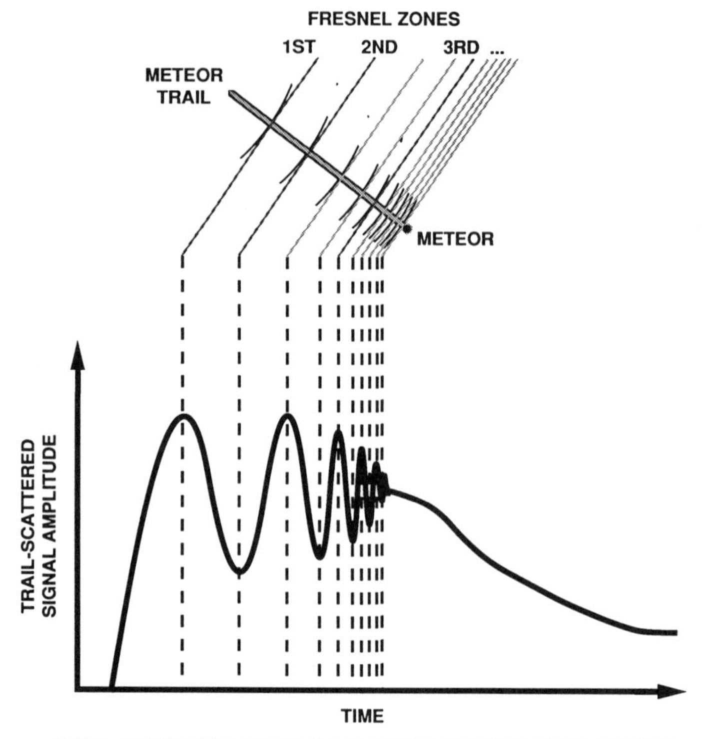

Figure 3.5. RSL showing diffraction effects.

As the trail electrons spread radially outward from the trail axis due to normal atmospheric diffusion, ray path lengths for a significant number of trail electrons contributing to the RSL violate the coherent trail-scattering criterion. This violation results in destructive interference of RSL contributions, resulting in the ultimate disappearance of the trail-scattered signal. These short-lived received signals, or meteor bursts, may last from several milliseconds to many seconds. Signal duration depends on trail line density, initial radius, diffusion rate, electron attachment and the state of upper atmospheric winds [20]. Digital communications is possible during these brief, intermittent meteor bursts, thus forming the basis for an over-the-horizon digital radio link. Since persistent meteoric ionization useful for trail-scatter occurs between 80 and 120 km above the earth's surface, maximum propagation distances of 2400 km are possible. The maximum usable distance for communications, however, is generally below 2000 km because of earth blockage, terrain obstructions, and antenna pattern ground tuck.

The time dependence of the trail-scattered RSL depends on the electron volume density in the meteor trail as well as atmospheric parameters within

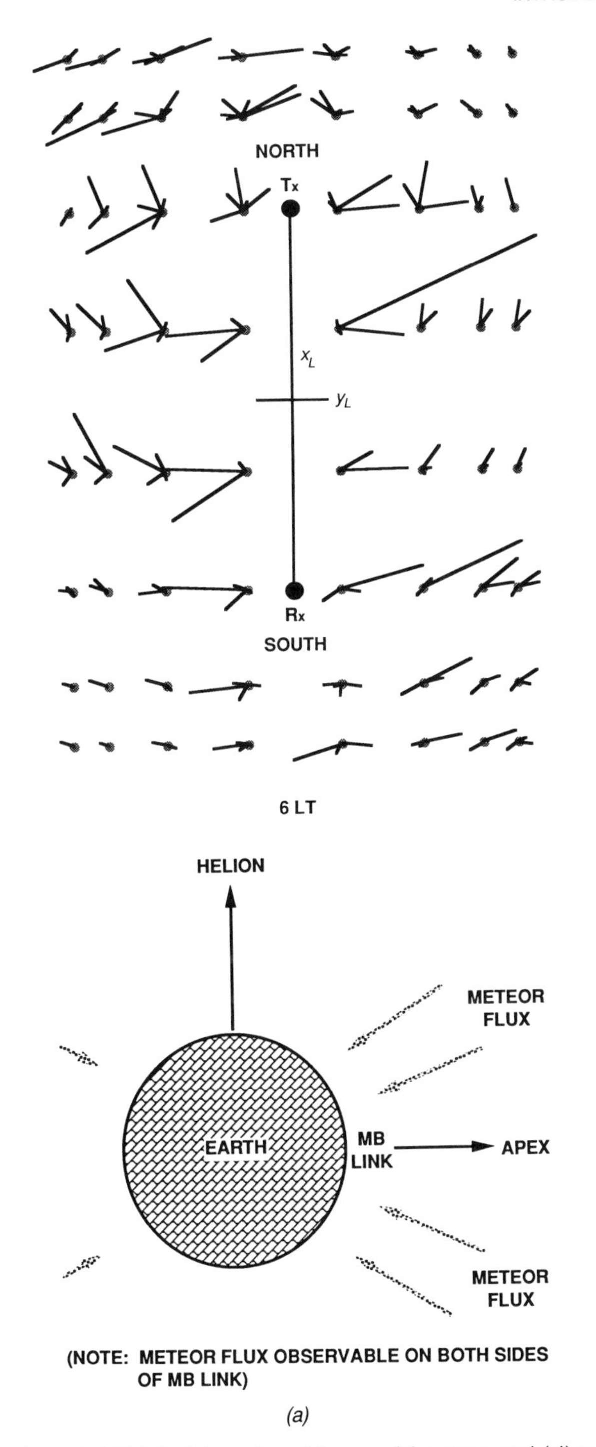

Figure 3.6. North-south MB link: (*a*) sunrise; (*b*) noon; (*c*) sunset, and (*d*) midnight.

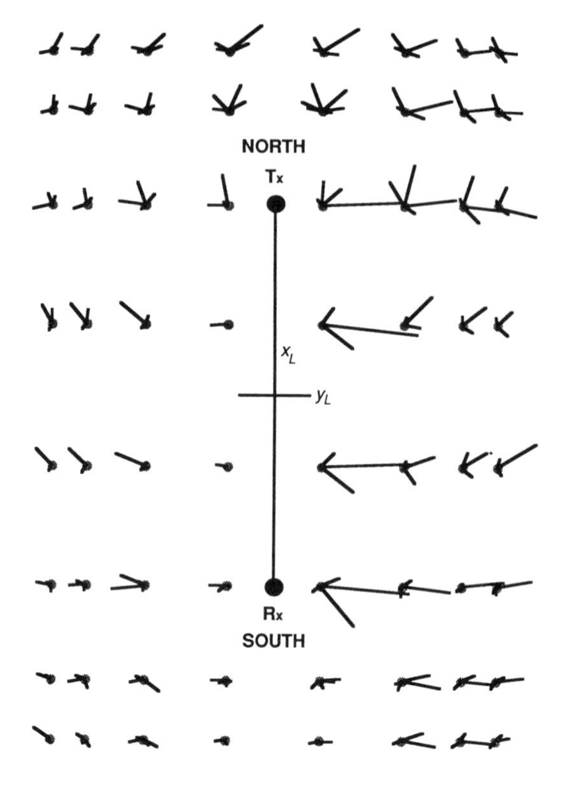

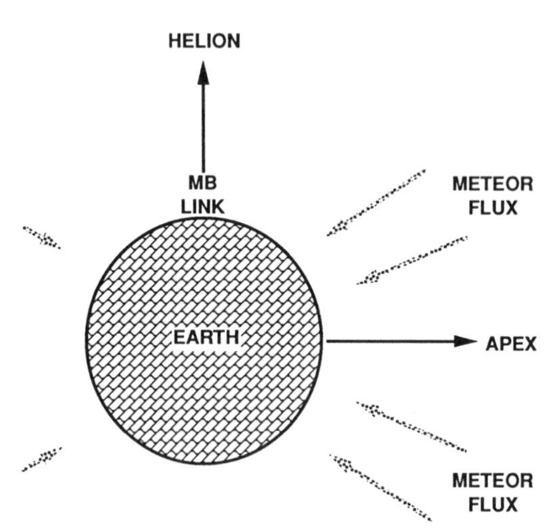

(NOTE: MOST OBSERVABLE METEOR FLUX TRAVERSES ACROSS LINK FROM WEST TO EAST)

(b)

Figure 3.6. (*Continued*)

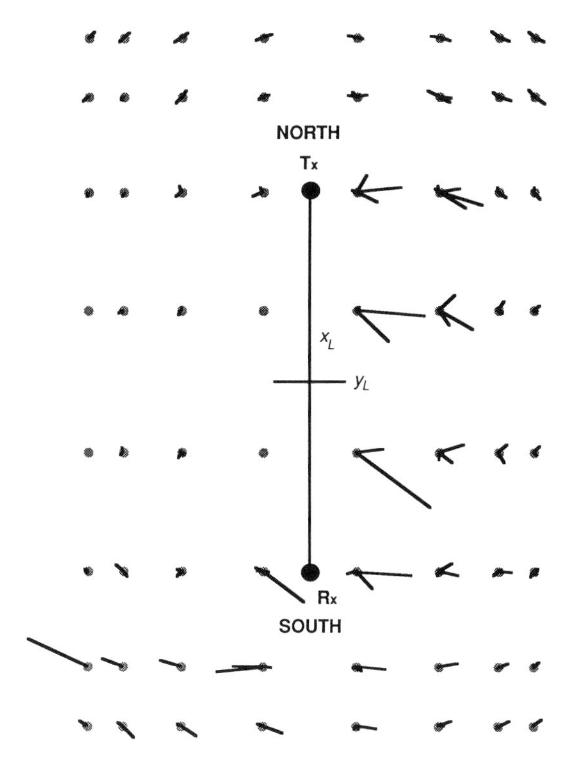

18 LT

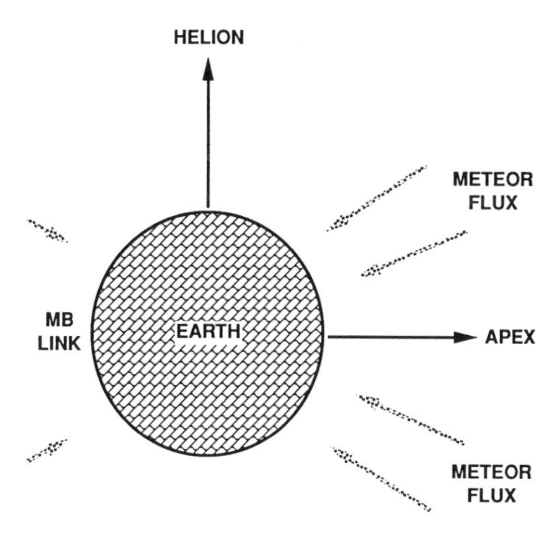

(NOTE: MINIMUM METEOR FLUX OBSERVABLE ON "BACK SIDE" OF EARTH)

(c)

Figure 3.6. (*Continued*)

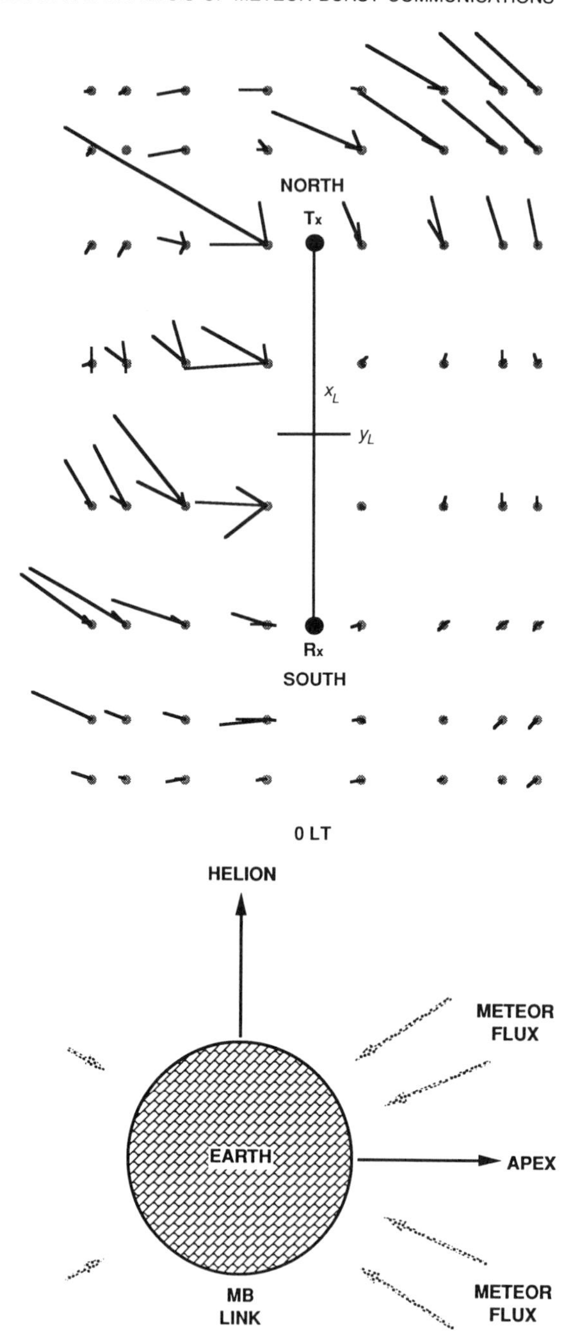

(d)

Figure 3.6. (*Continued*)

the central Fresnel zone. If the electrons in the trail are sufficiently sparse to permit single scattering, that is, if energy is scattered from single electrons between transmitter and receiver, then the trail is termed *underdense*. An underdense trail is depicted in Fig. 3.7*a* and the typical time dependence of the RSL in Fig. 3.7*b*. Peak power is achieved immediately after trail formation, when the phase dispersion of single-scattered RSL components is a minimum. The classic exponential decay characteristic of the underdense trail-scattered signal is evidenced in the measured RSL traces shown in Fig. 3.8. This decay is caused by the cancellation of phase-shifted RSL contributions from radially dispersing electron scatterers, that is, violation of the geometrical conditions for coherent trail-scatter.

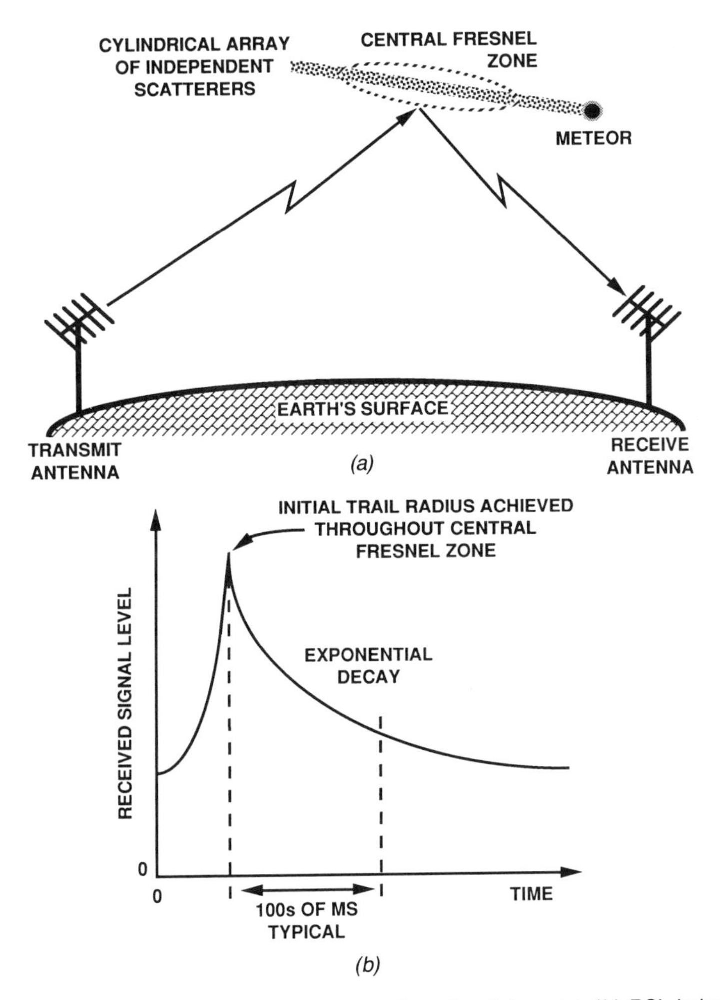

Figure 3.7. Long wavelength underdense trail-scatter: (*a*) model; (*b*) RSL behavior.

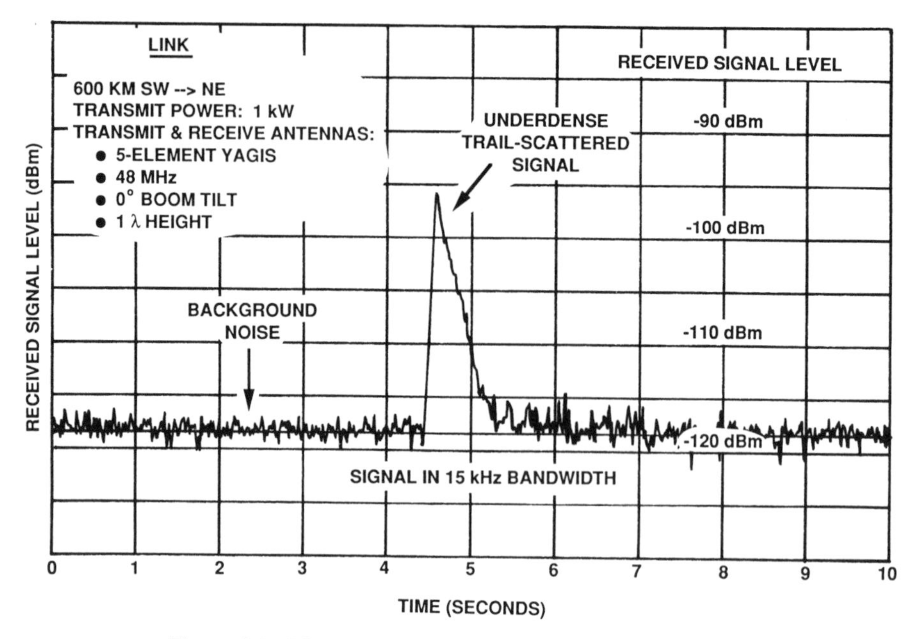

Figure 3.8. RSL trace from an underdense trail-scatter event.

The *long wavelength* trail model assumes that the duration of the underdense trail-scattered signal exceeds the time of trail formation. In other words, the trail-scatterer diameter is approximately constant throughout the central Fresnel zone relative to the incident signal wavelength. This assumption is not valid for all trail orientations, signal frequencies, meteor velocities, and link geometries, in which case the trail cannot be considered a cylindrical volume of electron scatterers. In this case, the RSL corresponds to scattering from a parabola-shaped ionized region formed around and immediately behind the meteor, as shown in Fig. 3.9*a* [21]. The resulting RSL time behavior, symmetric about the moment that the meteor passes through the middle of the central Fresnel zone, is shown in Fig. 3.9*b*.

The trail-scattered signal has been shown to exhibit significant sensitivity to the polarization of the incident signal [22]. Theoretical results [23–25] and radar measurements [26] show that the electric vector oriented perpendicular (transverse) to the trail axis yields the greatest scattered signal. A space-charge separation across the trail diameter supplies a restoring force, which, in conjunction with changing incident wave polarity, results in greater scattered signal amplitudes than produced by polarization parallel to the trail axis. Most researchers suggest an average ratio of transverse-to-parallel scattered signal amplitudes of two, although much larger values have been predicted [27], presumably due to inadequate model formulation [28]. Of course, this result is dependent on trail diameter and electron line

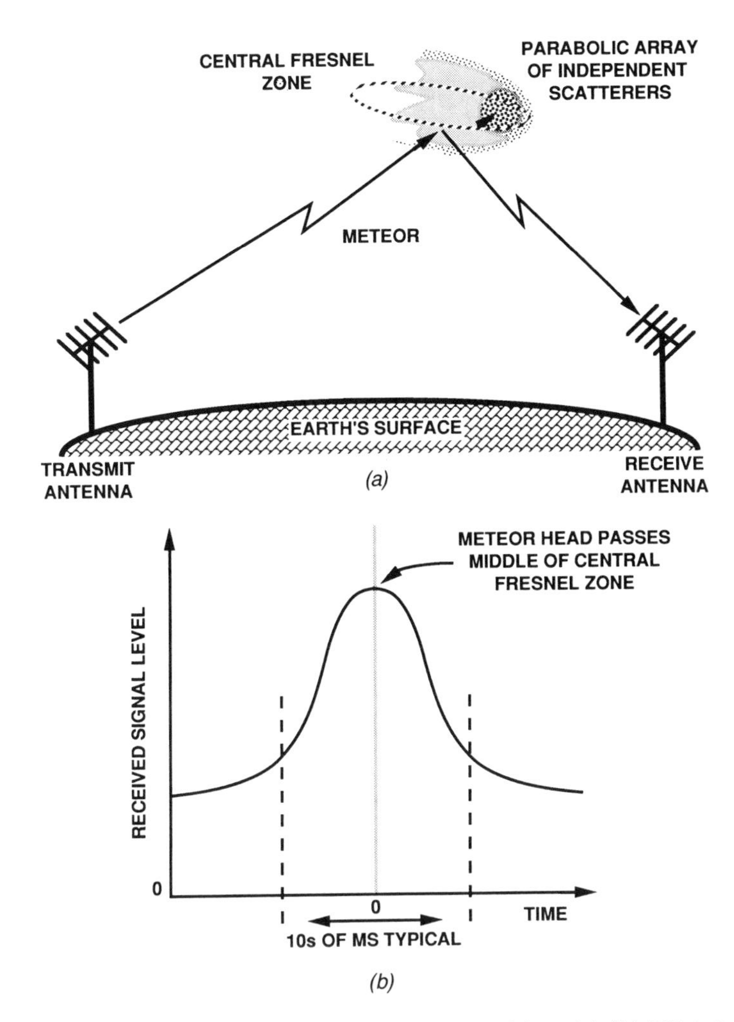

Figure 3.9. Short wavelength underdense trail-scatter: (*a*) model; (*b*) RSL behavior.

density, signal wavelength, and scattering angle as well as incident signal polarization. The resonance effect dissipates early in the trail-scattered signal lifetime, typically in less than 50 ms [29].

If the trail electron volume density exceeds a critical value, beyond which most of the RSL is the consequence of "multiple" scattering, the trail is described as *overdense*. Overdense trails are traditionally considered as metallic cylinders that completely reflect the incident signal (Fig. 3.10*a*) [30], although other overdense trail scattering models have been proposed [31]. The early history of the classic overdense trail-scattered RSL is marked by a gradual rise to a shallow maximum. As the dense trail expands, it presents maximum cross-sectional area relative to the incident signal wave-

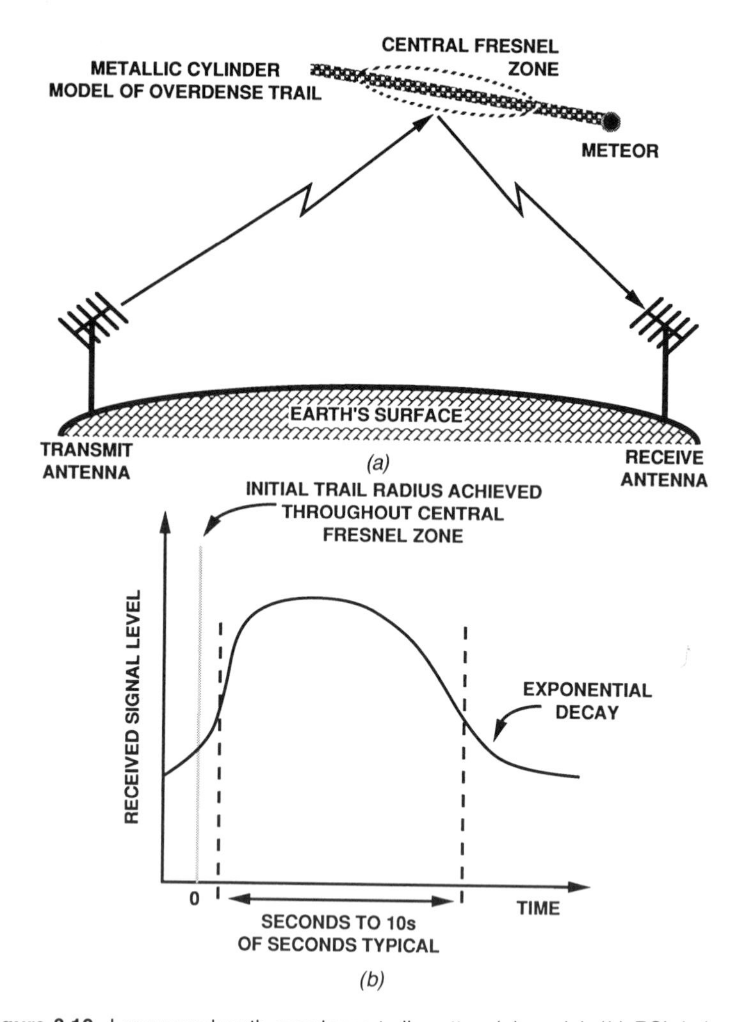

Figure 3.10. Long wavelength overdense trail-scatter: (*a*) model; (*b*) RSL behavior.

length and electron volume density (see Fig. 3.10*b*), producing maximum trail-scattered signal amplitude. Beyond the maximum point, the overdense trail diffuses and transitions to a single scattering volume density, ultimately producing an underdense RSL decay. As in the underdense case, a signal frequency threshold exists beyond which a cylindrical scattering volume assumption fails to adequately describe observed RSL behavior. In this case, the dominant signal is provided by the ionized volume close to the meteor itself.

The RSLs from classic overdense trails are shown in Figures 3.11*a* and 3.11*b*. The significant RSL fading apparent in these traces is caused by signal scattering from multiple trail reflection points [32]. These multiple

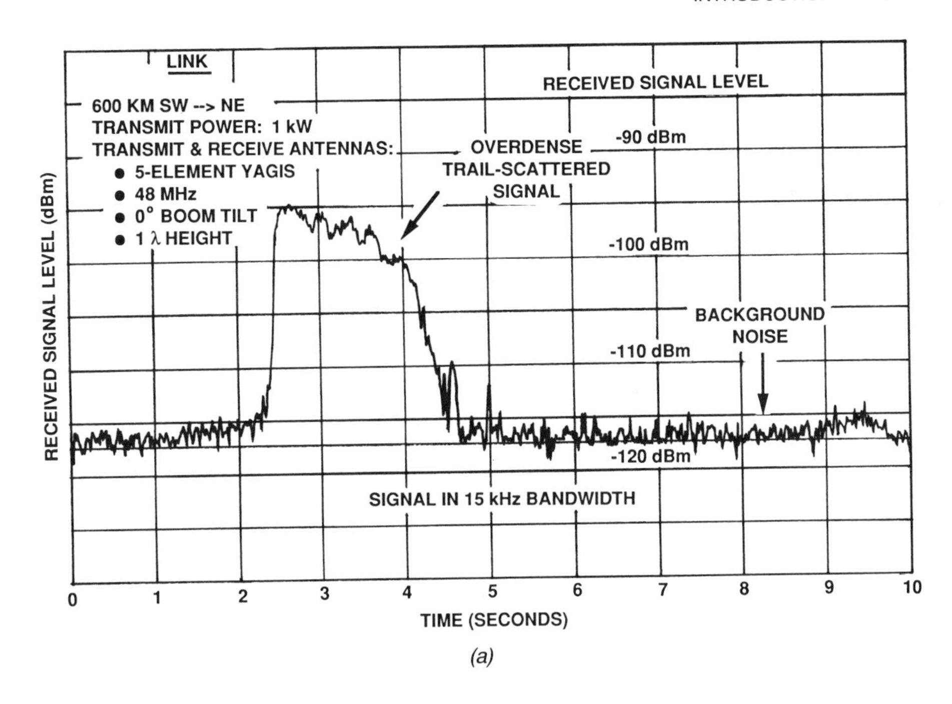

(*a*)

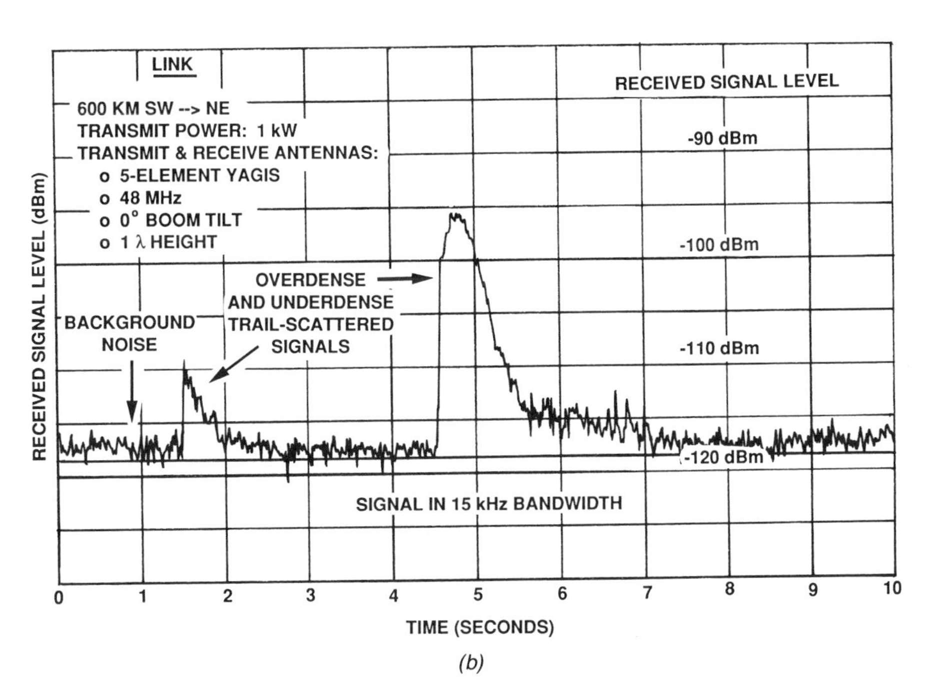

(*b*)

Figure 3.11. (*a*) Long wavelength overdense RSL trace; (*b*) overdense and underdense RSL traces.

reflection points are caused by upper atmospheric winds that force different parts of the trail into conformance with the trail-scattering geometry criterion are shown in Figure 3.12*a*. As a result, two or more significant RSL contributions arrive more than half a wavelength out of phase. The resulting multipath fading (see Figure 3.12*b*) reduces the effective trail RSL duration above threshold as well as the available channel bandwidth [33]. The dissipation of trail electron scatterers by ambipolar diffusion, electron recombination, electron attachment, and wind turbulence ultimately accounts for disappearance of the trail-scattered RSL [34].

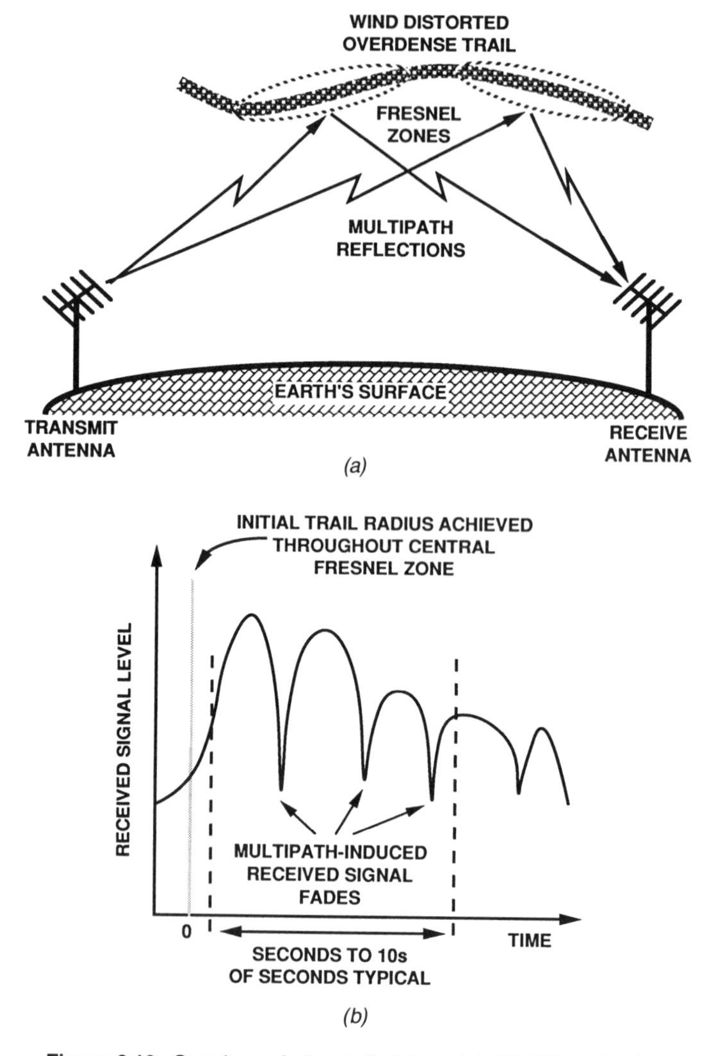

Figure 3.12. Overdense fading trail: (*a*) model; (*b*) RSL behavior.

3.1.1.3 Performance Measures

3.1.1.3.1 Meteor Rate and Duty Cycle. A meteor trail is said to be usable if it scatters sufficient incident (transmitted) signal to exceed a specified RSL threshold. Typically, this threshold is determined to provide some minimum average bit error rate (BER) at the demodulator output. This threshold may be the minimum RSL value required for carrier phase lock, but synchronization, or a minimum demodulator BER value for a specified transmission rate. The arrival rate of meteors with suitable trajectory and sufficient mass to exceed this RSL threshold is called the usable meteor rate, or simply the meteor rate (MR). The MR value is a fundamental performance measure for MB communication links.

Given that a meteor trail is usable, the duration of the RSL above some specified threshold determines the link duty cycle (DC). The DC is the percentage of time for which the desired RSL threshold is exceeded and a communication is possible. Short-duration trail-scattered RSLs show less fading than the long-duration signals, thus providing contiguous DC contributions. Of course, this phenomenon occurs because the likelihood of wind-induced multipath fading increases as the trail-scattered signal lifetime increases. For example, Fig. 3.13*a* shows the classic underdense trail behavior with horizontal lines indicating several trail DC values corresponding to different values of RSL threshold. Wind-induced fading can break up the DC provided by a single meteor trail, as shown in Figure 3.13*b*. In this case, the net trail DC value is the sum of all DC contributions. Nevertheless, fading statistics are required to determine link protocol performance given "drop outs" during the trail-scattered RSL lifetime.

3.1.1.3.2 Waiting Time. The classic performance measure for MB communications is waiting time. Waiting time may be defined as the interarrival time between the occurrence of usable meteors, that is, burst waiting time, or as the time between reception of complete messages, called message waiting time (MWT). If messages are sufficiently brief to permit complete transmission on single meteor bursts (single-burst messages), then burst waiting time and MWT are approximately equal to the inverse of the average MR value. Longer messages are completed by transmission of message packets on multiple bursts, called message piecing [35]. The MWT provides a network as well as link performance measure. In practice, the MWT includes both queue and channel transmission delays for each link between message source and destination. Queue delays arise from other network message and acknowledgment traffic vying for channel time.

MWT arguably provides the most natural measure of MB performance. Except for periods of continuous ionoscatter, for example, sporadic-E, the typical MB link will provide intermittent message transmission with periods of complete inactivity. These "quiet" periods may vary from a few seconds to many minutes and even hours depending on link parameters. This

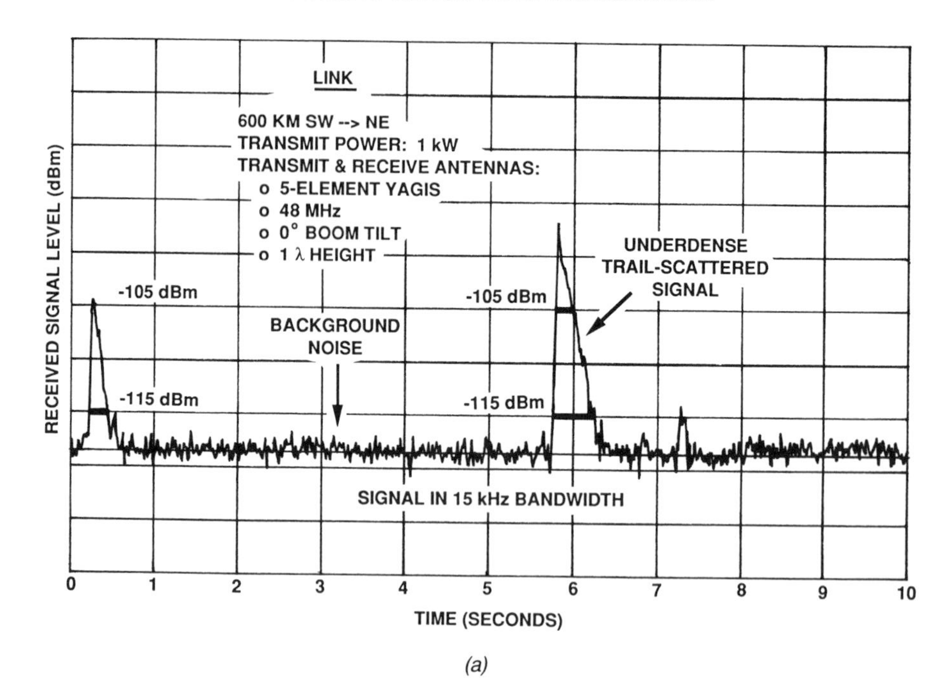

(a)

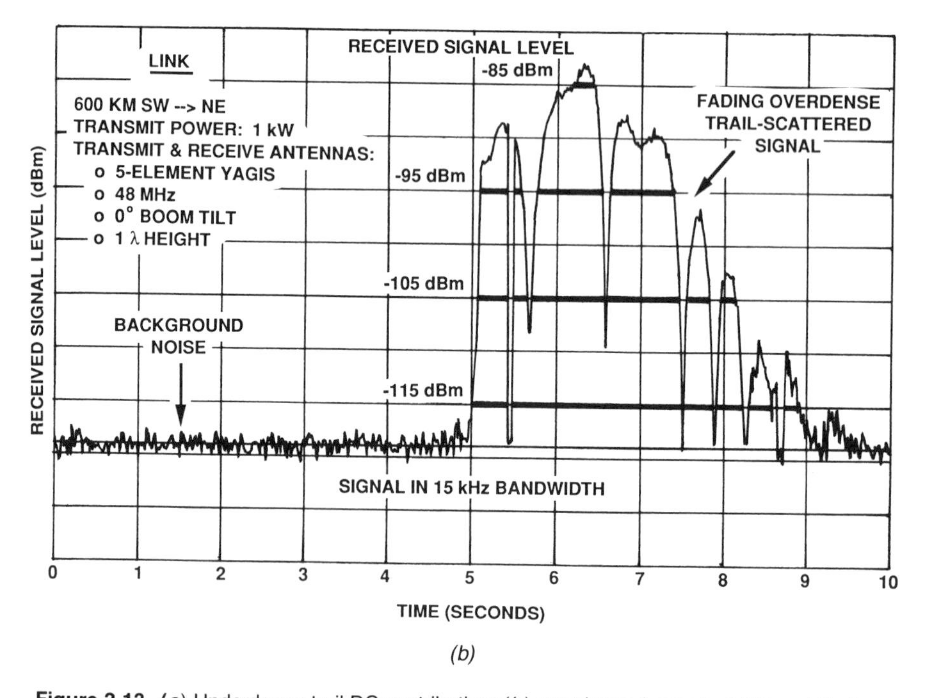

(b)

Figure 3.13. (*a*) Underdense trail DC contribution; (*b*) overdense fading trail DC contributions.

channel behavior must be contrasted with performance specification in terms of average throughput, typically measured in bps or words per minute (wpm). Throughput implies continuous transmission, not realized during operation of true MB communication links. Nevertheless, a well-designed MB link with adequate power margin can achieve sufficiently short MWT values to justify a throughput measure. Of course, the time interval for the throughput calculation must be specified, since hourly, daily, monthly, or even yearly-averaged throughput values may be used to specify the performance requirement.

3.1.1.3.3 Footprint. An important attribute of MB is the communication privacy associated with low probability of intercept (LPI) of the trail-scattered signal [36]. This privacy derives principally from low probability of signal detection (LPD), which results from the smaller geographic extent of trail-scattered RSL "footprints" compared to the larger footprints characteristic of satellite and HF skywave radio. In addition, the intermittent, short-lived meteor bursts combined with the typically weak RSL values further enhance the LPI attribute.

The MB signal footprint can be practically defined as the geographic region illuminated by sufficient scattered signal energy to simultaneously exceed the threshold RSL of a receiver located anywhere within the region. Instead of a single receiver, a receiver array dispersed over a wide area could be envisioned that monitors signal transmissions from a single transmitter. The portion of the array that simultaneously receives the same signal would then be considered within the same signal footprint. As the meteor trail forms at a rate proportional to the meteor's speed, multiple points along the trail are oriented to provide coherent scatter between the transmitter and one or more receivers in the array. In fact, the set of ground points receiving the same signal strength moves across the ground as the trail forms and moves with the upper atmospheric winds. Thus, the simultaneous signal footprint measure may be defined as the conditional probability that a receiver in the array uses a burst *while* the intended, or primary, receiver uses the same burst.

A trail footprint [37], defined by use of the same meteor trail, consists of the accumulation of moving "simultaneous" signal footprints. This definition expands the footprint to include all receivers that received a signal from each trail, although not necessarily simultaneous reception of the same signal. Therefore a useful extension of the definition of signal footprint to include this conditional probability permits link performance analysis from an LPD, and ultimately LPI, perspective. An example trail footprint for a 1000-km link is shown in Fig. 3.14, in which the contours of conditional reception are plotted for probabilities of 0.10, 0.25, 0.50, and 0.90.

3.1.1.3.4 Diversity. The limited geographic signal and trail footprints of meteor trail-scattered signals suggests that adequately spaced receivers can

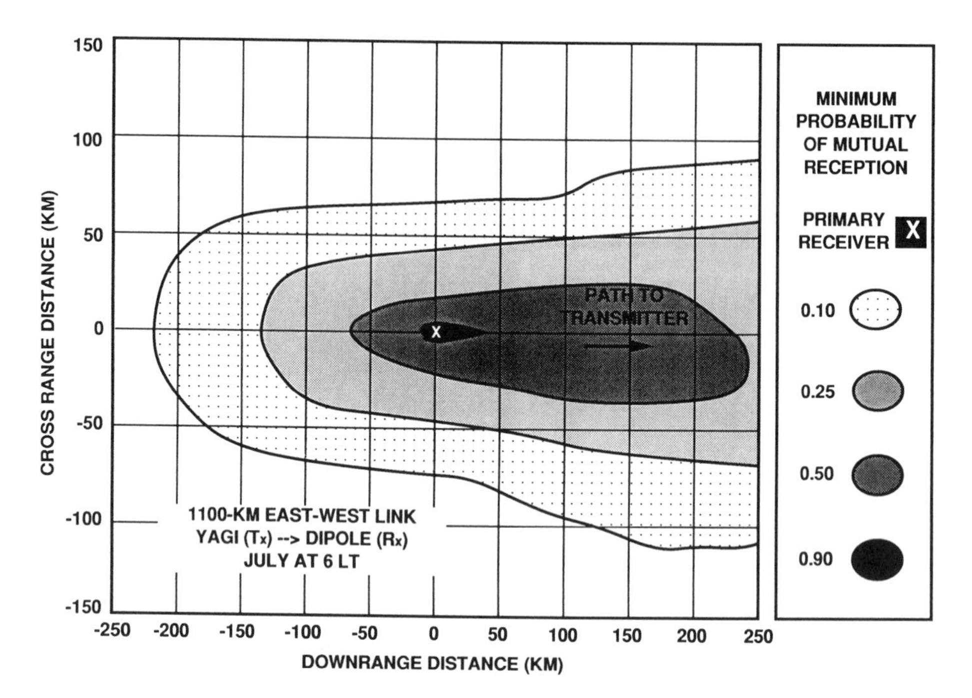

Figure 3.14. Example of a METEORTRAK-predicted trail footprint.

employ independent meteor trails as shown in Fig. 3.15*a*. This figure may be contrasted to Fig. 3.15*b*, in which all receivers lie within the same trail footprint probability contour. Thus multiple spaced receivers can increase the common volume and therefore increase the *effective* MR value from a single transmitter to the receiver array. Similarly, the use of adequately spaced transmitters can significantly increase the effective MR value to a single receiver. In general, space diversity increases the effective MR value, or array MR (AMR) value, between M_t ($M_t \geq 1$) spaced transmitters to M_r ($M_r \geq 1$) spaced receivers. This spatial diversity effect is inherent in the meteor scatter phenomenology and provides a means of improving link performance when inadequate power margin forces low link MR values, and consequently, long link MWT values.

In general, MB spatial diversity gain may be defined as the ratio of the array MR value (AMR), $\dot{N}_1$, of usable trail-scattered signals from *at least one* of M_t transmitters *to at least one* of M_r receivers *divided* by the average MR value, $\bar{\dot{N}}_L$, determined over all $M_t M_r$ individual links. Since diversity gain is a relative measure, it is somewhat insensitive to variation in link parameters [38]. In fact, diversity gain may provide a misleading measure of effective MB link performance. Consider the limiting case in which $M_r - 1$ of M_r receivers detect no usable trail-scattered signals. If the ith receiver

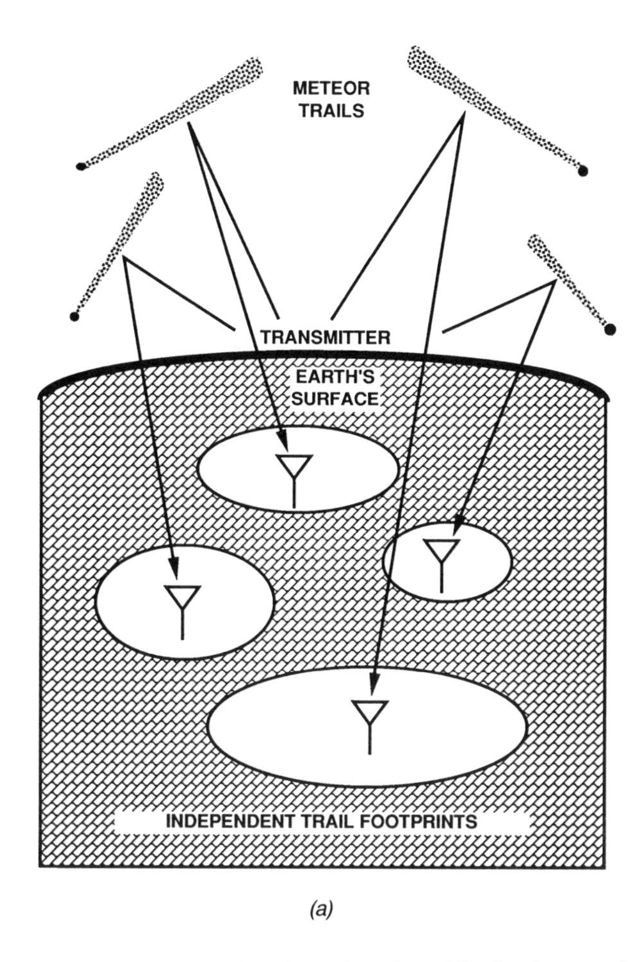

Figure 3.15. (*a*) Spaced receivers in unique footprints; (*b*) closely spaced receivers in common footprint.

measures trail-scattered signals at the rate $\dot{N}_{L_i} = \bar{\dot{N}}_L$, then the AMR value $\dot{N}_1 = \dot{N}_L$, the average link MR value becomes $\bar{\dot{N}}_L = \dot{N}_L / M_t M_r$, and the diversity gain value is given by $\dot{N}_1 / \bar{\dot{N}}_L = M_t M_r$. This diversity value, $M_t M_r$, is identical to the value obtained for an average link MR value of $\bar{\dot{N}}_L$ and an AMR value of $\dot{N}_1 = M_t M_r \bar{\dot{N}}_L$. Thus, diversity gain provides a reliable measure only when the distribution of individual link MR values, $\dot{N}_{L_i}$, $i = 1, 2, \ldots, M_t M_r$, is narrow, that is, concentrated near the mean value $\bar{\dot{N}}_L$. The AMR value combined with the distribution of individual link MR values provides an unambiguous measure of link performance and diversity gain.

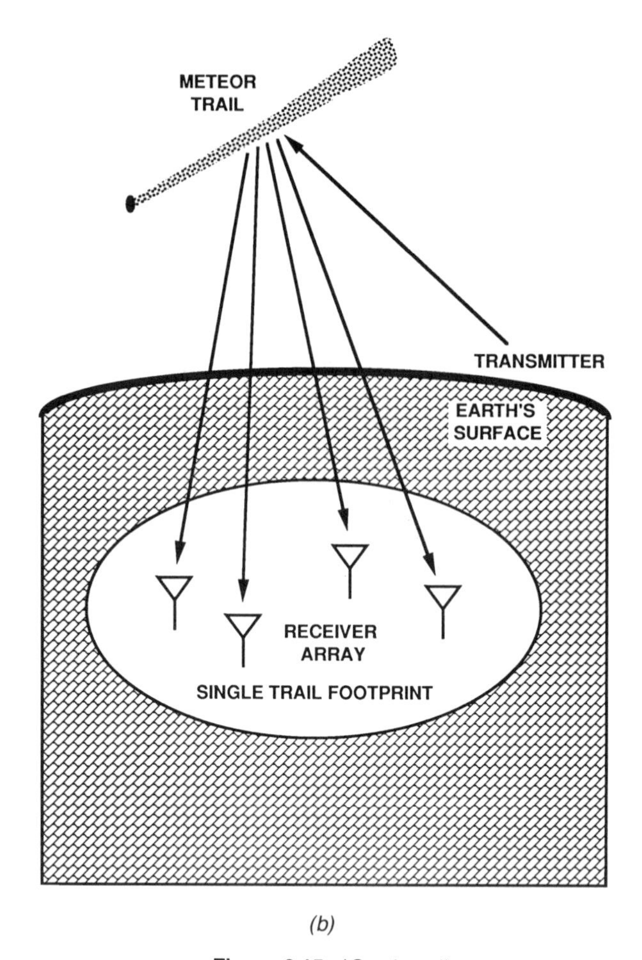

(b)

Figure 3.15. (*Continued*)

3.1.2 Modeling Applications

Computer modeling of MB communication systems has been applied for concept design and validation (CD & V) studies; equipment design, test, and evaluation (DT & E); link and network design; and associated trade-off studies. CD & V studies are intended to evaluate candidate communication architectures and provide performance estimates. In this regard, performance predictions determine the capability of MB radio to meet communication requirements. A broad range of link power budgets, candidate antenna designs, data communication, and protocol parameters may be used in these trade-off predictions. If these results justify use of MB radio, then preliminary system design parameters have been determined. Otherwise, unrealistic MB link design requirements or performance objectives should discourage the use of MB radio.

Use of computer models to support communication system design is particularly important for MB radio, given performance sensitivity to antenna pattern, power margin, and range [39]. MB terminal equipment and antenna designs may be evaluated by computer model predictions before incurring the costs of development. Moreover, trade studies permit optimization of various equipment parameters. For example, MWT can be predicted versus transmitter power, receiver sensitivity, signal frequency, antenna pattern, transmission rate, packet overhead and data frame lengths, and forward error correction (FEC) schemes. Prediction results would provide recommendations for prototype system design improvements necessary to meet specific performance requirements. In addition, the range of MWT performance achievable with adjustable or variable parameters, for example, data rate, packet length, antenna gain, can be predicted. In this way, performance improvements from such adaptive techniques may be evaluated without incurring the cost of development.

The procurement of MB terminal equipment and antennas requires independent test and evaluation to predict expected performance. Computer predictions permit estimates of test duration to achieve statistically significant results and provide a means for verifying proper test operation. For example, measured MR values significantly below predicted values may indicate equipment failure or test link configuration problems. Furthermore, performance prediction using manufacturer-specific equipment parameters can be used to rate products for different applications. Relating equipment cost to communications performance provides the cost-benefit trade-off necessary before final procurement decisions.

MB computer models provide a low-cost design tool for MB link and networks. Given specific MB terminal equipment and antenna parameters, models can predict performance for various link and network geographic configurations. For example, model predictions can be used to determine the best antennas as a function of fixed link range [40]. Antennas for mobile or transportable MB communications can be designed for omniazimuthal or limited azimuthal MB link coverage. In this context, the significant performance trade-off between antenna gain and coverage versus range can be quantified for antenna selection.

As a second example, computer models can quantify the impact of terminal functions and protocols designed to support network communications on link performance. Terminal functions may include message precedence and both link and end-to-end acknowledgment. Network-capable protocols may be encumbered by routing, destination and control information used to support network operations. These functional and protocol trade-offs are intimately connected to link and network antenna selection. For example, protocols designed to communicate using minimum burst DC values achievable with optimized antenna patterns at a single link range may be detrimental for other range/antenna combinations [41].

Consider a fixed network design consisting of a single MB base, or master, station and many remote MB terminal locations, as shown in Fig. 3.16. If the master station is centrally located relative to the remote sites (Fig. 3.16*a*), then master station antennas must provide omniazimuthal coverage. Better performance may be achieved by employing a high-gain, directive antenna at a suitable stand-off range, as shown in Figure 3.16*b*. In this case, the additional antenna gain and increased forward scattering angles improve link performance despite the reduction in potential common volume and increased path length. Computer modeling could be used to determine an optimized combination of range and antenna pattern to minimize master-remote MWT values.

3.1.3 Modeling Techniques

3.1.3.1 Reference Models. MB computer models have been classified as *arrival* and *reference*, or *scaling*, models [42]. Reference models compute the MR and DC as a product of independent factors designed to account for the effects of link range, power margin, and temporal variation [43]. These scaling factors are empirically determined from measurements performed on

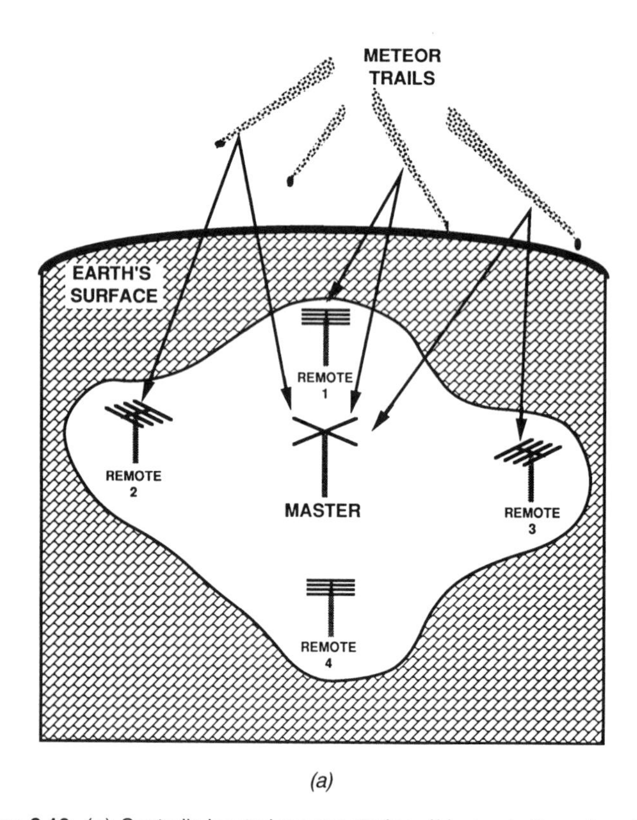

(*a*)

Figure 3.16. (*a*) Centrally-located master station; (*b*) stand-off master station.

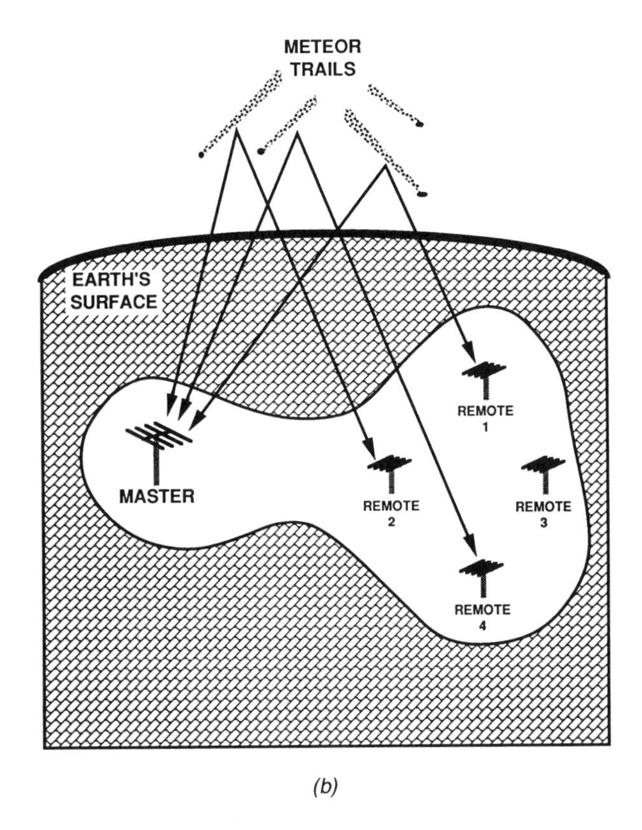

(b)

Figure 3.16. *(Continued)*

one or more reference MB radio links. For example, the complicated diurnal variation in MR values at mid-latitudes exhibits a pseudo-sinusoidal behavior. Reference model predictions are computed by scaling these factors to account for differences between the reference link (or links) and the link under evaluation. In general, reference models provide a simplified algorithm for rapid prediction of link MWT values. The absence of extensive footprint, diversity, and network performance measurements prevents the development of reference models for these model prediction objectives.

Link MWT predictions should be accurate for modeled links similar to the reference link(s) used in model development. As link parameters deviate from these parameters, however, model predictions may deviate significantly from actual measurement [44]. Therefore, reference models should be used with caution when predictions are the principal source for the evaluation of MB communication systems. For example, a reference model developed from a mid-latitude link would not be expected to show the diurnal variation characteristic of high-latitude links (see Section 3.2.2.7.1.1).

3.1.3.2 Physical Models. Arrival models determine the MR from the SMRD and corresponding velocity distributions [45]. The portion of arriving trails contributing to the MR and DC values is determined from link geometric factors, power budget parameters (including antenna patterns), and meteor astronomical and physical factors [46]. The motivation for arrival models is based on a desire to match the physical processes that constitute MB communications as closely as possible. In this sense, arrival models may also be designated as physical models [47].

The principal objective for physical models is to achieve accurate predictions independent of the MB link parameters. In this regard, the physical model not only seeks reference link independence, but also attempts to provide quantitative explanation for predicted performance. For example, a physical model may provide detailed calculations of the transmit–receive antenna gain product throughout the MB common volume [48]. This product describes the interaction of antenna patterns in the meteoric hot spots. In addition, these detailed predictions permit independent validation of the component physical models and provide a basis for the development of footprint and space diversity models. This basis consists of a trail-by-trail treatment of the MR prediction problem, permitting absolute and conditional MR predictions for diversity and footprint calculations, respectively.

3.1.3.3 Monte Carlo Simulation. MR, DC, footprint, and diversity performance predictions are readily computed by physical MB channel models. MWT prediction for arbitrary link protocols, network traffic loads, transmit/receive queue sizes, and so on, are most suited to Monte Carlo simulation. Exact solution of the general problem is improbable, although MWT predictions are possible given simplifying assumptions [49]. The principle advantage of the Monte Carlo approach is that complicated MB data exchange using half- or full-duplex packet protocols, as well as practical variable data rate models [50], may be modeled.

The Monte Carlo simulation generates meteoric trail-scatter events from MR and DC statistics measured or predicted for the modeled link. Predicted values may come from either reference or physical models. Alternatively, meteor arrivals and usable trail formation may be modeled as part of the Monte Carlo simulation [51] or recorded RSL time histories [52] may be "replayed" to generate realistic meteor trail-scatter events. Channel time for each event is subsequently apportioned to account for packet overhead, message characters, and transmit/receive delays. This approach is easily extended to network simulation by the addition of message queues, message generation rules, message precedence, and a time reference.

3.1.4 Contents

This chapter describes modeling and analysis of MB communications. Reference models, such as the BLINK model [53], have been thoroughly discussed in the available literature and will not be further addressed in this chapter. Section 3.2 provides an extensive description of the METEORCOM[1] family of commercially available computer programs [54], including link, footprint, and diversity models. This description includes coordinate transformations, trail-scattering equations, threshold line density calculations, and the determination of MB link-observable SMRD values. Model predictions are provided to illustrate MB link sensitivity to temporal parameters, antenna pattern, power margin, and range. The link model algorithm is extended to MB trail footprint and space diversity models. Example results from these models are also provided.

Section 3.3 describes a Monte Carlo simulation designed to predict MB link and network performance. This simulation uses the MR and DC values measured from MB links or computed using either reference of physical MB model components. Sample predictions for both link and network examples are provided and discussed. Section 3.4 addresses computer model validation. The purpose of validation is explained and a comparison is performed between METEORCOM model predictions and extensive field measurements. Finally, Section 3.5 summarizes the MB modeling and analysis results presented in this chapter and presents recommendations for empirical and theoretical studies designed to advance MB modeling and analysis capability.

3.2 PHYSICAL MODEL

3.2.1 Overview

A physical model called METEORLINK has been developed to predict MB link performance and provide a diagnostic analysis tool. METEORLINK is the link prediction program in the METEORCOM family of MB communications modeling and analysis programs. METEORLINK was initially created to predict the performance of MB link tests and was therefore designed to employ antenna electric field patterns determined from either numerical modeling or direct pattern measurements. This model uses terminal equipment parameters, antenna electric field patterns, link geographic parameters, and measured sporadic meteor radiant density (SMRD) distributions combined with the physics of meteoric phenomena to compute link MR and DC statistics. These statistics are used to generate MB channel openings in a Monte-Carlo MB link and network communications protocol simulator called METEORWAIT (see Section 3.3). In addition, the link model provides the fundamental algorithm employed by the METEORTRAK trail footprint and METEORDIV spatial diversity models.

[1] Proprietary computer programs developed by Science Applications International Corporation.

3.2.2 Link Model

3.2.2.1 Overall Approach. The principal physical link modeling objective is to determine those meteor radiants satisfying link geometric and power margin requirements. In this regard, the link serves as a filter that selects those meteors producing trails that contribute to the channel MR and DC. First, observable trail orientations are determined from link trail-scatter geometric considerations. Each orientation is used to reference the appropriate value of meteor radiant density from an empirical SMRD map. The link power budget, antenna polarization patterns, operating frequency, link/trail geometry, time of day, and day in the year determine the appropriate scaling of this SMRD value to compute the desired MR and DC values.

3.2.2.2 Observable Meteor Radiants

3.2.2.2.1 SMRD Map. The SMRD map is the principal component of any physical MB communications (MBC) model. SMRD maps should cover the entire celestial sphere and may be provided in either heliocentric or geocentric coordinate systems. The heliocentric coordinate system determines meteor velocity (speed and direction) relative to the sun. From an astronomical perspective, heliocentric distributions of meteoroids in space determine distributions of meteoroid orbital parameters that provide astronomers with information about the solar system in general [55].

Sporadic meteoroid orbital density (SMOD) maps have been determined from SMRD maps measured by earth-based radars [56–58]. Similar maps have also been measured for the major meteor streams [59], although streams are primarily important from a communications viewpoint for meteor scatter HAM radio enthusiasts [60] or as a source of SMRD enhancement during MB link tests. Figure 3.17*a*, *b* are plots of SMOD maps in ecliptic longitude and latitude, respectively, over the entire celestial sphere derived from radar measurements recorded in Mogadishu, Somalia (2°N) from December 1968 to May 1970 [61]. Ecliptic longitude is measured around the circle formed by the celestial sphere [62] and the plane of the earth's orbit with 0° corresponding to the direction of the sun (see Fig. 3.A.1). Ecliptic latitude is measured from the orbital plane to the ecliptic poles. These conventions define the METEORLINK ORBITAL coordinate system, or O system, as shown in Fig. 3.18.

Figure 3.17*a* shows the Mogadishu SMOD map in ecliptic longitude for ecliptic latitudes between ±20°. Flux density values are provided in units of meteoroids $\text{km}^{-2}\text{sr}^{-1}\text{hr}^{-1}$ for meteoroid masses estimated to lie between 10^{-3} and 10^{2} grams. The latitude distributions shown in Fig. 3.17*b* give the percentage of the observed meteors approaching from vertical angles measured relative to the ecliptic plane. SMOD values above 80° ecliptic latitude are not representative of the actual distribution because of Mogadishu

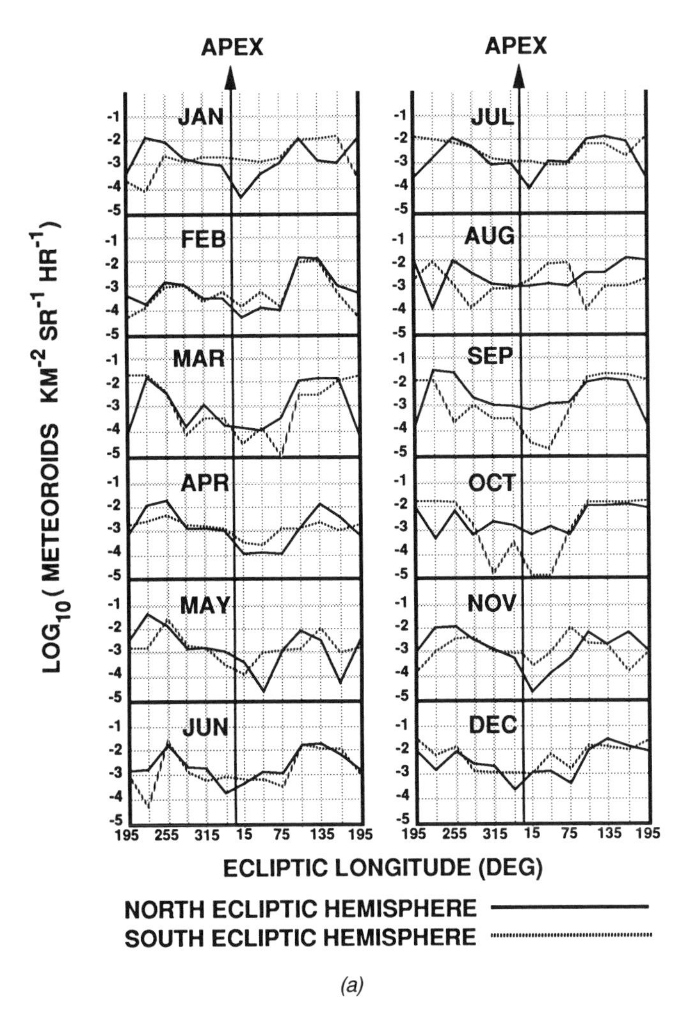

Figure 3.17. (*a*) SMOD ecliptic longitude distribution; (*b*) ecliptic latitude distribution.

(equatorial) radar selectivity which could not observe steeply angled radiants, that is, those meteoroids approaching the earth from high ecliptic latitudes.

These monthly distributions show that most meteoroids orbit the sun in the same direction as the earth; that is, most meteoroids are in direct motion. Fig. 3.17*a* shows that the principal SMOD contributions come from longitudes of 45° and 165°, forming two "petals" each about 60° wide. Latitudinal asymmetry is marked in February, April, and May, when the total measured flux is 1.7, 1.9, and 1.7, respectively, times greater in the northern ecliptic hemisphere than in the southern hemisphere. In October, however, total flux in the southern hemisphere is about 1.5 times greater

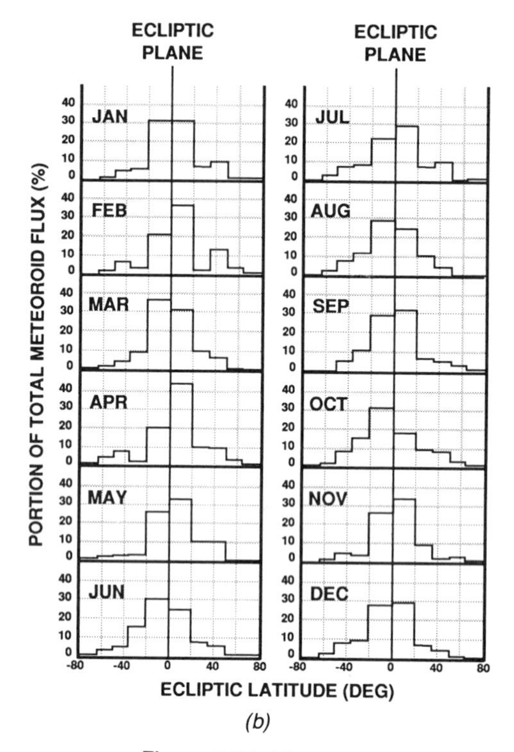

(b)

Figure 3.17. (*Continued*)

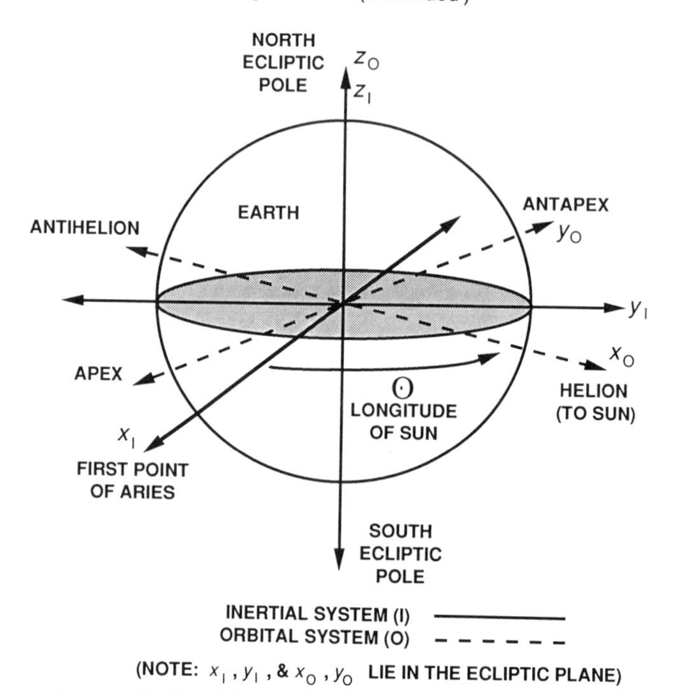

Figure 3.18. The INERTIAL and ORBITAL coordinate systems.

than the flux in the northern hemisphere. The seasonal variation in these distributions is apparent from these figures as well as the asymmetry in each distribution. Asymmetry in ecliptic longitude has been observed by independent measurements [63], although disagreement was apparent in the intensity of the antihelion (away from sun) source.

A MB link, however, "observes" meteors in the geocentric system. Thus, calculation of the MR and DC requires conversion of SMOD maps in the heliocentric system to SMRD maps in the geocentric system or conversion of link-observable (geocentric) radiants to the heliocentric system [64] in order to reference the appropriate SMOD values. These conversions must account for relative velocities of the earth and meteoroid zenith attraction, and a change in solid angle [65]. Both the conversions from geocentric radar measurements to heliocentric distributions and then back to geocentric distributions may introduce errors in the predicted MR and DC values if these conversions are not reciprocal. Therefore, it is arguably preferable to use geocentric flux density maps (SMRD maps) in a physical model.

The METEORLINK computer model employs a tabular map of radar measurement-derived SMRD values. Figure 3.19 shows the radar SMRD map for February, including both average radar SMRD values and the corresponding standard errors [66] with all values scaled by a factor of 100. These SMRD values were derived from $2\frac{1}{2}$ years of meteor radar measurements made during the mid 1960s at the Engelhardt Observatory of Kazan State University in the former Soviet Union [67] combined [68] with the Mogadishu radar measurements. Since the Kazan measurements were performed at 56° north geographic latitude, the combined radar data provides more accurate SMRD for the northerly radiants otherwise excluded from the equatorial Mogadishu measurements.

The SMRD values in Fig. 3.19 were determined from radar measurements of the observed meteor radiants, that is, they include the direction changes associated with zenith attraction and diurnal aberration on the true geocentric radiants (see Section 3.1.1.2). Since the MB link serves as an earth-bound observer of meteor radiants, SMRD values of the observed radiants provide the fundamental data base for MB link performance predictions. Let the observed radiant of a meteor whose trail supports meteoric trail-scatter be specified in the O system by $-\hat{\mathbf{v}}_m^{\mathrm{O}}$, where $\hat{\mathbf{v}}_m^{\mathrm{O}} = \langle \hat{v}_{m1}^{\mathrm{O}}, \hat{v}_{m2}^{\mathrm{O}}, \hat{v}_{m3}^{\mathrm{O}} \rangle$ is the unit vector (direction cosines) of the observed meteor velocity, $\mathbf{V}_o$. The unit vector $\hat{\mathbf{v}}_m^{\mathrm{O}}$ is determined from the geometric requirements for meteoric trail-scatter (see Section 3.1.1.2 and Eq. 3.A.22).

In general, it is not possible to determine the geocentric velocity vector $\mathbf{V}_G$ given only the unit direction vector, $\hat{\mathbf{v}}_m^{\mathrm{O}}$, of the observed velocity vector $\mathbf{V}_o$ but not the corresponding value of observed meteor speed, $v_o = |\mathbf{V}_o|$. It has been suggested [69], however, that since the maximum meteor deflection due to zenith attraction is only 17°, for the slowest meteors approaching normal to the zenith, their contributions to the link MR and DC values are negligible. If the small direction changes associated with zenith attraction and diurnal aberration are ignored, then

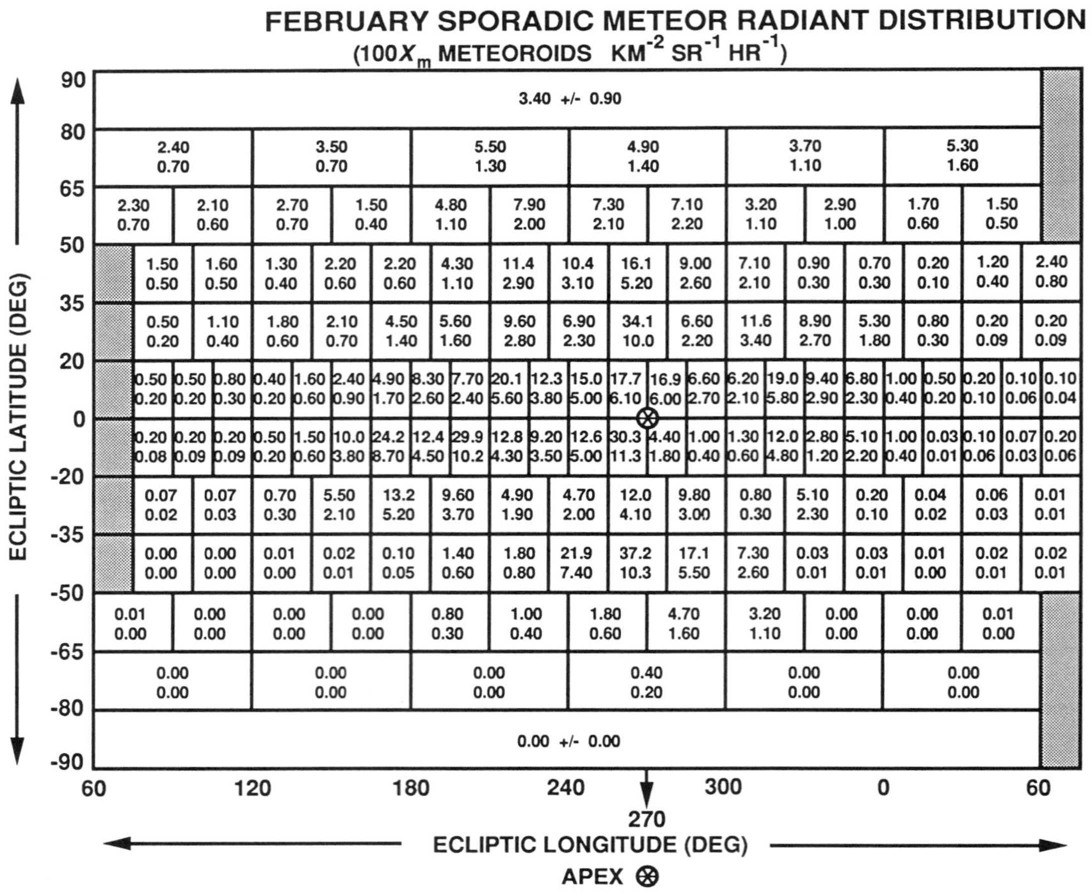

FEBRUARY SPORADIC METEOR RADIANT DISTRIBUTION

(100X_m METEOROIDS KM^{-2} SR^{-1} HR^{-1})

ECLIPTIC LATITUDE (DEG): 90, 80, 65, 50, 35, 20, 0, -20, -35, -50, -65, -80, -90

ECLIPTIC LONGITUDE (DEG): 60, 120, 180, 240, 270, 300, 0, 60

80 to 90: 3.40 +/- 0.90

65 to 80: 2.40 / 0.70; 3.50 / 0.70; 5.50 / 1.30; 4.90 / 1.40; 3.70 / 1.10; 5.30 / 1.60

50 to 65: 2.30 / 0.70; 2.10 / 0.60; 2.70 / 0.70; 1.50 / 0.40; 4.80 / 1.10; 7.90 / 2.00; 7.30 / 2.10; 7.10 / 2.20; 3.20 / 1.10; 2.90 / 1.00; 1.70 / 0.60; 1.50 / 0.50

35 to 50: 1.50 / 0.50; 1.60 / 0.50; 1.30 / 0.40; 2.20 / 0.60; 2.20 / 0.60; 4.30 / 1.10; 11.4 / 2.90; 10.4 / 3.10; 16.1 / 5.20; 9.00 / 2.60; 7.10 / 2.10; 0.90 / 0.30; 0.70 / 0.30; 0.20 / 0.10; 1.20 / 0.40; 2.40 / 0.80

20 to 35: 0.50 / 0.20; 1.10 / 0.40; 1.80 / 0.60; 2.10 / 0.70; 4.50 / 1.40; 5.60 / 1.60; 9.60 / 2.80; 6.90 / 2.30; 34.1 / 10.0; 6.60 / 2.20; 11.6 / 3.40; 8.90 / 2.70; 5.30 / 1.80; 0.80 / 0.30; 0.20 / 0.09; 0.20 / 0.09

0 to 20: 0.50 / 0.20; 0.50 / 0.20; 0.80 / 0.30; 0.40 / 0.20; 1.60 / 0.60; 2.40 / 0.90; 4.90 / 1.70; 8.30 / 2.60; 7.70 / 2.40; 20.1 / 5.60; 12.3 / 3.80; 15.0 / 5.00; 17.7 / 6.10; 16.9 / 6.00; 6.60 / 2.70; 6.20 / 2.10; 19.0 / 5.80; 9.40 / 2.90; 6.80 / 2.30; 1.00 / 0.40; 0.50 / 0.20; 0.20 / 0.10; 0.10 / 0.06; 0.10 / 0.04

-20 to 0: 0.20 / 0.08; 0.20 / 0.09; 0.20 / 0.09; 0.50 / 0.20; 1.50 / 0.60; 10.0 / 3.80; 24.2 / 8.70; 12.4 / 4.50; 29.9 / 10.2; 12.8 / 4.30; 9.20 / 3.50; 12.6 / 5.00; 30.3 / 11.3; 4.40 / 1.80; 1.00 / 0.40; 1.30 / 0.60; 12.0 / 4.80; 2.80 / 1.20; 5.10 / 2.20; 1.00 / 0.40; 0.03 / 0.01; 0.10 / 0.06; 0.07 / 0.03; 0.20 / 0.06

-35 to -20: 0.07 / 0.02; 0.07 / 0.03; 0.70 / 0.30; 5.50 / 2.10; 13.2 / 5.20; 9.60 / 3.70; 4.90 / 1.90; 4.70 / 2.00; 12.0 / 4.10; 9.80 / 3.00; 0.80 / 0.30; 5.10 / 2.30; 0.20 / 0.10; 0.04 / 0.02; 0.06 / 0.03; 0.01 / 0.01

-50 to -35: 0.00 / 0.00; 0.00 / 0.00; 0.01 / 0.00; 0.02 / 0.01; 0.10 / 0.05; 1.40 / 0.60; 1.80 / 0.80; 21.9 / 7.40; 37.2 / 10.3; 17.1 / 5.50; 7.30 / 2.60; 0.03 / 0.01; 0.03 / 0.01; 0.01 / 0.00; 0.02 / 0.01; 0.02 / 0.01

-65 to -50: 0.01 / 0.00; 0.00 / 0.00; 0.00 / 0.00; 0.00 / 0.00; 0.80 / 0.30; 1.00 / 0.40; 1.80 / 0.60; 4.70 / 1.60; 3.20 / 1.10; 0.00 / 0.00; 0.00 / 0.00; 0.01 / 0.00

-80 to -65: 0.00 / 0.00; 0.00 / 0.00; 0.00 / 0.00; 0.40 / 0.20; 0.00 / 0.00; 0.00 / 0.00

-90 to -80: 0.00 +/- 0.00

APEX ⊗

Figure 3.19. SMRD map for February.

$$\hat{\mathbf{v}}_m^{\mathrm{O}} = \hat{\mathbf{v}}_o^{\mathrm{O}} = \hat{\mathbf{v}}_\infty^{\mathrm{O}} = \hat{\mathbf{v}}_a^{\mathrm{O}} = \hat{\mathbf{v}}_G^{\mathrm{O}} \tag{3.1}$$

where, for example, $\mathbf{V}_G = |\mathbf{V}_G|\hat{\mathbf{v}}_G^{\mathrm{O}} = v_G\hat{\mathbf{v}}_m^{\mathrm{O}}$, and $\hat{\mathbf{v}}_o^{\mathrm{O}}$, $\hat{\mathbf{v}}_\infty^{\mathrm{O}}$, and $\hat{\mathbf{v}}_a^{\mathrm{O}}$ are the unit direction vectors for the observed, no-atmosphere, and apparent meteor velocities, respectively.

Given $\hat{\mathbf{v}}_m^{\mathrm{O}}$, the corresponding SMRD value from Fig. 3.19 is indexed by the O system angles

$$\phi_{\mathrm{O}} = \tan^{-1}\left(\frac{-\hat{v}_{m2}^{\mathrm{O}}}{-\hat{v}_{m1}^{\mathrm{O}}}\right) \quad \text{and} \quad \theta_{\mathrm{O}} = \tan^{-1}(-\hat{v}_{m3}^{\mathrm{O}}) \tag{3.2}$$

where ϕ_{O} is ecliptic longitude measured counter-clockwise from the longitude of the sun, ⊙, and θ_{O} is ecliptic latitude measured from the ecliptic equator. The apex direction is shown in the figure as ⊛ at $\phi_{\mathrm{O}} = 270°$, $\theta_{\mathrm{O}} = 0°$. The size of the $\Delta\phi_{\mathrm{O}}$ and $\Delta\theta_{\mathrm{O}}$ cells in Fig. 3.19 are varied to maintain approximately constant cell solid angle

$$\Omega_{\phi\theta}^{\mathrm{O}} = \Delta\phi_{\mathrm{O}}\,\Delta\theta_{\mathrm{O}}\sin(\theta_{\mathrm{O}})$$

over the entire celestial sphere.

Figure 3.20*a*, *b* are three-dimensional images of the radar-derived SMRD maps for February (Fig. 3.19) and July in the O system. In this figure, the length of each line segment depicts the arrival rate of meteors, the longer lengths representing greater arrival rates. Note that the largest values are produced in the apex direction in July while the smallest values occur in the antapex direction in February (see Section 3.1.1.2, Fig. 3.2). This result is apparent from the transformation of heliocentric distributions shown in Fig. 3.17 to the geocentric system, in which the vector difference of heliocentric meteoroid and earth velocities (see Fig. 3.3*a*) yields the geocentric meteor velocity. Thus, meteoroids are "swept up" by the earth in its forward motion, resulting in a concentration of the largest SMRD values in the earth's apex direction.

The SMRD values in Fig. 3.19 are provided in scaled units of radar-detected meteors $\mathrm{km}^{-2}\ \mathrm{sr}^{-1}\ \mathrm{hr}^{-1}$ and were determined by combining the measurements recorded at Kazan and Mogadishu [70]. The resulting flux values correspond to meteors whose trails produced a minimum electron line density detectable by both the Kazan and Mogadishu radars. Since the minimum detectable line densities for the Kazan and Mogadishu radars were 1.20×10^{13} electrons per meter (epm) and 0.36×10^{13} epm, respectively, the Mogadishu meteor counts were reduced to match the lower sensitivity of the Kazan radar. These line density values were computed from each radar's transmit power, required received power, and maximum antenna gain for a trail at the height of maximum meteoric ionization. Before these flux values can be used in a physical link model, however, they must be scaled to account for meteor velocity-selective effects inherent in the radar measurement technique.

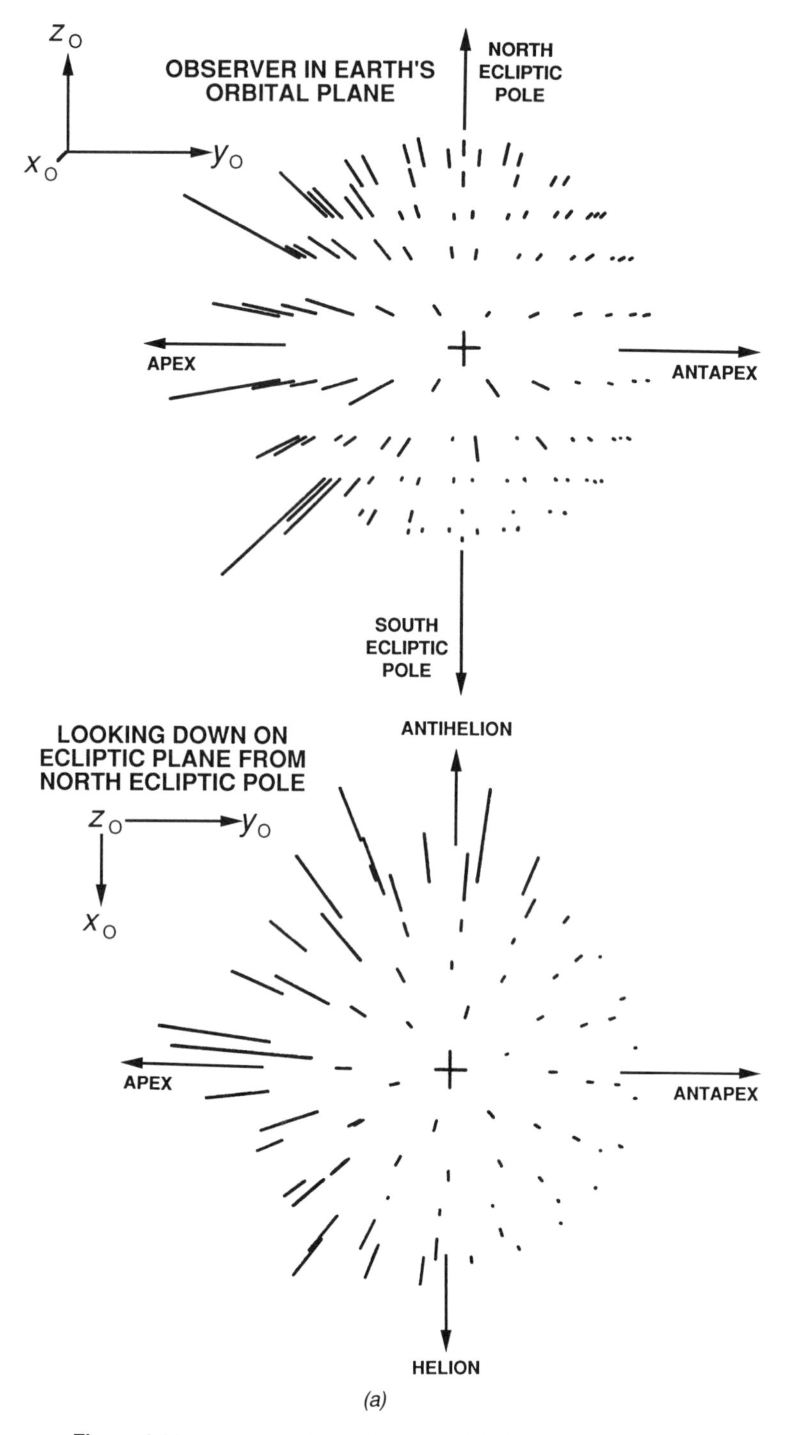

(a)

Figure 3.20. 3D image of SMRD values: (*a*) February; (*b*) July.

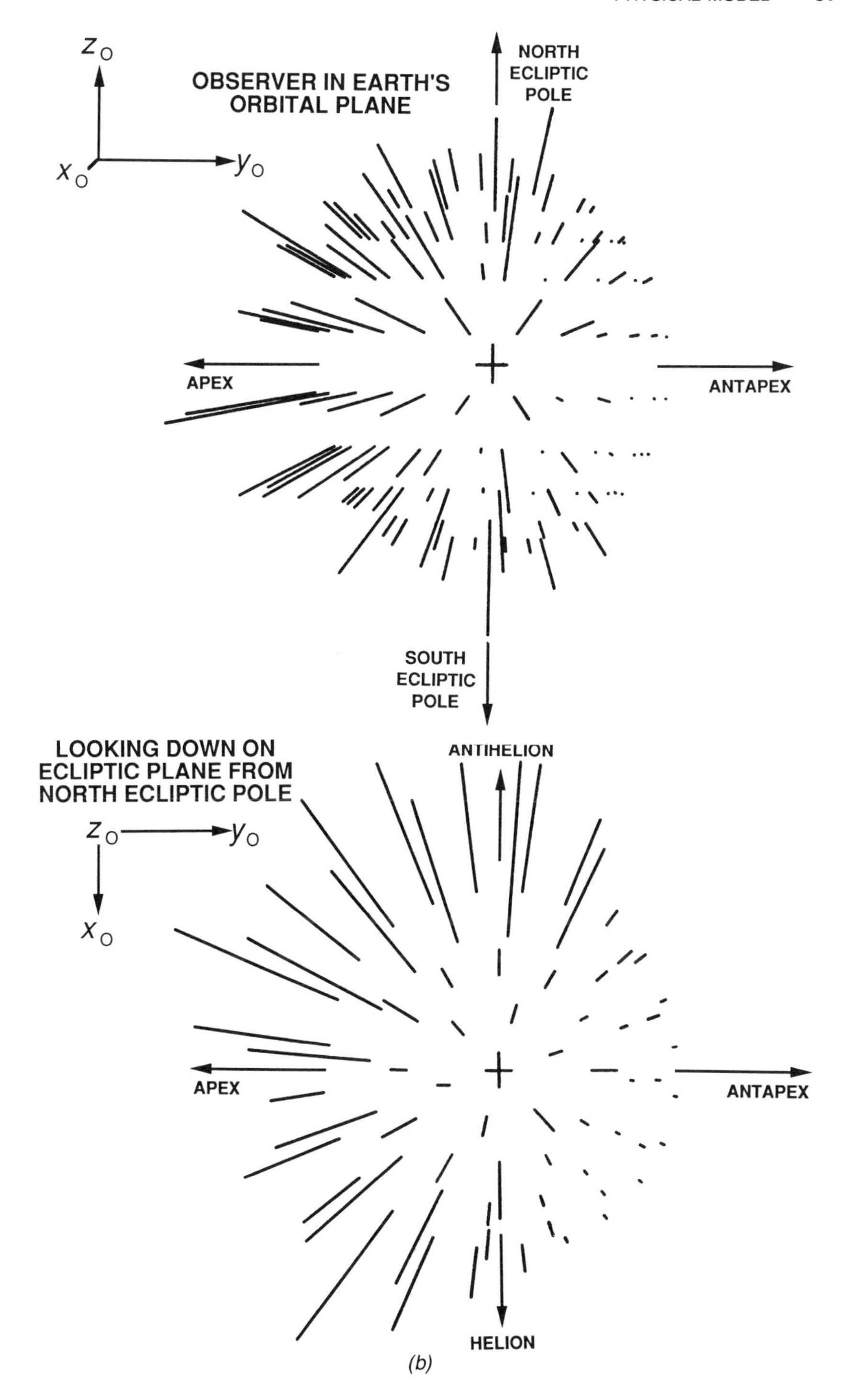

Figure 3.20. (*Continued*)

3.2.2.2.2 Radar-Observable SMRD Values. In general, accurate SMRD values cannot be determined directly from single radar measurements. The radar periodically transmits short-duration pulses and then monitors its receiver for backscattered signals from meteor trails. One of more returned pulses within a time interval that is short compared to the interarrival time of detectable trails would then be counted as a detected meteor. These pulses do not directly reveal the meteor's mass, speed, or direction. Meteor mass must be determined from a knowledge of the radar's power budget and the expected trail-scattered RSL versus trail electron line density for each trail vector $\hat{\mathbf{v}}_m^{\mathrm{O}}$. Meteor speed may be accurately determined from the time derivative of the pulse amplitude oscillations (see Fig. 3.5), albeit only for a percentage of detected trails. The meteor's direction vector, $\hat{\mathbf{v}}_m^{\mathrm{O}}$, may be determined from cross-referenced measurements taken at multiple receiver locations [71] or by statistical methods using a single radar site [72, 73]. Thus, the radar-selective effects implicit within the Kazan–Mogadishu SMRD map of Fig. 3.19 must be quantified before the corresponding SMRD values can be used to predict MB link performance.

The incremental number of sporadic meteors arriving with apparent velocity vector $\mathbf{V}_a = v_a \hat{\mathbf{v}}_m^{\mathrm{O}}$ between masses m and $m + dm$ and speeds $v = v_a$ and $v + dv$ crossing a plane area element dS within the solid angle element $d\Omega$ between times t and $t + dt$ is given by

$$d^5N = c_n m^{-s} p_{\mathbf{V}_a}(v, \hat{\mathbf{v}}_m^{\mathrm{O}})\, dm\, dv\, dS\, d\Omega\, dt \tag{3.3}$$

where

c_n = normalization constant

m = meteor mass

$p_{\mathbf{V}_a}(v, \hat{\mathbf{v}}_m^{\mathrm{O}})$ = density function for the apparent velocity of sporadic meteors

s = mass-rate exponent determined for each trail direction $\hat{\mathbf{v}}_m^{\mathrm{O}}$

v = apparent meteor speed between the physical limits $v_l \leq v \leq v_u$.

It has been assumed that the meteor mass m is independent of the meteor speed v [74].

The term $p_{\mathbf{V}_a}(v, \hat{\mathbf{v}}_m^{\mathrm{O}})\, dv\, d\Omega$ in Eq. 3.3 expresses the probability that a meteor will have a speed between v and $v + dv$ and trail unit vector $\hat{\mathbf{v}}_m^{\mathrm{O}}$ lying within the solid angle element $d\Omega$ [75]. Therefore, $p_{\mathbf{V}_a}(v, \hat{\mathbf{v}}_m^{\mathrm{O}})$ must satisfy

$$\int_{\Omega_{4\pi}} p_{\hat{\mathbf{v}}_m^{\mathrm{O}}}(\hat{\mathbf{v}}_m^{\mathrm{O}})\, d\Omega = \int_{\Omega_{4\pi}} \int_{v_l}^{v_u} p_{\mathbf{V}_a}(v, \hat{\mathbf{v}}_m^{\mathrm{O}})\, dv\, d\Omega = 1 \tag{3.4a}$$

where $\Omega_{4\pi}$ designates all possible trail orientations, $\hat{\mathbf{v}}_m^{\mathrm{O}}$, over the celestial sphere, and

$$p_{\hat{\mathbf{v}}_m^{\mathrm{O}}}(\hat{\mathbf{v}}_m^{\mathrm{O}}) = \int_{v_l}^{v_u} p_{\mathbf{v}_a}(v, \hat{\mathbf{v}}_m^{\mathrm{O}})\, dv \tag{3.4b}$$

The lower speed limit, $v_l \cong 11$ km/s, is experienced by a meteor with zero geocentric speed "falling" to earth from an infinite distance. The upper limit, $v_u \cong 72$ km/s, corresponds to the maximum speed possible for an object accelerated by earth's gravity while orbiting the sun and observed (head on) at the earth's orbital position [76].

Equation 3.3 gives the incremental number of meteors with trail vector $\hat{\mathbf{v}}_m^{\mathrm{O}}$ independent of the observational technique. Integrating Eq. 3.3 over all meteor speeds v with direction vector $\hat{\mathbf{v}}_m^{\mathrm{O}}$ yields the sporadic meteor flux with masses between m and $m + dm$ given by

$$\Phi_{sp}(m, \hat{\mathbf{v}}_m^{\mathrm{O}}) = \frac{d^4N}{dm\, dS\, d\Omega\, dt} = c_n m^{-s} p_{\hat{\mathbf{v}}_m^{\mathrm{O}}}(\hat{\mathbf{v}}_m^{\mathrm{O}})\, dm\, dS\, d\Omega\, dt \tag{3.5}$$

The normalization constant c_n is determined from the corresponding SMRD value, $\Theta_{sp}(m_0, \hat{\mathbf{v}}_m^{\mathrm{O}})$, by integrating Eq. 3.5 over all masses exceeding some minimum mass, m_0, that contribute to the sporadic meteor flux in the $\hat{\mathbf{v}}_m^{\mathrm{O}}$ direction. The SMRD value for meteors with direction $\hat{\mathbf{v}}_m^{\mathrm{O}}$ and minimum mass m_0 is then given by

$$\begin{aligned} \Theta_{sp}(m_0, \hat{\mathbf{v}}_m^{\mathrm{O}}) &= \int_{m_0}^{\infty} \Phi_{sp}(m, \hat{\mathbf{v}}_m^{\mathrm{O}})\, dm \\ &= \frac{d^3N}{dS\, d\Omega\, dt} \\ &= c_n p_{\hat{\mathbf{v}}_m^{\mathrm{O}}}(\hat{\mathbf{v}}_m^{\mathrm{O}}) \int_{m_0}^{\infty} m^{-s}\, dm \end{aligned} \tag{3.6}$$

The normalization constant for direction vector $\hat{\mathbf{v}}_m^{\mathrm{O}}$ is therefore given by

$$c_n = \frac{(s-1) m_0^{(s-1)}\, \Theta_{sp}(m_0, \hat{\mathbf{v}}_m^{\mathrm{O}})}{p_{\hat{\mathbf{v}}_m^{\mathrm{O}}}(\hat{\mathbf{v}}_m^{\mathrm{O}})}$$

Finally, the incremental number of meteors arriving with direction $\hat{\mathbf{v}}_m^{\mathrm{O}}$ is given by

$$d^5N = (s-1) m_0^{(s-1)}\, \Theta_{sp}(m_0, \hat{\mathbf{v}}_m^{\mathrm{O}}) m^{-s} \frac{p_{\mathbf{v}_a}(v, \hat{\mathbf{v}}_m^{\mathrm{O}})}{p_{\hat{\mathbf{v}}_m^{\mathrm{O}}}(\hat{\mathbf{v}}_m^{\mathrm{O}})}\, dm\, dv\, dS\, d\Omega\, dt \tag{3.7}$$

Thus, d^5N given by Eq. 3.7 is the total number of meteors with unit direction vector $\hat{\mathbf{v}}_m^O$, masses between m and $m + dm$, apparent speeds between v and $v + dv$, and crossing the area element dS at normal incidence within the solid angle element $d\Omega$ between times t and $t + dt$. It is tacitly assumed that the sporadic flux $\Theta_{sp}(m_0, \hat{\mathbf{v}}_m^O)$ is derived from meteor counts averaged over some time interval during which the meteor arrival statistics are stationary. In practice, this interval varies from several days to one month depending on the time required to complete a statistically significant sporadic flux measurement. Thus, d^5N is also a time-averaged value.

The incremental number of meteors d^5N given by Eq. 3.7 was independent of the measurement technique, that is, it was determined without the selective effects inherent in any radiant measurement technique. Now consider the use of an earth-bound radar to determine the meteor arrival rate from the detection of meteor trail-scattered signals. First, the radar cannot simultaneously illuminate the entire area S above the earth at the height of maximum meteoric ionization. Second, not all trail orientation vectors $\hat{\mathbf{v}}_m^O$ will be observable because of earth blockage, radar antenna pattern, and trail-scatter geometric effects. Thus, not all of the 4π steradians of solid angle of arriving meteor flux can be observed by the radar. Radar parameters therefore determine bounds on the available sky area and solid angle from which sporadic meteor radiants may be observed. These bounds provide integration limits for S and Ω in Eq. 3.7. In this context, a meteor trail is considered "observable" by a radar if its location and orientation permit radar detection.

Although a meteor trail may be observable by a specific radar, the electron line density in the trail may not be adequate given trail dimensions and orientation to produce a detectable signal in the radar's receiver. The received signal is considered detectable if it exceeds some minimum RSL threshold set just above maximum background noise levels. Since trail line density is proportional to meteor mass, there exists a minimum meteor mass $m_R(v, \hat{\mathbf{v}}_m^O)$ for which $p_{\mathbf{v}_a}(v, \hat{\mathbf{v}}_m^O)$ is non-zero that will produce a radar-detectable trail. The mass $m_R(v, \hat{\mathbf{v}}_m^O)$ depends on speed v through both the ionization efficiency and the initial dimensions of the trail. In this regard, the minimum detectable mass $m_R(v, \hat{\mathbf{v}}_m^O)$ is determined by the design parameters of the radar employed, such as transmitter power, antenna gain pattern, required RSL, and radio frequency as well as the height of radar detection.

The number of radar-detectable meteors, N_R, with trail vector $\hat{\mathbf{v}}_m^O$ crossing area dS in solid angle $d\Omega$ within time dt can then be determined by integrating Eq. 3.7 over all detectable meteor masses $m > m_R(v, \hat{\mathbf{v}}_m^O)$ and all meteor speeds v. In symbols,

$$d^4N_R = (s-1)m_0^{(s-1)}\Theta_{sp}(m_0, \hat{\mathbf{v}}_m^O)\,\frac{p_{\mathbf{v}_a}(v, \hat{\mathbf{v}}_m^O)}{p_{\hat{\mathbf{v}}_m^O}(\hat{\mathbf{v}}_m^O)}\int_{m_R(v,\,\hat{\mathbf{v}}_m^O)}^{\infty} m^{-s}\,dm\,dv\,dS\,d\Omega\,dt$$

$$= m_0^{(s-1)}\,\Theta_{sp}(m_0, \hat{\mathbf{v}}_m^O)\,\frac{p_{\mathbf{v}_a}(v, \hat{\mathbf{v}}_m^O)}{p_{\hat{\mathbf{v}}_m^O}(\hat{\mathbf{v}}_m^O)}\,[m_R(v, \hat{\mathbf{v}}_m^O)]^{(1-s)}\,dv\,dS\,d\Omega\,dt \qquad (3.8)$$

Define the velocity-dependent radar detection factor [77] to be

$$F_R(v, \hat{\mathbf{v}}_m^{\mathrm{O}}) = \frac{p_{\mathbf{v}_a}(v, \hat{\mathbf{v}}_m^{\mathrm{O}})}{p_{\hat{\mathbf{v}}_m^{\mathrm{O}}}(\hat{\mathbf{v}}_m^{\mathrm{O}})} \left[\frac{m_0}{m_R(v, \hat{\mathbf{v}}_m^{\mathrm{O}})} \right]^{(s-1)} \tag{3.9}$$

then Eq. 3.8 may be written as

$$d^3 N_R = \Theta_{sp}(m_0, \hat{\mathbf{v}}_m^{\mathrm{O}}) \int_{v_l}^{v_u} F_R(v, \hat{\mathbf{v}}_m^{\mathrm{O}})\, dv\, dS\, d\Omega\, dt \tag{3.10}$$

The scaled radar-detected SMRD values shown in Fig. 3.19 for February are then given by

$$100 X_m(\hat{\mathbf{v}}_m^{\mathrm{O}}) = 100 \Theta_{sp}(m_0, \hat{\mathbf{v}}_m^{\mathrm{O}}) \int_{v_l}^{v_u} F_R(v, \hat{\mathbf{v}}_m^{\mathrm{O}})\, dv \tag{3.11}$$

where the factor of 100 was used to reduce the number of digits needed for presentation in Fig. 3.19. Calculation of the actual SMRD value corresponding to trail direction $\hat{\mathbf{v}}_m^{\mathrm{O}}$ requires determination of the appropriate values for $m_R(v, \hat{\mathbf{v}}_m^{\mathrm{O}})$ and $p_{\mathbf{v}_a}(v, \hat{\mathbf{v}}_m^{\mathrm{O}})$.

The minimum value of the radar-detectable mass, $m_R(v, \hat{\mathbf{v}}_m^{\mathrm{O}})$, over all meteor speeds v is determined at the point of maximum electron line density created in the meteor trail. This value occurs for trail-scatter points, $\mathscr{P}_s^{\mathrm{S}}$, at the height of maximum meteoric ionization, $\widehat{H}_I$. In general, the maximum electron line density, $\widehat{q}_I(v, \hat{\mathbf{v}}_m^{\mathrm{O}})$, in a meteor trail with initial (meteoroid) mass m moving with speed v along direction $\hat{\mathbf{v}}_m^{\mathrm{O}}$ is determined from [78]

$$\widehat{q}_I(v, \hat{\mathbf{v}}_m^{\mathrm{O}}) = \frac{4\beta_I(v)\cos(\chi_{sm})}{9\mu_m \widehat{H}_P}\, m \tag{3.12}$$

where

χ_{sm} = the angle between the trail direction vector $\hat{\mathbf{v}}_m^{\mathrm{O}}$ and the zenith vector

$$\hat{\boldsymbol{\zeta}}_s^{\mathrm{O}} = \mathbf{T}_{\mathrm{SO}} \cdot \hat{\boldsymbol{\zeta}}_s^{\mathrm{S}} \tag{3.13}$$

at the trail-scatter point, $\mathscr{P}_s^{\mathrm{S}}$, located at the center of the principal Fresnel zone in the S system, the transformation matrix $\mathbf{T}_{\mathrm{SO}}$ from the S to the O system defined by Eq. 3.A.10, and $\hat{\boldsymbol{\zeta}}_s^{\mathrm{S}}$ given by Eq. 3.A.19

μ_m = the average mass of a meteoric atom

$\widehat{H}_P$ = the atmospheric pressure scale height, H_P, [79] at the height of maximum ionization, $\widehat{H}_I$

$\beta_I(v)$ = the average number of free electrons produced by a meteoric atom with speed v colliding with an atmospheric particle, called the ionization coefficient

An empirical expression for $\beta_I(v)$ [80] with v measured in kilometers per second is given by

$$\beta_I(v) = 4.36 \times 10^{-24}(10^5 v)^{3.42} \tag{3.14}$$

The ionization coefficient, $\beta_I(v)$, measures the mass-independent ionization efficiency of the trail-forming process.

The meteor mass m cannot be measured directly by the radar, so it must be transformed into a radar-observable quantity. Two such quantities are trail-scattered RSL and echo duration [81]. For example, the peak RSL value computed for a back-scattered signal (radar echo) from a trail formed by a meteor traveling with speed v in the $\hat{\mathbf{v}}_m^{\text{O}}$ direction determines the minimum radar-detectable line density

$$\widehat{q}_R(v, \hat{\mathbf{v}}_m^{\text{O}}) = \widehat{q}_I(v, \hat{\mathbf{v}}_m^{\text{O}})$$

at the height of maximum meteoric ionization, $\widehat{H}_I$. The corresponding value of meteor mass, $m_R(v, \hat{\mathbf{v}}_m^{\text{O}})$, is given by

$$\widehat{m}_R(v, \hat{\mathbf{v}}_m^{\text{O}}) = \frac{9\mu_m \widehat{H}_P}{4\beta_I(v)\cos(\chi_{sm})}\,\widehat{q}_R(v, \hat{\mathbf{v}}_m^{\text{O}}) = \frac{9\mu_m \widehat{H}_P}{4\beta_I(v)}\,\widehat{q}_{R_z}(v, \hat{\mathbf{v}}_m^{\text{O}}) \tag{3.15}$$

where $\widehat{q}_{R_z}(v, \hat{\mathbf{v}}_m^{\text{O}}) = \widehat{q}_R(v, \hat{\mathbf{v}}_m^{\text{O}})/\cos(\chi_{sm})$ is the equivalent minimum radar-detectable zenith line density for a meteor traveling vertically downward through $\mathscr{P}_s^{\text{S}}$ with speed v at height $\widehat{H}_I$.

The quantities $\widehat{q}_{R_z}(v, \hat{\mathbf{v}}_m^{\text{O}})$, $\widehat{q}_R(v, \hat{\mathbf{v}}_m^{\text{O}})$, and $\widehat{m}_R(v, \hat{\mathbf{v}}_m^{\text{O}})$ are dependent on both meteor velocity (v and $\hat{\mathbf{v}}_m^{\text{O}}$) as well as height $\widehat{H}_I$. This dependence is determined from the peak trail-scattered RSL using the radar link power budget, antenna gain pattern, and operating frequency. In general, $\widehat{q}_{R_z}(v, \hat{\mathbf{v}}_m^{\text{O}})$, and therefore $\widehat{m}_R(v, \hat{\mathbf{v}}_m^{\text{O}})$, decreases with increased transmitter power, reduced required received power (increased radar sensitivity), decreased radar frequency, and increased antenna gain.

If the meteor trail produces an underdense-type echo (see Section 3.1.1.2), the speed dependence of $\widehat{q}_{R_z}(v, \hat{\mathbf{v}}_m^{\text{O}})$ is realized through the initial trail radius, r_0, by the factor

$$\exp\left(\frac{4\pi^2 r_0^2}{\lambda_R^2}\right)$$

where the initial radius (meters) may be given by [82]

$$r_0(v, h_s) \cong \frac{7.35 \times 10^{11} v}{n_A(h_s)} \tag{3.16a}$$

or

$$r_0(v, h_s) \cong \frac{2.58 \times 10^{12} v^{0.8}}{n_A(h_s)} \tag{3.16b}$$

derived from different empirical studies. In these expressions, v is measured in km/s and

h_s = height of the trail-scatter point $\mathscr{P}_s^{\mathrm{S}}$ above the earth's surface ($h_s = \tilde{H}_I$ in Eqs. 3.16, measured in kilometers)

n_A = atmospheric particle density at height h_s (particles per cm^3) [83]

λ_R = radar signal wavelength (m).

The minimum detectable zenith underdense line density becomes

$$\widehat{q}^{\,U}_{R_z}(v, \hat{\mathbf{v}}^{\mathrm{O}}_m) = \widehat{q}^{\,U_0}_{R_z} \exp\left[\frac{4\pi^2 r_0(v, h_s)}{\lambda_R^2}\right] \tag{3.17}$$

where $\widehat{q}^{\,U_0}_{R_z}$ is the minimum radar-detectable zenith underdense line density for $r_0 = 0$.

In general, Eq. 3.17 includes a second exponential factor to account for trail diffusion (see Eqs. 3.47c and d) occurring after trail formation but before arrival of a radar pulse. This effect, not considered in this development, is eliminated when cw (bistatic) radars are employed.

If $\widehat{q}^{\,U}_{R_z}(v, \hat{\mathbf{v}}^{\mathrm{O}}_m)$ exceeds the velocity-dependent underdense-overdense threshold $\widehat{q}^{\,T}_{R_z}(v, \hat{\mathbf{v}}^{\mathrm{O}}_m)$ at height $\tilde{H}_I$, then an overdense trail-scattered signal will be detected by the radar. The threshold $\widehat{q}^{\,T}_{R_z}(v, \hat{\mathbf{v}}^{\mathrm{O}}_m)$ is an increasing function of initial trail radius r_0, therefore it is also an increasing function of meteor speed v. Since the maximum backscattered power from an overdense trail is independent of the initial trail radius [84], $\widehat{q}^{\,U}_{R_z}(v, \hat{\mathbf{v}}^{\mathrm{O}}_m) > \widehat{q}^{\,T}_{R_z}(v, \hat{\mathbf{v}}^{\mathrm{O}}_m)$ implies $\widehat{q}_{R_z}(v, \hat{\mathbf{v}}^{\mathrm{O}}_m) = \widehat{q}^{\,O}_{R_z}(v, \hat{\mathbf{v}}^{\mathrm{O}}_m)$ the minimum radar-detectable overdense trail line density at height $\widehat{H}_I$.

Since $\beta_I(v)$ is an increasing function of v in the denominator of Eq. 3.15 and $r_0(v, \widehat{H}_I)$ is an increasing function of v in the exponential of Eq. 3.17 (all other parameters fixed), there exists a value of $v = v^*$ such that

$$\widehat{m}_R(v^*, \hat{\mathbf{v}}^{\mathrm{O}}_m) = \frac{9\mu_m \widehat{H}_P}{4\beta_I(v^*)} \widehat{q}^{\,*}_{R_z}(v^*, \hat{\mathbf{v}}^{\mathrm{O}}_m) \leq \widehat{m}_R(v, \hat{\mathbf{v}}^{\mathrm{O}}_m) \tag{3.18}$$

for all values of meteor speed v, $v_l \leq v \leq v_u$. At $v = v^*$, the minimum radar-detectable line density $\widehat{q}^{\,*}_{R_z}(v^*, \hat{\mathbf{v}}^{\mathrm{O}}_m)$ corresponds to *maximum radar* *sensitivity*. The maximum sensitivity of the Kazan radar ($\lambda_R = 8$ m) was given as $\widehat{q}^*_{R_z}(v^*, \hat{\mathbf{v}}^{\mathrm{O}}_m) = 1.2 \times 10^{13}$ epm. This value was computed at the trail height $h_s = \widehat{H}_I = 96$ km/s for an initial radius $r_0 = 1.2$ m. This radius corresponds to a meteor speed $v^* \cong 45$ km/s, the approximate average atmospheric speed of sporadic meteors [85]. The 96 km trail height corresponds to the peak gain of the radar antenna used at both Kazan and Mogadishu (five-element Yagi) at an elevation of 35°. Using these values, the zero-radius q value, $\widehat{q}^{U_0}_{R_z}$, for the Kazan-Mogadishu results is given by

$$\widehat{q}^{U_0}_{R_z} = \frac{1.2 \times 10^{13}}{\exp(4\pi^2\, 1.2^2/8^2)} = 4.9 \times 10^{12} \text{ epm}$$

Given this value for $\widehat{q}^{U_0}_{R_z}$, $\widehat{q}_{R_z}(v, \hat{\mathbf{v}}^{\mathrm{O}}_m)$ determined from Eq. 3.17 will be less than $\widehat{q}^{T}_{R_z}(v, \hat{\mathbf{v}}^{\mathrm{O}}_m)$ for all meteor speeds of interest. Thus, Eq. 3.15 gives the minimum velocity-dependent radar-detectable mass for the combined Kazan-Mogadishu radar measurements as

$$\widehat{m}_R(v, \hat{\mathbf{v}}^{\mathrm{O}}_m) = \frac{9\mu_m \widehat{H}_P}{4\beta_I(v)}\, \widehat{q}^{U_0}_{R_z} \exp\left\{\frac{4\pi^2[r_0(v, \widehat{H}_P)]^2}{\lambda_R^2}\right\} \tag{3.19}$$

This m value is needed to compute the detection factor given by Eq. 3.9. The complete calculation of the detection factor, $F_R(v, \hat{\mathbf{v}}^{\mathrm{O}}_m)$, requires the density function, $p_{\mathbf{V}_a}(v, \hat{\mathbf{v}}^{\mathrm{O}}_m)$, for the apparent velocity of sporadic meteors over the celestral sphere.

Theoretical velocity density functions have been derived for both sporadic meteoroid orbital velocity [86] and sporadic meteor geocentric velocity. For example, a density function for the geocentric velocity of sporadic meteors, $p_{\mathbf{V}_G}(v_G, \hat{\mathbf{v}}^{\mathrm{O}}_m)$, was approximated by [87]

$$p_{\mathbf{V}_G}(v_G, \hat{\mathbf{v}}^{\mathrm{O}}_m) = c_G n_G(\epsilon_G) \exp\{-0.0363[v_G - \bar{v}_G(\epsilon_G)]^2\} \tag{3.20}$$

where

ϵ_G = geocentric elongation, the angle measured between the geocentric radiant, $-\hat{\mathbf{v}}^{\mathrm{O}}_m$, and the earth's apex direction vector, $\langle 0, -1, 0\rangle_{\mathrm{O}}$ in the O system, so $0 \le \epsilon_G \le \pi$, [88]

$\bar{v}_G(\epsilon_G)$ = average value of geocentric meteor speed, v_G, for elongation ϵ_G

$n_G(\epsilon_G)$ = probability density function for geocentric sporadic meteor radiants

c_G = normalization constant.

Since $\hat{\mathbf{v}}^{\mathrm{O}}_G = \hat{\mathbf{v}}^{\mathrm{O}}_m$ is assumed,

$$\epsilon = \epsilon_G = \epsilon_a = \cos^{-1}(-\hat{v}^{\mathrm{O}}_{m2}) \tag{3.21}$$

where $\hat{v}^{\mathrm{O}}_{m2}$ is the y_{O} component of the trail vector $\hat{\mathbf{v}}^{\mathrm{O}}_{m}$. Thus, $n_G(\epsilon_G) = n_a(\epsilon_a) = n(\epsilon)$, which is equivalent to $p_{\hat{\mathbf{v}}^{\mathrm{O}}_m}(\hat{\mathbf{v}}^{\mathrm{O}}_{m})$, the density function for apparent meteor radiants.

The geocentric velocity density function $p_{\mathbf{v}_G}(v_G, \hat{\mathbf{v}}^{\mathrm{O}}_{m})$ given by Eq. 3.20 must be converted to the apparent velocity density function $p_{\mathbf{v}_a}(v, \hat{\mathbf{v}}^{\mathrm{O}}_{m})$ for use in Eq. 3.9 in the physical factor, $F_R(v, \hat{\mathbf{v}}^{\mathrm{O}}_{m})$. The conversion from apparent to geocentric meteor speed at a height $\widehat{H}_I \cong 96$ km is given by [89]

$$v = [v_G^2 + 2G_E(R_E + \widehat{H}_i)]^{1/2} = (v_G^2 + 126.9)^{1/2}$$

where

G_E = the earth's gravitational acceleration (9.8 m/sec^2)

R_E = Earth's radius (6370 km).

Using Eq. 3.22 and the standard technique for computing the density of a continuous function of one variable [90], v_G, the apparent velocity density function becomes

$$p_{\mathbf{v}_a}(v, \hat{\mathbf{v}}^{\mathrm{O}}_{m}) \cong \frac{c_G n(\epsilon) v \exp\{-0.0363[\sqrt{v^2 - 126.9} - \sqrt{\bar{v}(\epsilon)^2 - 126.9}]^2\}}{|\sqrt{v^2 - 126.9}|} \tag{3.22}$$

where ϵ is determined from $\hat{\mathbf{v}}^{\mathrm{O}}_{m}$ by Eq. 3.21 and $\bar{v}(\epsilon)$ is the average apparent meteor speed. $\bar{v}(\epsilon)$ and the average geocentric meteor speed, $\bar{v}_G(\epsilon)$, are plotted versus ϵ in Fig. 3.21.

The portion of Eq. 3.22 comprising the meteor speed density function includes the term

$$\frac{v \exp\{-0.0363[\sqrt{v^2 - 126.9} - \sqrt{\bar{v}(\epsilon)^2 - 126.9}]^2\}}{|\sqrt{v^2 - 126.9}|}$$

which is significant only for meteor speeds within ±10 km/s of the average apparent speed, $\bar{v}(\epsilon)$. Therefore, consider replacing this term in Eq. 3.22 with the Dirac delta function, δ_D, centered at $\bar{v}(\epsilon)$. The density function for the apparent velocity of sporadic meteors may then be approximated as

$$p_{\mathbf{v}_a}(v, \hat{\mathbf{v}}^{\mathrm{O}}_{m}) \cong c_a n(\epsilon)\delta_D[v - \bar{v}(\epsilon)] \tag{3.23}$$

where c_a is a normalization constant determined so that the condition imposed by Eq. 3.4*a* is met. A significant condition inherent in this approximation is that $v^* = \bar{v}(\epsilon)$, independent of the radar wavelength.

Given the trail unit vector $\hat{\mathbf{v}}^{\mathrm{O}}_{m}$, ϵ can be computed from Eq. 3.21, and the average apparent meteor speed, $\bar{v}_\epsilon = \bar{v}(\epsilon)$, can be determined from a

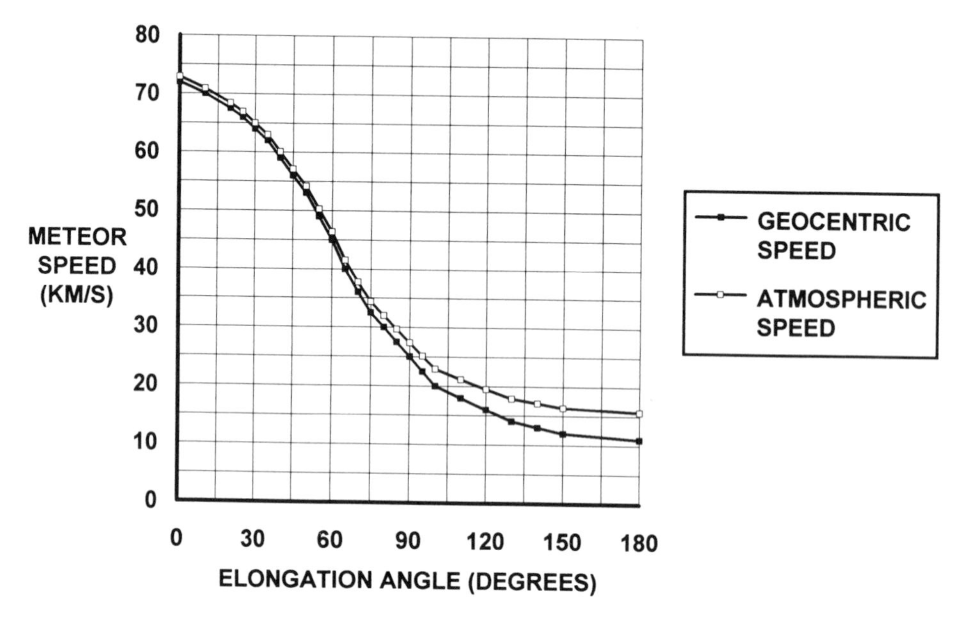

Figure 3.21. Geocentric and atmospheric meteor velocities.

curve-fit to the plot of Fig. 3.21. Thus, ignoring the direction changes imposed by zenith attraction and diurnal aberration, the integrated density function

$$p_{\hat{\mathbf{v}}_m^{\mathrm{O}}}(\hat{\mathbf{v}}_m^{\mathrm{O}}) = c_a n(\epsilon) \int_{v_l}^{v_u} \delta_D[v - \bar{v}_\epsilon] \, dv = c_a n(\epsilon)$$

so the detection factor defined by Eq. 3.9 can then be written as

$$F_R(v, \hat{\mathbf{v}}_m^{\mathrm{O}}) = \delta_D[v - \bar{v}_\epsilon] \left[\frac{m_0}{\widehat{m}_R(v, \hat{\mathbf{v}}_m^{\mathrm{O}})} \right]^{(s-1)}$$

The integration over v in Eq. 3.10 yields

$$\int_{v_l}^{v_u} F_R(v, \hat{\mathbf{v}}_m^{\mathrm{O}}) \, dv = \int_{v_l}^{v_u} \delta_D[v - \bar{v}_\epsilon] \left[\frac{m_0}{\widehat{m}_R(v, \hat{\mathbf{v}}_m^{\mathrm{O}})} \right]^{(s-1)} dv = \left[\frac{m_0}{\widehat{m}_R(\bar{v}_\epsilon, \hat{\mathbf{v}}_m^{\mathrm{O}})} \right]^{(s-1)}$$

The scaled SMRD value is then given by

$$X_m(\hat{\mathbf{v}}_m^{\mathrm{O}}) = \Theta_{sp}(m_0, \hat{\mathbf{v}}_m^{\mathrm{O}}) \left[\frac{m_0}{\widehat{m}_R(\bar{v}_\epsilon, \hat{\mathbf{v}}_m^{\mathrm{O}})} \right]^{(s-1)}$$

TABLE 3.1 Estimated Mass–Rate Exponent Versus Approximate Mass Interval

Mass (m) Interval (g)	Mass–Rate Exponent
$10^{-12} \leq m \leq 10^{-8}$	1.5
$10^{-8} \leq m \leq 10^{-4}$	2.0
$10^{-4} \leq m \leq 10^{1}$	2.4
$10^{1} \leq m \leq 10^{8}$	1.6

so the actual SMRD value corresponding to the trail direction vector $\hat{\mathbf{v}}_m^{\mathrm{O}}$ becomes

$$\Theta_{sp}(m_0, \hat{\mathbf{v}}_m^{\mathrm{O}}) = X_m(\hat{\mathbf{v}}_m^{\mathrm{O}})\left[\frac{\widehat{m}_R(\bar{v}_\epsilon, \hat{\mathbf{v}}_m^{\mathrm{O}})}{m_0}\right]^{(s-1)} \tag{3.24}$$

where values of 100 X_m have been tabulated [91] for every month as shown in Fig. 3.19 for February.

The exponential mass-rate relationship defined by c_n and s in Eq. 3.2 states that the number of meteors with masses between m and $m + dm$ is given by $c_n m^{-s}\, dm$. The parameter s varies over the celestial sphere according to [92]

$$s = s(\epsilon) = 1.90 + 0.12\epsilon + 0.01\epsilon^2 \tag{3.25}$$

More generally, s is a function of the mass interval as well as the elongation, ϵ. For example, estimates of the s parameter collected from several studies [93] are shown in Table 3.1. For MB communications, pre-burn meteor masses from 10^{-5} to 10^2 g span the range of values that produce usable trails. Eq. 3.25 yields values from 1.9 to 2.1, corresponding most closely to the mass interval from 10^{-8} to 10^{-4} g, rather than the somewhat higher value of 2.4 from the more relevant mass interval from 10^{-4} to 10 g (see Table 3.1). If the values in Table 3.1 are accurate, predictions using Eq. 3.25 would somewhat overestimate link MR values. Nevertheless, the Kazan-Mogadishu SMRD map used in METEORLINK employs s values computed from Eq. 3.25.

3.2.2.2.3 Link-Observable SMRD Values. Define the exact locations of the transmit and receive antennas as $(l_{\phi_j}, l_{\theta_j}, h_{a_j})$, where

- l_{ϕ_j} = earth longitude measured east of the Greenwich meridian ($-\pi \leq l_{\phi_j} \leq \pi$)
- l_{θ_j} = earth latitude measured north of the geographic equator ($\pi/2 \leq l_{\theta_j} \leq \pi/2$)

h_{a_j} = height of the site j antenna above sea level at the earth coordinates $(l_{\phi_j}, l_{\theta_j})$

where the index $j = 1$ references the transmit antenna site and $j = 2$ indexes the receive antenna site. These earth coordinates define the LINK coordinate system, or L system, shown in Fig. 3.22. The figure shows that the origin of the L system lies below the earth's surface for separated transmit and receive antenna sites (MB link or cw radar) and is located at the surface for collocated sites (monostatic radar). In the L system, the x_L axis lies in the plane containing the link GCP and the center of the earth with the transmit site ($j = 1$) assigned a positive x_L value. The z_L axis passes through the earth's center and intersects the x_L axis halfway between the surface

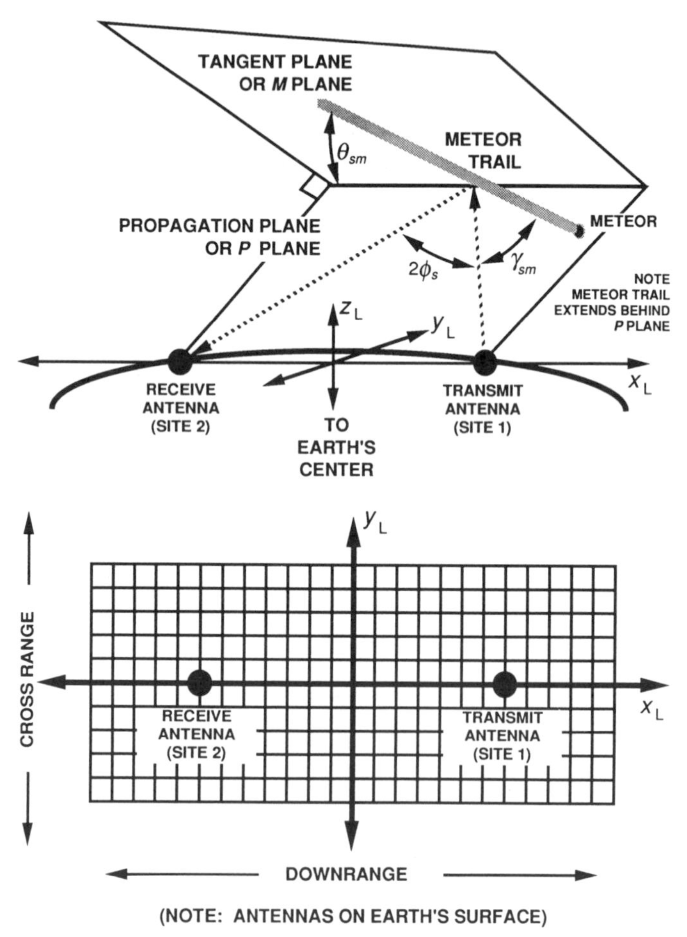

Figure 3.22. LINK coordinate system.

coordinates of the two sites, $(l_{\phi_j}, l_{\theta_j})$, $j = 1, 2$. Finally, the y_L axis is defined in the normal right-handed sense (see Section 3.A.2)

A meteor trail is considered "observable" by a MB link if it meets the geometrical requirements for coherent trail-scatter, although it may not produce a usable signal at the receiver. Consider the orientations of observable meteor trails passing through an arbitrary scatter point, $\mathcal{P}_s^L = \langle x_s^L, y_s^L, z_s^L \rangle$, in the L system. The height, h_s, of the point $\mathcal{P}_s^L$ above sea level is given by

$$h_s = R_s - R_E \tag{3.26a}$$

where

$$R_s = \sqrt{\left\{ z_s^L + \left[R_E^2 - \left(\frac{d_{12}}{2} \right)^2 \right]^{1/2} \right\}^2 + (x_s^L)^2 + (y_s^L)^2} \tag{3.26b}$$

and d_{12} is the distance along the x_L axis between the points $(l_{\phi_1}, l_{\theta_1})$ and $(l_{\phi_2}, l_{\theta_2})$ at sea level (see Eq. 3.A.6).

An observable trail through the point $\mathcal{P}_s^L$ lies in the meteor plane, or *M* plane [94], tangent to the surface of an ellipsoid of revolution containing the point $\mathcal{P}_s^L$ with foci located at the link antennas. If the antennas are sufficiently elevated above sea level (h_{a_j}), this ellipsoid is symmetric about the tilted SCATTER coordinate system (S system) shown in Fig. 3.23

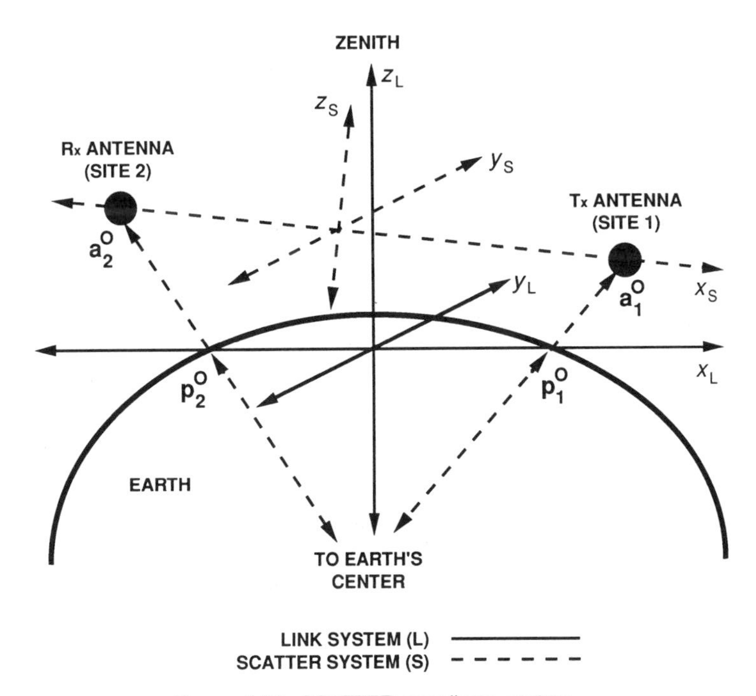

Figure 3.23. SCATTER coordinate system.

(same as Fig. 3.A.3). The point $\mathscr{P}_s^{\mathrm{L}}$ can be converted to the S system point $\mathscr{P}_s^{\mathrm{S}} = \langle x_s^{\mathrm{S}}, y_s^{\mathrm{S}}, z_s^{\mathrm{S}} \rangle$ by Eq. 3.A.15.

As shown in Fig. 3.24, the unit normal vector $\hat{\mathbf{n}}_s^{\mathrm{S}}$ to the scatter ellipsoid defines the M plane [94] at $\mathscr{P}_s^{\mathrm{S}}$. The M plane contains meteor trail orientations meeting the geometric criterion for specular reflection (see Fig. 3.4), i.e., observable trails are normal to $\hat{\mathbf{n}}_s^{\mathrm{S}}$. It is assumed that meteor trajectories do not gain height in the earth's atmosphere, so that limiting orientations are normal to the zenith vector $\hat{\boldsymbol{\zeta}}_s^{\mathrm{S}}$ through the point $\mathscr{P}_s^{\mathrm{S}}$ (see Eq. 3.A.17). Since limiting trail orientations are normal to both the $\hat{\mathbf{n}}_s^{\mathrm{S}}$ and $\hat{\boldsymbol{\zeta}}_s^{\mathrm{S}}$ vectors, these orientations are determined in the S system by the unit vector

$$\pm\hat{\mathbf{x}}_{\mathrm{M}}^{\mathrm{S}} = \pm\frac{\hat{\mathbf{n}}_s^{\mathrm{S}} \times \hat{\boldsymbol{\zeta}}_s^{\mathrm{S}}}{|\hat{\mathbf{n}}_s^{\mathrm{S}} \times \hat{\boldsymbol{\zeta}}_s^{\mathrm{S}}|}$$

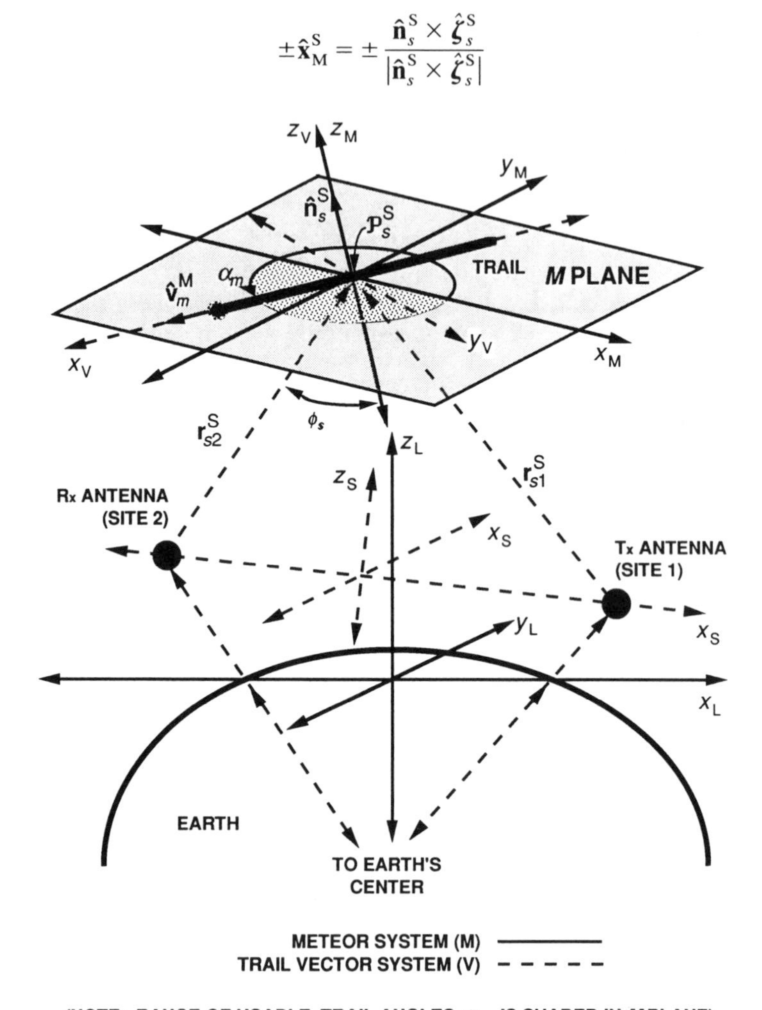

Figure 3.24. The METEOR and TRAIL VECTOR coordinate systems.

where the unit vectors $\hat{\mathbf{x}}_{\mathrm{M}}^{\mathrm{S}}$ and $\hat{\mathbf{z}}_{\mathrm{M}}^{\mathrm{S}} = \hat{\mathbf{n}}_s^{\mathrm{S}}$ define the x_{M} axis and z_{M} axis, respectively, of the METEOR coordinate system (M system) in the S system. The y_{M} axis is defined in the normal right-handed sense by the unit vector $\hat{\mathbf{y}}_{\mathrm{M}}^{\mathrm{S}}$. The M system and its relationship to the S system are shown in Fig. 3.24 and Fig. 3.A.4 at each trail-scatter point $\mathcal{P}_s^{\mathrm{S}}$.

In the M system, observable trail orientations may be specified by the unit vector $\hat{\mathbf{v}}_m^{\mathrm{M}}$ with components

$$\begin{aligned} \hat{v}_{m1}^{\mathrm{M}} &= \cos(\alpha_m) \\ \hat{v}_{m2}^{\mathrm{M}} &= \sin(\alpha_m) \\ \hat{v}_{m3}^{\mathrm{M}} &= 0 \end{aligned} \tag{3.27}$$

where α_m is the angle measured counterclockwise from the positive x_{M} axis to the trail vector $\hat{\mathbf{v}}_m^{\mathrm{M}}$. Observable trails lie in the range of α_m angles given by $\pi \le \alpha_m \le 2\pi$. Given α_m, the unit vector $\hat{\mathbf{v}}_m^{\mathrm{M}}$ determines the direction of meteor flight in the atmosphere that will produce a link-observable trail. The $\hat{\mathbf{v}}_m^{\mathrm{M}}$ vector may be converted to unit vectors in the S system, L system, and O system by

$$\hat{\mathbf{v}}_m^{\mathrm{S}} = \mathbf{T}_{\mathrm{MS}} \cdot \hat{\mathbf{v}}_m^{\mathrm{M}} \tag{3.28a}$$

$$\hat{\mathbf{v}}_m^{\mathrm{L}} = \mathbf{T}_{\mathrm{SL}} \cdot \hat{\mathbf{v}}_m^{\mathrm{S}} \tag{3.28b}$$

$$\hat{\mathbf{v}}_m^{\mathrm{O}} = \mathbf{T}_{\mathrm{MO}} \cdot \hat{\mathbf{v}}_m^{\mathrm{M}} \tag{3.28c}$$

respectively, where the corresponding transformation matrices were derived in Appendix A. These equivalent direction vectors will be used interchangeably throughout this development. Similarly, the trail-scatter point $\mathcal{P}_s^{\mathrm{S}}$ in the S system is coincident with the point $\mathcal{P}_s^{\mathrm{L}}$ in the L system. The cosine of the zenith angle χ_{sm} (see Eq. 3.13) of this trail is then given by

$$\cos(\chi_{sm}) = \cos(\hat{\boldsymbol{\zeta}}_s^{\mathrm{O}} \cdot \hat{\mathbf{v}}_m^{\mathrm{O}}) \tag{3.29}$$

where $\hat{\boldsymbol{\zeta}}_s^{\mathrm{O}} = \mathbf{T}_{\mathrm{SO}} \cdot \hat{\boldsymbol{\zeta}}_s^{\mathrm{S}}$, $\mathbf{T}_{\mathrm{SO}}$, is the transformation matrix from the S system to the O system, and the $(\cdot)$ is the vector dot product. The α_m angles π and 2π correspond to meteor direction vectors $\hat{\mathbf{v}}_m^{\mathrm{O}}$ oriented normal to the zenith vector $\hat{\boldsymbol{\zeta}}_s^{\mathrm{O}}$.

Given an observable meteor trail with unit direction vector $\hat{\mathbf{v}}_m^{\mathrm{O}}$, the corresponding orientation angles $(\phi_{\mathrm{O}}, \theta_{\mathrm{O}})$ needed to index the appropriate SMRD values, that is

$$\Theta_{sp}(m_0, \hat{\mathbf{v}}_m^{\mathrm{O}}) = \text{meteors}(m \ge m_0)\ \mathrm{km}^{-2}\ \mathrm{sr}^{-1}\ \mathrm{hr}^{-1}$$

are found from Eq. 3.2. Although a trail with direction $\hat{\mathbf{v}}_m^{\mathrm{O}}$ may be observable on a given MB link, it will not produce a usable trail-scattered signal unless the trail's electron line density exceeds a threshold value. This threshold is determined by the scattering efficiency of the trail as well as the link power budget.

Let q^*_{sm} be the minimum line density detectable by a MB link for a meteor with trail orientation $\hat{\mathbf{v}}^{\mathrm{S}}_m$ and speed v through the point $\mathcal{P}^{\mathrm{S}}_s$ in the S system. The corresponding line density $\widehat{q}^*_{sm}$ at the height of maximum ionization, $\widehat{H}_I$, from a solid meteor body exhibiting little or no fragmentation is given by

$$\widehat{q}^*_{sm} = \frac{4q^*_{sm}}{9\exp(H_e)[1-\exp(H_e)/3]^2} \tag{3.30a}$$

where

$$H_e = \frac{\widehat{H}_I - h_s}{H_P} \tag{3.30b}$$

where H_P is the scale height at the trail-scatter point height h_s. The minimum mass $\widehat{m}^*_{sm}$ of the corresponding meteor at the height of maximum ionization, $\widehat{H}_I$, in the arriving flux $\Theta_{sp}(m_0, \hat{\mathbf{v}}^{\mathrm{O}}_m)$ is then given by

$$\widehat{m}^*_{sm} = \frac{9\mu_m \widehat{H}_P}{4\beta_I(v)\cos(\chi_{sm})}\,\widehat{q}^*_{sm} \tag{3.31}$$

where $\widehat{q}^*_{sm}$, and therefore $\widehat{m}^*_{sm}$, are functions of meteor velocity (speed v and direction $\hat{\mathbf{v}}^{\mathrm{O}}_m$).

Let the total number of usable meteor trails detected by a MB link over some time interval during which the SMRD statistics are stationary be designated by N_L. The differential MR value d^3N_L contributed by meteors passing through the point $\mathcal{P}^{\mathrm{S}}_s$ at height h_s with unit direction vector $\hat{\mathbf{v}}^{\mathrm{M}}_m$ (or, equivalently, $\hat{\mathbf{v}}^{\mathrm{O}}_m$) that have the minimum mass $\widehat{m}^*_{sm}$ at height $\widehat{H}_I$ can now be written in terms of the Kazan-Mogadishu measurements as

$$\begin{aligned}
d^3N_L &= (s-1)m_0^{(s-1)}\Theta_{sp}(m_0, \hat{\mathbf{v}}^{\mathrm{O}}_m)\int_{v_l}^{v_u} \frac{p_{\mathbf{v}_a}(v, \hat{\mathbf{v}}^{\mathrm{O}}_m)}{p_{\hat{\mathbf{v}}^{\mathrm{O}}_m}(\hat{\mathbf{v}}^{\mathrm{O}}_m)} \\
&\quad \times \int_{\widehat{m}^*_{sm}(v, \hat{\mathbf{v}}^{\mathrm{O}}_m)}^{\infty} m^{-s}\, dm\, dv\, dS\, d\Omega\, dt \\
&= m_0^{(s-1)} X_m(\hat{\mathbf{v}}^{\mathrm{O}}_m)\left[\frac{\widehat{m}_R(\bar{v}_\epsilon, \hat{\mathbf{v}}^{\mathrm{O}}_m)}{m_0}\right]^{(s-1)} \widehat{m}^*_{sm}(\bar{v}_\epsilon, \hat{\mathbf{v}}^{\mathrm{O}}_m)^{(1-s)}\, dS\, d\Omega\, dt \\
&= X_m(\hat{\mathbf{v}}^{\mathrm{O}}_m)\left[\frac{\widehat{m}_R(\bar{v}_\epsilon, \hat{\mathbf{v}}^{\mathrm{O}}_m)}{\widehat{m}^*_{sm}(\bar{v}_\epsilon, \hat{\mathbf{v}}^{\mathrm{O}}_m)}\right]^{(s-1)} dS\, d\Omega\, dt
\end{aligned} \tag{3.32}$$

Since μ_m, $\widehat{H}_P$, and $\beta_I(\bar{v}_\epsilon)$ appear in both $\widehat{m}_R(\bar{v}_\epsilon, \hat{\mathbf{v}}_m^{\mathrm{O}})$ and $\widehat{m}_{sm}^*(\bar{v}_\epsilon, \hat{\mathbf{v}}_m^{\mathrm{O}})$, the mass ratio in Eq. 3.32 reduces to the ratio of electron line densities

$$\left[\frac{\widehat{q}_{R_z}(\bar{v}_\epsilon, \hat{\mathbf{v}}_m^{\mathrm{O}})\cos(\chi_{sm})}{\widehat{q}_{sm}^*(\bar{v}_\epsilon, \hat{\mathbf{v}}_m^{\mathrm{O}})}\right]^{(s-1)}$$

where $\bar{v}_\epsilon$ is the average observed meteor speed for trail direction $\hat{\mathbf{v}}_m^{\mathrm{O}}$. Finally, the differential link MR value may be written

$$d^3N_L = X_m(\hat{\mathbf{v}}_m^{\mathrm{O}})\left[\frac{\widehat{q}_{R_z}(\bar{v}_\epsilon, \hat{\mathbf{v}}_m^{\mathrm{O}})\cos(\chi_{sm})}{\widehat{q}_{sm}^*(\bar{v}_\epsilon, \hat{\mathbf{v}}_m^{\mathrm{O}})}\right]^{(s-1)} dS\, d\Omega\, dt \qquad (3.33)$$

The element d^3N_L is the MB link MR contribution through the point $\mathcal{P}_s^{\mathrm{S}}$ oriented along $\hat{\mathbf{v}}_m^{\mathrm{M}}$ to produce a usable trail on the MB radio link. Next, the minimum detectable line density q_{sm}^* is computed from the MB link power budget and minimum required burst duration at $\mathcal{P}_s^{\mathrm{S}}$ for a trail with orientation $\hat{\mathbf{v}}_m^{\mathrm{M}}$. Finally, q_{sm}^* is converted to $\widehat{q}_{sm}^*$ at the height of maximum ionization using Eq. 3.30*a*. Substitution of $\widehat{q}_{sm}^*$ into Eq. 3.33 yields the usable MR contribution d^3N_L along $\hat{\mathbf{v}}_m^{\mathrm{M}}$ through $\mathcal{P}_s^{\mathrm{S}}$.

3.2.2.3 Usable Meteor Rate

3.2.2.3.1 Link Power Budget. The minimum required line density q_{sm}^* is computed for each modeled trail-scatter point $\mathcal{P}_s^{\mathrm{L}}$ and trail vector $\hat{\mathbf{v}}_m^{\mathrm{L}}$ from the link antenna gain pattern product G_{sm}^{12} (dB) at $\mathcal{P}_s^{\mathrm{L}}$, the transmitted power P_t (dBW), line losses L_t and L_r (dB), and the required received power P_r^* (dBm). P_r^* may be determined from various criteria, including the following:

- minimum bit error probability P_e given modulation type and data rate, that is, $P_e^*(SNR_b^*, R_c, N_0, BW_n)$, where SNR_b^* is the required SNR per bit or E_b/N_0, R_c is the bit transmission rate, BW_n is the modulation bandwidth, and N_0 is the noise power spectral density
- minimum detection threshold used to prompt the link communications processor to monitor the output of the demodulator
- minimum threshold required for signal identification to trigger time-logged RSL sampling as part of an MB test bed [95].

Values of P_t for experimental or operational MB links and radars range from a minimum of about 10 dBW (10 W) to as high as 50 dBW (100 kW), although off-the-shelf MB communications equipment typically operates between 27 dBW (500 W) and 30 dBW (1 kW). The *effective* antenna gain product, G_{sm}^{12}, varies over a wide range, with peak values of about 20 dB for links employing five-element Yagi antennas. The subscript m denotes a possible dependence of G_{sm}^{12} on trail orientation when plasma resonance effects are modeled for underdense trail-scattered signals. Typical values for P_r^* vary between -120 and -90 dBm for nominal terminal equipment and noise environments. Sensitivity studies using P_r^* as the parameter are typically employed in model validation efforts and link design studies. In general, all link power budget parameters are functions of the link operating frequency $f_L = c/\lambda_L$, where c is the speed of light and λ_L is the signal wavelength for the MB link.

The classic link power budget equation determines the maximum allowable path loss L^*(dB) for each point $\mathcal{P}_s^{\mathrm{L}}$ from

$$L^* = P_t + G_{sm}^{12} - L_t - L_r - P_r^* + 30 \tag{3.34a}$$

where

$$P_r^* = SNR_b^* + R_b + N_0 + BW_n \tag{3.34b}$$

with the noise spectral density N_0 referenced to the receiver input. Typically, N_0 is measured or computed from antenna temperature T_a (K), receive line temperature T_r (K), and receiver noise factor, n_f, from the expression

$$N_0 = 10 \log(kT_e) \tag{3.35a}$$

with

$$T_e = \frac{T_a}{l_r} + \left(1 - \frac{1}{l_r}\right) T_r + (n_f - 1) T_0 \tag{3.35b}$$

where

$$T_a = T_0 10^{(F_a/10)}, \quad l_r = 10^{(L_r/10)}, \quad n_f = 10^{(N_f/10)} \tag{3.35c}$$

and

k = Boltzmann's constant (1.38×10^{-23} J/K)

T_e = effective temperature of the receive system (K)

l_r = receive line attenuation factor

T_0 = reference temperature (290 K)

F_a = antenna noise factor (dB above thermal noise, kT_0)

N_f = receiver noise figure (dB)

Values of the antenna noise factor F_a may be measured with a sensitive receiver or computed from integration of the incident noise field over the receive antenna pattern. Typical values of F_a for the omnipresent galactic background noise using nominal MB antennas are between 10 and 15 dB above kT_0 and show a marked diurnal variation. The peak value of this variation occurs when the Milky Way Galaxy lies in the main beam. Of course, this variation is most significant for highly directive antenna patterns and least significant for broadbeamed antenna patterns.

In general, the variety of antennas employed for meteor scatter, diversity in the antenna environments, and polarization coupling loss at the trail-scatter point $\mathcal{P}_s^{\mathrm{L}}$, determine the need for detailed pattern data to accurately model G_{sm}^{12} throughout the common volume of the link antennas. In METEORLINK, antenna electric field patterns are used to determine the antenna gain product G_{sm}^{12} at each of hundreds of trail-scatter points, $\mathcal{P}_s^{\mathrm{L}}$, throughout the common volume. These patterns can be computed by the Numerical Electromagnetics Code (NEC) developed by Lawrence Livermore Laboratories [96]. NEC has been used to calculate electric field patterns for arbitrary antennas mounted on towers, buildings, vehicles, ships and aircraft positioned above perfect or real ground. The antenna electric field (E field) patterns computed by NEC consist of the ϕ and phasor θ component magnitudes and their relative phase angle specified for arbitrary $\Delta\phi_j$ and $\Delta\theta_j$ increments at antenna j for $j = 1, 2$.

METEORLINK normalizes the NEC-generated E field vectors by the field strength, E_i, from an isotropic radiator computed at some far-field range r_a (m) and transmitted power P_a (W) used by the NEC calculation. In the PATTERN coordinate system at antenna site j (P_j system), this normalized complex E field vector is given by

$$\tilde{\mathbf{e}}_s^{\mathrm{A}_j} = \frac{\langle \tilde{E}_{s_\psi r}^{\mathrm{P}_j}(=0), \tilde{E}_{s_\psi \phi}^{\mathrm{P}_j}, \tilde{E}_{s_\psi \theta}^{\mathrm{P}_j} \rangle}{E_i} \tag{3.36a}$$

where $\langle \tilde{E}_{s_\psi r}^{\mathrm{P}_j}(=0), \tilde{E}_{s_\psi \phi}^{\mathrm{P}_j}, \tilde{E}_{s_\psi \theta}^{\mathrm{P}_j} \rangle$ (see Eq. 3.A.35) is the *complex* (˜) electric phasor vector at a distance r_a from the antenna at site j, $j = 1, 2$, and the subscript ψ indicates that the phasors have been rotated to account for the effects of Faraday rotation in the lower ionosphere. The magnitude of the corresponding isotropic E field vector $\tilde{\mathbf{E}}_i$ (V/m) at the same distance r_a is given by

$$E_i = \|\tilde{\mathbf{E}}_i\| = \sqrt{\left(\frac{P_a}{2\pi e_0 r_a^2}\right)} \tag{3.36b}$$

where

c = speed of light (3.0×10^8 m/s)

ϵ_0 = free space permittivity constant (8.85×10^{-12} F/m) and ($\|\ \|$) denotes complex magnitude. The normalized electric vectors, $\tilde{\mathbf{e}}_s^{\mathrm{A}_j}$, are determined from the transmit ($j=1$) and receive ($j=2$) antenna patterns. These patterns are provided in the ANTENNA coordinate system for site j (A_j system) as described in Appendix A, Section 3.A.4. These $\tilde{\mathbf{e}}_s^{\mathrm{A}_j}$ vectors must be transformed into a common coordinate system for calculation of the magnitude of the complex vector dot product, that is, the antenna gain product G_{sm}^{12} at $\mathcal{P}_s^{\mathrm{L}}$. Let $\tilde{\mathbf{e}}_s^{\mathrm{A}_j}$, $j=1,2$, represent the normalized electric vectors in the TRAIL VECTOR or V system (see Section 3.A.3), then

$$\tilde{\mathbf{e}}_{sj}^{\mathrm{V}} = \mathbf{T}_{\mathrm{SV}} \cdot \mathbf{T}_{\mathrm{A}_j\mathrm{S}} \cdot \tilde{\mathbf{e}}_s^{\mathrm{A}_j} \tag{3.37}$$

where $\mathbf{T}_{\mathrm{A}_j\mathrm{S}}$ is the transformation from the A_j system to the S system (see Eq. 3.A.39b) and $\mathbf{T}_{\mathrm{SV}}$ transforms the S system coordinates into the V system (Eq. 3.A.26). The E field vector incident on the meteor trail, $\tilde{\mathbf{e}}_{s1}^{\mathrm{V}}$, has the same polarization as the transmit antenna in the direction of the scatter point $\mathcal{P}_s^{\mathrm{L}}$. In addition, the scattered field vector, $\tilde{\mathbf{e}}_{s2}^{\mathrm{V}}$, is polarization-matched to the receive antenna from the direction of $\mathcal{P}_s^{\mathrm{L}}$.

The $\tilde{\mathbf{e}}_{sj}^{\mathrm{V}}$ vectors provide the incident and scattered electric vector orientations in the V system. Thus, $\tilde{e}_{s11}^{\mathrm{V}}$ is the component of the transmitted, normalized E vector incident along the trail axis (parallel with $\hat{\mathbf{v}}_m^{\mathrm{M}}$). Both $\tilde{e}_{s12}^{\mathrm{V}}$ and $\tilde{e}_{s13}^{\mathrm{V}}$ are normally incident on the trail. Similarly, $\tilde{e}_{s21}^{\mathrm{V}}$ is the component of the scattered E vector parallel to the trail axis and $\tilde{e}_{s22}^{\mathrm{V}}$ *and* $\tilde{e}_{s23}^{\mathrm{V}}$ are the scattered components normal to the trail axis.

Resolution of the E vectors into trail parallel and normal components permits modeling of plasma resonance effects [97] (see Section 3.1.1.2). The resonance phenomena produces an amplification of the component of the scattered E vector [98] oriented normal, or transverse, to the trail axis by a factor of ρ. The effective dot product of the antenna patterns for this case would then be given by

$$g_{sm}^{12}(\rho) = \|\tilde{e}_{s11}^{\mathrm{V}}(\tilde{e}_{s21}^{\mathrm{V}})^{\mathrm{c}} + \rho[\tilde{e}_{s12}^{\mathrm{V}}(\tilde{e}_{s22}^{\mathrm{V}})^{\mathrm{c}} + \tilde{e}_{s13}^{\mathrm{V}}(\tilde{e}_{s23}^{\mathrm{V}})^{\mathrm{c}}]\| \tag{3.38a}$$

with

$$G_{sm}^{12} = G_{sm}^{12}(\rho) = 20\log[g_{sm}^{12}(\rho)] \tag{3.38b}$$

where the (^{c}) denotes complex conjugate. Thus, $G_{sm}^{12}(\rho)$ represents an *effective* value of the gain product to be used in the calculation of the maximum allowable path loss, L^*, given by Eq. 3.34a. As will be shown, L^* is a key parameter in the calculation of the minimum required line density, q_{sm}^*, which ultimately determines the link MR and DC values.

The value $\rho=1$ in Eq. 3.38 models G_{sm}^{12} for overdense trail-scattered signals or when the effects of plasma resonance are to be ignored in the calculation of underdense signals. If $\rho=2$, the resulting value of G_{sm}^{12} may be used to include the effects of plasma resonance in the calculation of underdense trail line densities, q_{sm}^{U} [99, 100]. More specifically, $\rho = 2$ increases G_{sm}^{12} which increases L^* for trail-scatter at $\mathscr{P}_s^{\mathrm{L}}$ above the corresponding value for no resonance, that is, $\rho = 1$. As a result, the corresponding value of underdense line density, q_{sm}^{U}, required to achieve L^* would be reduced. Resonance effects occur when the dielectric constant of the ionized trail approaches

$$\kappa = 1 - \frac{q\lambda_L^2 r_e}{\pi[r_q(t)]^2} \cong -1$$

where λ_L is the signal wavelength, r_e is the classical electron radius, and $r_q(t)$ is the *effective* radius of the trail with line density q. The function $r_q(t)$ is defined such that $q/[r_q(t)]^2$ is the volume density of electrons in the meteor trail. As the trail expands due to normal diffusion, $r_q(t)$ increases so κ increases. By at most 50 ms after trail formation, $\kappa > -1$ and the effects of resonance are no longer observed. Since κ decreases with increasing wavelength, λ_L, resonance effects also diminish with increasing signal frequency.

In general, the maximum allowable path loss L^* (Eq. 3.34*a*) must be achieved for some minimum time τ^* for the trail $\hat{\mathbf{v}}_m^{\mathrm{L}}$ through $\mathscr{P}_s^{\mathrm{L}}$ to be usable for MB communications. In practice, there is always some minimum time required for signal detection, bit and bit field synchronization. MB test bed operation also requires some minimum burst duration for detection and signal identification. At a minimum, τ^* is on the order of 10 ms. For MB systems employing short messages intended to be transmitted during a single trail-scattering event, τ^* would correspond to the minimum time required for a "single-burst" message. For half-duplex communications of single-burst messages, τ^* includes both the channel probe and message duration. Monte Carlo simulation of MB link performance, described in Section 3.3, requires the MR and DC values for all trail-scattered signals corresponding to a maximum path loss, L^*, independent of duration.

3.2.2.3.2 Minimum Line Density

3.2.2.3.2.1 Approach. The fundamental philosophy of the METEOR-LINK physical model is to determine the "filter" that transforms the radar-measured SMRD map into predicted MR and DC values on an arbitrary MB link. This filter specifies both geometric (observability) and power budget (usability) requirements met by meteor trails that support radio communication. The combination of observability $(\mathscr{P}_s^{\mathrm{L}}, \hat{\mathbf{v}}_s^{\mathrm{L}})$ and usability (L^*, τ^*) requirements determine the minimum trail electron line density q_{sm}^{*} yielding P_r^* at the receiver input for duration τ^*.

The most accurate approach for calculating q_{sm}^{*} is to rely on well-developed trail-scattering theory [101, 102], which minimizes reliance on simplifying assumptions and approximations. These techniques solve simultaneous differential equations for the E and H fields internal and external to the trail. Solutions of these equations determine the trail reflection coefficients for arbitrary trail orientation, electron line density q, and both incident and scattered E and H field vectors. Unfortunately, the nonlinear q dependence of the reflection coefficient prevents a direct solution.

An iterative trial and error approach could be employed to create the trail reflection coefficient versus time. The desired q_{sm}^* value would provide a maximum path loss of L^* for the minimum required burst duration τ^*. The computer time needed to implement this approach is excessive, given that thousands of observable trails are modeled for a typical link prediction. Furthermore, the accuracy of the classic approximations [103, 104] combined with uncertainty associated with radio propagation predictions in general reduces the need for the more exhaustive approach. For example, many meteors break into fragments that produce trails with unpredictable line density distributions. A more useful application of the exact solutions would be the enhancement of the classic equations to better match the exact solution. This approach would be particularly useful in handling the transition region between underdense and overdense trail-scattered signals.

An alternative approach to the common volume integration method employed by METEORLINK is a Monte Carlo simulation of the meteor arrival process [105]. This simulation generates observable meteor trails from probability density functions computed versus link range. The preburn meteor mass, m_∞, and the corresponding line density, q, at the simulated trail height are then determined from the appropriate probability distributions. Conditions are imposed on the selection of this minimum q value to maximize the probability that the maximum allowable path loss, L^*, is achieved for the simulated trail orientation. Next, the E field vectors incident at the trail from the transmit antenna and the scattered E vector component accepted by the receive antenna are determined relative to the orientation of each simulated observable trail. At this point, the relative orientation of E field vectors and each observable trail are combined with the line density q to determine the trail-scattered signal amplitude from a look-up table. This table was derived from the exact solution of Maxwell's equations applied to the trail-scattering problem [106]. Finally, the interarrival times between usable trail-scattered signals are accumulated into a distribution function and the link MR value is determined.

Section 3.1.1.2 described four classic trail-scattering models categorized by the applicable range of electron line density q_{sm}^* and the signal wavelength λ. In particular, trail-scattered signals are modeled as long wavelength underdense and overdense or short wavelength underdense and overdense. Expressions have been derived relating $(\mathscr{P}_s^{\mathrm{L}}, \hat{\mathbf{v}}_s^{\mathrm{L}})$, or $(\mathscr{P}_s^{\mathrm{S}}, \hat{\mathbf{v}}_m^{\mathrm{S}})$, and (L^*, τ^*) to q_{sm}^* for both forward and backscatter from long wavelength underdense and overdense trail models as well as the corresponding short wavelength models [107]. An expression has also been derived for maximum power from the backscatter echo of a short wavelength overdense trail and modified to yield an applicable forward scatter formula. These equations may be solved either explicitly or numerically to yield

$$q_{sm}^{*} = q[(\mathscr{P}_s^{\mathrm{S}}, \hat{\mathbf{v}}_m^{\mathrm{S}}), (L^*, \tau^*)] \tag{3.39}$$

which gives the minimum value of line density q_{sm}^{*} for meteor trails passing through the point $\mathscr{P}_s^{\mathrm{S}}$, oriented by $\hat{\mathbf{v}}_s^{\mathrm{S}}$, and providing at most L^* path loss for burst duration τ^*.

The point $\mathscr{P}_s^{\mathrm{S}}$ and unit vector $\hat{\mathbf{v}}_s^{\mathrm{S}}$ determine the dependence of q_{sm}^{*} on trail location and orientation, respectively. First, define the R vectors, $\mathbf{r}_{sj}^{\mathrm{S}}$, that point from the antenna locations at $\mathscr{P}_j^{\mathrm{S}} = (x_j^{\mathrm{S}}, 0, 0)$, $j = 1, 2$, to the center of the principal Fresnel zone at the scatter point $\mathscr{P}_s^{\mathrm{S}}$. These vectors determine the free space propagation distances

$$r_j = |\mathbf{r}_{sj}^{\mathrm{S}}| = |\langle \mathscr{P}_s^{\mathrm{S}} - \mathscr{P}_j^{\mathrm{S}} \rangle| \tag{3.40a}$$

as well as the unit direction numbers

$$\hat{\mathbf{r}}_{sj}^{\mathrm{S}} = \frac{\mathbf{r}_{sj}^{\mathrm{S}}}{|\mathbf{r}_{sj}^{\mathrm{S}}|} = \langle \hat{r}_{sj1}^{\mathrm{S}}, \hat{r}_{sj2}^{\mathrm{S}}, \hat{r}_{sj3}^{\mathrm{S}} \rangle \quad \text{and} \quad \hat{\mathbf{r}}_s^{\mathrm{A}_j} = \frac{\mathbf{T}_{\mathrm{SA}_j} \cdot \hat{\mathbf{r}}_{sj}^{\mathrm{S}}}{|\mathbf{T}_{\mathrm{SA}_j} \cdot \hat{\mathbf{r}}_{sj}^{\mathrm{S}}|} = \langle \hat{r}_{s1}^{\mathrm{A}_j}, \hat{r}_{s2}^{\mathrm{A}_j}, \hat{r}_{s3}^{\mathrm{A}_j} \rangle \tag{3.40b}$$

where

$$\mathbf{T}_{\mathrm{SA}_j} = (\mathbf{T}_{\mathrm{A}_j\mathrm{S}})^{-1}$$

in the A_j systems. The unit vectors $\hat{\mathbf{r}}_s^{\mathrm{A}_j}$ determine the angles $\phi_s^{\mathrm{A}_j}$ and $\theta_s^{\mathrm{A}_j}$

$$\phi_s^{\mathrm{A}_j} = \tan^{-1}\left(\frac{\hat{r}_{s2}^{\mathrm{A}_j}}{\hat{r}_{s1}^{\mathrm{A}_j}}\right) \tag{3.41a}$$

$$\theta_s^{\mathrm{A}_j} = \cos^{-1}(\hat{r}_{s3}^{\mathrm{A}_j}) \tag{3.41b}$$

that locate $\mathscr{P}_s^{\mathrm{S}}$ relative to the antenna E field pattern at each link site (see Eq. 3.A.33). The angles $\phi_s^{\mathrm{A}_j}$ and $\theta_s^{\mathrm{A}_j}$ determine the normalized E field vectors used to compute G_{sm}^{12} in Eq. 3.38 and ultimately L^* from Eq. 3.34a.

The unit direction numbers $\hat{\mathbf{r}}_{sj}^{\mathrm{S}}$ of the R vectors in the S system also determine the forward scattering angle $2\phi_s$ included between $\hat{\mathbf{r}}_{s1}^{\mathrm{S}}$ and $\hat{\mathbf{r}}_{s2}^{\mathrm{S}}$. In symbols

$$\phi_s = \tfrac{1}{2}\cos^{-1}(\hat{\mathbf{r}}_{s1}^{\mathrm{S}} \cdot \hat{\mathbf{r}}_{s2}^{\mathrm{S}}) = \cos^{-1}(\hat{\mathbf{n}}_s^{\mathrm{S}} \cdot \hat{\mathbf{r}}_{s1}^{\mathrm{S}}) = \cos^{-1}(\hat{\mathbf{n}}_s^{\mathrm{S}} \cdot \hat{\mathbf{r}}_{s2}^{\mathrm{S}}) \tag{3.42}$$

as shown in Fig. 3.22, where $\hat{\mathbf{n}}_s^{\mathrm{S}}$ is the normal vector at $\mathscr{P}_s^{\mathrm{S}}$. A second orientation angle, θ_{sm}, is measured from the trail to the M plane intersection with the propagation or P plane as shown in Fig. 3.22. It has been shown [108] that given the trail unit vector $\hat{\mathbf{v}}_s^S$, ϕ_s, and θ_{sm} are related by

$$\sin^2(\gamma_{sm}) = 1 - \sin^2(\phi_{sm})\cos^2(\theta_{sm}) \tag{3.43}$$

where γ_{sm} is the angle between the meteor trail and the $\mathbf{r}_{s1}^S$ vector, that is,

$$\gamma_{sm} = \cos^{-1}(-\hat{\mathbf{v}}_m^S \cdot \hat{\mathbf{r}}_{s1}^S) \tag{3.44}$$

The distances r_1 and r_2, combined with the angles $(\phi_s^{A_j}, \theta_s^{A_j})$, ϕ_s, and θ_{sm} (or ϕ_s and γ_{sm}) completely determine the dependence of q_{sm}^* on the observability criteria $(\mathcal{P}_s^S, \hat{\mathbf{v}}_m^S)$.

Once the meteor passes point $\mathcal{P}_s^S$, electrons within the newly formed trail diffuse radially outward at a rate

$$D(h_s, v_m) = \frac{v_m}{3n_A(h_s)\sigma_D(v_m)} \tag{3.45}$$

where

v_m = mean velocity of meteoric particles relative to atmospheric particles at height h_s [109]

σ_D = effective diffusion cross section for a particle traveling with velocity v_m

n_A = density of atmospheric particles at height h_s (see Eq. 3.16)

The cross section $\sigma_D(v_m)$ may be given by [110]

$$\sigma_D(v_m) = 5.6 \times 10^{-15} v_m^{-0.8}(1.09 - 0.004 v_m) \tag{3.46}$$

where relative particle speed v_m is in km/s. During trail formation, $v_m \cong \bar{v}_\epsilon$ and Eqs. 3.45 and 3.46 are used in the calculation of the initial trail radius r_0 given by Eq. 3.16 [111] and the initial diffusion rate is $D_0 = D(h_s, \bar{v}_\epsilon)$.

Once the trail has formed, however, meteoric particle speeds rapidly slow to the ambient atmospheric particle speed v_A. v_A is below about 0.5 km/s for altitudes in the height band of enduring meteoric ionization [112]. Substituting Eq. 3.46 in Eq. 3.45 with $v_m \cong v_A$ determines the ambipolar diffusion coefficient $D = D(h_s, v_A)$, which dominates the lifetime of meteoric trail-scattered signals for up to about 10 s [113]. Since both n_A and v_A vary with atmospheric temperature and pressure, r_0 and D are subject to diurnal, seasonal, and solar cycle variations. Classical expressions for r_0 and D, varying with only height, are given by

$$r_0(h_s) = 10^{(0.075h_s - 7.9)} \tag{3.47a}$$

$$r_0(h_s) = 10^{(0.075h_s - 7.2)} \quad [114] \tag{3.47b}$$

$$D(h_s) = 10^{(0.067h_s - 5.6)} \quad [115] \tag{3.47c}$$

where h_s is measured in kilometers. Diffusion rate was the independent variable in the regression analysis leading to Eq. 3.47*c*. An alternative expression [116] for D using h_s as the independent variable in the regression yields

$$D(h_s) = 10^{(0.0775h_s - 1.225)} \tag{3.47d}$$

Given h_s, r_0 and D are the principal model parameters controlling the relationship between τ^* and q^*_{sm}. The rapid decrease in underdense trail-scattered signal amplitude due to the termination of the plasma resonance enhancement may produce an *apparent* or *effective* diffusion rate $D_U > D$ [117] while for overdense trails $D_O = D(h_s, v_A)$. Values for r_0 and D derived from these expressions are plotted in Figs. 3.25*a* and *b*, respectively.

3.2.2.3.2.2 Long Wavelength Underdense Trail Model. The minimum value of line density $q^*_{sm} = q^U_{sm}$ required for the long wavelength underdense (LU) trail model to provide a maximum path loss L^* for a duration τ^* is given by

$$q^{LU}_{sm} = \sqrt{a^{LU}_1 \exp(a^{LU}_2 \tau^* + a^{LU}_3)/l^*} \tag{3.48}$$

where

$$a^{LU}_1 = \frac{16\pi^2 r_1 r_2 (r_1 + r_2) \sin^2(\gamma_{sm})}{r_e^2 \lambda_L^3} \tag{3.49a}$$

$$a^{LU}_2 = \frac{32\pi^2 D_U \cos^2(\phi_s)}{\lambda_L^2} \tag{3.49b}$$

$$a^{LU}_3 = \frac{8\pi^2 r_0^2 \cos^2(\phi_s)}{\lambda_L^2} \tag{3.49c}$$

where

$$l^* = 10^{(L^*/10)} \tag{3.50}$$

and r_e is the classical electron radius 2.8178×10^{-15} m. τ^* in Eq. 3.48 is measured from $t = 0$, when the LU trail model produces the minimum path loss value

$$l^{LU}_{sm}(t)\big|_{t=0} = l^{LU}_{sm}(0) = \frac{a^{LU}_1 \exp(a^{LU}_3)}{(q^{LU}_{sm})^2}$$

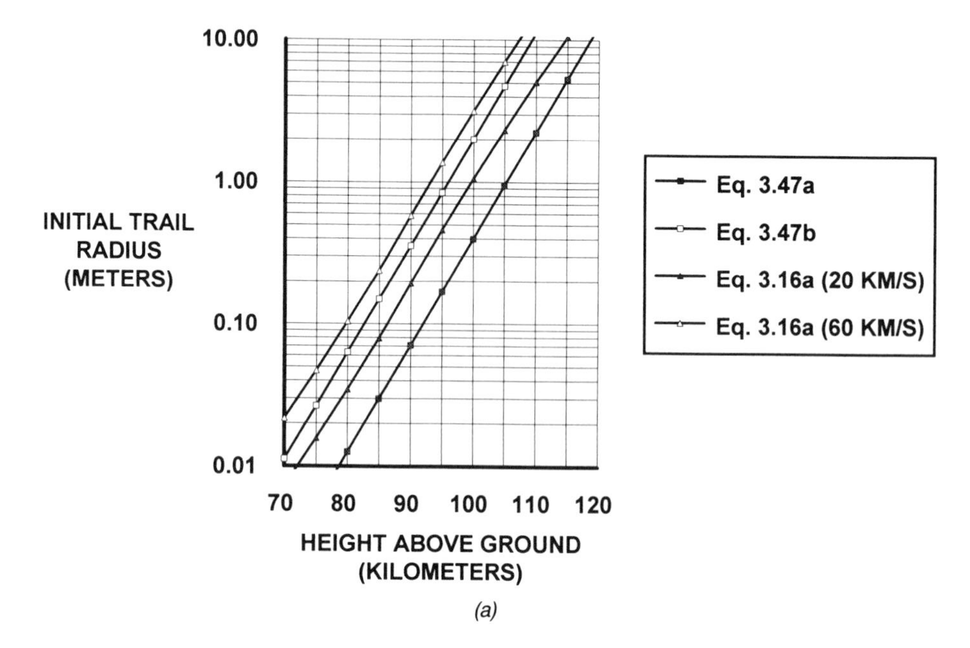

(a)

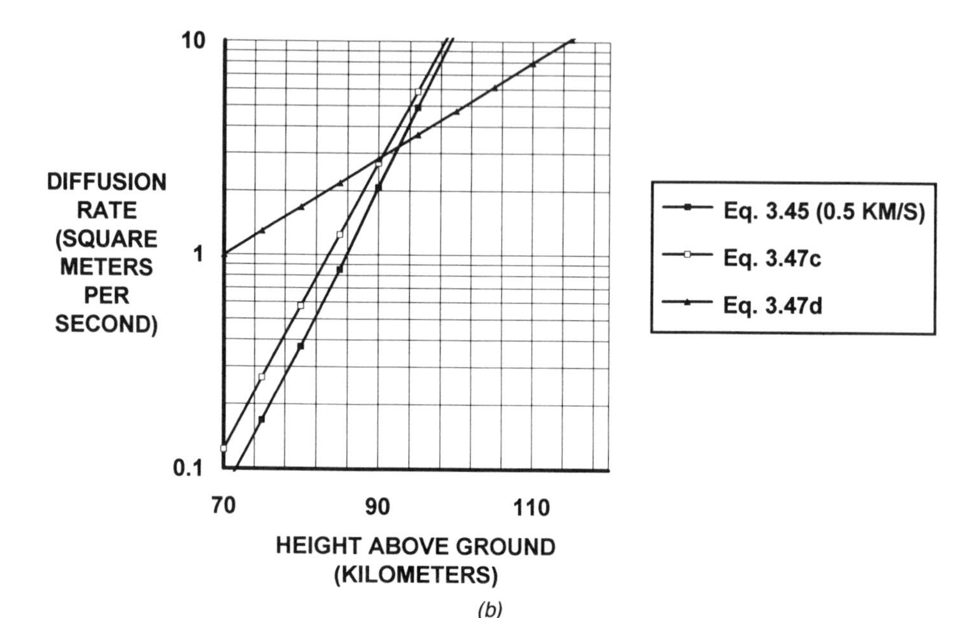

(b)

Figure 3.25. Examples of computed values: (a) initial radius r_0; (b) diffusion rate D.

until $t = \tau^*$, where the path loss determined by the LU model exceeds l^* (see Fig. 3.7b).

This formulation assumes that the meteor trail may be accurately modeled as a right circular cylinder during the trail lifetime (see Figs. 3.8 and 3.11b). Since the trail begins diffusing with time t immediately after formation, the trail radius

$$r(t) = \sqrt{4\pi^2 D_U t + r_0^2} \tag{3.51}$$

is smaller near the meteor head than further back on the trail. If this size difference is significant relative to the signal wavelength, then the LU model no longer applies and the short wavelength underdense (SU) trail model must be considered.

3.2.2.3.2.3 Short Wavelength Underdense Trail Model. The LU model applies if the time required for trail formation in the principal Fresnel zone is small relative to the nominal LU burst duration

$$\tau_s^{LU} = \frac{\lambda_L^2 \sec^2(\phi_s)}{32\pi^2 D_U} \tag{3.52}$$

where τ_s^{LU} is the time required for the signal to fall to $1/e$ of its maximum amplitude. The length of the principal Fresnel zone in the forward scatter case is given by

$$2\ell_{sm} = 2\sqrt{\frac{\lambda_L r_1 r_2}{(r_1 + r_2)\cos^2(\gamma_{sm})}}$$

where ℓ_{sm} is the half length. If the meteor is moving with speed $\bar{v}_\epsilon$, then the signal wavelength threshold λ_{sm}^{UT} for which the LU model applies is determined from

$$\frac{2\ell_{sm}}{\bar{v}_\epsilon} = \frac{2}{\bar{v}_\epsilon}\sqrt{\frac{\lambda_{sm}^{UT} r_1 r_2}{(r_1 + r_2)\sin^2(\gamma_{sm})}} \le \tau_s^{LU}$$

or, using Eq. 3.52, the wavelength threshold becomes

$$\lambda_{sm}^{UT} = \left[\frac{4096\pi^4 D_U^2 r_1 r_2 \cos^4(\phi_s)}{\bar{v}_\epsilon^2 (r_1 + r_2)\sin^2(\gamma_{sm})}\right]^{1/3} \tag{3.53}$$

Thus, if the signal wavelength $\lambda_L > \lambda_{sm}^{UT}$, the SU rather than the LU model is applicable to the calculation of q_{sm}^* from the observability ($\mathcal{P}_s^{\mathrm{L}}$, $\hat{v}_m^{\mathrm{L}}$) and usability (L^*, τ^*) requirements.

The SU model considers the trail-scattering electrons to be configured in a paraboloid (see Fig. 3.9a) such that solution of the corresponding transmission equation [118] for $q_{sm}^{*} = q_{sm}^{U}$ yields

$$q_{sm}^{SU} = a_1^{SU}\sqrt{[1 + (a_2^{SU}\tau^*)^2]\exp(a_3^{SU})/l^*} \tag{3.54}$$

where

$$a_1^{SU} = \frac{256\pi^3 r_1 r_2 D_U \cos^2(\phi_s)}{r_e \lambda_L^3 \bar{v}_\epsilon} \tag{3.55a}$$

$$a_2^{SU} = \frac{\lambda_L(r_1 + r_2)\sin^2(\gamma_{sm})\bar{v}_\epsilon^2}{8\pi r_1 r_2 D_U \cos^2(\phi_s)} \tag{3.55b}$$

$$a_3^{SU} = a_3^{LU} = \frac{8\pi^2 r_0^2 \cos^2(\phi_s)}{\lambda_L^2} \tag{3.55c}$$

The peak power from the SU model occurs at time $t = 0$, when the meteor passes the center of the principal Fresnel zone. The maximum allowable path loss l^* is achieved at $t = -\tau^*/2$, as the meteor approaches this point, and again at $t = \tau^*/2$, after the meteor passes this point (see Fig. 3.9b).

For $\tau^* = 0$, increasing meteor speed, $\bar{v}_\epsilon$, or increasing the wavelength, λ_L, produces an *effective* reduction in the diffusion rate, D_U (see Eq. 3.55a). This reduction decreases q_{sm}^{SU} from Eq. 3.54 and produces a corresponding increase in the MR contribution from trail meeting the observability requirements (L^*, τ^*). Physically, the electrons remain more densely packed near the meteor head relative to the signal wavelength and therefore produce a greater trail-scattered signal than occurs for lower meteor speeds.

The underlying assumption of both the LU and SU models is that only single scattering occurs. In other words, each ray path from the transmit antenna to the receive antenna encounters at most one electron scatterer and much of the incident signal passes through the underdense trail unaffected. This assumption is valid throughout the time history of any usable underdense trail-scattered signal. Plasma resonance effects may be approximated in the LU and SU trail models by using Eq. 3.38a to compute G_{sm}^{12} in the calculation of L^* from Eq. 3.34a. Consider the approximate conditions for resonance in the underdense trail. It has been stated [119] that resonance can occur in the backscatter case ($\phi_s = 0$) only when the trail radius

$$r(t) = \sqrt{4\pi^2 D_U t + r_0^2} < \frac{\lambda_L}{2\pi}$$

Generalizing this constraint to the forward scatter case, λ_L is replaced with $\lambda_L \sec^2(\phi_s)$. Thus, if $\mathrm{r}_q(\tau^*) < \lambda_L \sec^2(\phi_s)/2\pi$ in the LU trail model or $\mathrm{r}_q(\tau^*/2) < \lambda_L \sec^2(\phi_s)/2$ in the SU trail model, then $p = 2.0$ in Eq. 3.38a; otherwise, $\rho = 1.0$. This algorithm provides an approximate treatment of plasma resonance effects of calculation of q_{sm}^{*}. If the $(\mathscr{P}_s^{\mathrm{L}}, \hat{\mathbf{v}}_m^{\mathrm{L}})$ or (L^*, τ^*)

requirements force increasing values of q_{sm}^* (for example, increasing noise density N_0 decreases L^*), the electron volume density becomes so great that none of the incident signal can pass through the trail unaltered. In this case, the electron density is so great that the single scattering assumption required for applicability of the *LU* and *SU* models becomes invalid. This condition defines the overdense meteor trail.

3.2.2.3.2.4 Long Wavelength Overdense Trail Model. The transition from underdense to overdense trail-scattering behavior with increasing electron volume density in the trail is a continuous phenomena. In fact, all overdense trails ultimately diffuse to an underdense volume density. In practice, however, the METEORLINK computer model employs trail-scattering equations designed explicitly for one or the other of these line density regimes. Calculations of a suitable line density threshold, q_{sm}^T, serving as an upper bound for q values in the underdense regime, have yielded values ranging from 0.75×10^{14} epm to 2.4×10^{14} epm [120]. Intuitively, the signal wavelength should affect whether single or multiple scattering behavior is observed. Similarly, the orientation of the trail relative to the P plane (see Fig. 3.22) should also affect the scattering path through the trail and therefore impact the calculation of q_{sm}^T.

One attempt at deriving a more accurate expression for q_{sm}^T, including signal wavelength and trail orientation effects, focused on the phase shift, Φ_{sm}^T, undergone by the incident wave as it penetrates the meteor trail [121]. If the Φ_{sm}^T value is small, say 25°, then the index of refraction in the ionized gas is close to unity and the underdense assumption is valid. As the electron density increases, however, this phase shift increases beyond some small but subjective threshold and the trail-scattered signal is classified overdense. Thus, the scattered signal is considered underdense in the backscatter case when the following conditions are met:

$$B(q_{sm}^U) \le K(r_0, \lambda_L) \quad \text{for} \quad B(q_{sm}^U) < 1$$

and (3.56*a*)

$$[B(q_{sm}^U)]^2 \le K(r_0, \lambda_L) \quad \text{for} \quad B(q_{sm}^U) \ge 1$$

where

$$B(q_{sm}^U) = 2\sqrt{r_e q_{sm}^U} \quad \text{and} \quad K(r_0, \lambda_L) = 2\pi r_0/\lambda_L \qquad (3.56b)$$

for angles γ_{sm} (see Eq. 3.44) close to 90° (backscatter) and small values of Φ_{sm}^T.

The maximum q_{sm}^T values implied by Eq. 3.56 are shown plotted versus signal frequency, f_L, and trail height, h_s, in Fig. 3.26*a*. The initial trail radius, r_0, used for these calculations was given by Eq. 3.47*b*. The irregular

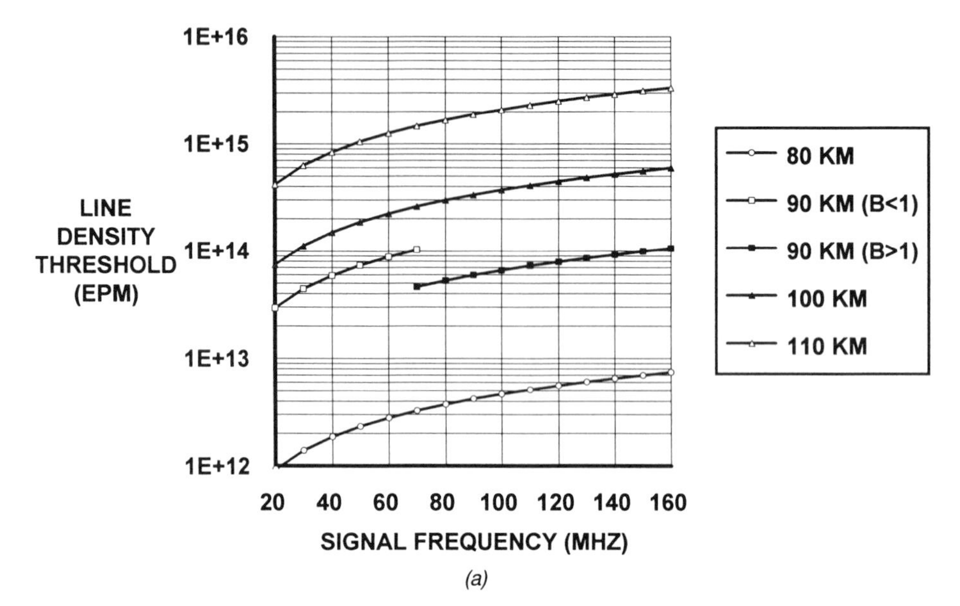

(a)

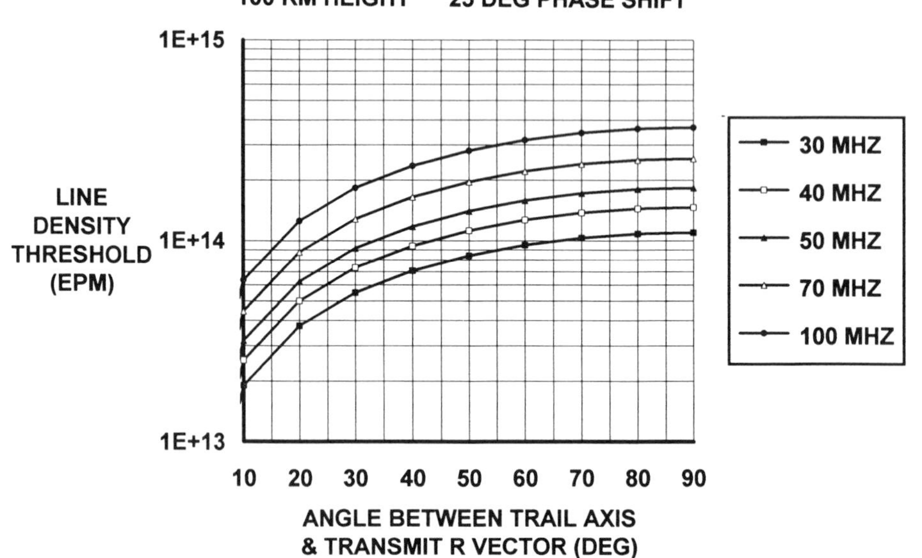

(b)

Figure 3.26. (*a*) Threshold variation with frequency and height; (*b*) threshold variation with trail orientation and frequency.

behavior of the 90-km curve is due to the change in the appropriate inequality expressed by Eq. 3.56 at $B(q_{sm}^{LU}) = 1$. The threshold values in Fig. 3.26*a* increase with increasing values of trail height, h_s, because the initial trail radius, r_0, increases with height and reduces the electron volume density in the trail. Since there is a minimum electron volume density required to achieve overdense scattering conditions in the trail, increased r_0 values raise the minimum line density needed to achieve this volume density threshold.

Increasing signal frequency, f_L, reduces the corresponding signal wavelength λ_L and causes the trail dielectric constant to approach unity ($\kappa \rightarrow 1$), thus decreasing the phase shift Φ_{sm}^T imparted to the signal. Again, increased trail line density is required to maintain a specified Φ_{sm}^T value delineating the onset of overdense trail-scatter behavior. From a scattering perspective, the likelihood of single scattering increases as the signal wavelength is reduced relative to the nominal separation distance between individual trail scatterers. Thus, use of a single value for the underdense-overdense transition line density, such as 2.4×10^{14} epm, is not appropriate for all trail heights or signal frequencies.

The q_{sm}^T values plotted in Fig. 3.26*a* were computed for meteor trails oriented normal to the direction of the propagating wave (transmit R vector, Eq. 3.40). If the incident signal approaches the trail at an angle $\gamma_{sm} < 90°$, then the signal travels a greater distance within the trail and encounters more scatterers. The derivation of Eqs. 3.56 can be manipulated to compute a close upper bound for q_{sm}^T from the maximum allowable phase shift Φ_{sm}^T, that is

$$q_{sm}^T \leq \frac{2\pi^{1/2} \sin(\gamma_{sm}) r_0 \Phi_{sm}^T}{\lambda_L r_e} \tag{3.57}$$

For example, Fig. 3.26*b* is a plot of an upper bound for the line density threshold q_{sm}^T versus trail orientation angle γ_{sm} and signal frequency at a height of 100 km and a maximum phase shift Φ_{sm}^T of 25°. The figure shows a threshold line density of about 1.1×10^{14} epm for a 30 MHz signal at normal trail incidence ($\gamma_{sm} = 90°$). This threshold corresponds to the value shown in Fig. 3.26*a* for 30 MHz, 100 km trail height, and normal incidence.

In practice, an underdense line density $q_{sm}^U = q_{sm}^{LU}$ or $q_{sm}^U = q_{sm}^{SU}$ is first computed from the $(\mathscr{P}_s^{\mathrm{L}}, \hat{\mathbf{v}}_m^{\mathrm{L}})$ and (L^*, τ^*) requirements. If $q_{sm}^U > q_{sm}^T$ given by Eq. 3.57, then an overdense trail-scatter model will be considered. The standard expression for the long wavelength overdense trail-scattered signal models the trail as a metallic cylinder [122–124]. The resulting expression for the time-dependent path loss from a LO trail is given by

$$l_{sm}^{LO}(t) = \left[a_1^{LO} \sqrt{r_s(t) \ln\left(\frac{a_4^{LO} q_{sm}^{LO}}{r_s(t)} \right)} \right]^{-1} \tag{3.58}$$

where

$$r_s(t) = a_2^{LO} t + a_3^{LO} \tag{3.59}$$

$$a_1^{LO} = \frac{32\pi^2 r_1 r_2 (r_1 + r_2) \sin^2(\gamma_{sm})}{(r_e \lambda_L)^2} \tag{3.60a}$$

$$a_2^{LO} = 4D_O \cos^2(\phi_s) \tag{3.60b}$$

$$a_3^{LO} = r_0^2 \cos^2(\phi_s) \tag{3.60c}$$

$$a_4^{LO} = \frac{\lambda_L^2 r_e}{\pi^2} \tag{3.60d}$$

Although Eq. 3.58 can be solved for q_{sm}^{LO}, the resulting function is transcendental in t and cannot be put in terms of the minimum required duration τ^*. In this case, a second order Taylor series expansion of $l_{sm}^{LO}(t)$ in powers of $t - t_{\min}^{LO} \cong \tau^*/2$ yields the parabolic approximation

$$l_{sm}^{LO}(\tau^*) \cong \left\{ a_1^{LO} \left[\sqrt{\left(\frac{a_4^{LO} q_{sm}^{LO}}{e} \right)} - \frac{(a_2^{LO} \tau^*)^2}{16} \left(\frac{e}{a_4^{LO} q_{sm}^{LO}} \right)^{3/2} \right] \right\}^{-1} \tag{3.61}$$

where

$$t_{\min}^{LO} = \frac{a_4^{LO} q_{sm}^{LO} - e a_3^{LO}}{e a_2^{LO}} \tag{3.62a}$$

at which time the scatter path loss has reached its minimum value

$$\min\{l_{sm}^{LO}(t)\} = l_{sm}^{LO}(t_{\min}^{LO}) = \frac{\sqrt{e / a_4^{LO} q_{sm}^{LO}}}{a_1^{LO}} \tag{3.62b}$$

Note that the minimum path loss, $l_{sm}^{LO}(t_{\min}^{LO})$, corresponding to maximum received power P_r^*, is independent of the initial trail radius r_0 contained in the coefficient a_3^{LO}. Finally, Eq. 3.61 is solved numerically for q_{sm}^{LO} using a root finding algorithm to determine the two times for which $l_{sm}^{LO}(t_i) = l_{sm}^{LO}(t_f) = l^*$. If $\tau^* = 0$, then $t_i = t_f = t_{\min}^{LO}$.

Plots of the exact and approximate expressions for RSL versus time, t, from the LO trail model are shown in Fig. 3.27. These plots show that the approximation achieves the same maximum power as the exact equation but with slightly greater duration for the same q values. More accurate calculation of the two times at which $l_{sm}^{LO}(t) = l^*$, $\tau^* > 0$, could be achieved using higher order approximations or root finding with $l_{sm}^{LO}(t)$ given by Eq. 3.58. This improved accuracy does not justify the corresponding increase in computer time required. More importantly, this improvement is small compared to the inaccuracies and uncertainties inherent in employing tractable models to explain the overdense trail-scatter phenomena.

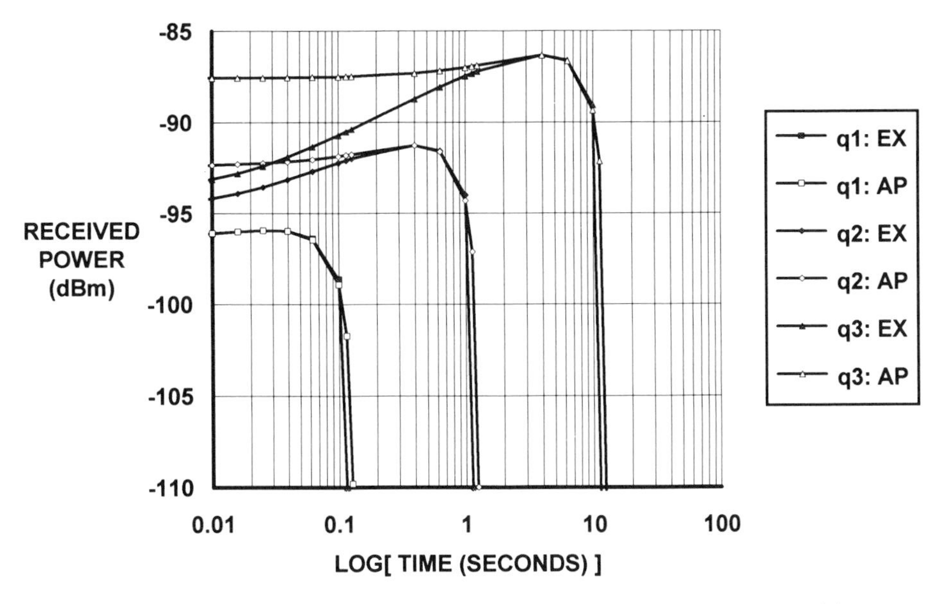

Figure 3.27. Time behavior of classic and approximate overdense trail-scatter models.

3.2.2.3.2.5 SHORT WAVELENGTH OVERDENSE TRAIL MODEL. As in the underdense case, the cylindrical trail model applies if the time required for trail formation is small compared to the usable lifetime of the trail-scattered signal. The nominal duration of an overdense trail-scattered signal for the classic LO trail model is given by [125]

$$\tau_{sm}^{LO} = \frac{a_4^{LO} q_{sm}^{LO} - a_3^{LO}}{a_2^{LO}} \tag{3.63}$$

where the initial radius r_0 has been included in the a_3^{LO} term. The duration τ_{sm}^{LO} must exceed the time required for the meteor to traverse the central Fresnel zone. This criterion translates into the requirement

$$\tau_{sm}^{LO} \geq \frac{2}{\bar{v}_\epsilon} \sqrt{\frac{\lambda_L r_1 r_2}{(r_1 + r_2)\sin^2(\gamma_{sm})}} \tag{3.64}$$

When Eqs. 3.60b, c, and d are substituted for a_2^{LO}, a_3^{LO}, and a_4^{LO}, respectively, in Eq. 3.63 and Eq. 3.63 is substituted for τ_{sm}^{LO} in Eq. 3.64, the transition wavelength λ_{sm}^{TO} for overdense trails is expressed as a quartic equation. Since r_0 has no significant impact on the majority of overdense trails, setting $r_0 = 0$ yields the transition wavelength

$$\lambda_{sm}^{TO} = \left[\frac{64\pi^4 D_O r_1 r_2 \cos^4(\phi_s)}{\bar{v}_\epsilon^{-2}(r_e q_{sm}^{LO})^2 (r_1 + r_2)\sin^2(\gamma_{sm})} \right]^{1/3} \tag{3.65}$$

Unlike the underdense transition wavelength λ_{sm}^{TU}, λ_{sm}^{TO} is a function of line density q_{sm}^{LO}.

The minimum path loss produced by the short wavelength overdense (SO) trail-scatter model (see Section 3.1.1.2) occurs at $t = 0$ and is given by

$$l_{sm}^{SO} = \frac{a_1^{SO}}{\sqrt{q_{sm}^{SO}}(q_{sm}^{SO} - a_2^{SO})^2} \tag{3.66}$$

where

$$a_1^{SO} = \frac{512\pi^7 \sqrt{e} D_O^2 r_1^2 r_2^2 \cos^6(\phi_s)}{\bar{v}_\epsilon^{-2} \lambda_L^6 r_e^{5/2}} \tag{3.67a}$$

$$a_2^{SO} = \frac{\pi^2 r_0^2 \cos^2(\phi_s)}{\lambda_L^2 r_e} \tag{3.67b}$$

If the initial radius $r_0 = 0$ at $t = 0$, then $a_2^{SO} = 0$ and Eq. 3.66 reduces to

$$l_{sm}^{SO}(0) = \frac{a_1^{SO}}{(q_{sm}^{SO})^{5/2}} \tag{3.68}$$

These results at $t = 0$ apply when the meteor passes through the center of the principal Fresnel zone and the path loss achieves its minimum value. It has been postulated [126] that the time behavior for the SO model is similar to that of the SU model. In this case, the expression for path loss versus time for the SO model would be approximated by

$$l_{sm}^{SO}(t) = l_{sm}^{SO}(0)[1 + (a_3^{SO} t)^2] \tag{3.69}$$

where $a_3^{SO} \cong a_2^{SU}$. In general, Eq. 3.66 must be solved numerically. Setting $r_0 = 0$, however, simplifies the expression so that the minimum required line density is given by

$$q_{sm}^{SO} = \{a_1^{SO}(a_2^{SO})^2[1 + (a_3^{SO}\tau^*)^2]\}^{2/5} \tag{3.70}$$

where

$$a_3^{SO} = \frac{\bar{v}_\epsilon^{-2} \lambda_L \sin^2(\gamma_{sm})(r_1 + r_2)}{16\pi D_O r_1 r_2 \cos^2(\phi_s)} \tag{3.71}$$

Equation 3.70 gives the minimum line density needed to meet the observability $(\mathscr{P}_s^{L}, \hat{\mathbf{v}}_m^{L})$ and usability (L^*, τ^*) requirements for the SO trail model.

3.2.2.3.2.6 Overdense Trail-Scatter Model Modifications. A modification to the classic LO trail expression given by Eq. 3.58 has been proposed [127] to account for refraction effects from overdense trail electrons outside the critical radius. The critical radius defines that portion of the overdense trail in which the electron volume density is adequate to produce multiple scattering. These refraction effects have been associated with an observed decrease in the duration of overdense trail-scattered signals. A suggested modification of Eq. 3.58 to model refraction effects is the replacment of the $\sin^2(\gamma_{sm})$ term with $1 - \sin^2(\phi_s)\cos^{\vartheta}(\theta_{sm})$, where the exponent $\vartheta = 2.0$ for $\theta_{sm} = 0°$ and $\vartheta = 0.3$ for $\theta_{sm} = 90°$ [128]. The variation between these values was not defined and little comparison with measured data has been presented. Nevertheless, simultaneous forward and backscatter measurements used to determine trail orientations found that a value of 2.0 was not observed in practice, with ϑ estimated at 1.13 across all overdense trails observed for which the backscatter duration exceeded 3.5 s [129]. In this context, refraction deflects signal energy away from the critical density contour of the meteor trail, effectively reducing the scattered signal amplitude.

The minimum LO line density q_{sm}^{LO} required to satisfy the trail usability criteria (L^*, τ^*) was determined from Eq. 3.61. The q_{sm}^{LO} value increases with increasing minimum burst duration τ^* because of the ambipolar diffusion constant D_O in Eq. 3.60*b*. In fact, it has been empirically determined that diffusion D_O is the principal physical mechanism that limits LO trail-scattered signal durations less than about 10 s [130]. In practice, however, Eq. 3.61 yields predicted LO RSL durations above P_r^* far in excess of observed values [131]. Therefore, diffusion alone cannot explain the ultimate deterioration of the LO trail-scattered signal.

Multipath fading of the trail-scattered RSL from long-lasting bursts is routinely observed on MB links. Although fading-induced RSL drop-outs certainly reduce burst duration, this reduction is insignificant compared to the difference between predicted and measured values. For this reason, three physical mechanisms other than ambipolar diffusion have been considered: recombination, attachment, and turbulent diffusion. The first mechanism, the recombination of trail electrons with positive ions, has been shown [132] to account for only a small part of the reduction in trail-scattered RSL lifetime.

Electron attachment to neutral atmospheric particles is believed to limit burst durations for τ^* values between about 10 and 50 seconds [133] after trail formation. Although these long duration signals are infrequent, they nevertheless contribute significantly to predicted link DC values. In addition, these long duration signals often provide the maximum link RSL values needed to provide minimal connectivity on a "disadvantaged" link or high data rate transmission (e.g., 500 kbps) on a well-designed link.

In METEORLINK, the attachment effect is modeled by replacing the line density q_{sm}^{LO} in Eq. 3.61 with

$$q_{sm}^{LO} \exp(-\beta_A n_A t) \tag{3.72}$$

where

β_A = attachment rate determined from atmospheric measurements at the trail scatter point height above ground h_s

n_A = volume density of neutral particles involved in the attachment process at height h_s.

In general, both β_A and n_A are strong functions of height and temperature as well as geographic and temporal parameters. Nominal values for $\beta_A n_A$ at meteoric ionization heights have been estimated between about 0.005 and 0.010 [134].

After substitution of Eq. 3.72 into Eq. 3.61, a second order Taylor series expansion of the resulting expression for q_{sm}^{LO} is derived. This expression is then used in a root-finding algorithm to determine the q_{sm}^{LO} value satisfying the usability requirements (L^*, τ^*). Unfortunately, the $\exp(-\beta_A n_A t)$ term changes the time $t_{\min}^{LO}$ at which the minimum path loss, $l_{sm}^{LO}(t_{\min}^{LO})$, is achieved, so Eq. 3.62*b* is no longer valid. As a result, $t_{\min}^{LO}$ must be determined using a second root-finding algorithm for each *LO* modeled q value used in the iterative solution for q_{sm}^{LO}. Use of the $\exp(-\beta_A n_M t)$ term in METEORLINK to account for attachment has the desired effect, it increases the q values needed to meet trail usability requirements (L^*, τ^*) and therefore decreases the corresponding MR values for long duration bursts.

The decrease in burst duration due to attachment may be estimated from the trail's electron volume density at the moment the trail ceases to satisfy the overdense criteria. Ignoring the initial trail radius (i.e., $r_0 = 0$), unimportant for long duration echoes, and considering only ambipolar diffusion, this volume density is given by

$$N_c = \frac{q_{sm}^{LO}}{4\pi D_O t_D} \tag{3.72a}$$

where

N_c = critical electron volume density at the transition between overdense and underdense trail-scatter behavior

t_D = time from overdense trail formation at $t = 0$ to the transition from overdense to underdense trail-scatter behavior, that is, the duration of the overdense state, assuming only ambipolar diffusion.

If both attachment and ambipolar diffusion are considered, then this volume density becomes

$$N_c = \left(\frac{q_{sm}^{LO}}{4\pi D_O t_A}\right)\exp(-\beta_A n_A t_A) \tag{3.72b}$$

where

t_A = time from overdense trail formation at $t = 0$ to the transition from overdense to underdense trail-scatter behavior, assuming both ambipolar diffusion and attachment

Solving Eqs. 3.72*a* and 3.72*b* for t_D yields

$$t_D = t_A \exp(\beta_A n_A t_A)$$

Since both t_D and t_A were measured approximately from the time of overdense trail formation at $t = 0$, they may be replaced with the corresponding burst durations τ_D and τ_A, respectively, so

$$\tau_D = \tau_A \exp(\beta_A n_A \tau_A)$$

Defining $\tau_D = \tau_D^*$ when $\tau_A^* = \tau^*$ yields

$$\tau_D^* = \tau^* \exp(\beta_A n_A \tau^*) \tag{3.73}$$

Equation 3.73 gives the burst duration τ_D^* needed to provide a required burst duration τ^* accounting for attachment effects while modeling ambipolar diffusion as the only duration-limiting mechanism. For example, Fig. 3.28 plots the necessary τ_D^* value to achieve a specified τ^* value for three

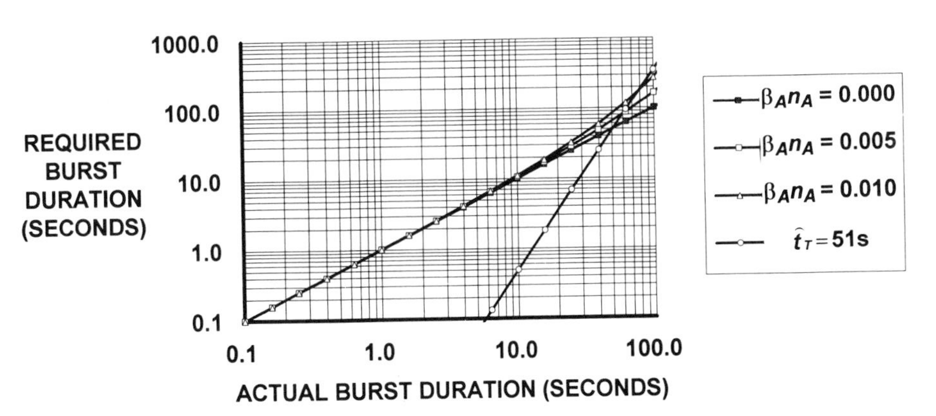

Figure 3.28. Reduction in trail-scattered signal lifetime due to attachment and turbulent diffusion.

representative values of $\beta_A n_A$. The slope of the (τ_D^*, τ^*) curve is near unity until $\tau^* = 10$ s, for $\tau^* > 10$ s, the τ_D^*/τ^* slope increases rapidly, requiring ever increasing values of τ_D^* to meet the duration requirement τ^*.

Turbulent diffusion from upper atmospheric wind motion has been proposed as a principal mechanism for the disintegration of meteor trails, and termination of the trail-scattered signal, for durations exceeding about 50 s [135]. This duration is unusual for sporadic meteors but is nevertheless characteristic of high speed meteors in major showers such as the Perseids ($\bar{v}_\epsilon = 54$ km/s) and the Leonids ($\bar{v}_\epsilon = 72$ km/s). Equating the effective trail radius due to turbulent diffusion with the corresponding radius achieved from ambipolar diffusion alone yields [136]:

$$[r_q(t)]^2 = 4\widehat{D}\exp(H_e)t_D = 4\widehat{M}\exp(\eta H_e)t_T^\varphi$$

where

$\widehat{D}$ = ambipolar diffusion rate at the height of maximum ionization (m²/s)

$\widehat{M}$ = turbulent diffusion rate at the height of maximum ionization (m²/s³)

η, φ = empirically determined constants

H_e = normalized difference between the height of maximum ionization and the scatter point height given by Eq. 3.13*b*

Solving for t_D yields

$$t_D = (\widehat{M}/\widehat{D})\exp[(\eta - 1)H_e]t_T^\varphi$$

Thus, the effective trail duration τ_D^* required to assure a minimum burst duration τ^* in the presence of turbulent diffusion can be approximated by:

$$\tau_D^* = (\widehat{M}/\widehat{D})\exp[(\eta - 1)H_e](\tau^*)^\varphi \tag{3.74}$$

At the height of maximum ionization, $\widehat{H}_I$, where enduring trail-scattered signals are observed, $H_e = 0$ and

$$\tau_D^* = \widehat{t}_T^{(1-\varphi)}(\tau^*)^\varphi \tag{3.75}$$

where

$$\widehat{t}_T = (\widehat{D}/\widehat{M})^{1/(\varphi-1)}$$

is the time at which turbulence effects become dominant.

The constants $\eta = 2.85$, $\varphi = -1.11$, $\widehat{t}_T = 51$ s, and the turbulent diffusion coefficient $\widehat{M} = 3.88 \times 10^5$ m²/s³ were derived from measurements performed for the Leonid and Perseid meteor showers [137]. These values were substituted into Eq. 3.75 and the resulting expression for τ_D^* versus τ^* was plotted in Fig. 3.28. The $(\tau_D^* 1 \tau^*)$ curve shows the constant slope expected from the constant exponential in Eq. 3.75 and crosses the $\beta_A n_A = 0.0611$ curve at 51 s. This example calculation has been included primarily for illustrative purposes. The values used to generate the (τ_D^*, τ^*) curve in Fig. 3.28 may not be appropriate for most sporadic meteors, however, whose average speed is about 40 km/s. Given appropriate values, however, the maximum of τ_D^* from Eq. 3.73 or Eq. 3.75 should be used to determine the specified RSL duration to achieve the required burst duration τ^*.

3.2.2.3.2.7 SUMMARY. The observability $(\mathcal{P}_s^{\mathrm{L}}, \hat{\mathbf{v}}_m^{\mathrm{L}})$ and usability (L^*, τ^*) requirements have been used to compute the minimum detectable line density q_{sm}^*. Using these requirements and the four approximate trail-scattering models, the minimum line densities q_{sm}^{LU}, q_{sm}^{SU}, q_{sm}^{LO}, and q_{sm}^{SO} can be computed. In addition, the underdense/overdense transition line density q_{sm}^T, is computed and the transition wavelengths λ_{sm}^{UT} and λ_{sm}^{OT} are determined for underdense and overdense trails, respectively. If $q_{sm}^{LU} \le q_{sm}^T$, then the minimum required line density corresponds to an underdense trail and is given by

$$q_{sm}^* = q_{sm}^U = q_{sm}^{LU} \qquad \text{if } \lambda_L \le \lambda_{sm}^{UT}$$

or

$$q_{sm}^* = q_{sm}^U = q_{sm}^{SU} \qquad \text{if } \lambda_L > \lambda_{sm}^{UT} \tag{3.76}$$

If $q_{sm}^{LU} > q_{sm}^T$, then the minimum required line density may be overdense. If the overdense trail duration (Eq. 3.63) is less than the time it takes to achieve its initial radius (Eqs. 3.16, 3.47*a*, *b*), that is,

$$\tau_{sm}^{LO} \le \frac{r_0^2}{4D_0}$$

then only an underdense signal would be observed and q_{sm}^* would be determined from Eqs. 3.48 or 3.54. Otherwise, the minimum line density becomes

$$q_{sm}^* = q_{sm}^O = q_{sm}^{LO} \qquad \text{if } \lambda_L \le \lambda_{sm}^{OT}$$

or

$$q_{sm}^* = q_{sm}^O = q_{sm}^{SO} \qquad \text{if } \lambda_L > \lambda_{sm}^{OT} \tag{3.77}$$

Under some anomalous conditions, $q_{sm}^{LU} > q_{sm}^T$ while $q_{sm}^{LO} < q_{sm}^T$. In this case, choosing $q_{sm}^* = q_{sm}^T$ provides a simple, albeit approximate, solution.

3.2.2.3.3 **Density-Rate Transformation.** The incremental MR value of usable meteors passing through the area element dS_s within the small solid angle $d\Omega_{sm}$ is given by

$$d^2\dot{N}^*_{sm} = X_m(\hat{\mathbf{v}}^{\mathrm{O}}_m)\left[\frac{\overline{q}_{R_z}(\bar{v}_\epsilon, \hat{\mathbf{v}}^{\mathrm{O}}_m)\cos(\chi_{sm})}{\overline{q}^*_{sm}(\bar{v}_\epsilon, \hat{\mathbf{v}}^{\mathrm{O}}_m)}\right]^{(s-1)} dS_s\, d\Omega_{sm}$$

$$= Q^*_{sm}(\bar{v}_\epsilon, \hat{\mathbf{v}}^{\mathrm{O}}_m)\, dS_s\, d\Omega_{sm} \tag{3.78}$$

where $\hat{\mathbf{v}}^{\mathrm{O}}_m$ is normal to the differential area element dS_s, and the function

$$Q^*_{sm}(\bar{v}_\epsilon, \hat{\mathbf{v}}^{\mathrm{O}}_m) = X_m(\hat{\mathbf{v}}^{\mathrm{O}}_m)\left[\frac{\overline{q}_{R_z}(\bar{v}_\epsilon, \hat{\mathbf{v}}^{\mathrm{O}}_m)\cos(\chi_{sm})}{\overline{q}^*_{sm}(\bar{v}_\epsilon, \hat{\mathbf{v}}^{\mathrm{O}}_m)}\right]^{(s-1)} \tag{3.79}$$

where the ϵ dependence of the s value is implied. The trail direction vector $\hat{\mathbf{v}}^{\mathrm{O}}_m$ corresponds to $\hat{\mathbf{v}}^{\mathrm{M}}_m$ in the M system and $\hat{\mathbf{v}}^{\mathrm{L}}_m$ in the L system.

Each usable trail vector $\hat{\mathbf{v}}^{\mathrm{M}}_m$ lies in the $x_{\mathrm{M}}y_{\mathrm{M}}$ plane (M plane) in the M system with origin at $\mathcal{P}^{\mathrm{S}}_s$ in the S system and a specified orientation angle α_m (see Eq. 3.A.21 in Section 3.A.3 and Fig. 3.A.4). The α_{M} angle is assigned values $\pi \le \alpha_m \le 2\pi$. The point $\mathcal{P}^{\mathrm{S}}_s$ corresponds to the point $\mathcal{P}^{\mathrm{L}}_s = (x^{\mathrm{L}}_s, y^{\mathrm{L}}_s, z^{\mathrm{L}}_s)$ in the L system defined in Section 3.A.2, $\hat{\mathbf{v}}^{\mathrm{M}}_m$ corresponds to $\hat{\mathbf{v}}^{\mathrm{L}}_m$ in the L system, and the M plane makes an angle β_s with the $x_{\mathrm{L}}y_{\mathrm{L}}$ plane as pictured in Fig. 3.29. In reality, the location and slope of the M plane is determined by the location of the link sites 1 and 2 as well as the scatter point $\mathcal{P}^{\mathrm{L}}_s$. Thus the M plane tilt angle β_s may be written as

$$\beta_s = \beta(\mathcal{P}^{\mathrm{L}}_j, \mathcal{P}^{\mathrm{L}}_s) \tag{3.80}$$

where $\mathcal{P}^{\mathrm{L}}_j = (x^{\mathrm{L}}_j, 0, 0)$, $j = 1, 2$, are the transmit and receive antenna locations projected onto the L system ($x^{\mathrm{L}}_2 = -x^{\mathrm{L}}_1$, see Section 3.A.3). The function $\beta(\mathcal{P}^{\mathrm{L}}_j, \mathcal{P}^{\mathrm{L}}_s)$ can be written as the dot product of the normal vector defining the M plane, that is

$$\hat{\mathbf{n}}^{\mathrm{L}}_s = \mathbf{T}_{\mathrm{SL}} \cdot \hat{\mathbf{n}}^{\mathrm{S}}_s \tag{3.81}$$

and the unit vector $\langle 0, 0, 1\rangle_{\mathrm{L}}$ for the z_{L} axis defining the $x_{\mathrm{L}}y_{\mathrm{L}}$ plane. The resulting dot product is merely the z_{L} component (direction cosine) of the $\hat{\mathbf{n}}^{\mathrm{L}}_s$ vector (see Section 3.A.3), that is, $\hat{n}^{\mathrm{L}}_{s3}$.

Now consider a second scatter point in the L system defined by

$$\mathcal{P}^{\mathrm{L}}_{\delta z} = (x^{\mathrm{L}}_s, y^{\mathrm{L}}_s, z^{\mathrm{L}}_s + dz^{\mathrm{L}}_s)$$

where dz^{L}_s is a small increment in the z_{L} direction. Following the calculations for the determination of observable trail orientations (Section 3.A.3) would yield a second M plane tangent to the scattering ellipsoid at $\mathcal{P}^{\mathrm{L}}_{\delta z}$ rather than at the point $\mathcal{P}^{\mathrm{L}}_s$ as shown in Fig. 3.29. This second plane would make an angle $\beta_s + d\beta_s$ with the $x_{\mathrm{L}}y_{\mathrm{L}}$ plane determined by a second normal vector $\hat{\mathbf{n}}^{\mathrm{L}}_\delta$. The included angle between these two M planes is given by

$$d\beta_s = \cos^{-1}(\hat{\mathbf{n}}^{\mathrm{L}}_s \cdot \hat{\mathbf{n}}^{\mathrm{L}}_\delta) \tag{3.82}$$

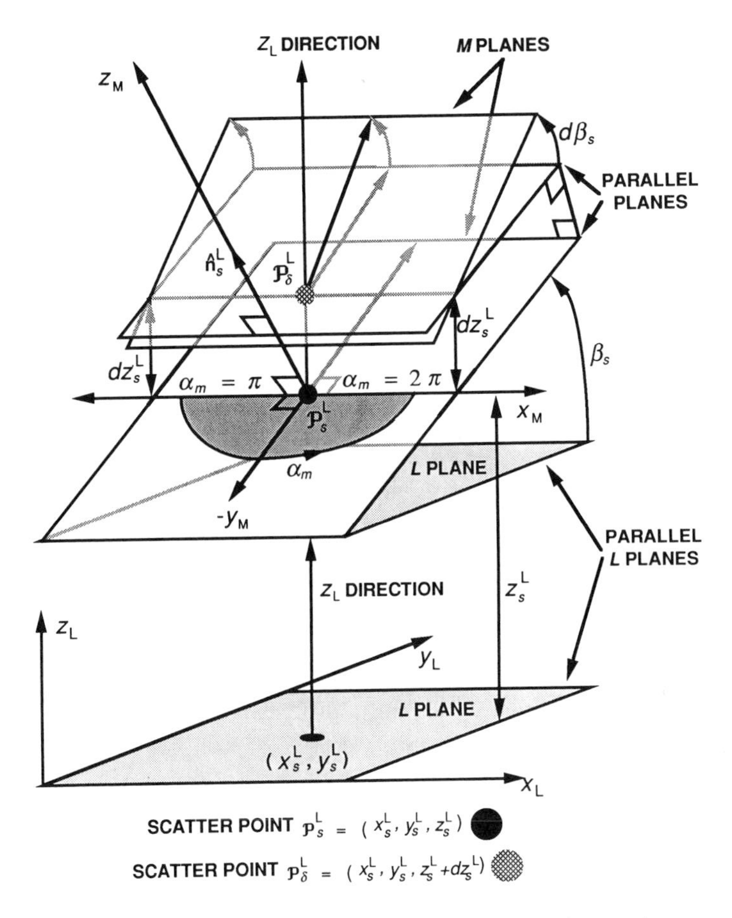

Figure 3.29. *M* planes at the trail-scatter points $\mathcal{P}_s^L$ and $\mathcal{P}_\delta^L$.

If dz_s^L is sufficiently small, the two M planes intersect along a line close to the x_M axis in the original M plane and define the included angle $d\beta_s$. Thus, two trails with the same orientation angle α_m in their respective M planes should have essentially the same value of MR density $d^2\dot{N}_{sm}^*$. None of the link geometric, atmospheric, or link budget parameters will change significantly if dz_s^L is small. In fact, since these $d^2\dot{N}_{sm}^*$ values are indistinguishable, any trail vectors with β angles between β_s and $\beta_s + d\beta_s$ would then account for the same MR contributions $d^2\dot{N}_{sm}^*$.

Similarly, consider a third meteor trail in either the $\hat{\mathbf{n}}_s^L$ or $\hat{\mathbf{n}}_\delta^L$ M plane with orientation angle $\alpha_m + d\alpha_m$. The same MR contribution $d^2\dot{N}_{sm}^*$ should also be observed for the corresponding trail vector. Thus, the MR contribution $d^2\dot{N}_{sm}^*$ corresponding to the trail vector $\hat{\mathbf{v}}_m^L$ applies across the element of solid angle

$$d\Omega_{sm} = -\sin(\alpha_m)d\alpha_m\, d\beta_s = |\sin(\alpha_m)|\,d\alpha_m\, d\beta_s \tag{3.83}$$

where the $\sin(\alpha_m)$ term is required to account for the decreasing values of solid angle, or observable SMRD values, approaching $\mathcal{P}_s^{\mathrm{L}}$ with observable orientations near π and 2π (see Fig. 3.29). The usable MR contributed along $\hat{\mathbf{v}}_m^{\mathrm{L}}$ through $\mathcal{P}_s^{\mathrm{L}}$ is then given by

$$d^3\dot{N}_{sm}^* = Q_{sm}^*(\bar{v}_\epsilon, \hat{\mathbf{v}}_m^{\mathrm{O}})|\sin(\alpha_m)|\,d\alpha_m\,d\beta_s\,dS_s$$

Since the angles α_m and β_s ultimately determine $\hat{\mathbf{v}}_m^{\mathrm{L}}$, the Q function may be written more succinctly as

$$Q_{\alpha\beta}^* = Q_{sm}^*(\bar{v}_\epsilon, \hat{\mathbf{v}}_m^{\mathrm{O}})$$

where both $Q_{\alpha\beta}^*$ and $Q_{sm}^*(\bar{v}_\epsilon, \hat{\mathbf{v}}_m^{\mathrm{O}})$ are long-term (e.g., monthly) average values.

Using arguments analogous to those for z_s^{L}, the MR contributions given by Eq. 3.78 will be essentially constant at the two L system points

$$\mathcal{P}_{\delta x}^{\mathrm{L}} = (x_s^{\mathrm{L}} + dx_s^{\mathrm{L}},\ y_s^{\mathrm{L}},\ z_s^{\mathrm{L}})$$

and

$$\mathcal{P}_{\delta y}^{\mathrm{L}} = (x_s^{\mathrm{L}},\ y_s^{\mathrm{L}} + dy_s^{\mathrm{L}},\ z_s^{\mathrm{L}})$$

The points $\mathcal{P}_{\delta x}^{\mathrm{L}}$, $\mathcal{P}_{\delta y}^{\mathrm{L}}$, and $\mathcal{P}_s^{\mathrm{L}}$ define a differential area $dx_s^{\mathrm{L}}\,dy_s^{\mathrm{L}}$ making an angle $\cos^{-1}(\hat{v}_{m3}^{\mathrm{L}})$ with the trail unit vector

$$\hat{\mathbf{v}}_m^{\mathrm{L}} = \mathbf{T}_{\mathrm{ML}} \cdot \hat{\mathbf{v}}_m^{\mathrm{M}}$$

Thus, if the MR contribution is measured as it passes through $dx_s^{\mathrm{L}}\,dy_s^{\mathrm{L}}$, then the area element

$$dS_s = |\hat{v}_{m3}^{\mathrm{L}}|\,dx_s^{\mathrm{L}}\,dy_s^{\mathrm{L}} \tag{3.84}$$

The total MR contribution passing through $\mathcal{P}_s^{\mathrm{L}}$, measured across $dx_s^{\mathrm{L}}\,dy_s^{\mathrm{L}}$ with orientation $\hat{\mathbf{v}}_m^{\mathrm{L}}$ is then given by

$$d^4\dot{N}_{sm}^* = Q_{\alpha\beta}^*|\hat{v}_{m3}^{\mathrm{L}}\sin(\alpha_m)|\,d\alpha_m\,d\beta_s\,dx_s^{\mathrm{L}}\,dy_s^{\mathrm{L}} \tag{3.85}$$

where $Q_{\alpha\beta}^* = Q_{sm}^*(\bar{v}_\epsilon, \hat{\mathbf{v}}_m^{\mathrm{O}})$ given by Eq. 3.79.

Summarizing the motivation for Eq. 3.83 and Eq. 3.84, a small deviation from specular orientation by a trail passing though $\mathcal{P}_s^{\mathrm{L}}$ will not appreciably effect the scattering geometry, the corresponding SMRD value, or the ultimate MR contribution. The differential $d^4\dot{N}_{sm}^*$ is the cumulative MR contribution for all meteors satisfying the observability criteria $(\mathcal{P}_s^{\mathrm{L}}, \hat{\mathbf{v}}_m^{\mathrm{O}})$ and producing trails with line densities in excess of q_{sm}^*, that is, meeting the usability criteria (L^*, τ^*). Integration of Eq. 3.85 over all contributing meteor radiants determined by $(\mathcal{P}_s^{\mathrm{L}}, \hat{\mathbf{v}}_m^{\mathrm{O}})$ and area elements $dx_s^{\mathrm{L}}\,dy_s^{\mathrm{L}}$ yields the link MR value, $\dot{N}_L$.

3.2.2.4 *Duty Cycle.* Link DC is the second standard measure of MB link performance and is measured, for example, in units of seconds of burst duration per second of elapsed time. The MR value, DC value, and DC/MR ratio fully characterize the MB communications channel. In the MR calculation, the usability requirements (L^*, τ^*) were specified for each observable trail meeting the observability criteria $(\mathscr{P}_s^{\mathrm{L}}, \hat{\mathbf{v}}_m^{\mathrm{O}})$ and the corresponding value of q_{sm}^* was sought. In this case, the net DC value contributed from meteors arriving at the incremental rate $d^4\dot{N}_{sm}^*$ with orientation $\hat{\mathbf{v}}_m^{\mathrm{L}}$ and line density exceeding q_{sm}^* must be determined. Setting $q = \widehat{q}_{sm}^*$ in Eq. 3.79 and differentiating $Q_{\alpha\beta}^*$ with respect to q yields

$$dQ_{\alpha\beta}^* = (1-s)X(\hat{\mathbf{v}}_m^{\mathrm{O}})[\widehat{q}_{R_z}\cos(\chi_{sm})]^{(s-1)}q^{-s}\,dq \tag{3.86}$$

where the $\bar{v}_\epsilon$ and $\hat{\mathbf{v}}_m^{\mathrm{O}}$ dependence is implicit in the symbol $\widehat{q}_{R_z}$. Since $1-s<0$, $dQ_{\alpha\beta}^*<0$, so $Q_{\alpha\beta}^*$ and the MR contribution $d^4\dot{N}_{sm}^*$ decrease with increasing values of q, thus

$$\begin{aligned} d^5\dot{N}_{sm}^* &= dQ_{sm}^*|\hat{v}_{m3}^{\mathrm{L}}\sin(\alpha_m)|\,d\alpha_m\,d\beta_s\,dx_s^{\mathrm{L}}\,dy_s^{\mathrm{L}} \\ &= n_t(q)\,dq\,d\alpha_m\,d\beta_s\,dx_s^{\mathrm{L}}\,dy_s^{\mathrm{L}} \end{aligned} \tag{3.87}$$

where

$$\begin{aligned} n_t(q) &= n_t(q|\alpha_m, \beta_s, x_s^{\mathrm{L}}, y_s^{\mathrm{L}}) \\ &= (1-s)X(\hat{\mathbf{v}}_m^{\mathrm{O}})[\widehat{q}_{R_z}\cos(\chi_{sm})]^{(s-1)}|\hat{v}_m^{\mathrm{L}}\sin(\alpha_m)|q^{-s} \end{aligned} \tag{3.88}$$

Since the MR value decreases with increasing required q values, $n_t(q)$ constitutes the *reduction* in the MR contribution for usable trails with line densities between q and $q+dq$. The appropriate value needed to compute the DC increment $d^5T_{sm}^*$ is the *increment* to the MR contribution, $m_t(q) = -n_t(q)$, that is, the MR contribution for trails with line densities between q and $q+dq$. The link DC contribution from the trail with orientation angles (α_m, β_s) and line densities between q and $q+dq$ can then be written

$$d^5\dot{T}_{sm}^* = m_t(q)\tau_q(q)\,dq\,d\alpha_m\,d\beta_s\,dx_s^{\mathrm{L}}\,dy_s^{\mathrm{L}} \tag{3.89}$$

where $\tau_q(q) = \tau_q(q|\alpha_m, \beta_s, x_s^{\mathrm{L}}, y_s^{\mathrm{L}})$ is the burst duration exceeding P_r^* for a trail with line density q.

The burst duration $\tau_q(q) = \tau_{sm}^{LU}(q)$ derived from the LU trail model defined by Eq. 3.48 is given by

$$\tau_{sm}^{LU}(q) = \frac{\ln(l^*q^2/a_1^{LU}) - a_3^{LU}}{a_2^{LU}} \tag{3.90}$$

and for the short wavelength underdense case by

$$\tau_{sm}^{SU}(q) = \frac{\sqrt{l^* q^2/[a_1^{SU}\exp(a_3^{SU})] - 1}}{a_2^{SU}} \tag{3.91}$$

where l^* is the maximum allowable path loss defined by Eq. 3.50 and the *LU* and *LO* model coefficients are given by Eqs. 3.48 and 3.61, respectively.

In the long wavelength overdense trail case, the burst duration for line density q has the form

$$\tau_{sm}^{LO} = \frac{4}{a_2^{LO}} \sqrt{\left[\sqrt{\frac{a_4^{LO}q}{e}} - \frac{1}{a_1^{LO}l^*}\right]\left(\frac{a_4^{LO}q}{e}\right)^{3/2}} \tag{3.92}$$

The *LO* trail contribution to the link DC value must account for the effects of attachment and turbulent diffusion. Otherwise the DC contribution from the infrequent large meteors will be significantly overestimated. Thus, if the *LO* trail expression for $\tau_q(q)$ given by Eq. 3.92 yields a value τ_{sm}^{LO}, it must be decreased to a realistic value $\tau_q^{LO}(\tau_{sm}^{LO}) = \tau_A^{LO}$ or $\tau_q^{LO}(\tau_{sm}^{LO}) = \tau_T^{LO}$ from consideration of attachment or turbulent diffusion, respectively. Attachment effects are modeled by the transcendental expression of Eq. 3.73, that is

$$\tau_{sm}^{LO} = \tau_A^{LO}\exp(\beta_A n_A \tau_A^{LO}) \tag{3.93}$$

This expression cannot be explicitly solved for τ_A^{LO} as a function of τ_{sm}^{LO}. Instead, a curve fit was generated for the inverse relationship $\tau_A^{LO}(\tau_{sm}^{LO})$ from a set of sample points, $(\tau_A^{LO}, \tau_{sm}^{LO})$. In the case of turbulent diffusion, Eq. 3.74 can be inverted to yield

$$\tau_q^T = \left(\frac{\widehat{D}}{\widehat{M}}\tau_{sm}^{LO}\exp[(1-\eta)H_e]\right)^{1/\varphi} \tag{3.94}$$

where η and φ are empirically determined constants (see Section 3.2.2.3.2.6) and H_e is given by Eq. 3.30*b*. The final value of $\tau_q^{LO}(\tau_{sm}^{LO})$ to be used in the DC calculation from the *LO* trail model is given by

$$\tau_q^{LO} = \min\{\tau_A^{LO}, \tau_T^{LO}\} \tag{3.95}$$

The expression for $\tau_q(q)$ derived from the *SO* trail model with $r_0 = 0$ is given by

$$\tau_{sm}^{SO} = \frac{\sqrt{l^* q^{5/2}/a_1^{SO} - 1}}{a_3^{SO}} \tag{3.96}$$

with the coefficients a_1^{SO} and a_3^{SO} given by Eq. 3.67a and 3.71, respectively. The *SO* burst durations are much shorter than observed for the *LO* trail, so attachment and turbulent diffusion effects are not considered.

The total DC contributed by meteor trails meeting the observability requirements $(\mathscr{P}_s^{\mathrm{L}}, \hat{\mathbf{v}}_m^{\mathrm{O}})$ and the usability requirements (L^*, τ^*) is given by

$$d^4\dot{T}_{sm}^* = \left[\int_{q_{sm}^*}^{\infty} m_t(q)\tau_q(q)\,dq\right] d\alpha_m\,d\beta_s\,dx_s^{\mathrm{L}}\,dy_s^{\mathrm{L}} \tag{3.97}$$

The function $\tau_q(q)$ in Eq. 3.97 must be determined from the appropriate burst duration in Eqs. 3.90 through 3.96. If the minimum required line density $q_{sm}^* < q_{sm}^T$, the increment $d^4\dot{T}_{sm}^*$ given by Eq. 3.97 becomes

$$d^4\dot{T}_{sm}^* = \left[\int_{q_{sm}^*}^{q_{sm}^T} m_t(q)\tau_q^U(q)\,dq + \int_{q_{sm}^T}^{\infty} m_t(q)\tau_q^O(q)\,dq\right] d\alpha_m\,d\beta_s\,dx_s^{\mathrm{L}}\,dy_s^{\mathrm{L}} \tag{3.98}$$

where $\tau_q^U(q)$ and $\tau_q^O(q)$ are the underdense and overdense trail-scattered RSL durations, respectively. If the minimum line density $q_{sm}^* \geq q_{sm}^T$, then only the second integral in Eq. 3.98 contributes to $d^4\dot{T}_{sm}^*$ and the DC contribution becomes

$$d^4\dot{T}_{sm}^* = \left[\int_{q_{sm}^*}^{\infty} m_t(q)\tau_q^O(q)\,dq\right] d\alpha_m\,d\beta_s\,dx_s^{\mathrm{L}}\,dy_s^{\mathrm{L}} \tag{3.99}$$

In either Eq. 3.98 or 3.99, the $\tau_q(q)$ expressions are determined from the signal wavelength, λ_L, dependence as well as the threshold q value, q_{sm}^T. For underdense trail-scattered signals

$$\tau_q^U(q) = \tau_{sm}^{LU} \quad \text{if } \lambda_L \leq \lambda_{sm}^{TU} \quad \text{(Eq. 3.90)}$$

or

$$\tau_q^U(q) = \tau_{sm}^{SU} \quad \text{if } \lambda_L > \lambda_{sm}^{TU} \quad \text{(Eq. 3.91)} \tag{3.100}$$

and

$$\tau_q^O(q) = \tau_q^{LO} \quad \text{if } \lambda_L \leq \lambda_{sm}^{TO} \quad \text{(Eq. 3.92 with Eq. 3.95)}$$

or

$$\tau_q^O(q) = \tau_{sm}^{SO} \quad \text{if } \lambda_L > \lambda_{sm}^{TO} \quad \text{(Eq. 3.96)}$$

The DC contribution $d^4\dot{T}_{sm}^*$ must be integrated over α_m, β_s, x_s^{L}, and y_s^{L} to produce the net link DC value.

3.2.2.5 *Link Performance Predictions*

3.2.2.5.1 Integral Formulations. The link MR and DC values are the accumulations of contributions from thousands of modeled trail orientations

$\mathbf{v}_m^{\mathrm{L}}$ at representative scatter points $\mathcal{P}_s^{\mathrm{L}}$ located throughout the meteor burn layer illuminated by both link antennas. This layer, or common volume, is between 80 and 120 km above the earth's surface within radio line-of-sight (LOS) of both link antennas. The MR and DC contributions $d^4\dot{N}_{sm}^*$ and $d^4\dot{T}_{sm}^*$, respectively, were determined from the observability criteria $(\mathcal{P}_s^{\mathrm{L}}, \hat{\mathbf{v}}_m^{\mathrm{O}})$ and the usability requirements (L^*, τ^*) for each modeled meteor trail. Integrating these contributions over all usable trails yields the link MR value

$$\dot{N}_L = \int_{x_{\mathrm{L}}^l}^{x_{\mathrm{L}}^u} \int_{y_{\mathrm{L}}^l}^{y_{\mathrm{L}}^u} \int_{\beta_s^l}^{\beta_s^u} \int_{\alpha_m^l}^{\alpha_m^u} d^4\dot{N}_{sm}^*$$

$$= \int_{x_{\mathrm{L}}^l}^{x_{\mathrm{L}}^u} \int_{y_{\mathrm{L}}^l}^{y_{\mathrm{L}}^u} \int_{\beta_s^l}^{\beta_s^u} \int_{\alpha_m^l}^{\alpha_m^u} Q_{\alpha\beta}^* |\hat{v}_{m3}^{\mathrm{L}} \sin(\alpha_m)| \, d\alpha_m \, d\beta_s \, dx_{\mathrm{L}} \, dy_{\mathrm{L}} \tag{3.101a}$$

and the link DC value

$$\dot{T}_L = \int_{x_{\mathrm{L}}^l}^{x_{\mathrm{L}}^u} \int_{y_{\mathrm{L}}^l}^{y_{\mathrm{L}}^u} \int_{\beta_s^l}^{\beta_s^u} \int_{\alpha_m^l}^{\alpha_m^u} d^4\dot{T}_{sm}^*$$

$$= \int_{x_{\mathrm{L}}^l}^{x_{\mathrm{L}}^u} \int_{y_{\mathrm{L}}^l}^{y_{\mathrm{L}}^u} \int_{\beta_s^l}^{\beta_s^u} \int_{\alpha_m^l}^{\alpha_m^u} \int_{q_{sm}^*}^{\infty} \left[\int_{q_{sm}^*}^{q_{sm}^T} m_t(q) \tau_q^U(q) \, dq \right.$$

$$\left. + \int_{q_{sm}^T}^{\infty} m_t(q) \tau_q^O(q) \, dq \right] d\alpha_m \, d\beta_s \, dx_{\mathrm{L}} \, dy_{\mathrm{L}} \tag{3.101b}$$

for $q_{sm}^* \le q_{sm}^T$ or

$$\dot{T}_L = \int_{x_{\mathrm{L}}^l}^{x_{\mathrm{L}}^u} \int_{y_{\mathrm{L}}^l}^{y_{\mathrm{L}}^u} \int_{\beta_s^l}^{\beta_s^u} \int_{\alpha_m^l}^{\alpha_m^u} \int_{q_{sm}^*}^{\infty} \left[\int_{q_{sm}^T}^{\infty} m_t(q) \tau_q^O(q) \, dq) \right] d\alpha_m \, d\beta_s \, dx_{\mathrm{L}} \, dy_{\mathrm{L}} \tag{3.101c}$$

when $q_{sm}^* > q_{sm}^T$. In Eqs. 3.101

$$Q_{\alpha\beta}^* = Q_{sm}^*(\bar{v}_\epsilon, \hat{\mathbf{v}}_m^{\mathrm{O}}) = X_m(\hat{\mathbf{v}}_m^{\mathrm{O}}) \left[\frac{\widehat{q}_{R_z} \cos(\chi_{sm})}{\widehat{q}_{sm}^*} \right]^{(s-1)}$$

and

$$m_t(q) = -n_t(q) = (1-s) X(\hat{\mathbf{v}}_m^{\mathrm{O}}) [\widehat{q}_{R_z} \cos(\chi_{sm})]^{(s-1)} |\hat{v}_{m3}^{\mathrm{L}} \sin(\alpha_m)| q^{-s}$$

with

x_{L}^l = minimum value of x_{L} for modeled scatter points above $(x_{\mathrm{L}}, y_{\mathrm{L}})$ in the L plane (see Fig. 3.30), serving as the lower limit for down-range integration of link-usable SMRD values

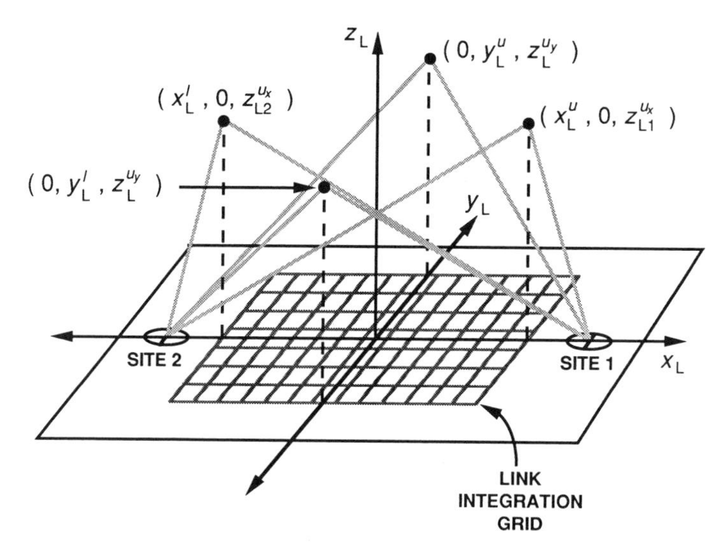

Figure 3.30. Integration limits in the L system.

x_L^u = maximum value of x_L for modeled scatter points above (x_L, y_L) in the L plane, serving as the upper limit for down-range integration of link-usable SMRD values

y_L^l = minimum value of y_L for modeled scatter points above (x_L, y_L), serving as the lower limit on cross-range integration of link-usable SMRD values

y_L^u = maximum value of y_L, serving as the upper limit on cross-range integration of link-usable SMRD values

β_s^l = minimum value of M-plane tilt angles β_s measured relative to the L plane above the point (x_L, y_L) as shown in Fig. 3.30

β_s^u = maximum value of β_s measured relative to the L plane above the point (x_L, y_L)

α_m^l = minimum value of trail orientation angle α_m in the M plane determined by β_s ($\alpha_m^l = \pi$)

α_m^u = maximum value of α_m determined by β_s ($\alpha_m^u = 2\pi$)

q_{sm}^* = minimum required trail line density meeting the observability requirement ($\mathscr{P}_s^L = \langle x_L, y_L, z_L \rangle$, $\hat{\mathbf{v}}_m^O = \hat{\mathbf{f}}(x_L, y_L, z_L, \alpha_m)$), where $\hat{\mathbf{f}}$ is a vector-valued function, as well as the usability requirements (L^*, τ^*)

q_{sm}^T = trail line density threshold corresponding to the observability requirement ($\mathscr{P}_s^L = \langle x_L, y_L, z_L \rangle$, $\hat{\mathbf{v}}_m^O = \hat{\mathbf{f}}(x_L, y_L, z_L, \alpha_m)$) as well as the signal wavelength λ_L.

with the definitions $x_L = x_s^L$, $y_L = y_s^L$, and $z_L = z_s^L$, where the subscript s has been deleted for simplicity. The calculation of $\dot{N}_L$ and $\dot{T}_L$ from Eqs. 3.101 requires specification of the integration limits x_{L}^l, x_{L}^u, y_{L}^l, y_{L}^u, β_s^l, and β_s^u.

3.2.2.5.2 Integration Limits. Consider the determination of β_s^l and β_s^u at the L-system trail-scatter point $\mathcal{P}_s^{\mathrm{L}} = (x_{\mathrm{L}}, y_{\mathrm{L}}, z_{\mathrm{L}})$. Given the antenna locations at points $\mathcal{P}_j^{\mathrm{L}}$, $j = 1, 2$, and the point $\mathcal{P}_s^{\mathrm{L}}$ in the L system (see Eq. 3.A.14), the angle β_s (Eq. 3.80) becomes a function of z_{L} alone. In symbols,

$$\beta_s = \beta(\mathcal{P}_j^{\mathrm{L}}, \mathcal{P}_s^{\mathrm{L}}) = \beta_{z_{\mathrm{L}}}(z_{\mathrm{L}}) \tag{3.102}$$

so that finding β_s^l and β_s^u corresponds to finding the minimum and maximum values of z_{L} (z_{L}^l and z_{L}^u, respectively) in the height band of meteoric ionization above the L-plane point $(x_{\mathrm{L}}, y_{\mathrm{L}})$. Values for z_L^l and z_{L}^u are determined from the corresponding trail-scatter point height limits $h_s^l = 80$ km and $h_s^u = 120$ km above $(x_{\mathrm{L}}, y_{\mathrm{L}})$ are given by

$$z_{\mathrm{L}}^l = [(R_E + h_s^l)^2 - x_{\mathrm{L}}^2 - y_{\mathrm{L}}^2]^{1/2} - \left[R_E^2 - \left(\frac{d_{12}^{\mathrm{L}}}{2}\right)^2\right]^{1/2} \tag{3.103a}$$

and

$$z_{\mathrm{L}}^u = [(R_E + h_s^u)^2 - x_{\mathrm{L}}^2 - y_{\mathrm{L}}^2]^{1/2} - \left[R_E^2 - \left(\frac{d_{12}^{\mathrm{L}}}{2}\right)^2\right]^{1/2} \tag{3.103b}$$

with

$$d_{12}^{\mathrm{L}} = |x_2^{\mathrm{L}} - x_1^{\mathrm{L}}| \tag{3.104}$$

where $\mathcal{P}_1^{\mathrm{L}} = (x_1^{\mathrm{L}}, 0, 0)$ and $\mathcal{P}_2^{\mathrm{L}} = (x_2^{\mathrm{L}}, 0, 0) = (-x_1^{\mathrm{L}}, 0, 0)$ are the x_{L}-axis points corresponding to antenna sites 1 and 2 (see Eq. 3.A.6).

The expression for $\beta(\mathcal{P}_j^{\mathrm{L}}, \mathcal{P}_s^{\mathrm{L}})$, $j = 1, 2$, is simply the inverse sine of the z_{L} component of the unit vector $\hat{\mathbf{n}}_{\mathrm{L}}$ normal to the M plane at the L-system point $(x_{\mathrm{L}}, y_{\mathrm{L}}, z_{\mathrm{L}})$. In symbols,

$$\beta_{z_{\mathrm{L}}}(z_{\mathrm{L}}) = \sin^{-1}(\hat{\mathbf{n}}_{\mathrm{L}} \cdot \langle 0, 0, 1 \rangle_{\mathrm{L}}) = \sin^{-1}(\hat{n}_{\mathrm{L}3})$$

where

$$\hat{\mathbf{n}}_{\mathrm{L}} = \langle \hat{n}_{\mathrm{L}1}, \hat{n}_{\mathrm{L}2}, \hat{n}_{\mathrm{L}3} \rangle = \mathbf{T}_{\mathrm{SL}} \cdot \hat{\mathbf{n}}_{\mathrm{S}}$$

The unit normal vector $\hat{\mathbf{n}}_{\mathrm{S}}$ in the S system is the normalized sum of the unit $\hat{\mathbf{r}}_{sj}^{\mathrm{S}}$ vectors, $j = 1, 2$, as given by Eq. 3.A.16. Equivalently, the normal vector $\hat{\mathbf{n}}_{\mathrm{L}}$ in the L system is the normalized sum of the unit vectors $\hat{\mathbf{r}}_{\mathrm{L}j}$ in the L system, that is

$$\hat{\mathbf{n}}_{\mathrm{L}} = \frac{\hat{\mathbf{r}}_{\mathrm{L1}} + \hat{\mathbf{r}}_{\mathrm{L2}}}{|\hat{\mathbf{r}}_{\mathrm{L1}} + \hat{\mathbf{r}}_{\mathrm{L2}}|}$$

where

$$\hat{\mathbf{r}}_{\mathrm{L}j} = \langle \hat{r}_{\mathrm{L}j1}, \hat{r}_{\mathrm{L}j2}, \hat{r}_{\mathrm{L}j3} \rangle = \hat{\mathbf{r}}_{sj}^{\mathrm{L}} = \mathbf{T}_{\mathrm{SL}} \cdot \hat{\mathbf{r}}_{sj}^{\mathrm{S}}$$

for $j = 1, 2$. Thus, the angle between the M plane and the L plane becomes

$$\beta_{z_{\mathrm{L}}}(z_{\mathrm{L}}) = \sin^{-1}\left(\frac{\hat{\mathbf{r}}_{\mathrm{L1}} + \hat{\mathbf{r}}_{\mathrm{L2}}}{|\hat{\mathbf{r}}_{\mathrm{L1}} + \hat{\mathbf{r}}_{\mathrm{L2}}|} \cdot \langle 0, 0, 1 \rangle_{\mathrm{L}} \right) = \sin^{-1}\left(\frac{\hat{r}_{\mathrm{L13}} + \hat{r}_{\mathrm{L23}}}{|\hat{\mathbf{r}}_{\mathrm{L1}} + \hat{\mathbf{r}}_{\mathrm{L2}}|} \right)$$

so that

$$\sin[\beta_{z_{\mathrm{L}}}(z_{\mathrm{L}})] = \frac{\hat{r}_{\mathrm{L13}} + \hat{r}_{\mathrm{L23}}}{|\hat{\mathbf{r}}_{\mathrm{L1}} + \hat{\mathbf{r}}_{\mathrm{L2}}|} \tag{3.105}$$

with

$$\hat{r}_{\mathrm{L11}} = \frac{x_{\mathrm{L}} - x_1^{\mathrm{L}}}{r_1} \qquad \hat{r}_{\mathrm{L21}} = \frac{x_{\mathrm{L}} - x_2^{\mathrm{L}}}{r_2}$$

$$\hat{r}_{\mathrm{L12}} = \frac{x_{\mathrm{L}}}{r_1} \qquad \hat{r}_{\mathrm{L22}} = \frac{y_{\mathrm{L}}}{r_2}$$

$$\hat{r}_{\mathrm{L13}} = \frac{z_{\mathrm{L}}}{r_1} \qquad \hat{r}_{\mathrm{L23}} = \frac{z_{\mathrm{L}}}{r_2}$$

and the R-vector magnitudes

$$r_j = \sqrt{(x_{\mathrm{L}} - x_1^{\mathrm{L}})^2 + y_{\mathrm{L}}^2 + z_{\mathrm{L}}^2} \qquad j = 1, 2$$

Let

$$b_r = |\hat{\mathbf{r}}_{\mathrm{L1}} + \hat{\mathbf{r}}_{\mathrm{L2}}|$$

and

$$\cos[\beta_{z_{\mathrm{L}}}(z_{\mathrm{L}})] = \sqrt{1 - \sin^2[\beta_{z_{\mathrm{L}}}(z_{\mathrm{L}})]}$$

then the differential element $d\beta_s$ is given by

$$d\beta_s = B_s(z_{\mathrm{L}})\, dz_{\mathrm{L}} = (\beta_1 - \beta_2)\, dz_{\mathrm{L}} \tag{3.106}$$

where

$$\beta_1 = \frac{\beta_1^n}{b_r} \quad \text{and} \quad \beta_2 = \frac{(\hat{r}_{L13} + \hat{r}_{L23})(b_3 - b_2 - b_1)}{b_r^3}$$

with

$$\beta_1^n = \frac{1 - (\hat{r}_{L13})^2}{r_1} + \frac{1 - (\hat{r}_{L23})^2}{r_2}$$

$$b_1 = (\hat{r}_{L11} + \hat{r}_{L21})\left(\frac{\hat{r}_{L11}\hat{r}_{L13}}{r_1} + \frac{\hat{r}_{L21}\hat{r}_{L23}}{r_2}\right)$$

$$b_2 = (\hat{r}_{L12} + \tilde{r}_{L22})\left(\frac{\hat{r}_{L12}\hat{r}_{L13}}{r_1} + \frac{\hat{r}_{L22}\hat{r}_{L23}}{r_2}\right)$$

$$b_3 = (\hat{r}_{L13} + \hat{r}_{L23})\beta_1^n$$

Using Eq. 3.106 for $d\beta_s$ and defining z_L^l by $\beta_s^l = \beta_{z_L}(z_L^l)$ and z_L^u by $\beta_s^u = \beta_{z_L}(z_L^u)$, the integration over β_s is changed to integration over z_L. Writing $d\beta_s = B_s(z_L)\, dz_L$, the expressions for $\dot{N}_L$ and $\dot{T}_L$ become

$$\dot{N}_L = \int_{x_L^l}^{x_L^u} \int_{y_L^l}^{y_L^u} \int_{z_L^l}^{z_L^u} \int_{\alpha_m^l}^{\alpha_m^u} d^4\dot{N}_L \tag{3.107a}$$

and the link DC value

$$\dot{T}_L = \int_{x_L^l}^{x_L^u} \int_{y_L^l}^{y_L^u} \int_{z_L^l}^{z_L^u} \int_{\alpha_m^l}^{\alpha_m^u} d^4\dot{T}_L \tag{3.107b}$$

where

$$d^5\dot{N}_L = B_s(z_L)Q_{\alpha\beta}^*|\hat{v}_{m3}^L \sin(\alpha_m)|\, d\alpha_m\, dz_L\, dy_L\, dx_L \tag{3.108a}$$

$$d^4\dot{T}_L = B_s(z_L) \times \left[\int_{q_{sm}^*}^{q_{sm}^T} m_t(q)\tau_q^U(q)\, dq + \int_{q_{sm}^T}^{\infty} m_t(q)\tau_q^O(q)\, dq\right] d\alpha_m\, dz_L\, dy_L\, dx_L \tag{3.108b}$$

for $q_{sm}^* \leq q_{sm}^T$ and

$$d^4\dot{T}_L = B_s(z_L)\left[\int_{q_{sm}^*}^{\infty} m_t(q)\tau_q^O(q)\, dq\right] d\alpha_m\, dz_L\, dy_L\, dx_L \tag{3.108c}$$

for $q_{sm}^* > q_{sm}^T$, with the z_L-integration limits given by Eq. 3.103.

The integration limits x_L^l, x_L^u, y_L^l, and y_L^u are determined so that each modeled trail-scatter point $\mathcal{P}_s^L = (x_s^L, y_s^L, z_s^L) = (x_L, y_L, z_L)$ is within the radio line-of-sight (LOS) of both link antennas located at the points $\mathcal{P}_1^L$ and

$\mathcal{P}_2^L$. More specifically, the maximum down-range extent for x_L integration occurs between the potential trail-scatter points $(x_L^l, 0, z_{L1}^{u_x})$ and $(x_L^u, 0, z_{L2}^{u_x})$ while the maximum cross-range extent occurs between $(0, y_L^l, z_L^{u_y})$ and $(0, y_L^u, z_L^{u_y})$. These points are shown in Fig. 3.30, in which $(x_L^l, 0, z_{L1}^{u_x})$ is closest to the transmit antenna (site 1) and $(x_L^u, 0, z_{L2}^{u_x})$ is closest to the receive antenna (site 2). The z_L values $z_{L1}^{u_x}$ and $z_{L2}^{u_x}$ differ if the antenna heights h_{a1} and h_{a2} differ. The distance from site 1 to the maximum point $(x_L^u, 0, z_{L2}^{u_x})$ is given by

$$r_1^u = \sqrt{((x_1^L - x_L^u)^2 + (z_{L2}^{u_x})^2} \tag{3.109a}$$

where

$$z_{L2}^u = \sqrt{(r_{\text{eff}} - h_s^u)^2 - (x_L^u)^2} - \sqrt{r_{\text{eff}}^2 - \left(\frac{d_{12}^L}{2}\right)^2} \tag{3.109b}$$

The maximum radio LOS range from the transmit antenna at height h_{a1} to the maximum trail-scatter point height h_s^u is given by

$$r_1^u = \sqrt{h_s^u(2r_{\text{eff}} + h_s^u)} + \sqrt{h_{a1}(2r_{\text{eff}} + h_{a1})} \tag{3.109c}$$

where

$$r_{\text{eff}} = k_E R_E$$

and

k_E = effective earth radius factor accounting for the effects of atmospheric refraction, e.g., 4/3.

The integration limit x_L^u is found by setting Eq. 3.109*a* equal to r_1^u, substituting Eq. 3.109*b* into 3.109*a*, and then solving for the proper root of the resulting quadratic equation in x_L^u. An analogous procedure can be used to determine the remaining limits x_L^l, y_L^l, and y_L^u. In general, the potential scatter point, $\mathcal{P}_s^L$, is within radio LOS if $r_j < r_j^u$ for both antenna sites j, $j = 1, 2$.

3.2.2.5.3 Quadrature Formula. The integrals in Eqs. 3.107 cannot be solved explicitly, so numerical quadrature is required to compute the link MR and DC values. In general, the definite integral

$$I\{f\} = \int_{u_1}^{u_2} f(u)\, du$$

can be approximated by quadrature formula of the form

$$I_N\{f\} = \sum_{i}^{N_u} w_i f(u_i) \tag{3.110}$$

where the N_u coefficients w_i and corresponding points u_i are chosen to minimize the quadrature error $|I\{f\} - I_N\{f\}|$. Applying this quadrature formula to approximate the integrals in Eq. 3.107 yields

$$\dot{N}_L = \sum_{i}^{N_x} \sum_{j}^{N_y} \sum_{k}^{N_z} \sum_{l}^{N_\alpha} w_i w_j w_k w_l \, \delta^4 \dot{N}_L(x_i, y_j, z_k, \alpha_l) \tag{3.111a}$$

and

$$\dot{T}_L = \sum_{i}^{N_x} \sum_{j}^{N_y} \sum_{k}^{N_z} \sum_{l}^{N_\alpha} w_i w_j w_k w_l \, \delta^4 \dot{T}_L(x_i, y_j, z_k, \alpha_l) \tag{3.111b}$$

where

$$\delta^4 \dot{N}_L(x_i, y_j, z_k, \alpha_l) = B_s(z_k) Q^*_{\alpha\beta} |\hat{v}^{\mathrm{L}}_{m3} \sin(\alpha_m)| \tag{3.112a}$$

and

$$\delta^4 \dot{T}_L(x_i, y_j, z_k, \alpha_l) = B_s(z_k)\left[\sum_{n_U}^{N_U} w_{n_U} m_t(q_{n_U}) \tau^U_q(q_{n_U}) + \sum_{n_O}^{N_O} w_{n_O} m_t(q_{n_O}) \tau^O_q(q_{n_O})\right] \tag{3.112b}$$

for $q^*_{sm} \le q^T_{sm}$ and

$$\delta^4 \dot{T}_L(x_i, y_j, z_k, \alpha_l) = B_s(z_k)\left[\sum_{n_O}^{N_O} w_{n_O} m_t(q_{n_O}) \tau^O_q(q_{n_O})\right] \tag{3.112c}$$

for $q^*_{sm} > q^T_{sm}$. The L-system notation for the values x_i, y_j, and z_k is assumed as well as M-system notation for α_l. The elements $Q^*_{\alpha\beta}$, $B_s(z_k)$, $m_t(q_{n_U}$ or $q_{n_O})$, $\tau^U_q(q_{n_U})$, $\tau^O_q(q_{n_O})$, and $\hat{v}^{\mathrm{L}}_{m3}$ in Eqs. 3.112 are functions of x_i, y_j, z_k, and α_l. The calculation values x_i, y_j, z_k, α_l, q_{n_U}, and q_{n_O} for quadrature are defined by

x_i: $x^l_{\mathrm{L}} \le x_i \le x^u_{\mathrm{L}}$ for $i = 1, 2, \ldots, N_x$

y_j: $y^l_{\mathrm{L}} \le y_j \le y^u_{\mathrm{L}}$ for $j = 1, 2, \ldots, N_y$

z_k: $z^l_{\mathrm{L}} \le z_k \le z^u_{\mathrm{L}}$ for $k = 1, 2, \ldots, N_z$

α_l: $\alpha^l_m(=\pi) \le \alpha_l \le \alpha^u_m(=2\pi)$ for $l = 1, 2, \ldots, N_\alpha$

q_{n_U}: $q^*_{sm} \le q_{n_U} \le q^T_{sm}$ for $n_U = 1, 2, \ldots, N_U$

and

$$q_{n_O}: q_{sm}^T \leq q_{n_O} \leq q_{\max}^O \qquad \text{for } n_O = 1, 2, \ldots, N_O$$

if $q_{sm}^* \leq q_{sm}^T$, and

$$q_{n_O}: q_{sm}^* \leq q_{n_O} \leq q_{\max}^O \qquad \text{for } n_O = 1, 2, \ldots, N_O$$

if $q_{sm}^* > q_{sm}^T$, where $q_{\max}^O$ is a suitably large value of electron line density such that the corresponding MR contribution is negligible, for example, $q_{\max}^O = 1 \times 10^{20}$ epm.

The values for x_i, y_j, z_k, α_l, q_{n_U}, and q_{n_O} and the corresponding coefficient weights w_i, w_j, w_k, w_l, w_{n_U}, and, w_{n_O}, are specified according to the requirements imposed by the quadrature technique, or techniques, employed. For example, a depiction of quadrature values x_i, y_j, and z_k forming the calculation points $\mathcal{P}_s^{\mathrm{L}} = \mathcal{P}_{ijk}^{\mathrm{L}} = (x_i, y_j, z_k)$ is shown at the intersection of lines in the LINK grid of Fig. 3.22. Ultimately, computation of sufficiently accurate values for $\dot{N}_L$ and $\dot{T}_L$ requires that the appropriate numbers of points N_x^*, N_y^*, N_z^*, N_α^*, N_U^*, and N_O^* be determined to achieve acceptable convergence of the quadratures in Eqs. 3.111. This determination is based on minimizing the change in predicted values for $\dot{N}_L$ and $\dot{T}_L$ as the numbers of quadrature points are increased.

The α_l quadrature performed at each modeled trail-scatter point $\mathcal{P}_{ijk}^{\mathrm{L}} = (x_i, y_j, z_k)$ is primarily dependent on the range of $X_m(\hat{\mathbf{v}}_m^{\mathrm{O}})$ values corresponding to α_l values between π and 2π. This *range* is not strongly sensitive to the location of $\mathcal{P}_s^{\mathrm{L}}$, although the *absolute* values of $\dot{N}_L$ and $\dot{T}_L$ vary significantly with $\mathcal{P}_s^{\mathrm{L}}$. For this reason, the number of α_l quadrature values can be estimated for any point $\mathcal{P}_s^{\mathrm{L}}$, independent of the values chosen for N_x, N_y, and N_z. Thus, nominal values of N_x, N_y, and N_z are fixed for each calculation of $\dot{N}_L$ as the number of points N_α is increased until

$$\left| \dot{N}_L|_{N_\alpha} - \dot{N}_L|_{N_\alpha^*} \right| < \epsilon(\dot{N}_L|_{N_\alpha^*}) \tag{3.113}$$

where $\dot{N}_L|_{N_\alpha}$ designates the calculation of $\dot{N}_L$ using N_α values α_l in the corresponding quadrature and $\epsilon(\dot{N}_L|_{N_\alpha^*})$ is some small percentage of $\dot{N}_L|_{N_\alpha^*}$ for any number of values $N_\alpha > N_\alpha^*$. This procedure can be repeated to determine N_z, with $N_\alpha = N_\alpha^*$ and both N_x and N_y fixed and, similarly, to determine N_y with $N_\alpha = N_\alpha^*$, $N_z = N_z^*$, and N_x held constant. Finally, the value $N_z x = N_x^*$ is determined using the chosen values for N_α^*, N_z^*, and N_y^*. In practice, the upper limit on the number of quadrature values is a trade-off between prediction accuracy and CPU time. In general, acceptable accuracy can be achieved before processing time becomes excessive on modern PC equipment and minicomputers.

Once the values N_x^*, N_y^*, N_z^*, and N_α^* have been determined for the calculation of $\dot{N}_L$ using Eq. 3.111, the optimal values for N_U and N_O can be determined for the calculation of $\dot{T}_L$ from Eq. 3.111b. The values of N_U and

N_O can then be chosen for nominal values of q^*_{sm} and q^T_{sm}, such as $q^*_{sm} = 1 \times 10^{11}$ epm and $q^T_{sm} = 2.4 \times 10^{14}$ epm. In METEORLINK, the q quadratures are performed by first transforming the variable of integration, q, in the corresponding integrations into $Q = \ln(q)$ and then defining the quadrature points

$$Q_{n_U}: Q^*_{sm} = \ln(q^*_{sm}) \le Q_{n_U} \le Q^T_{sm} = \ln(q^T_{sm}) \quad \text{for } n_U = 1, 2, \ldots, N_U$$

and

$$Q_{n_O}: Q^T_{sm} \le Q_{n_O} \le Q^O_{\max} = \ln(q^O_{\max}) \quad \text{for } n_O = 1, 2, \ldots, N_O$$

if $q^*_{sm} \le q^T_{sm}$, and

$$Q_{n_O}: Q^*_{sm} \le Q_{n_O} \le Q^O_{\max} \quad \text{for } n_O = 1, 2, \ldots, N_O$$

for $q^*_{sm} > q^T_{sm}$. This transformation reduces the range of the integration variable from nine orders of magnitude (q) to one order of magnitude (Q). Using the same condition imposed in Eq. 3.113 for both N_U and N_O yields the ultimate values N^*_U and N^*_O.

The MR and DC quadrature formulas in Eqs. 3.111 may be used to estimate the variability in the short-term link MR and DC values, defined by the random variables $\underset{\approx}{\dot{N}}_L$ and $\underset{\approx}{\dot{T}}_L$. In general, the SMRD-values $\Theta_{sp}(m_0, \hat{\mathbf{v}}^{\mathrm{O}}_m)$ contributing to the MR value, $\dot{N}_L$, and DC value, $\dot{T}_L$, are short-term average values of the stochastic processes, $\underset{\approx}{\Theta}_{sp}(m_0, \hat{\mathbf{v}}^{\mathrm{O}}_m)$. For a short interval of time (days to weeks), however, the statistics of each $\underset{\approx}{\Theta}_{sp}(m_0, \hat{\mathbf{v}}^{\mathrm{O}}_m)$ value in the SMRD map may be considered stationary. In other words, the SMRD value from cell n (see Fig. 3.19) is a random variable $\underset{\approx}{\Theta}_n$ corresponding to radiant position $(\phi_{\mathrm{O}}, \theta_{\mathrm{O}})$ in the O system, with mean value $\bar{\underset{\approx}{\Theta}}_n = \Theta_{sp}(m_0, \hat{\mathbf{v}}^{\mathrm{O}}_m)$ for some trail vector $\hat{\mathbf{v}}^{\mathrm{O}}_m$ and standard deviation $\sigma_{\underset{\approx}{\Theta}_n}$, $n = 1, 2, \ldots, M_\Theta$. M_Θ is the total number of SMRD cells in Fig. 3.19 containing values for both $\bar{\underset{\approx}{\Theta}}_n$ and $\sigma_{\underset{\approx}{\Theta}_n}$.

The values of $\dot{N}_L$ and $\dot{T}_L$ approximated by the summations in Eqs. 3.111 may be rewritten as

$$\dot{N}_L = \mathrm{E}\left\{\sum_{n=1}^{M_\Theta} K_n^{\dot{N}_L} \underset{\approx}{\Theta}_n\right\} = \sum_{n=1}^{M_\Theta} K_n^{\dot{N}_L} \mathrm{E}\{\underset{\approx}{\Theta}_n\} = \sum_{n=1}^{M_\Theta} K_n^{\dot{N}_L} \bar{\underset{\approx}{\Theta}}_n$$

and

$$\dot{T}_L = \mathrm{E}\left\{\sum_{n=1}^{M_\Theta} K_n^{\dot{T}_L} \underset{\approx}{\Theta}_n\right\} = \sum_{n=1}^{M_\Theta} K_n^{\dot{T}_L} \mathrm{E}\{\underset{\approx}{\Theta}_n\} = \sum_{n=1}^{M_\Theta} K_n^{\dot{T}_L} \bar{\underset{\approx}{\Theta}}_n$$

where

$K_n^{\dot{N}_L}$ = sum of factors derived from the observability criteria $(\mathscr{P}_s^{\mathrm{L}}, \hat{\mathbf{v}}_m^{\mathrm{O}})$ and the usability criteria (L^*, τ^*) multiplying the SMRD value $\underset{\approx}{\Theta}_n$ in Eq. 3.111*a*

$K_n^{\dot{T}_L}$ = sum of factors derived from the observability criteria $(\mathscr{P}_s^{\mathrm{L}}, \hat{\mathbf{v}}_m^{\mathrm{O}})$ and the usability criteria (L^*, τ^*) multiplying the SMRD value $\underset{\approx}{\Theta}_n$ in Eq. 3.111*b*

and E{ } denotes expected value. The variance of the link MR value is then given by

$$\sigma_{\dot{N}_L}^2 = \sum_{i=1}^{M_\Theta} \sum_{j=1}^{M_\Theta} K_i^{\dot{N}_L} K_j^{\dot{N}_L} (\mathrm{E}\{\underset{\approx}{\Theta}_i \underset{\approx}{\Theta}_j\} - \bar{\underset{\approx}{\Theta}}_i \bar{\underset{\approx}{\Theta}}_j)$$

and the variance of the corresponding link DC value is given by

$$\sigma_{\dot{T}_L}^2 = \sum_{i=1}^{M_\Theta} \sum_{j=1}^{M_\Theta} K_i^{\dot{T}_L} K_j^{\dot{T}_L} (\mathrm{E}\{\underset{\approx}{\Theta}_i, \underset{\approx}{\Theta}_j\} - \bar{\underset{\approx}{\Theta}}_i \bar{\underset{\approx}{\Theta}}_j)$$

The term within the square brackets in these expressions can be replaced with $r_{ij}^{\Theta} \sigma_{\underset{\approx}{\Theta}_i} \sigma_{\underset{\approx}{\Theta}_j}$ to yield

$$\sigma_{\dot{N}_L}^2 = \sum_{i=1}^{M_\Theta} \sum_{j=1}^{M_\Theta} K_i^{\dot{N}_L} K_j^{\dot{N}_L} r_{ij}^{\Theta} \sigma_{\underset{\approx}{\Theta}_i} \sigma_{\underset{\approx}{\Theta}_j} \quad \text{and} \quad \sigma_{\dot{T}_L}^2 = \sum_{i=1}^{M_\Theta} \sum_{j=1}^{M_\Theta} K_i^{\dot{T}_L} K_j^{\dot{T}_L} r_{ij}^{\Theta} \sigma_{\underset{\approx}{\Theta}_i} \sigma_{\underset{\approx}{\Theta}_j}$$

where r_{ij}^{Θ} is the correlation coefficient of $\underset{\approx}{\Theta}_i$ and $\underset{\approx}{\Theta}_j$. No estimate of r_{ij}^{Θ} is readily available, although the apparent similarity between values in nearby cells of the SMRD map in Fig. 3.19 suggests a significant correlation exists. Heuristically, adjacent radiants may be derived from a common orbital source, such as a decaying meteor stream or comet, which suggests that r_{ij}^{Θ} may be a function of cell spacing, decreasing with increasing radiant separation.

3.2.2.6 LINK Grid Diagnostic Analysis. The LINK grid is a useful diagnostic tool for METEORLINK model predictions. It consists of the display of z_{L}-quadrature values at each $(x_{\mathrm{L}}, y_{\mathrm{L}})$ quadrature point scaled by the corresponding difference area

$$\Delta x_i \, \Delta y_j = (x_i - x_{i-1})(y_j - y_{j-1})$$

For example, Fig. 3.31 shows the LINK grid of MR contributions on a 1200-km south-to-north high-latitude MB link at 3 AM local time in March, 1-kW transmit power, five-element Yagi transmit and receive antenna patterns, and -120-dBm required received power [138]. Each value in the grid is the estimated contribution to the link MR value measured in usable

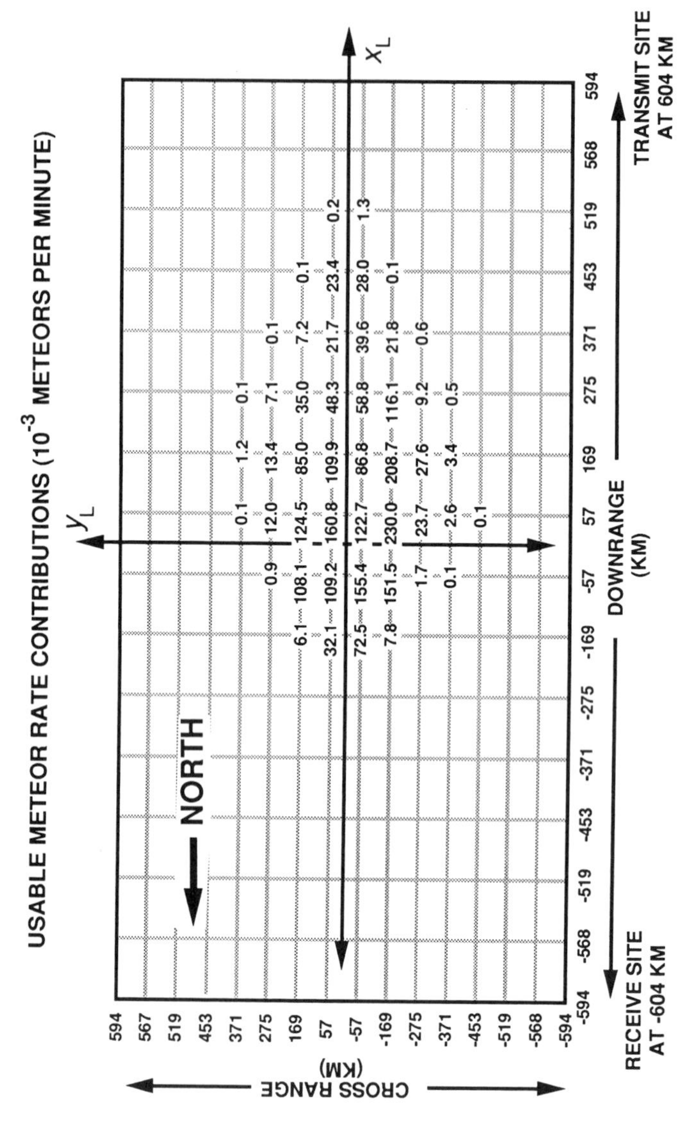

Figure 3.31. MR LINK grid for a 1200-km south-to-north MB link.

meteors per minute over all heights z_L and angles α_m above the indicated (x_L, y_L) point. The numbers of quadrature points for this example were $N_x^* = N_y^* = 16$, $N_z^* = 10$, $N_\alpha^* = 10$, $N_U^* = 4$, and $N_O^* = 8$. The unequal spacing of x_L and y_L values shown on the LINK grid is characteristic of the Gauss quadrature technique used by METEORLINK.

Fig. 3.31 shows that the only nonzero contributions ($>10^{-4}$ meteors per minute) extend downrange from the x_L value at about -170 km to the value at 520 km. Cross-range y_L values exceed 10^{-4} meteors per minute between -450 and 370 km. The peak value of 0.23 meteors per minute occurs at about -60 km downrange and -170 km cross range on the east side of the link GCP. At 3 AM, the link is exposed to the greatest meteor flux moving from west to east. Thus, the peak MR contribution should come from the eastern side of the link as predicted by METEORLINK. This result can be deemed from inspection of Figs. 3.6*a* and 3.6*b*, showing dominant contributions moving from the east side at midnight to balanced east–west contributions at 6 AM. The MR grid plot in Fig. 3.31 shows significant contribution to the west of the GCP as well, since the 3 AM results show characteristics of both the midnight and 6 AM distributions of MR contributions over the link.

Similar grid plots can be generated for the percentage of MR and DC values found to be underdense, effective antenna takeoff angles, antenna gain product, maximum allowable path loss, and Faraday rotation angles (see Appendix B). The LINK grid of antenna takeoff angles is useful in the evaluation of antenna designs for both point-to-point and mobile MB link applications. Figures 3.32*a*, *b* show the height-averaged transmit antenna elevation and azimuth angles, respectively, that were responsible for MR contributions exceeding 1% of the link MR in Fig. 3.31. The antenna boresight for this example was pointed along the GCP to the receive antenna (site 2). The peak MR contribution at $x_L = -60$ km, $y_L = -170$ km corresponds to an antenna pattern elevation angle of about 8° and an azimuth angle of 17° from the antenna boresight. Contributing elevation angles range from 6 to 30°, and the usable azimuth angles range from about -30 to 30°.

The LINK grid may also be plotted as a three-dimensional image or as a contour (intensity) plot. The relative MR contributions for a mid-latitude, east-west link at sunrise employing omnidirectional antenna patterns is shown in Fig. 3.33 plotted for three path lengths. Plot conventions are shown in Fig. 3.33*a*. Figure 3.33*b* shows the overhead, backscatter, and mid-link contributions at short range. At mid-ranges, earth blockage begins to reduce these contributions, as shown in Fig. 3.33*c*. Finally, earth blockage at long range produces the classic meteor "hot spots" on either side of the GCP at mid-link shown in Fig. 3.33*d*. Although numeric data cannot be easily determined from these relief plots, they provide a tool for rapid evaluation of the relative contributions throughout meteoric common volume.

The classic null zone is evident in Fig. 3.33*d* as negligible MR contribu-

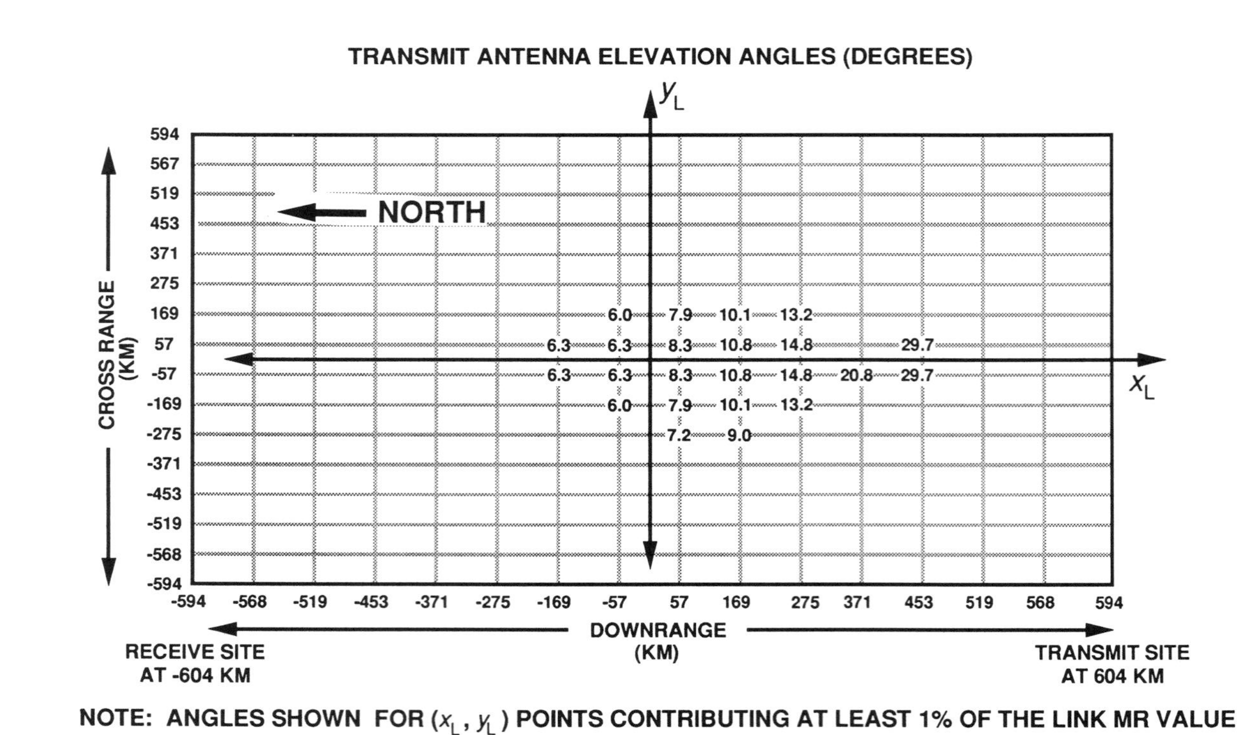

NOTE: ANGLES SHOWN FOR (x_L, y_L) POINTS CONTRIBUTING AT LEAST 1% OF THE LINK MR VALUE

(a)

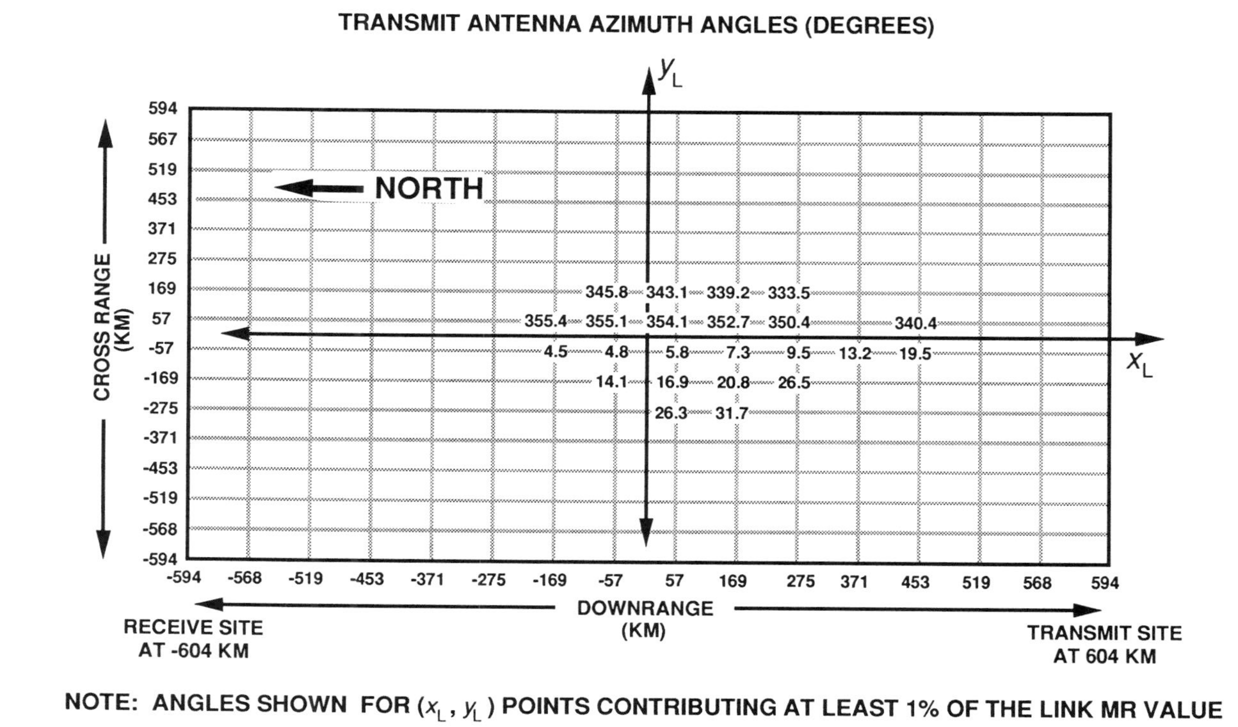

(*b*)

Figure 3.32. LINK grid of usable ray take-off angles: (*a*) elevation angles; (*b*) azimuth angles.

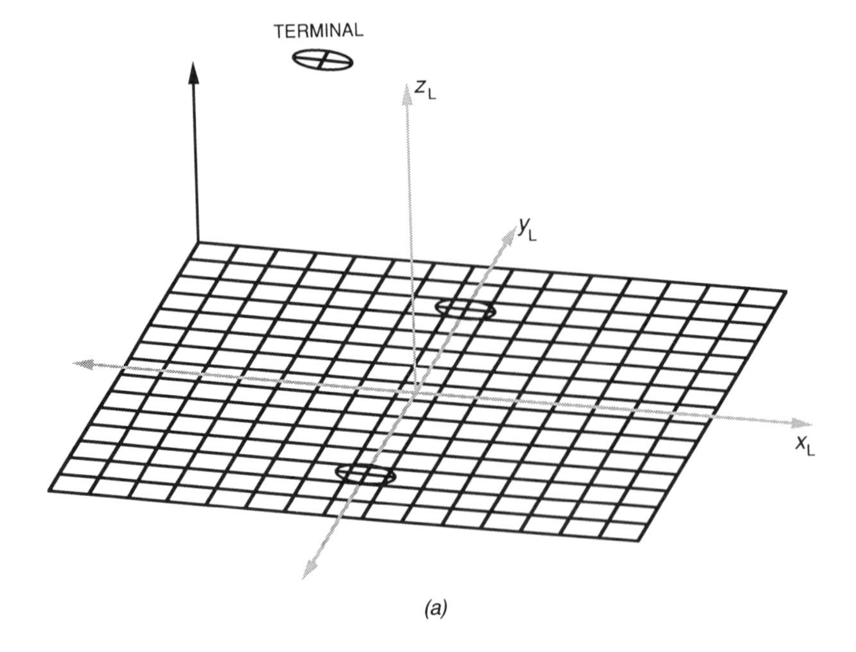

(a)

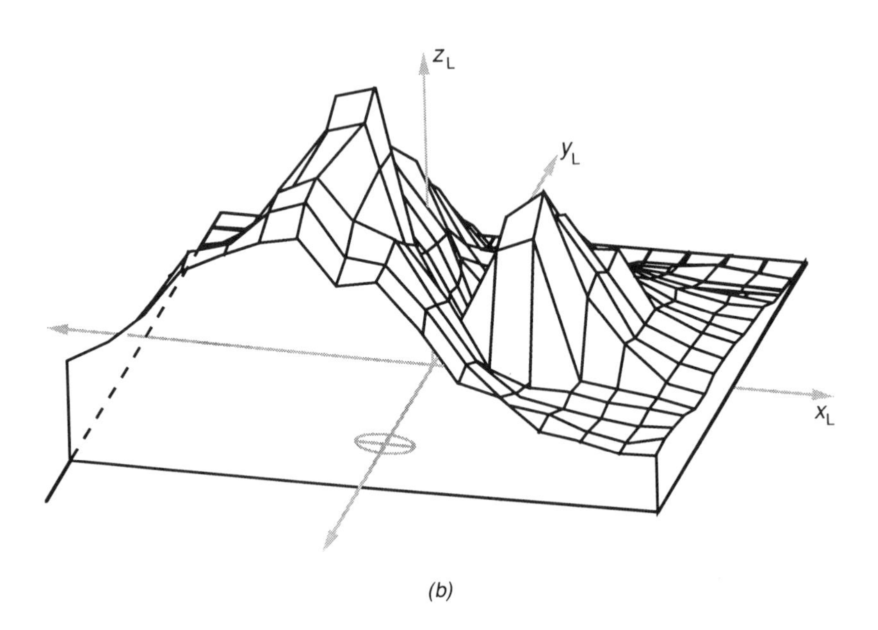

(b)

Figure 3.33. LINK MR grid displayed as a 3D image: (*a*) plot conventions; (*b*) short range; (*c*) mid-range; (*d*) long range.

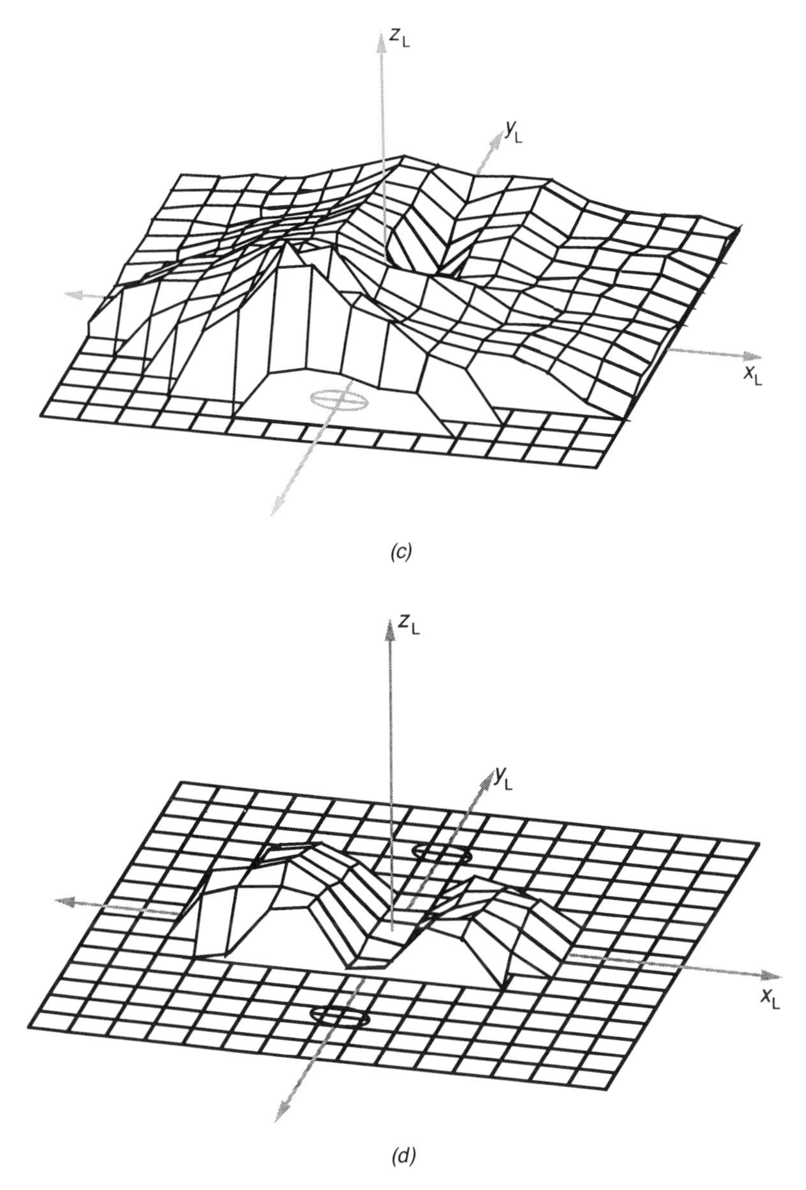

Figure 3.33. (*Continued*)

tions along the GCP between transmit and receive sites. This null zone occurs because the trail zenith angle $\chi_{sm} = 90°$, so the mass required to produce adequate trail line density from Eq. 3.15 becomes infinite. This mathematical result models the effect that a meteor with trajectory perpendicular to the zenith must pass through the maximum amount of atmosphere before reaching the trail-scatter point, $\mathscr{P}_s^L$, along the link GCP. In other words, the pre-burn meteor mass required to produce adequate trail line density for meteor trajectories tangent to the earth's surface is so large that their occurrence rate is negligible. Nevertheless, such "earth-grazing" meteoroids have been observed in practice [139].

3.2.2.7 Applications

3.2.2.7.1 Sensitivity Analysis

3.2.2.7.1.1 Temporal Variation. The position and orientation of an MB link on the earth's surface determines link exposure to meteor flux from the SMRD map. This exposure changes with the earth's rotation, resulting in the well-known diurnal variation in MR and DC values. The amount of diurnal variation is dependent on the latitude of the meteoric hot spots. Thus, equatorial MB links should show the greatest diurnal MR and DC swings, and transpolar links should show little, if any, diurnal variation.

Diurnal variation is also affected by the seasonal variation in both the SMRD map and the earth's tilt. In March, the north pole is tilted away from the earth's apex direction and the south pole is tilted forward, while in September these positions are reversed [140]. As the forward tilt of the link's latitude increases, that is, as the earth's apex rises above the MB link, the link's exposure to larger SMRD values will also increase. This effect, combined with variation in the magnitude of the SMRD throughout the year, produces the classic seasonal variation in link MR and DC performance.

To illustrate these points, consider the temporal variation in MR for north–south 1200-km MB links at 45 MHz in February and July, two transpolar links (north and south along Greenwich longitude), two north–south links centered at 45° latitude (north and south, also 0° longitude), and a single north–south link centered about the celestial equator (0° longitude). The ordinate is the MR value in meteors per minute and the abscissa is Universal Time (UT), which in this case is identical to local time (LT) at the transmitter. Each link employed single five-element Yagi antennas at a 3/2 wavelength height above ground. Antenna boresights were directed along the GCP between transmit and receive sites. The transmitter output power (P_t) was 1 kW and the required received power (P_r^*). No other system losses were modeled. A minimum required burst duration (τ^*) of 100 ms was specified.

Bi-hourly MR predictions for the transpolar links are shown in Fig. 3.34*a*. The first apparent characteristic of the transpolar MR is the lack of significant diurnal variation, due to link exposure to the available SMRD throughout the day without the earth blockage produced at mid-latitudes. The greatest swing in MR values occurs for the south pole link in July, in which two peak rates are observed near 10 and 22 UT and the minima are near 2 and 14 UT. These peaks are also apparent at the south pole in

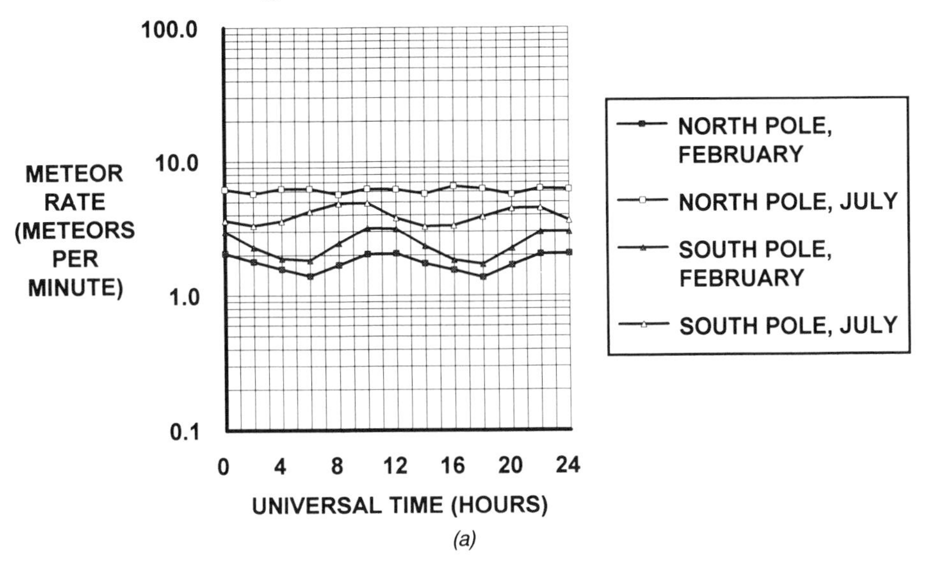

(a)

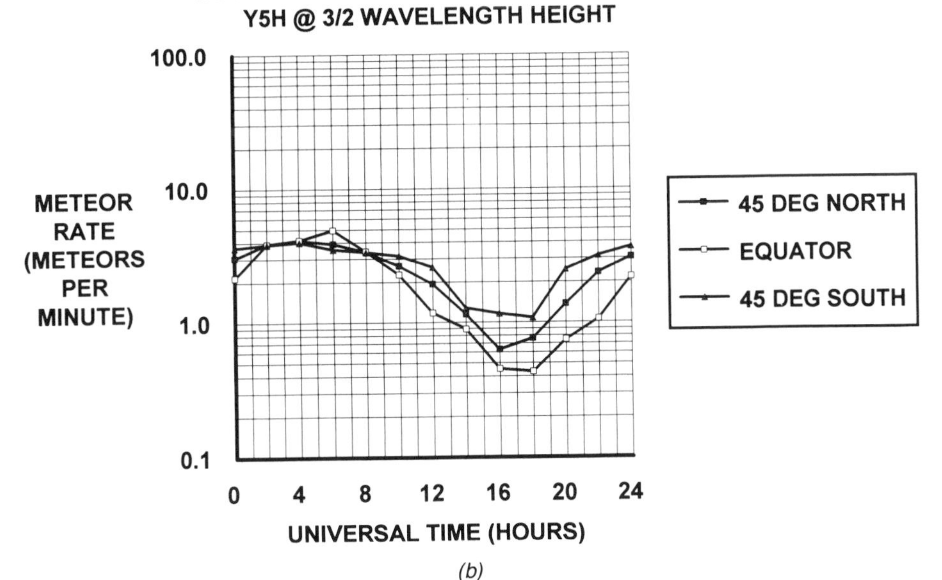

(b)

Figure 3.34. (*a*) Transpolar MB link performance in February and July; (*b*) mid-latitude MB link performance in February; (*c*) mid-latitude MB link performance in July.

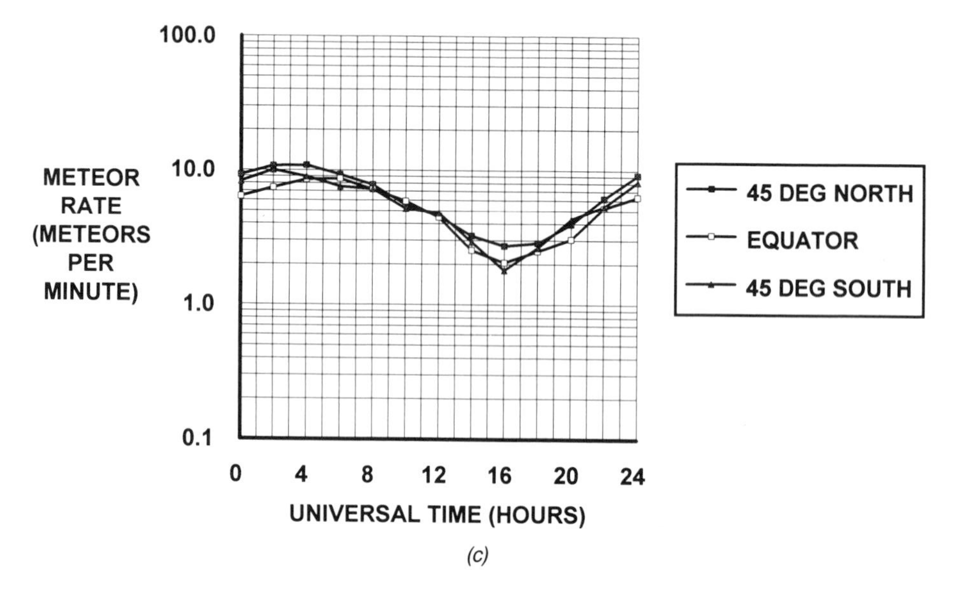

(c)

Figure 3.34. (*Continued*)

February, although the peaks have shifted to 0 and 12 UT. Note that the minimum MR values now occur not only at 18 UT, but also at 6 UT (6 AM LT). These peak MR values correspond to times when the link is approximately transverse to the direction of earth's motion and the greatest SMRD values are available. The minima occur when the link is "end-on" to the majority of incident meteor flux, thus creating a reduction in observable trail orientations.

The north transpolar link is tilted more toward the apex direction in July and shows about twice the average daily MR value as for the same link in February. The north link in July therefore shows greater MR values than the south pole link in July, principally because earth tilt favors the northern latitudes in July. The north pole link shows no significant diurnal variation, due to its nearly continuous exposure to observable trails throughout a July day. Note, however, that the south pole link does not produce greater MR values than the north pole link in February. This result is due in part to failure of the Mogadishu radar to observe extreme southern radiants in February, when the north pole is tilted away from the apex. This result can be verified by inspection of the southernmost SMRD cells in Fig. 3.19, showing several cells to be zero-valued.

Figures 3.34*b* and *c* show the classic mid-latitude diurnal variation in

February and July, respectively, with maximum MR values near 6 UT and minimum values near 18 UT. The greatest maximum-to-minimum ratio of MR values is apparent on the equatorial link in February (12-to-1). In July, the equatorial link shows approximately a 5-to-1 variation from 6 to 18 UT and all three links show similar MR values throughout the day. This result occurs because the equatorial link experiences greatest earth blockage of the largest SMRD values when it is positioned on the antapex side of the earth.

This analysis of temporal variation did not consider the observed variation in radar MR values with the solar cycle, although a "reference-type" model for this variation is included in METEORLINK. This variation achieves peak rates near the solar minimum, while the lowest rates occur near the solar maximum, with an observed swing of MR values by about a factor of two. The cause of this effect has been hypothesized to be the solar flux-induced change in the atmospheric density variation with height [141]. The higher solar flux expands the atmosphere and decreases its density in the height band of 80 to 120 km. Thus, the meteor encounters a longer path of lower particle density, which translates into an increased scale height H_P. At any trail-scatter point height h_s, the electron line density in the trail of a meteor of mass m moving with speed v decreases as H_P increases. This effect may be quantified for the height of maximum meteoric ionization, $\widehat{H}_I$, since H_P appears in the denominator on the right side of Eq. 3.12. This effect has been observed not only for time scales on the order of the solar cycle, but also during short-term enhancements to the incident solar flux [142].

3.2.2.7.1.2 Range, Power, and Burst Duration Effects. The effects of power and range variation on MB link performance can be significant, particularly for low-gain antennas at long range. The use of directional antennas complicates these effects because the effective meteoric common volume is determined by antenna directivity as well as blockage by the earth's bulge. To minimize pattern effects on range variation, consider the use omnidirectional antennas at both transmit and receive sites. As range increases, the maximum common volume available despite earth blockage is illuminated by both antennas and the distortion due to pattern effects is eliminated.

Figure 3.35 is a plot of link MR versus range, parametric in required received power, for an equatorial north–south link at local noon at the transmit site (12 UT). Transmit and receive antennas consisted of crossed dipoles mounted horizontally at a height of 3 m above a 6-m-radius ground screen. A transmit power (P_t) of 100 kW (50 dBW) was employed with minimum required received powers (P_r^* values) of −120, −110, and −100 dBm for minimum burst durations (τ^*) of 100 and 400 ms. This value of transmit power was required to compensate for the lower power gain of the crossed dipole antennas as compared to more directive antennas, such as five-element Yagis.

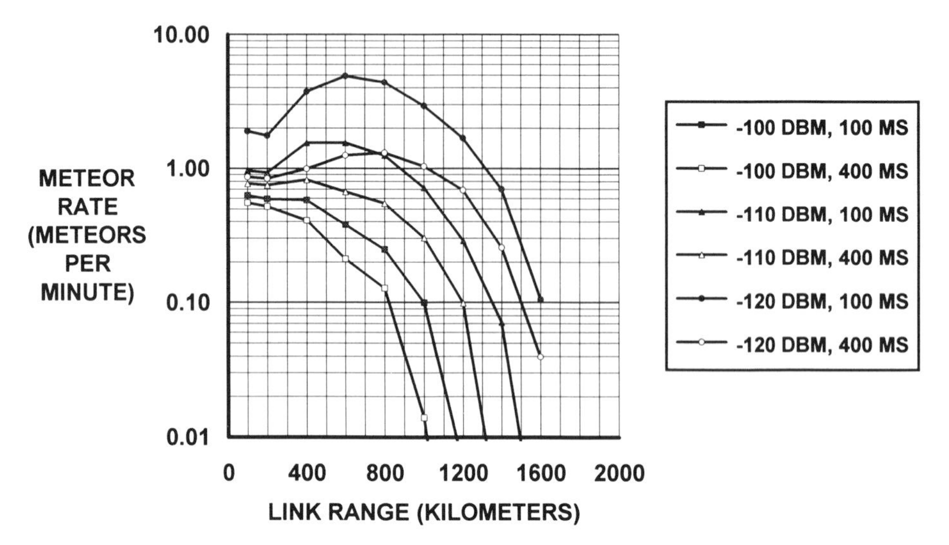

Figure 3.35. Equatorial MB link performance versus power, burst duration, and range.

As expected, the highest MR values are achieved for $P_r^* = -120$ dBm and $\tau^* = 100$ ms, while the lowest values correspond to $P_r^* = -100$ dBm and $\tau^* = 400$ ms. At short ranges, that is, below about 400 km, overdense trails are most capable of supporting both the smaller forward scattering angles, ϕ_s, and the higher required P_r^* values. These results show little MR variation with minimum burst duration τ^*. The greatest τ^* effects on the link MR value occur for -120 dBm at 400 km, where the 100-ms MR value is almost a factor of five times the corresponding 400-ms value. This result is due to the dominant effects of underdense trail-scattered signals, which show greater sensitivity to τ^* than overdense trail-scattered signals, used almost exclusively at the -100-dBm P_r^* value.

3.2.2.7.1.3 Antenna Type Versus Range. The results plotted in Fig. 3.35 were achieved using a high transmit power to compensate for the low antenna gain available with crossed dipole antennas. Consider instead the use of directional antennas designed to achieve similar performance with a lower powered transmitter. For example, Fig. 3.36 is a plot of link performance versus range parametric in the transmit–receive antenna configuration. Four horizontally polarized antenna combinations were used:

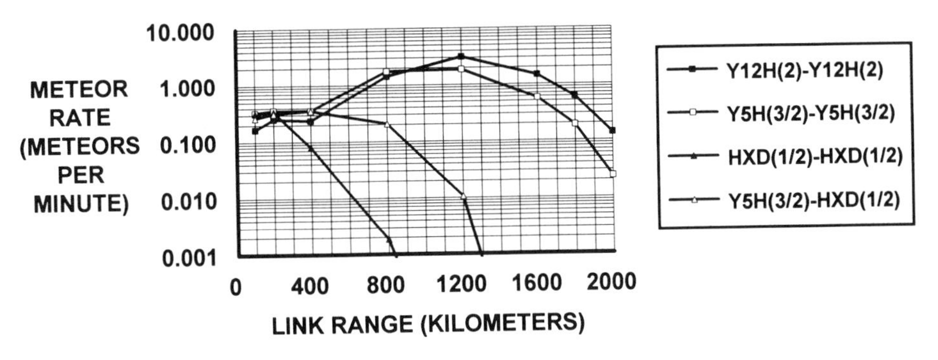

Figure 3.36. Antenna type versus range.

- Twelve-element Yagis, 3/2-wavelength height, 0° boom tilt, used for both transmit and receive (Y12H–Y12H)
- Five-element Yagis, 3/2-wavelength height, 0° boom tilt, used for both transmit and receive (Y5H–Y5H)
- The 3-m-height crossed dipole used in the plots of Fig. 3.35 for both transmit and receive (HXD–HXD)
- Five-element Yagis, 3/2-wavelength height, 0° boom tilt, for the transmit antenna and the 3-m height crossed dipole used in the plots of Fig. 3.35 for the receive antenna (Y5H–HXD)

All directional antennas were pointed along the link GCP between transmit and receive sites. A transmit power (P_t) of 1 kW, required received power (P_r^*) of -110 dBm, and a minimum burst duration (τ^*) of 100 ms were assumed for each link antenna combination.

Figure 3.36 shows that the Y5H-HXD and HXD-HXD antenna pairs yielded the highest MR values below about 400-km range, followed by the MR values determined for the Y5H-HXD and Y12H-Y12H combinations. This result occurred because the larger common volume illuminated by the broad-beamed combinations compensated for the corresponding lower power gain values. The MR values computed for the HXD-HXD combination diminished rapidly with range, because of the reduction in available common volume due to earth blockage combined with lower gain values. This effect was somewhat mitigated by replacing one of the HXD antennas with the higher-gain Y5H antenna to provide MR values for the Y5H-HXD antenna combination. The Y5H-Y5H combination yields the highest MR values between about 400 and 800 km. The Y12H-Y12H pair, with greater gain and greater height above ground than the Y5H-Y5H combination, provides a higher MR value at ranges of 1000 km and beyond. Even at 2000 km, the Y12H-Y12H pair yields MR values in excess of 0.1 meteors per minute.

The superior range performance of the directional antennas in this case was anticipated. However, this directionality may be inappropriate for some applications, such as mobile or network communications. Nevertheless, the significant performance advantage of directional gain is apparent and may be achieved by placing key sites at sufficient standoff distances so that directional antennas become feasible. Of course, a viable compromise between omnidirectionality and directive gain is an adaptive array of omnidirectional or broad-beamed elements [143].

3.2.2.7.2 Optimized Antenna Designs

3.2.2.7.2.1 MOTIVATION. MB link performance is highly dependent on the directive gain and polarization of the transmit and receive antenna patterns throughout the meteoric common volume. Increased antenna power gain, in the absence of adaptive beam-forming techniques [144], may degrade rather than improve performance. Degradation results because the increased gain is achieved with a concomitant reduction in main lobe beamwidth and, consequently, link common volume (see Fig. 3.37*a*, *b*). This decrease in common volume reduces the number of meteor trails available to support communications. On the contrary, illuminating sky region outside the common volume wastes transmitted power or receive sensitivity. These concepts suggest that "matching" the link common volume to the usable MB sky volume would maximize link performance [145].

The classical hot spots for long-range links are prominent on either side of the link GCP and are located approximately halfway between transmit and receive antennas [146]. At this range, a high-gain antenna is "matched" to the channel if it focuses all gain toward the meteoric hot spots. In fact, a movable, high-gain (narrow-beam) pattern designed to track these hot spots as they move from one side of the GCP to the other side has been considered [147]. At short range, an omnidirectional antenna may be expected to provide best performance by spreading available power gain over a much larger sky volume. The illumination pattern may differ, however, depending on whether link MR or DC values are to be maximized [148]. Although individual trails are illuminated by reduced signal strength, sufficiently more trails are illuminated that the net usable meteor rate is maximized. This tradeoff suggests that optimum patterns may be matched to the MB channel as a function of link range. These "matched" designs would be derived from predictions of the MR values, DC values, and antenna takeoff angles from the METEORLINK computer program. The resulting antenna patterns would be optimized either for single antennas or as the 'element' pattern for adaptive antennas arrays.

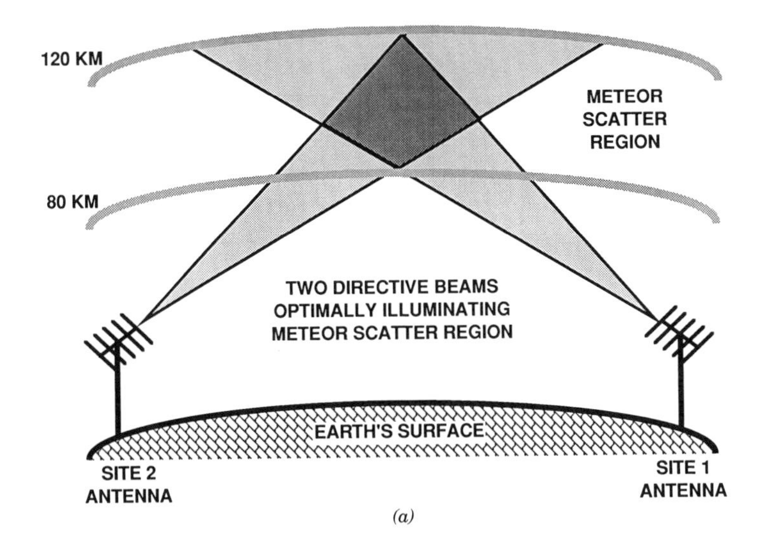

(a)

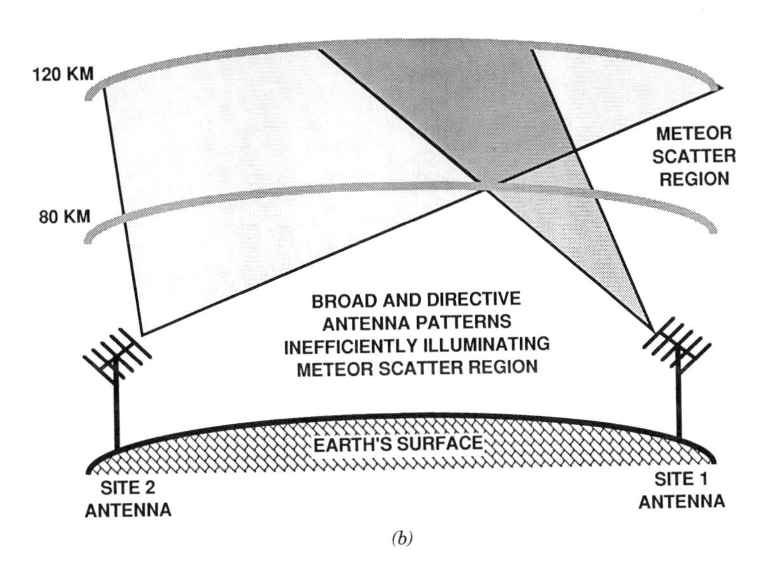

(b)

Figure 3.37. Antenna beamwidth versus common volume: (*a*) Directional antennas; (*b*) Directional and broadbeamed antennas.

3.2.2.7.2.2 Design Approach. The METEORLINK computer program determines the appropriate azimuth and elevation angles from each antenna to each quadrature point (x_i, y_j, z_k). The MR contribution $d^2\dot{N}_L$ from this point, that is, after α_M quadrature, can then be compiled into a two-dimensional probability density function as shown for a 200-km link in the relief plot of Fig. 3.38. This density weights each elevation–azimuth angle

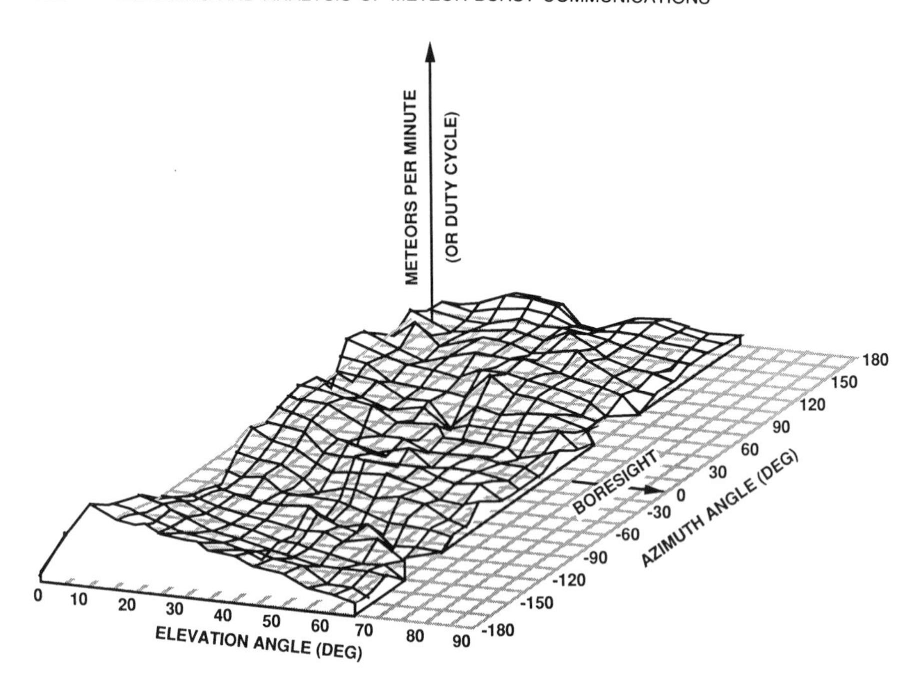

Figure 3.38. Contour plot of usable MR contributions versus omniazimthal antenna take-off angles for a 200-km link.

pair by the relative contribution to the link MR value from the corresponding point in the antenna pattern. For this example, assume that this density can be factored into separate elevation and azimuth angle densities.

Antenna elevation and azimuth angle densities were created from the MR contributions for omnidirectional antenna patterns (no ground effects) with perfect horizontal polarization ($\|\tilde{e}_{s2}^{P_j}\| = 1$, $\|\tilde{e}_{s3}^{P_j}\| = 0$), perfect vertical polarization ($\|\tilde{e}_{s2}^{P_j}\| = 0$, $\|\tilde{e}_{s3}^{P_j}\| = 1$), and perfect circular polarization

$$\left(\left\|\tilde{e}_{s2}^{P_j}\right\| = \frac{1}{\sqrt{2}}, \left\|\tilde{e}_{s3}^{P_j}\right\| = \frac{1}{\sqrt{2}}\right).$$

These densities were taken from METEORLINK predictions using these patterns made for link ranges of 200-, 1000- and 1800-km, east–west and north–south link bearings at mid-latitudes and local noon for a typical day in May. Plasma resonance, Faraday rotation, and atmospheric refraction effects were also modeled. In addition, all link predictions employed 10-kW transmit powers (P_t), -120-dBm required received power (P_r^*), and a 0-s minimum burst duration (τ^*), and an operating frequency (f_L) of 45 MHz. These values led to a predominance of underdense trail-scattered MR and DC contributions for most of the predicted results.

Table 3.2 gives the MR and DC predicted for all link ranges, bearings,

TABLE 3.2 MR and DC Values for Idealized Omnidirectional Antenna Patterns

		East-to-west			North-to-South		
		Polarization			Polarization		
Range (km)	MR[a] DC[b]	Horizontal	Vertical	Circular	Horizontal	Vertical	Circular
160	MR	5.8	6.6	2.3	8.5	8.4	2.3
	DC	2.1	2.2	0.5	2.4	2.4	0.4
1000	MR	4.8	7.0	5.6	4.9	6.3	5.9
	DC	3.0	3.9	3.8	3.3	4.2	3.4
1800	MR	1.9	2.0	1.8	1.7	1.8	2.0
	DC	1.8	2.0	1.8	1.7	1.9	2.0

[a]Meteors per minute.
[b]Duty cycle in percent time.

and antenna polarizations. The elevation and azimuth angle densities for the perfect horizontally polarized antenna and the north-south link configuration at each range are shown in Figs. 3.39*a* and *b*, respectively. The densities for the other antenna polarizations, link bearings, and times of day did not deviate from these values sufficiently to alter the design process, particularly at 1000 and 1800 km. The elevation angle densities were used to specify the required elevation of the antenna's main lobe and first null for each of the three link ranges and polarizations. The azimuthal densities were used to specify the number of Yagi antenna elements that could be used without excessively narrowing the azimuthal beam.

To demonstrate the effects of antenna height on pattern performance, optimum heights were determined for each link range (200, 1000, and 1800 km) using horizontally, vertically, and circularly polarized dipole antennas. Circular polarization was achieved with a crossed dipole oriented such that its boresight pointed at the horizon that is, its elements were in a plane perpendicular to the ground. Electric field patterns for these antennas at each height were computed with NEC. Ground effects were computed in NEC using the Sommerfeld–Norton approximation with medium dry ground parameters (conductivity $\sigma = 4 \times 10^{-3}$ Seimens/m, relative permittivity $\epsilon_r = 4.0$).

The relative MR performance of the dipole antennas at each height is plotted in Figs. 3.40*a–c* for ranges of 200, 1000, and 1800 km, respectively. The predicted MR and DC values were scaled by the corresponding values predicted for the horizontal dipole at 1/2 wavelength height for each range; thus, the normalized value for the horizontal dipole at 1/2 wavelength height at each range is unity. Values are provided for each dipole polarization, link bearing, and range. Figure 3.40 shows that the horizontal dipole provides the greatest relative MR at all ranges. At 200 and 1000 km, the optimum dipole height is 3/2 wavelengths. At 1800 km, the optimum dipole height moves to four wavelengths. The crossed dipole, whose pattern departs significantly from circular polarization over much of the common

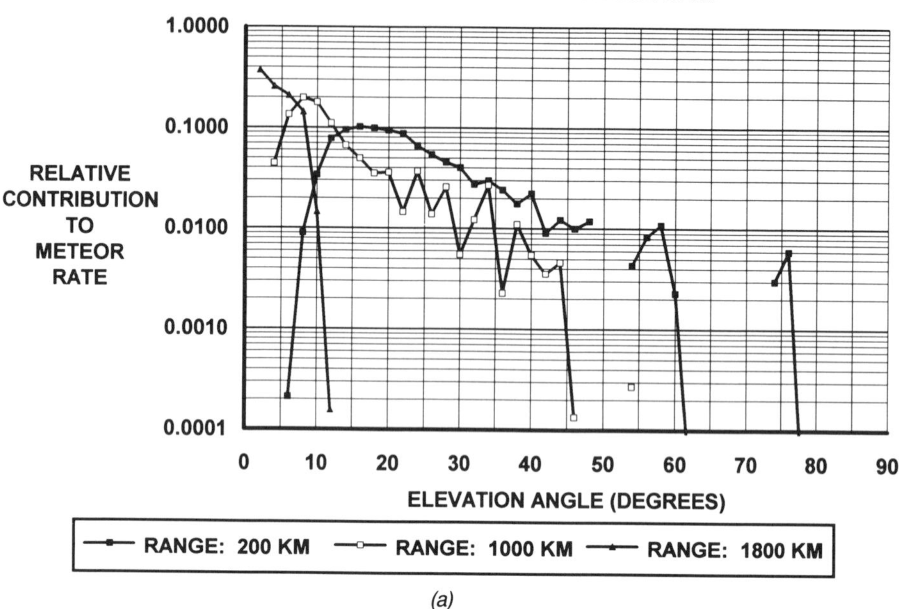

(a)

AZIMUTH ANGLE DENSITY VS LINK RANGE
NORTH-SOUTH LINK MAY @ 12 HRS
HORIZONTAL POLARIZATION

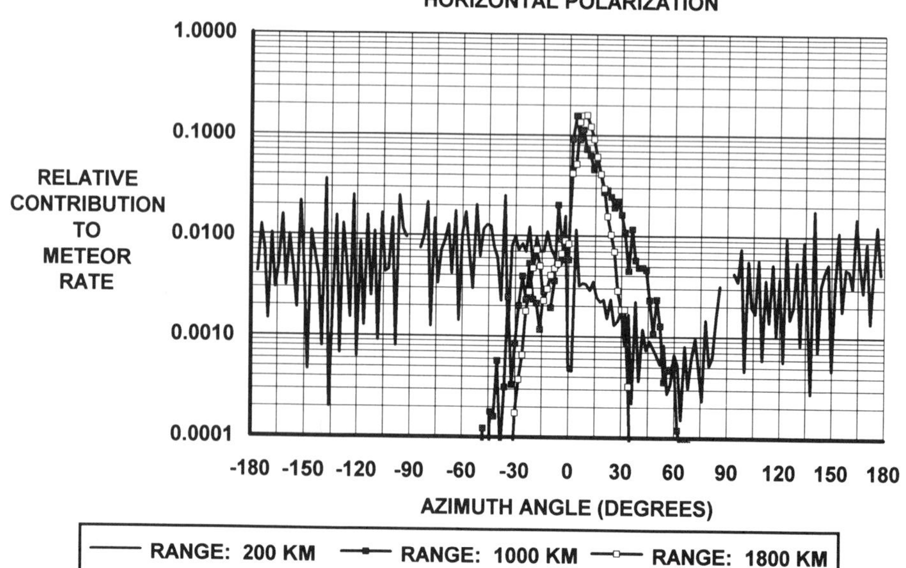

(b)

Figure 3.39. (*a*) Distribution of usable antenna elevation angles; (*b*) distribution of usable antenna azimuth angles.

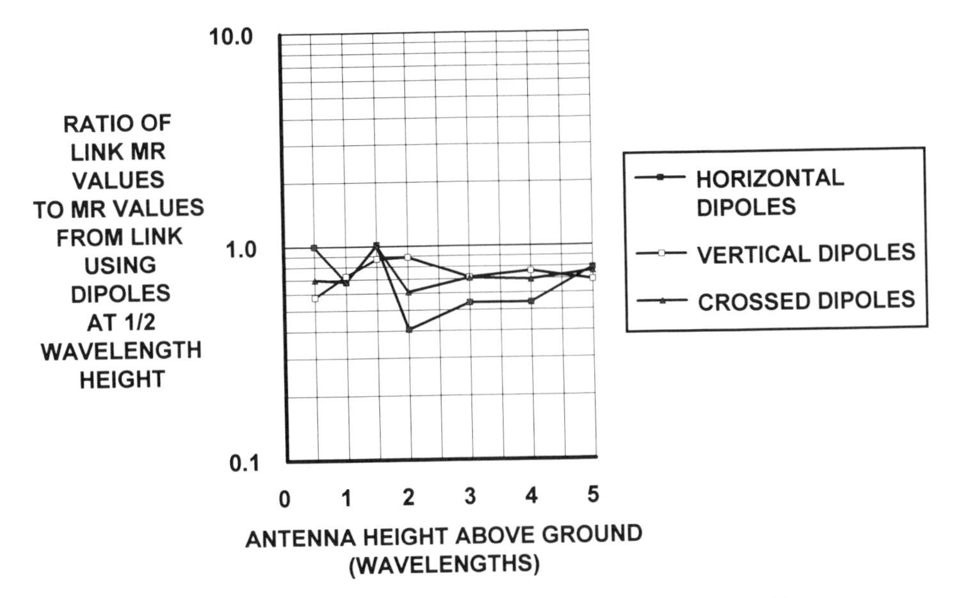

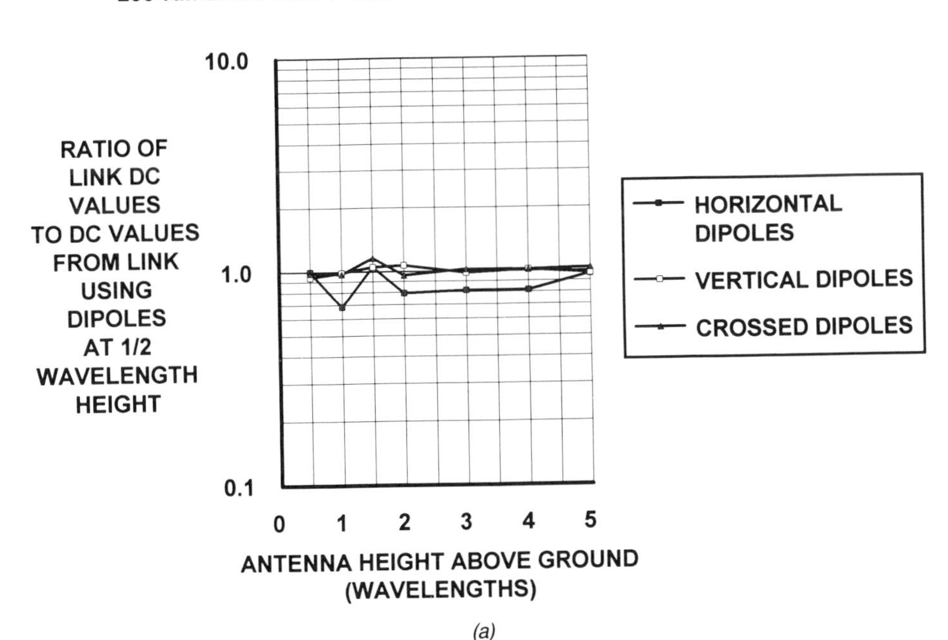

(a)

Figure 3.40. MR and DC versus dipole height: (*a*) 200-km range; (*b*) 1000-km range; (*c*) 1800-km range.

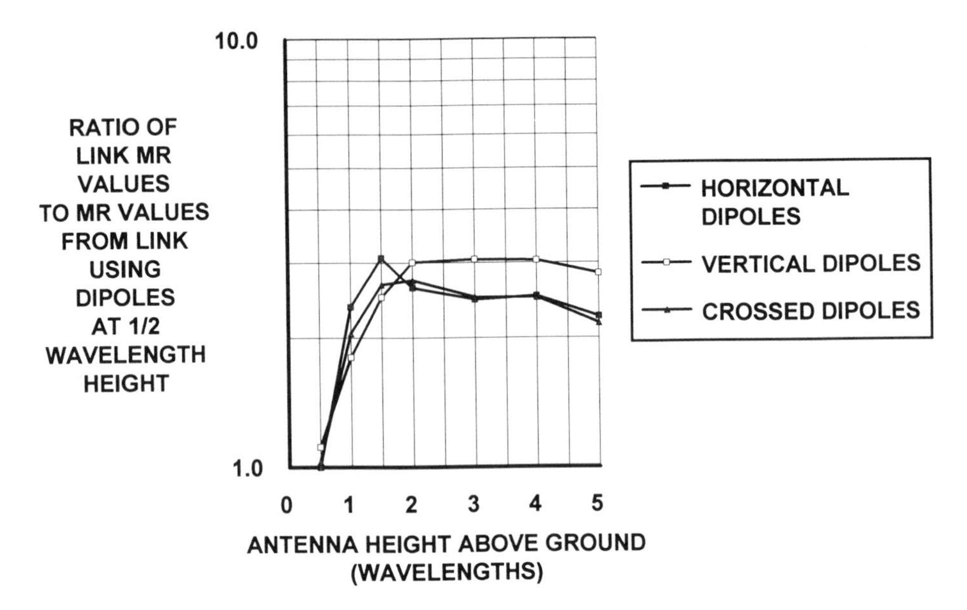

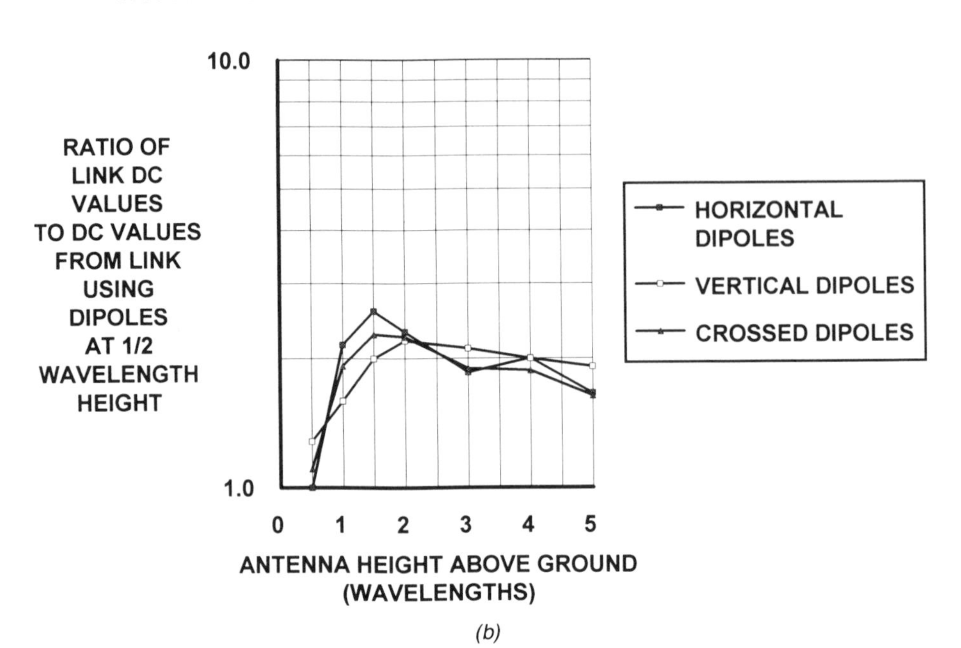

(b)

Figure 3.40. *(Continued)*

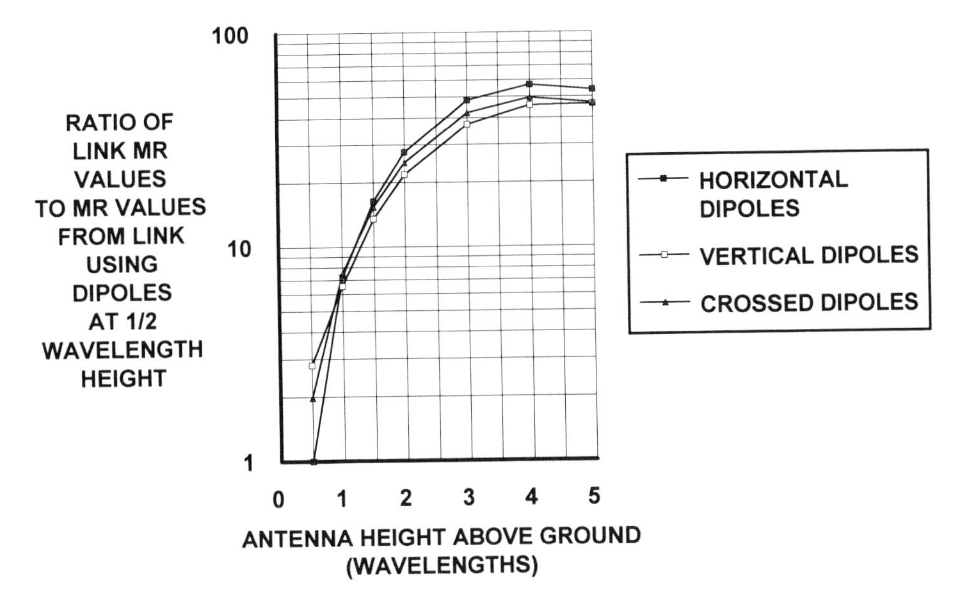

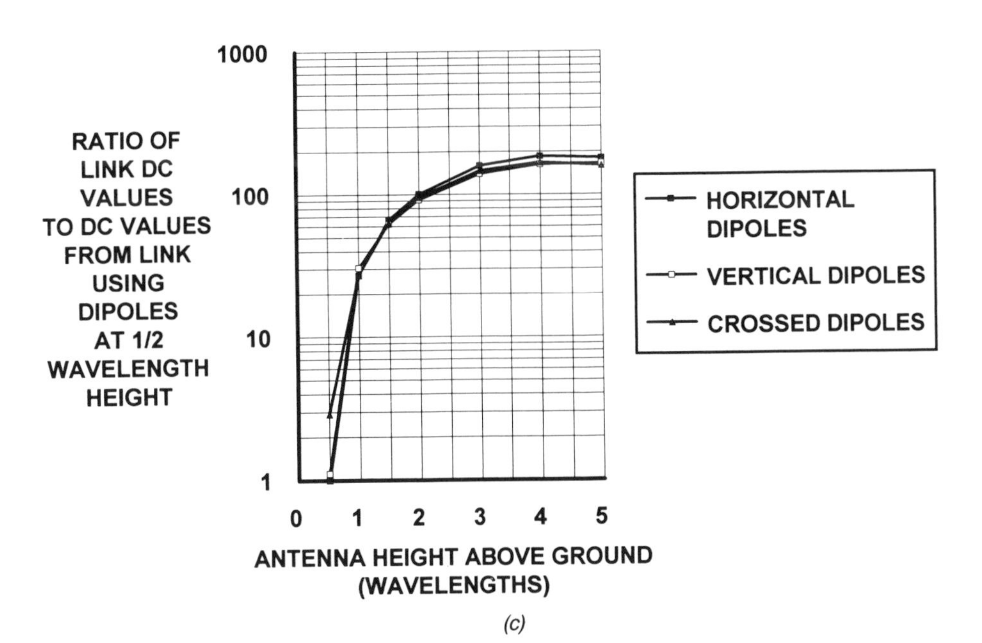

(c)

Figure 3.40. (*Continued*)

TABLE 3.3 MB Link Performance Versus Height and Range

Antenna Type	Polarization	Height wavelengths	Peak Gain (dBi)	Elevation of Peak Gain (°)	Average Gain (dBi)	3-dB Azimuthal Beamwidth (°)	Normalized MR value	Normalized DC value
				Range: 200 km				
Dipole	Horz[a]	1.5	7.8	10.0	1.5	70.0	1.00	1.00
	Vert[b]	2.0	4.8	6.0	1.1	360.0	0.87	1.02
	Circ[c]	1.5	6.2	8.0	1.3	110.0	0.99	1.10
3-el Yagi	Horz	1.5	14.9	10.0	1.7	50.0	1.16	1.00
	Vert	2.0	11.9	6.0	1.1	70.0	0.62	0.91
	Circ	0.5	10.5	20.0	1.1	50.0	0.71	0.88
5-el Yagi	Horz	0.5	13.3	20.0	1.5	45.0	1.00	0.92
	Vert	2.0	12.9	6.0	1.1	50.0	0.64	0.88
	Circ	1.5	14.1	8.0	1.4	45.0	0.88	0.97
12-el Yagi	Horz	—	—	—	—	—	—	—
	Vert	—	—	—	—	—	—	—
	Circ	—	—	—	—	—	—	—
				Range: 1000 km				
Dipole	Horz	1.5	7.8	10.0	1.5	70.0	1.00	1.00
	Vert	3.0	5.7	4.0	1.1	360.0	0.99	0.85
	Circ	2.0	6.5	6.0	1.3	110.0	0.88	0.88
3-el Yagi	Horz	1.5	14.9	10.0	1.7	50.0	4.40	4.77
	Vert	2.0	11.9	6.0	1.1	70.0	3.15	3.04
	Circ	1.5	13.3	8.0	1.1	50.0	3.32	3.53

5-el Yagi	Horz	1.5	15.7	10.0	1.6	45.0	4.66	5.34
	Vert	2.0	12.9	6.0	1.1	50.0	3.36	3.30
	Circ	1.5	14.1	8.0	1.4	45.0	3.55	3.90
12-el Yagi	Horz	1.5	17.8	8.0	1.7	35.0	6.77	7.94
	Vert	2.0	15.2	6.0	1.1	40.0	4.58	4.71
	Circ	—	—	—	—	—	—	—
				Range: 1800 km				
Dipole	Horz	4.0	7.9	4.0	1.5	70.0	1.00	1.00
	Vert	5.0	6.1	4.0	1.1	360.0	0.81	0.89
	Circ	4.0	7.0	4.0	1.3	110.0	0.87	0.90
3-el Yagi	Horz	—	—	—	—	—	—	—
	Vert	—	—	—	—	—	—	—
	Circ	—	—	—	—	—	—	—
5-el Yagi	Horz	4.0	16.1	4.0	1.6	45.0	6.11	6.03
	Vert	5.0	14.3	4.0	1.2	50.0	4.86	4.99
	Circ	4.0	15.2	4.0	1.4	45.0	5.00	5.07
12-el Yagi	Horz	4.0	18.5	4.0	1.7	35.0	9.10	9.62
	Vert	4.0	16.5	4.0	1.2	40.0	7.22	8.05
	Circ	4.0	17.6	4.0	1.4	35.0	7.27	7.81

[a]Horizontal polarization.
[b]Vertical polarization.
[c]Circular polarization.

volume, yields the best link DC value at 200 km. The horizontally polarized dipole gave the greatest DC values at 1000 and 1800 km. Note that the best height for the vertical dipole exceeds the best height of the horizontal dipole in all cases. This result is due to the 180° phase shift incurred by the horizontally polarized ground-reflected wave which is not experienced by the vertical reflected wave.

Based on inspection of the azimuth angle densities, antenna types selected for this analysis were the dipole, three- and five-element Yagi antennas for the 200-km link; three-, five-, and 12-element Yagis for the 1000-km link; and five- and 12-element Yagis for the 1800-km link. NEC models of these antenna types were configured for horizontal, vertical, and circular (crossed) polarization at several heights, including the optimum height found for the dipole antenna with the corresponding polarization.

The resulting NEC-generated electric field patterns were input to the METEORLINK program to predict MR and DC values versus antenna height and range on the east–west link at noon in May. The dipole antennas were used at all ranges to provide a common reference. Once it was verified that polarization trends matched the dipole case, circular polarization was not employed with the 12-element Yagi at the 1000-km range. These predictions showed that the optimum heights determined for the dipole results in Fig. 3.40 were, in fact, close to the optimum height for the corresponding Yagi antennas at each range.

3.2.2.7.2.3 Results. Table 3.3 gives the MR and DC values for all antenna types, heights, and link ranges normalized by the corresponding values predicted for the horizontal dipole at its optimum height. In addition, the table gives the antenna height for best performance, maximum power gain, elevation angle for peak gain, average power gain, and the 3-dB beamwidth for each antenna. The average power gain was included to provide a measure of the antenna's susceptibility to interference. In an omnidirectional noise field, for example, the antenna with the highest average power gain would produce the highest received noise signal at its terminals.

The horizontally polarized antennas provided the greatest MR at all link ranges irrespective of the antenna type, indicating best performance for short, or single-burst, messages. This result was significant only at the 1000-km range, in which the horizontally polarized antennas provided a 30% improvement in MR and about a 40% improvement in DC over the vertically polarized antennas. This result is due to the greater power gain of the horizontally polarized antennas despite their narrower azimuthal beamwidths. The greater gain of the horizontally polarized antennas is due to the larger amplitude predicted for the horizontally polarized (tangential) ground-reflected signal than predicted for the vertically polarized reflection. The greater azimuthal beamwidth of the vertically polarized antennas somewhat mitigated this effect by illuminating more common volume at the

200-km range, where the MR and DC values for all modeled polarizations were more nearly equal than at the longer ranges.

The horizontal three-element Yagi at 3/2 wavelengths height (6.7 m at 45 MHz) showed the best MR performance at 200 km. The five-element Yagi at 1/2 wavelength height was second best at 200 km, probably because the relatively low height directed the main beam upward and illuminated the usable MB sky volume. The 12-element Yagi (3/2 wavelength height) performed best at 1000 km, yielding almost 50% higher MR values than the five-element Yagi. Apparently, the gain/azimuthal beamwidth tradeoff favors the 12-element Yagi over the five-element Yagi at 1000 km. As expected, the 12-element Yagi at four wavelengths (28 m) height yielded maximum MR values at the 1800-km range. The performance penalty associated with use of these antennas at nonoptimum antenna heights can be estimated from Fig. 3.40.

As in Fig. 3.40, Table 3.3 shows that the crossed dipole provided greatest DC value at 200 km, while the horizontal antennas, particularly the 12-element Yagi, yielded maximum DC values at 1000 and 1800 km. Note, however, that the performance advantage of horizontal over vertical or circular polarization at the optimum heights does not take into account difference in received noise levels. Since the average power gains of the horizontal antennas were also greater than for the other polarizations, its relative SNR performance would be diminished in an omnidirectional noise field.

These antenna design results show that antenna pattern shape as well as maximum gain combine to determine link performance at each range. In fact, a retrodirective array [149] designed to focus its main beam in the direction of each link-usable trail may provide an optimum MB antenna design. This array consists of M_t transmit antennas or M_r receive antennas which may be independently phased to obtain maximize sensitivity in a desired direction. One approach to focus the retrodirective array employs a *narrowband* pilot signal to phase each antenna element in the array for maximum gain in the direction of the trail-scatter point, $\mathcal{P}_s^{\mathrm{L}}$, within a minimum time, τ^*, after trail formation. The narrowband pilot signal permits detection, and array focusing, on more trail-scattered signals at the detection threshold, P_d^*, than would be detected at the higher receiver threshold $P_r^* > P_d^*$, required for the wider signaling bandwidth needed for effective MB communications. The receive antenna for the pilot signal would necessarily have the same pattern shape and pointing angles as each element in the retrodirective array.

Before trail detection, the maximum path loss supported by the pilot link is given by

$$L_{\mathrm{pilot}}^* = P_t + G_{sm}^{12} - L_t - L_r - P_d^* + 30$$

which determines the pilot link detected MR value, $\dot{N}_L^d$. Once a trail-scattered signal is detected on the pilot link, the transmit and receive arrays focus peak gain at the trail-scatter point $\mathscr{P}_s^{\mathrm{L}}$. At this point, the MR value, $\dot{N}_L^{\mathrm{array}}$, for the communications link is determined from the maximum path loss value

$$L^*_{\mathrm{array}} = \log(2M_tM_r) + P_t + G^{12}_{sm} - L_t - L_r - P^*_r + 30$$

where the gain product from one transmit and receive antenna pair at $\mathscr{P}_s^{\mathrm{L}}$ is G^{12}_{sm}. The factor of two in the logarithm is required because an additional transmitter with output power P_t is assumed to be added for each additional transmit antenna. It was also assumed that the antenna elements in each array are spaced adequately so coupling losses between array elements are negligible and the maximum possible array gain is achieved.

Thus, the difference between the maximum path loss values for the pilot link and the communications link is then given by

$$L^*_{\mathrm{pilot}} - L^*_{\mathrm{array}} = \log(2M_tM_r) + P^*_r - P^*_d \;.$$

If $L^*_{\mathrm{pilot}} > L^*_{\mathrm{array}}$, then $\dot{N}_L^{\mathrm{pilot}} > \dot{N}_L^{\mathrm{array}}$, and trail-scattered signals detected by the pilot link could not support the communications link. If $L^*_{\mathrm{pilot}} < L^*_{\mathrm{array}}$, then $\dot{N}_L^{\mathrm{pilot}} < \dot{N}_L^{\mathrm{array}}$, and trail-scattered signals usable for communications are not detected by the pilot link. Therefore, $L^*_{\mathrm{pilot}} \cong L^*_{\mathrm{array}}$ provides the most efficient balance between pilot and communications link performance. Thus the number of antennas in the transmit and receive arrays should be chosen such that

$$L^*_{\mathrm{pilot}} = L^*_{\mathrm{array}} \Rightarrow \log(2M_tM_r) = P^*_r - P^*_d$$

so the detected MR value $\dot{N}_L^{\mathrm{pilot}} = \dot{N}_L^{\mathrm{array}}$.

These results are demonstrated in Figs. 3.41*a-d*, which show METEORLINK-predicted MR and DC values plotted versus the number of antennas in the transmit and receive arrays. In this example, horizontal crossed dipole (HXD) antennas were used in both transmit and receive arrays for link ranges of 350 and 1200 km. A transmit power $P_t = 20$ dBW (100 W) was assumed at the input of each antenna element in the transmit array and a required received power $P^*_r = -115$ dBm was required at the communications link receiver. A detection threshold $P^*_d = -132$ dBm was specified for the *narrowband* pilot signal. No other system losses were modeled. The resulting MR and DC values are plotted in Figs 3.41*a* and *b*, respectively, for the 350-km link range. Both plots show that the incremental MR improvement decreases rapidly as the number of antenna elements used in the array is increased. This result occurs because each 3-dB increase in L^*_{array} requires a doubling of the number of antennas in the transmit or receive arrays. More importantly, no MR improvement is apparent when $\log(2M_tM_r) > P^*_r - P^*_d = 17$ dB. This result occurs because the additional antennas increase L^*_{array} but P^*_d is fixed, so L^*_{pilot} is fixed and $L^*_{\mathrm{pilot}} < L^*_{\mathrm{array}}$.

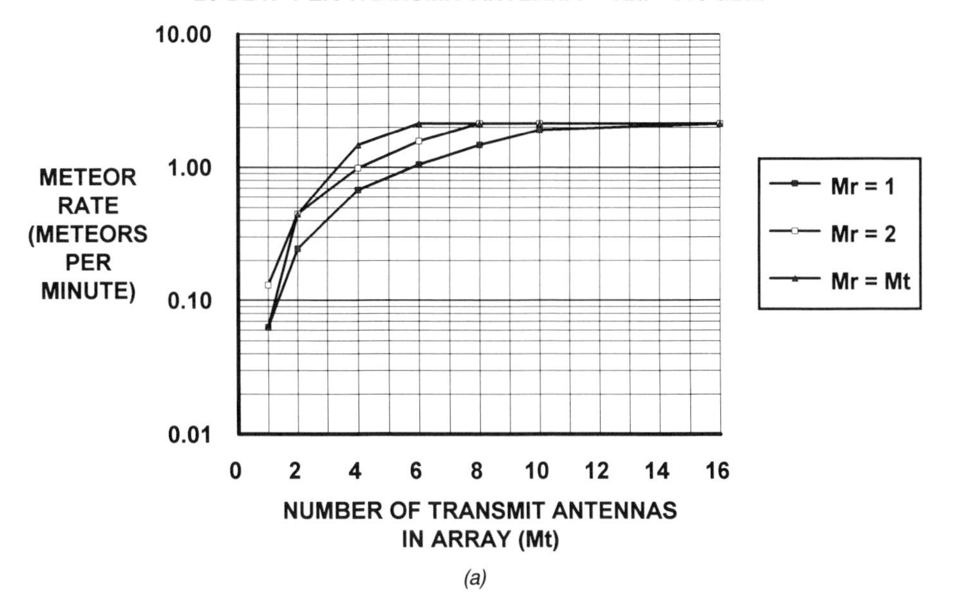

(a)

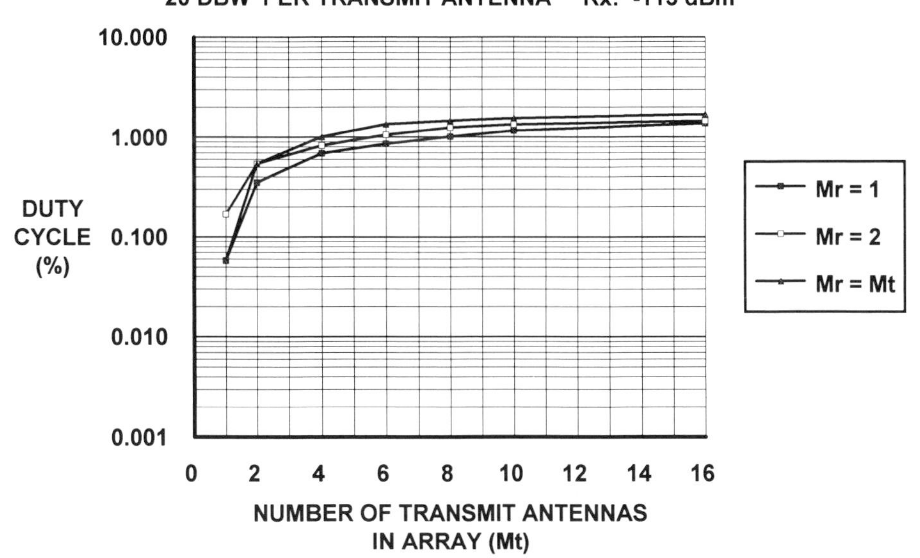

(b)

Figure 3.41. (*a*) MR performance of a retrodirective array at 350 km; (*b*) DC performance of a retrodirective array at 350 km; (*c*) MR performance of a retrodirective array at 1200 km; (*d*) DC performance of a retrodirective array at 1200 km.

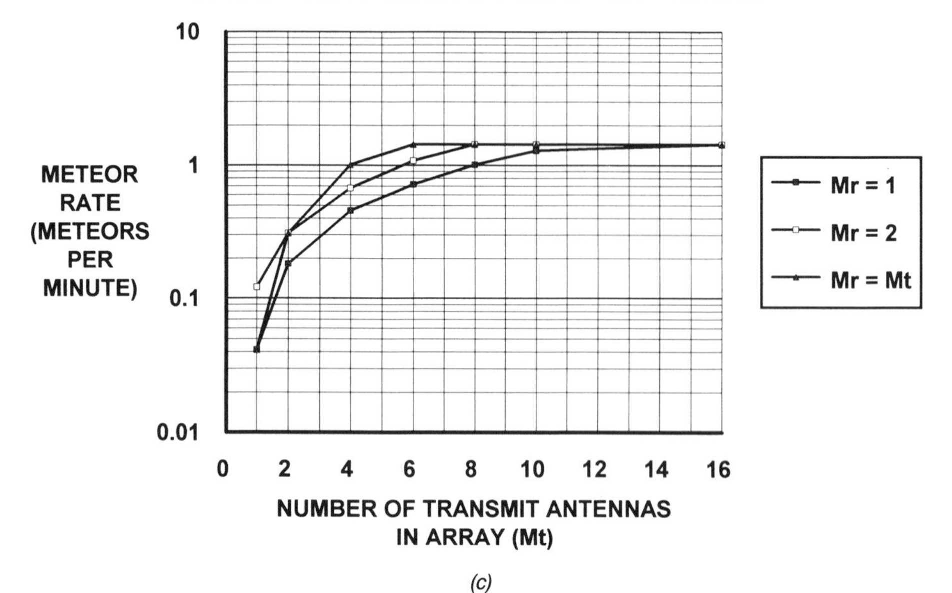

(c)

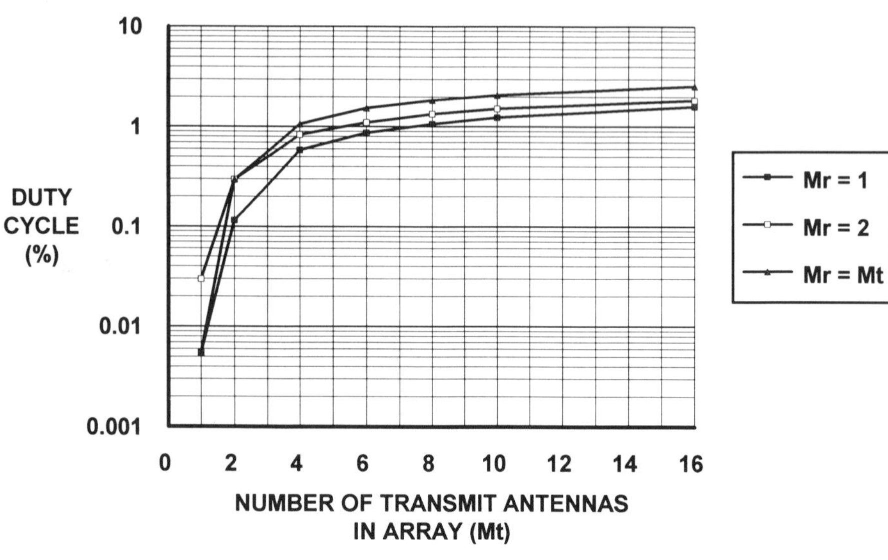

(d)

Figure 3.41. (*Continued*)

These results are mimicked at the 1200-km range in Figs. 3.41*c* and *d*, although the DC improvement increases faster with the number of antennas in the array than at 350-km range due to the increased forward scattering angle characteristic of the longer-link range.

This example shows how a physical model, such as METEORLINK, can be used as part of an antenna design technique to optimize performance of MB communications links. The results indicate that horizontal polarization provides best overall performance at lower required antenna heights than other polarizations. At short ranges, performance is not significantly affected by polarization or directivity, as long as directional antennas are placed within 3/2 wavelengths of the ground. Horizontal polarization performs significantly better than other polarizations at medium range for antennas with equivalent free space power gains and beamwidths. In addition, employing Yagi antennas with as many as 12 elements can significantly out-perform the five-element Yagi irrespective of polarization with a concomitant increase in antenna size and weight. For path lengths approaching the MB-range limit, horizontal polarization performs best although antenna height, and not polarization, is the dominant mechanism. The results also show that the difference in link MR values between horizontal and vertical polarizations decreases at the maximum ranges. Since low take-off angles may be more easily achieved with vertical polarization combined with a suitable ground screen, vertical polarization may provide better link performance near the MB-range limit.

These results ignore the effects of terrain blockage and irregular foreground. Path-interposed obstructions can obscure the link common volume, particularly for MB links approaching the range limit. In addition, irregular foreground terrain can change the arrival angles of ground-reflected rays and distort the antenna pattern. In practice, obstruction of the radio LOS is modeled as a minimum effective elevation angle in the antenna pattern, while foreground terrain effects are included in the calculation of ground reflection coefficients.

The results presented in this section are intended to illustrate use of a physical MB model in analysis and provide representative results. A more complete treatment would investigate other antenna types, including arrayed versions of the antennas employed in this study and links with differing transmit and receive antenna configurations. Study results would be extended to marginal MB links for which the MR value is provided primarily by overdense trails. This treatment should also consider temporal variation of the MR value as well as the received noise level, which is essential in actual MB link design efforts. The performance of adaptive arrays, particularly useful for short to medium range point-to-point links and mobile applications, should be more extensively evaluated. Additional adaptive beam-forming techniques, such as dynamic arrays that track the meteoric hot spots with time of day or multiple fixed-beam (or split-beam) techniques [150] that simultaneously monitor all hot-spots, should also be considered. This latter technique avoids beam focusing delays and makes the maximum use of the available link MR and DC values while avoiding the need for variable phase-adjustment circuitry. A successful practical application of the multiple-beam technique has been reported [151].

3.2.3 Footprint Model

***3.2.3.1* Overview.** The LPI feature of MB communications [152] derives from the low probability of detection (LPD), which results from the limited geographic extent of the footprint of meteor trail-scattered signals. In addition, the intermittent and brief channel openings associated with the short-lived ionized trail phenomenon as well as the typically weak RSL values are believed to further enhance this feature. However, the conditions under which a MB communication system can achieve LPD must be stated to avoid misrepresenting this capability.

Practical MB communication links are susceptible to anomalous propagation mechanisms, including sporadic-E, ducting, and ionospheric skywave. Sporadic-E and ducted propagation can provide significant RSL values over geographic areas much larger than the typical meteor scatter footprint, persist for many tens of minutes, and occur without warning [153]. Furthermore, frequencies well above 30 MHz will be useful for skywave propagation during peaks in the 11-year solar cycle. During these times, MB links may be enhanced by skywave modes and extend propagation range to thousands of kilometers. A strong, continuous RSL, combined with the expanded signal footprint, may compromise LPI for MB communications.

In general, meteor trail-scattered signals produce short-lived footprints whose shapes vary greatly with meteor trail orientation and ionization. However, infrequent large meteors will produce trails with dense, enduring ionization that can produce viable received signals for hundreds of seconds [154]. These trails are ultimately distorted by upper atmospheric winds, resulting in a significant decrease in aspect sensitivity and an expanded signal footprint, after about 10 s [155]. In practice, however, the trail footprints (see Section 3.1.1.3.3) from the more frequent specular trail-scattered signals are of more interest for most model applications. From an LPD perspective, an important measure of LPI susceptibility is the probability that at least one receiver in a receiver array will detect a signal from a meteor trail given that the intended, or primary, receiver detected a signal from the same trail.

3.2.3.2 Footprint Calculation. A MB trail footprint model METEORTRAK was developed as an adaptation of the METEORLINK computer model to compute the conditional probability that a meteor trail will produce a usable scattered signal to one or more peripheral receivers, given that it produced a usable signal at an intended, or primary, receiver. Receptions at the primary and peripheral receivers are not necessarily simultaneous. This conditional probability is the ratio of the MR value computed for the link between the transmitter and a peripheral receiver *to* the MR value computed for the link to the primary receiver. The term "conditional" describes the use of the observability criteria $(\mathscr{P}_s^{\mathrm{L}}, \dot{\mathbf{v}}_m^{\mathrm{L}})$ derived from the locations of the transmitter and *primary* receiver to determine the trail-scatter points $\mathscr{P}_s^{\mathrm{L}}$ and trail orientations $\hat{\mathbf{v}}_m^{\mathrm{L}}$ for the MB links between the transmitter and each of the *peripheral* receivers.

Let the transmitter be located at the point $\mathscr{T}_0^{\mathrm{L}}$, the primary receiver located at the point $\mathscr{R}_0^{\mathrm{L}}$, and the peripheral receivers positioned at the points $\mathscr{R}_i^{\mathrm{L}}$, $i = 1, 2, \ldots, M_r$, in the L plane shown in Fig. 3.42. As shown in the figure, potential trail-scatter points $\mathscr{P}_{s_0}^{\mathrm{L}}$ are selected for the $\mathscr{T}_0^{\mathrm{L}} - \mathscr{R}_0^{\mathrm{L}}$ link within the meteor burn later between 80 and 120 km above the earth's surface. Next, meteor trails are selected according to the observability criteria $(\mathscr{P}_{s_0}^{\mathrm{L}}, \hat{\mathbf{v}}_m^{\mathrm{L}})$ imposed by the $\mathscr{T}_0^{\mathrm{L}} - \mathscr{R}_0^{\mathrm{L}}$ link geometry. METEORTRAK considers each of the observable orientations $\hat{\mathbf{v}}_m^{\mathrm{L}}$ through the point $\mathscr{P}_{s_0}^{\mathrm{L}}$ to determine if the corresponding meteor trail would meet the observability criteria $(\mathscr{P}_{s_i}^{\mathrm{L}}, \hat{\mathbf{v}}_m^{\mathrm{L}})$ for all MB links, $\mathscr{T}_0^{\mathrm{L}} - \mathscr{R}_i^{\mathrm{L}}$, $i = 1, 2, \ldots, M_r$. In other words, the algorithm searches for points $\mathscr{P}_{s_i}^{\mathrm{L}}$, with trail unit vector $\hat{\mathbf{v}}_m^{\mathrm{L}}$ providing coherent trail-scatter on the $\mathscr{T}_0^{\mathrm{L}} - \mathscr{R}_i^{\mathrm{L}}$ link, $i = 1, 2, \ldots, M_r$. If the height, h_{s_k}, of the point $\mathscr{P}_{s_k}^{\mathrm{L}}$ for some $i = k$ does not lie between 80 and 120 km above sea level, then meteors with atmospheric trajectories defined by $\hat{\mathbf{v}}_m^{\mathrm{L}}$ passing through the point $\mathscr{P}_{s_k}^{\mathrm{L}}$ are not observable on the $\mathscr{T}_0^{\mathrm{L}} - \mathscr{R}_k^{\mathrm{L}}$ MB link.

The principal calculation of the METEORTRAK model is the determination of scatter points along meteor trails meeting both the observability

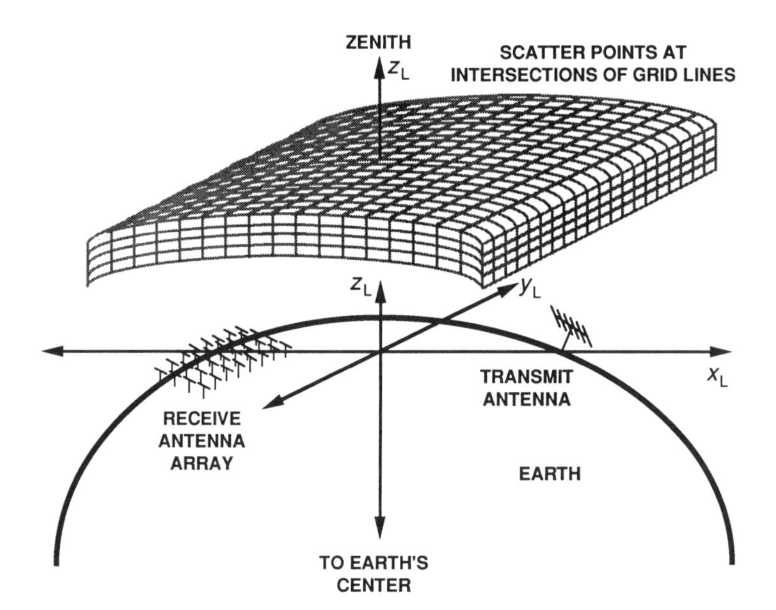

Figure 3.42. LINK grid showing receiver array for trail footprint prediction.

criteria $(\mathscr{P}^{\mathrm{L}}_{s_0}, \hat{\mathbf{v}}^{\mathrm{L}}_m)$ and $(\mathscr{P}^{\mathrm{L}}_{s_i}, \hat{\mathbf{v}}^{\mathrm{L}}_m)$ for each receiver location $\mathscr{R}^{\mathrm{L}}_i$, $i = 1, 2, \ldots, M_r$. With this objective, let the coordinates of the transmit antenna be given by $\mathscr{T}^{\mathrm{L}}_0 = (x^{\mathrm{L}}_{t_0}, y^{\mathrm{L}}_{t_0}, z^{\mathrm{L}}_{t_0})$, the primary receiver at $\mathscr{R}^{\mathrm{L}}_0 = (x^{\mathrm{L}}_{r_0}, y^{\mathrm{L}}_{r_0}, z^{\mathrm{L}}_{r_0})$, and the peripheral receivers at $\mathscr{R}^{\mathrm{L}}_i = (x^{\mathrm{L}}_{r_i}, y^{\mathrm{L}}_{r_i}, z^{\mathrm{L}}_{r_i})$, for $i = 1, 2, \ldots, M_r$. Let $\hat{\mathbf{v}}^{\mathrm{L}}_m = \{\hat{v}^{\mathrm{L}}_{m1}, \hat{v}^{\mathrm{L}}_{m2}, \hat{v}^{\mathrm{L}}_{m3}\rangle$ define a specular trail orientation on the $\mathscr{T}^{\mathrm{L}}_0 - \mathscr{R}^{\mathrm{L}}_0$ link passing through the L-system point $\mathscr{P}^{\mathrm{L}}_{s_0} = (x^{\mathrm{L}}_{s_0}, y^{\mathrm{L}}_{s_0}, z^{\mathrm{L}}_{s_0})$ in the meteoric ionization layer. Given these definitions, compute the parameters

$$
\begin{aligned}
\Delta x^{\mathrm{L}}_{t_0} &= x^{\mathrm{L}}_{t_0} - x^{\mathrm{L}}_{s_0} \quad \Delta x^{\mathrm{L}}_{r_i} = x^{\mathrm{L}}_{r_i} - x^{\mathrm{L}}_{s_0} \\
\Delta y^{\mathrm{L}}_{t_0} &= y^{\mathrm{L}}_{t_0} - y^{\mathrm{L}}_{s_0} \quad \Delta y^{\mathrm{L}}_{r_i} = y^{\mathrm{L}}_{r_i} - y^{\mathrm{L}}_{s_0} \\
\Delta z^{\mathrm{L}}_{t_0} &= z^{\mathrm{L}}_{t_0} - z^{\mathrm{L}}_{s_0} \quad \Delta z^{\mathrm{L}}_{r_i} = z^{\mathrm{L}}_{r_i} - z^{\mathrm{L}}_{s_0} \\
\alpha_{t_0} &= \hat{v}^{\mathrm{L}}_{m1}\, \Delta x^{\mathrm{L}}_{t_0} + \hat{v}^{\mathrm{L}}_{m2}\, \Delta y_{t_0} + \hat{v}^{\mathrm{L}}_{m3}\, \Delta z^{\mathrm{L}}_{t_0} \\
\alpha_{t_i} &= \hat{v}^{\mathrm{L}}_{m1}\, \Delta x^{\mathrm{L}}_{r_i} + \hat{v}^{\mathrm{L}}_{m2}\, \Delta y^{\mathrm{L}}_{r_i} + \hat{v}^{\mathrm{L}}_{m3}\, \Delta z^{\mathrm{L}}_{r_i} \\
\beta_{t_0} &= (\Delta x^{\mathrm{L}}_{t_0})^2 + (\Delta y^{\mathrm{L}}_{t_0})^2 + (\Delta z^{\mathrm{L}}_{t_0})^2 \\
\beta_{t_i} &= (\Delta x^{\mathrm{L}}_{r_i})^2 + (\Delta y^{\mathrm{L}}_{r_i})^2 + (\Delta z^{\mathrm{L}}_{r_i})^2
\end{aligned}
\tag{3.115}
$$

The specular point along the trail determined by the observability criteria $(\mathscr{P}^{\mathrm{L}}_{s_0}, \hat{\mathbf{v}}^{\mathrm{L}}_m)$ is then given by

$$
\mathscr{P}^{\mathrm{L}}_{s_i} = \mathscr{P}^{\mathrm{L}}_{s_0} + d_i \hat{\mathbf{v}}^{\mathrm{L}}_m \tag{3.116}
$$

where d_i is the appropriate root of the quadratic equation

$$
d_i = \frac{b_i^2 \pm \sqrt{b_i^2 - 4a_i c_i}}{2a_i} \tag{3.117}
$$

with

$$
\begin{aligned}
a_i &= (\beta_{r_i} - \beta_{t_0}) + (\alpha^2_{t_0} - \alpha^2_{r_i}) \\
b_i &= 2\alpha_{r_i}(\beta_{t_0} + \alpha_{t_0}\alpha_{r_i}) - 2\alpha_{t_0}(\beta_{r_i} + \alpha_{i_0}\alpha_{r_i}) \\
c_i &= \alpha^2_{t_0}\beta_{r_i} - \alpha^2_{r_i}\beta_{t_0}
\end{aligned}
\tag{3.118}
$$

The appropriate root of Eq. 3.117 to be used in Eq. 3.116 must meet *both* of the following criteria:

- the dot products of the unit R vectors pointing from the transmit antenna at $\mathscr{T}^{\mathrm{L}}_0$ to the new scatter point $\mathscr{P}^{\mathrm{L}}_{s_i}$, and from the receive antenna at $\mathscr{R}^{\mathrm{L}}_0$ to $\mathscr{P}^{\mathrm{L}}_{s_i}$, must be of equal magnitude and opposite sign

- the height of the new trail-scatter point $\mathscr{P}_{s_i}^{\mathrm{L}}$ above ground, as determined from Eq. 3.109*b*, must lie between h_s^l and h_s^u, the minimum and maximum heights of meteoric ionization, respectively.

If both criteria are not satisfied for some receiver at $\mathscr{R}_k^{\mathrm{L}}$, then the trail $\hat{\mathbf{v}}_m^{\mathrm{L}}$ through $\mathscr{P}_{s_i}^{\mathrm{L}}$ does not support the $\mathscr{T}_0^{\mathrm{L}} - \mathscr{R}_k^{\mathrm{L}}$ MB link.

Let $\mathscr{R}_{i_j}^{\mathrm{L}}, j = 1, 2, \ldots, M_r^*$, be the M_r^* peripheral receivers locations with specular points $\mathscr{P}_{s_{ij}}^{\mathrm{L}}$ in the meteor burn layer and direction vector $\hat{\mathbf{v}}_m^{\mathrm{L}}$ through $\mathscr{P}_{s_0}^{\mathrm{L}}$. A link budget calculation is performed for the $\mathscr{T}_{s_0}^{\mathrm{L}} - \mathscr{R}_0^{\mathrm{L}}$ link and each $\mathscr{T}_0^{\mathrm{L}} - \mathscr{R}_{i_j}^{\mathrm{L}}$ link to determine the minimum electron line densities $q_{sm_0}^*$ and $q_{sm_j}^*$ (at $\mathscr{P}_{s_{i_j}}^{L}$), respectively, needed to produce a usable trail-scattered signal at the corresponding receivers at points $\mathscr{R}_0^{\mathrm{L}}$ and $\mathscr{R}_{i_j}^{\mathrm{L}}$. This link budget calculation in METEORTRAK uses independent antenna patterns for the primary receiver at $\mathscr{R}_0^{\mathrm{L}}$ as well as the peripheral receivers at the points $\mathscr{R}_{i_j}^{\mathrm{L}}$, $j = 1, 2, \ldots, M_r^*$. In general, the usability criteria (L_0^*, τ_0^*) on the $\mathscr{T}_0^{\mathrm{L}} - \mathscr{R}_0^{\mathrm{L}}$ link may differ from the usability criteria $(L_{i_j}^*, \tau_{i_j}^*)$ on each $\mathscr{T}_0^{\mathrm{L}} - \mathscr{R}_{i_j}^{\mathrm{L}}$ link.

To implement the conditional nature of the trail footprint, the minimum line density value, $q_{sm_j}^*$, computed for each $\mathscr{T}_0^{\mathrm{L}} - \mathscr{R}_{i_j}^{\mathrm{L}}$ link, is compared with the corresponding minimum line density, $q_{sm_0}^*$, computed for the $\mathscr{T}_0^{\mathrm{L}} - \mathscr{R}_0^{\mathrm{L}}$ link. If $q_{sm_0}^* < q_{sm_j}^*$, then $q_{sm_j}^*$ is set equal to $q_{sm_0}^*$. In other words, $q_{sm_j}^*$ is the minimum line density which satisfies the trail-scatter criteria $(\mathscr{P}_{s_0}^{\mathrm{L}}, \hat{\mathbf{v}}_m^{\mathrm{L}})$ and (L_0^*, τ_0^*) for the $\mathscr{T}_0^{\mathrm{L}} - \mathscr{R}_0^{\mathrm{L}}$ link as well as the criteria $(\mathscr{P}_{s_{i_j}}^{\mathrm{L}}, \hat{\mathbf{v}}_m^{\mathrm{L}})$ and $(L_{i_j}^*, \tau_{i_j}^*)$ for the $\mathscr{T}_0^{\mathrm{L}} - \mathscr{R}_{i_j}^{\mathrm{L}}$ link. Finally, the $\mathscr{T}_0^{\mathrm{L}} - \mathscr{R}_0^{\mathrm{L}}$ and $\mathscr{T}_0^{\mathrm{L}} - \mathscr{R}_{i_j}^{\mathrm{L}}$ link MR contributions, $d^4\dot{N}_{00}^*$ and $d^4N_{0i_j}^*$, respectively, are computed from these q values for each j, $j = 1, 2, \ldots, M_r^*$. The MR contributions for the observability criteria $(\mathscr{P}_{s_0}^{\mathrm{L}}, \hat{\mathbf{v}}_m^{\mathrm{L}})$ for the remaining peripheral receivers $\mathscr{R}_k^{\mathrm{L}}$, $k \neq i_j$ for $j = 1, 2, \ldots, M_r^*$, are considered negligible. The incremental MR contributions $d^4\dot{N}_{00}^*$ and $d^4\dot{N}_{0i}^*$ are then integrated over all usable trail-scatter points, $\mathscr{P}_{s_i}^{\mathrm{L}}$, with trail vectors $\hat{\mathbf{v}}_m^{\mathrm{L}}$, using the quadrature expressions described in Section 3.2.2.5 to yield the link MR values $\dot{N}_{00}$ and $\dot{N}_{0i}$, $i = 1, 2, \ldots, M_r$. $i = 1, 2, \ldots, M_r$.

The final step in the trail footprint calculation is to create the conditional probabilities

$$p_i = \Pr\{\text{link} - \text{usable meteor on } \mathscr{T}_0^{\mathrm{L}} - \mathscr{R}_i^{\mathrm{L}} | \text{link} - \text{usable meteor on } \mathscr{T}_0^{\mathrm{L}} - \mathscr{R}_0^{\mathrm{L}}\}$$

$$= \frac{\dot{N}_{0i}}{\dot{N}_{00}}$$

for $i = 1, 2, \ldots, M_r$. These ratios approximate the conditional probability that a meteor which produces a trail-scattered signal for the primary link will also be useful for one or more peripheral links. The interpolated probabilities provide sufficient points on the LINK grid to plot conditional probability contours that define the trail footprint.

3.2.3.3 Sensitivity Analysis. METEORTRAK was used to calculate meteor trail footprint variation with range, time of day, and link power margin [156]. These sample results were computed for a five-element Yagi transmit antenna 3/2 wavelengths above lossy ground and pointed along the line GCP to the primary receiver in the center of the receiver array. A transmit power (P_t) of 1 kW was employed with a minimum required received power (P_r^*) of −110 dBm for the $\mathcal{T}_0^{\mathrm{L}} - \mathcal{R}_0^{\mathrm{L}}$ link as well as each $\mathcal{T}_0^{\mathrm{L}} - \mathcal{R}_i^{\mathrm{L}}$ link, $i = 1, 2, \ldots, M_r$. The $\mathcal{T}_0^{\mathrm{L}} - \mathcal{R}_0^{\mathrm{L}}$ link operating frequency (f_L) was 45 MHz and the minimum required burst duration (τ^*) was 0 s. The primary receiver and each peripheral receiver employed a horizontal dipole at 1/2 wavelength above lossy ground. Each dipole was oriented perpendicular to the GCP to the transmit antenna. All predictions were performed for 21 September, the autumnal equinox, which orients the earth's celestial pole in the direction of the earth's motion in its orbit about the sun. Contours were generated for trail footprint (conditional) probabilities of 0.90, 0.50, 0.25, and 0.10 at the peripheral receiver locations. The absolute MR value at the primary receiver is labeled on each plot.

Figures 3.43*a–d* are trail footprints computed for east-west links at ranges of 500, 800, 1100, and 1600 km, respectively, between the transmitter at $\mathcal{T}_0^{\mathrm{L}}$ and the primary receiver at $\mathcal{R}_0^{\mathrm{L}}$. In each contour plot, the transmitter location is in the direction indicated. These figures show that the footprint size increases with path length, becoming wider and more elongated as the scattered signal spreads over larger and larger geographic areas. At ranges of 500 and 800 km, usable meteor trails are formed between, around and behind the points $\mathcal{T}_0^{\mathrm{L}}$ and $\mathcal{R}_0^{\mathrm{L}}$. The resulting propagation path geometries create trail footprints that are approximately symmetric about the point $\mathcal{R}_0^{\mathrm{L}}$. As link range increases, usable meteor trails become concentrated at midlink. The resulting propagation paths favor illumination of the region behind $\mathcal{R}_0^{\mathrm{L}}$ as shown in Figs. 3.43*c* and *d*. Although the 1600-km trail footprint extends several hundred kilometers in length, the MR value $\dot{N}_{00}$ is only six meteors per hour.

Diurnal variation in the trail footprint is illustrated in Figs. 3.44*a*, *b* on an 1100-km north-south link at 0 and 12 LT. At 12 LT (Fig. 3.44*b*), the sun is directly above the link and the usable meteor flux is dominant on the eastern side of the scatter ellipsoid as the link moves west into the greatest SMRD values. This position for the meteoric hot spots explains the apparent sway of the southern end of the footprint to the west. Similarly, the principle flux contributions at 0 LT are positioned on the western side of the north-south link. Propagation paths from the transmitter would therefore sway the trail footprint to the southeast as shown in Fig. 3.44*b*. Footprints at 6 and 18 LT (not shown) appear symmetric about the link GCP because the rate of

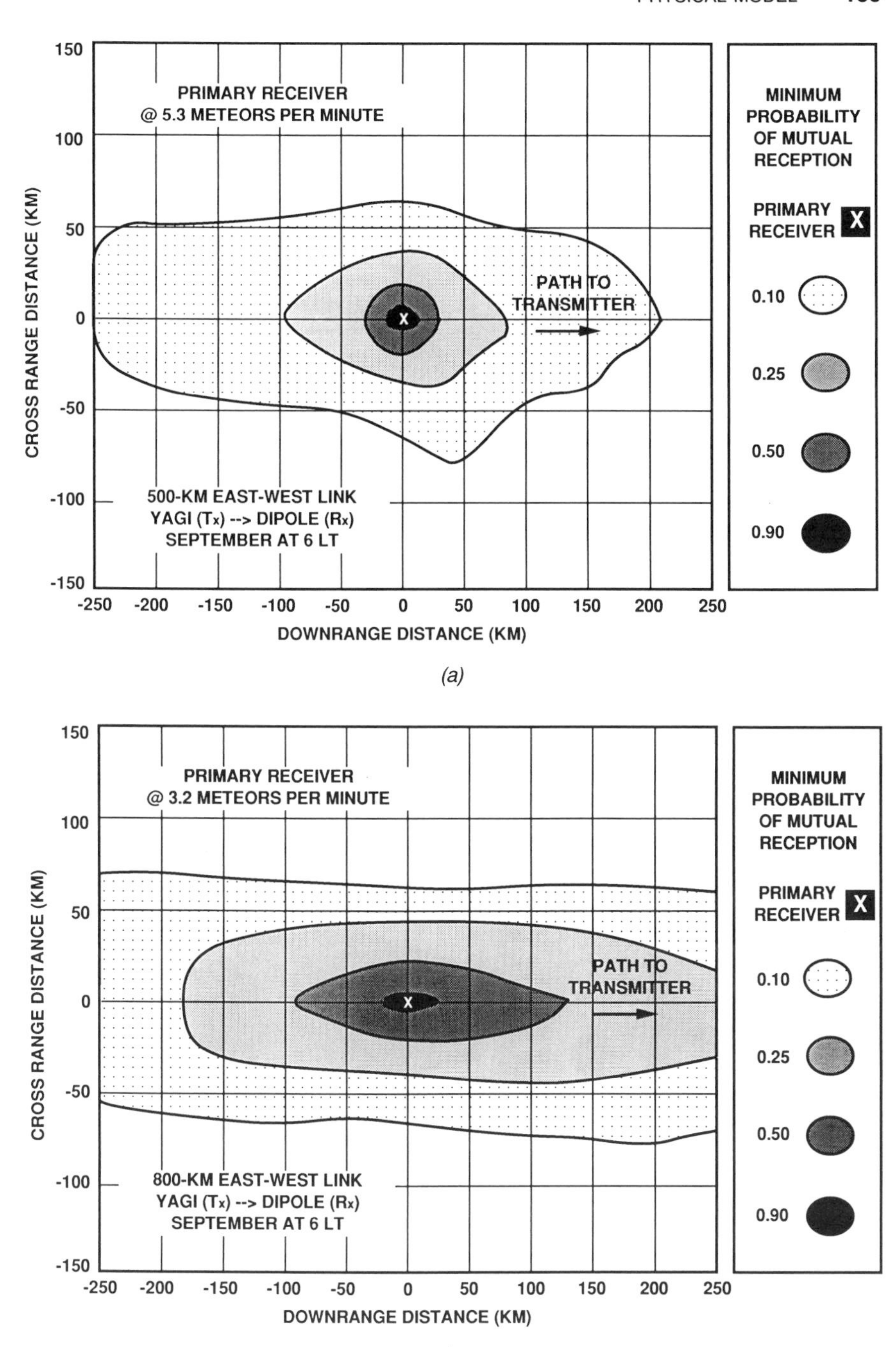

Figure 3.43. Trail footprint versus transmitter range to the primary receiver: (*a*) 500-km range; (*b*) 800-km range; (*c*) 1100-km range; (*d*) 1600-km range.

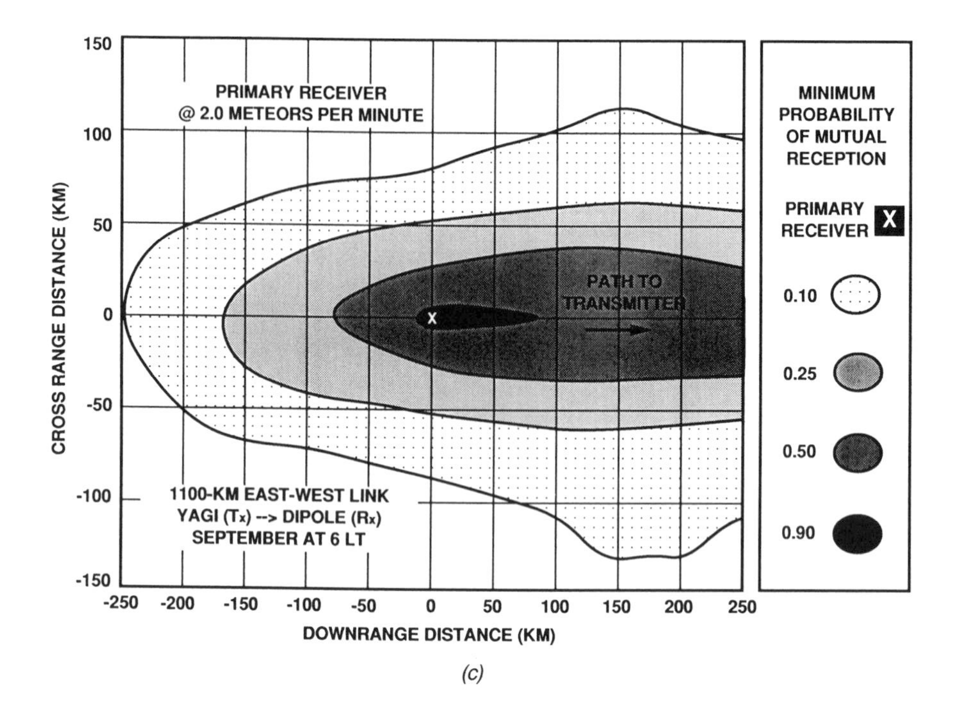

(c)

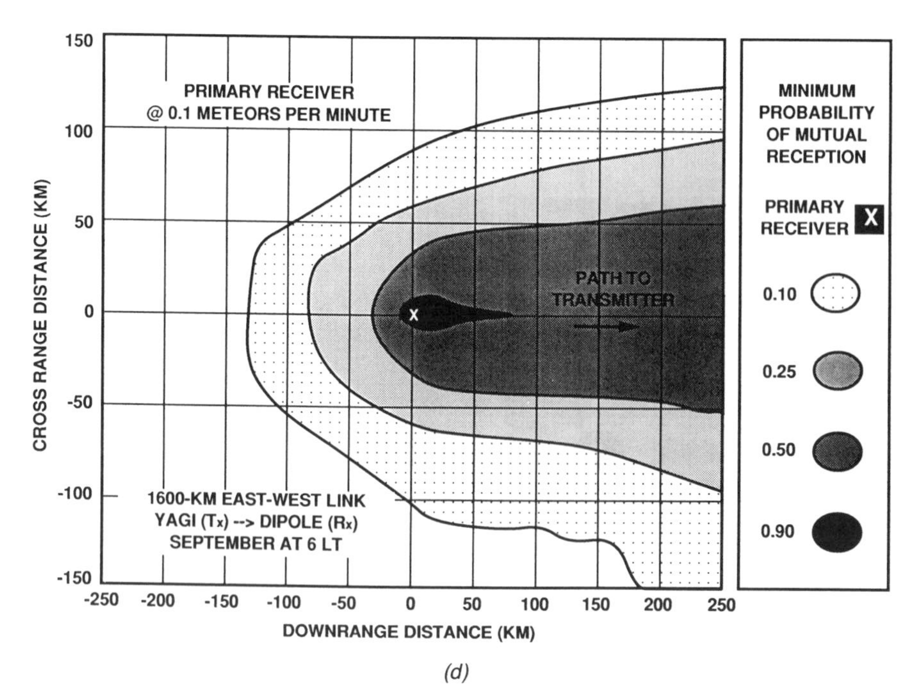

(d)

Figure 3.43. *(Continued)*

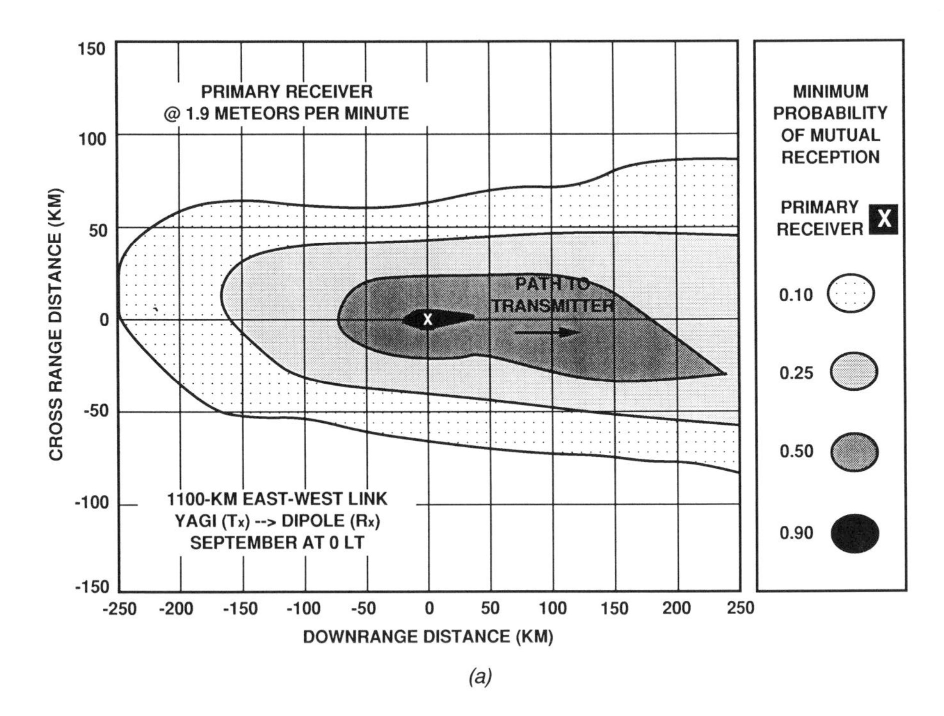

(a)

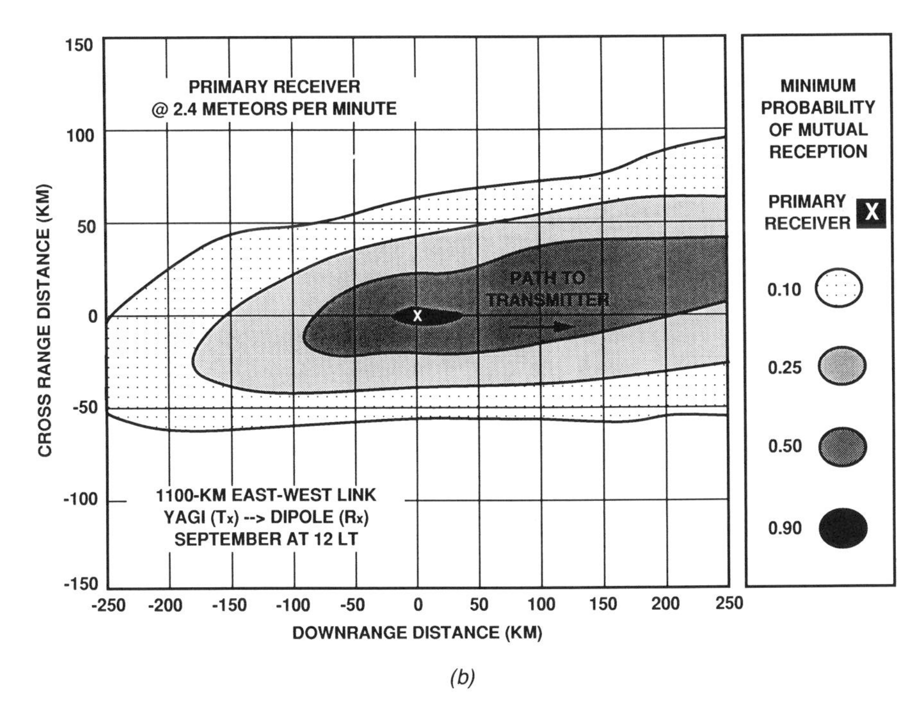

(b)

Figure 3.44. Trail footprint versus time of day on a north-south link: (*a*) 0 LT; (*b*) 12 LT.

usable trails is somewhat symmetric about the scatter ellipsoid for a mid-latitude link positioned at the earth's apex or antapex.

Figures 3.45*a*, *b* show the trail footprints for an 800 km east-west link for which the transmitted powers, P_t (at $\mathcal{T}_0^L$), were 250 W and 4 kW, respectively. The 12 dB increase in radiated power produced a noticeable increase in the footprint coverage area. Since trail footprints are relative measures, increasing the P_t value might be expected to yield equal MR increases on the $\mathcal{T}_0^L - \mathcal{R}_i^L$ links, $i = 1, 2, \ldots, M_r$, as well as the $\mathcal{T}_0^L - \mathcal{R}_0^L$ link. In this case, the ratio of these MR values would remain constant and the footprint contour would be unaffected by the increased P_t value. The trail footprint contours are expanded at the higher P_t value, however, because the relative percentage of underdense trail-scattered signals on the $\mathcal{T}_0^L - \mathcal{R}_0^L$ and $\mathcal{T}_0^L - \mathcal{R}_i^L$ links is different for $P_t = 250$ W than for $P_t = 4$ kW. Apparently, the non-linear relationship between minimum required q values and MR values produced a larger increase in $\dot{N}_{0i}$, $i = 1, 2, \ldots, M_r$, than in the corresponding $\dot{N}_{00}$ value as the transmitted power was increased.

3.2.4 Spatial Diversity Model

3.2.4.1 Overview. Consider the link MR value from a single transmitter at $\mathcal{T}_0^L$ to an array of receivers at the LINK grid points $\mathcal{R}_i^L$, $i = 1, 2, \ldots, M_r$, as shown in Figs. 3.46. If these receivers are closely spaced such that they lie within a single nonzero trail footprint probability contour (see Section 3.2.3) as shown in Fig. 3.46*a*, then all receivers have some probability of using the same meteor trail-scatterer. On the contrary, the receivers may be spaced such that each lies within statistically independent footprints as shown in Fig. 3.46*b*. In this case, different receivers detect independent meteor trail-scattered signals, that is, no two receivers detect the same trail-scatterers, so the links $\mathcal{T}_0^L - \mathcal{R}_i^L$, $i = 1, 2, \ldots, M_r$, yield independent MR values $\dot{N}_{0i}$. In this limiting case, the effective MR value from the point $\mathcal{T}_0^L$ to at least one receiver in the M_r-receiver array, called the array MR (AMR) value, is given by $\dot{N}_1 = \Sigma_{i=1}^{M_r} \dot{N}_{0i}$. The ratio of the AMR value, $\dot{N}_1$, to the average MR value of a typical receiver, $\dot{\bar{N}}_L$, is called the diversity improvement factor (DIF), $F_{M_r} = \dot{N}_1/\dot{\bar{N}}_L$ [157]. If some receivers lie within the same nonzero trail footprint contours, then the individual MR values $\dot{N}_{0i}$, $i = 1, 2, \ldots, M_r$, employ some number of meteor trail-scatterers in common. In this case, the AMR value $\dot{N}_1$ cannot be given by $\Sigma_{i=1}^{M_r} \dot{N}_{0i}$ because of multiple counting of single trail-scatterers.

The distribution of MR values observed on the individual links $\mathcal{T}_0^L - \mathcal{R}_i^L$, $i = 1, 2, \ldots, M_r$, may vary significantly in practice. Two limiting cases were described in Section 3.1.1.3.4 that demonstrated the ambiguity of the DIF value as a measure of MB diversity improvement. First, let $\dot{N}_{0i} = \dot{\bar{N}}_L$, for $i = 1, 2, \ldots, M_r$, then the DIF value $F_{M_r} = M_r$. On the contrary, let $\dot{N}_{0k} =$

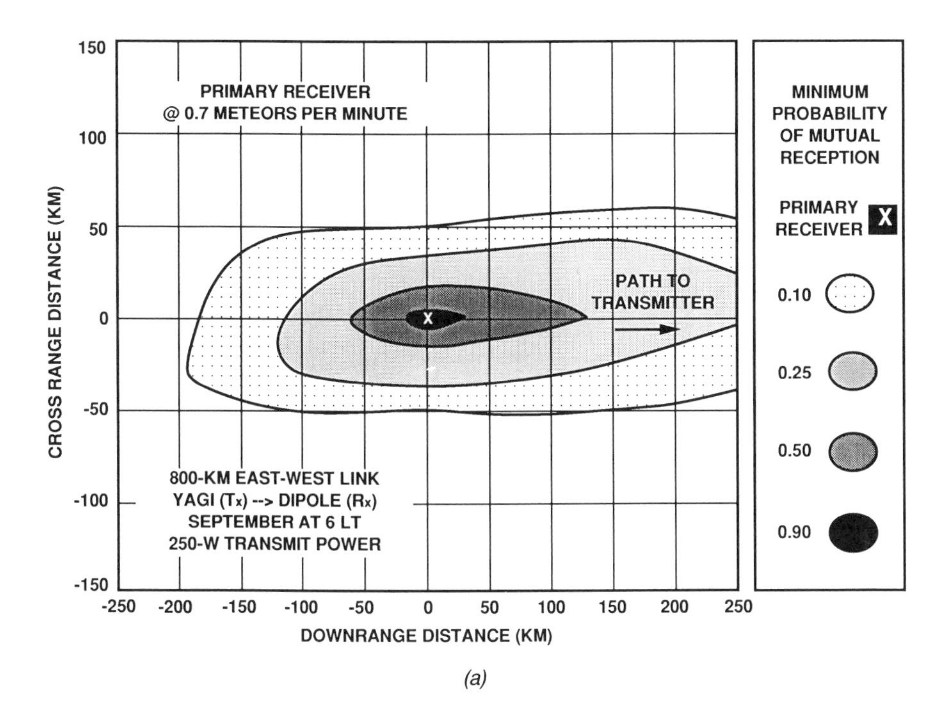

(a)

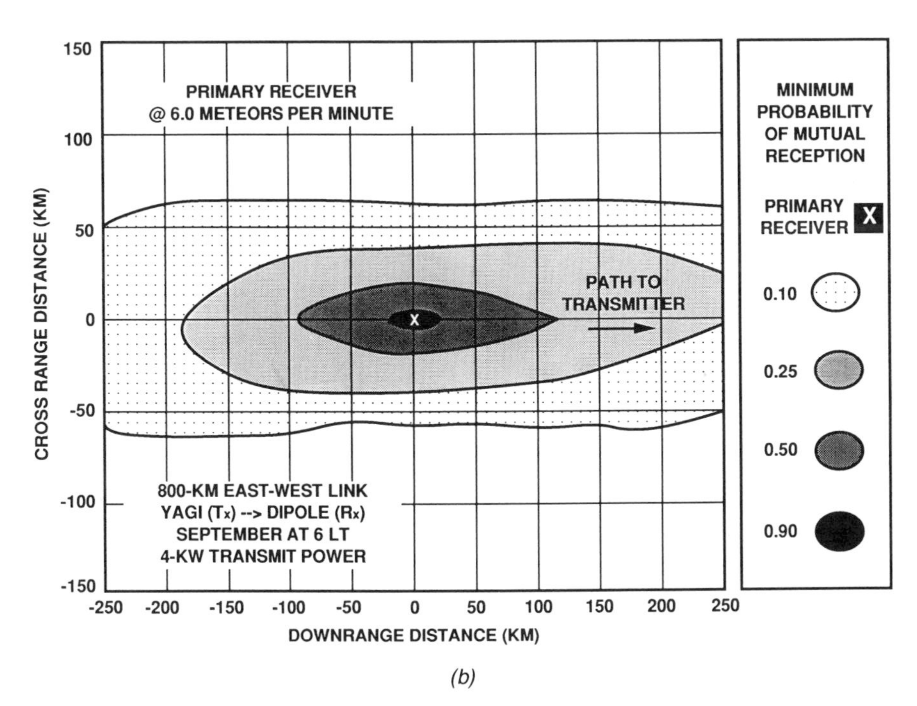

(b)

Figure 3.45. Trail footprint versus transmitter power at an 800-km range: (*a*) 250 W; (*b*) 4 kW.

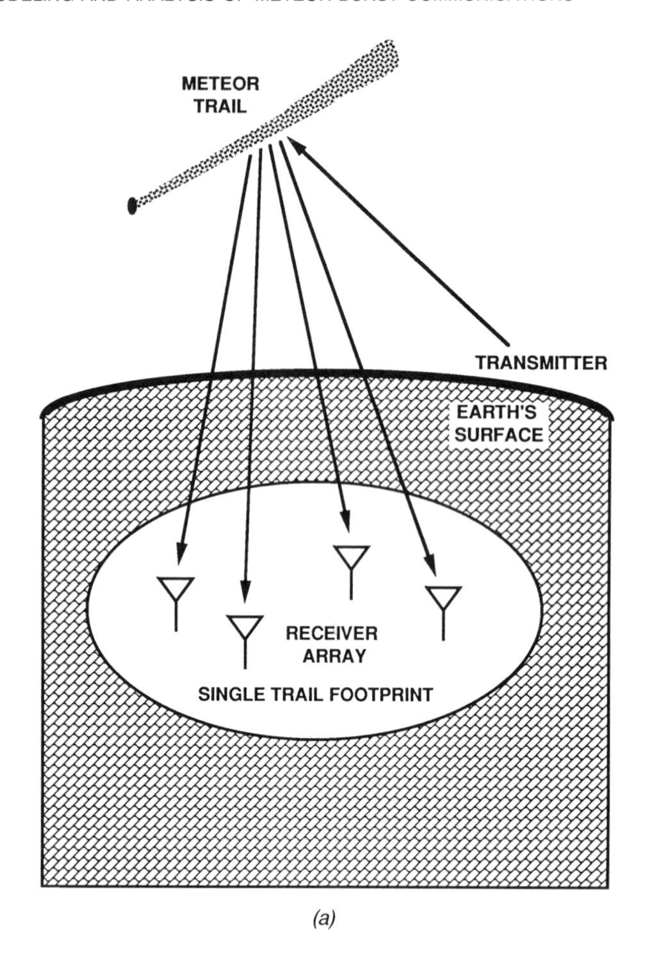

(a)

Figure 3.46. Receiver spacing relative to area covered by trail footprints: (*a*) Receivers in same footprint; (*b*) Receivers in separate footprints.

$\bar{\dot{N}}_L$ for some $k \leq M_r$ and $\dot{N}_{0i} \cong 0$, for $i = 1, 2, \ldots, M_r$ and $i \neq k$, then the DIF value $F_{M_r} = M_r$, as the in the first case. The DIF value $F_{M_r} = M_r$ whether the AMR value $\dot{N}_1 = M_r\bar{\dot{N}}_L$ *or* $\dot{N}_1 = \bar{\dot{N}}_L$. Thus the AMR value, rather than the DIF value, provides an unambiguous measure of spatial diversity improvement when compared with the distribution of individual link MR values.

Meteoric spatial diversity may be compared with classical diversity techniques used for tropospheric scatter (troposcatter) communications. Space diversity, for example, uses multiple spaced receive antennas to provide independently fading signals for a diversity combiner [158]. These independently fading signals are received from the illumination of the same common volume but employing different electrical path lengths. For MB

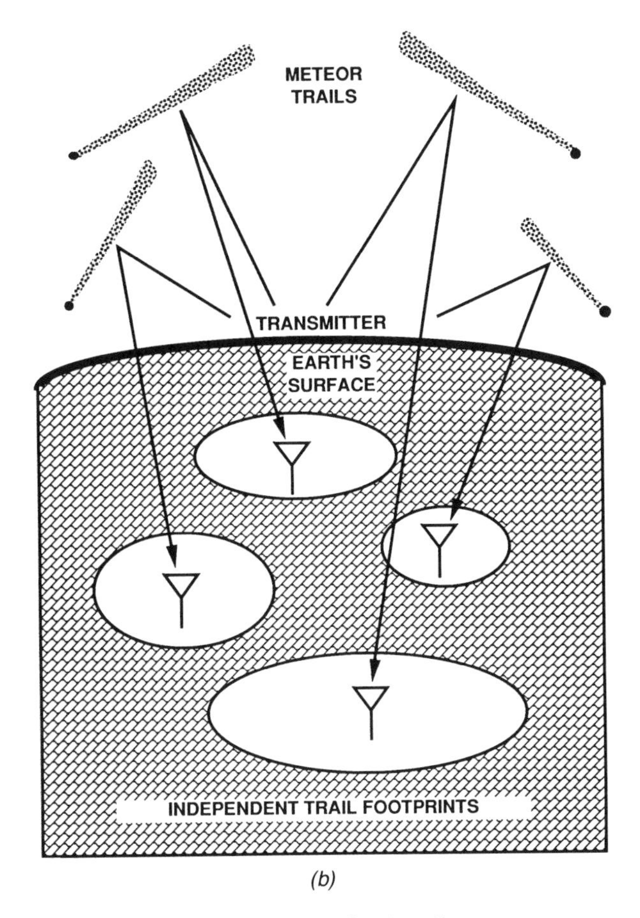

(b)

Figure 3.46. (*Continued*)

communications, space diversity may be used to minimize fading outages which are characteristic of long duration overdense trail-scattered signals and thus increase the effective channel DC contribution per meteor burst.

Unlike space diversity, meteoric spatial diversity employs non-overlapping, or partially overlapping, common scattering volumes much like angle diversity [159] used in troposcatter communications. Angle diversity provides independently fading received signals from disjoint common volumes to produce fewer deep fades at the output of a diversity combiner. Spatial diversity increases the effective MR value to at least one receiver in an array of spaced receivers. If the receivers are interconnected by a wireline, optical fiber, LOS radio, or groundwave radio network, then the array is equivalent to a single wide-area receiver. Spatial diversity then becomes a useful meteor engineering technique to increase the usable common volume and improve MB link performance between the transmitter, $\mathscr{T}_0^{\mathrm{L}}$, and an *effective receiver set*, $\mathfrak{R}_{\mathrm{eff}}^{\mathrm{L}}$, consisting of the receiver locations $\{\mathscr{R}_i^{\mathrm{L}},\ i = 1, 2, \ldots,$

$M_r\}$. This approach may be easily applied to an array, or set, of M_t transmitters ($\mathfrak{T}_{\text{eff}}^{\text{L}}$) and a single receiver at $\mathscr{R}_0^{\text{L}}$, defining the $\mathfrak{T}_{\text{eff}}^{\text{L}} - \mathscr{R}_0^{\text{L}}$ link. The approach is easily extended to the case of M_t transmitters at points $\mathscr{T}_i^{\text{L}}$ and M_r receivers at points $\mathscr{R}_j^{\text{L}}$, thus defining the $\mathfrak{T}_{\text{eff}}^{\text{L}} - \mathfrak{R}_{\text{eff}}^{\text{L}}$ MB link.

Several important questions arise in the consideration of spatial diversity for MB link performance improvement. Most importantly, what is the DIF value as a function of receiver spacing in the array? If this spacing is too great, it might be impossible to provide adequate intra-array communication links or include the contribution of outlying receivers due to transmit antenna pattern or MB link geometry limitations. Clearly, the relationship between the AMR value and receiver spacing is significantly affected by MB link range. This range is measured from the transmitter, $\mathscr{T}_0^{\text{L}}$ to a central receiver in the $\mathfrak{R}_{\text{eff}}^{\text{L}}$ array located at $\mathscr{R}_0^{\text{L}}$. This section describes the diversity calculation implemented by the METEORDIV program, a computer model developed to calculate spatial diversity improvement and estimate the probability of simultaneous reception by two or more receivers, a by-product of the diversity calculation. In addition, results are presented from a sensitivity analysis of the AMR value to receiver spacing and MB link range.

3.2.4.2 Diversity Modeling Approach. The METEORDIV computer model uses the geographic coordinates of the transmit antenna at the point $\mathscr{T}_0^{\text{L}}$ and a single receiver at the point $\mathscr{R}_0^{\text{L}}$ to define the L system. The remaining receivers at $\mathscr{R}_i^{\text{L}}$, $i = 1, 2, \ldots, M_r$, in the receive array are located at chosen points in the L system. A grid of potential trail-scatter points, $\mathscr{P}_s^{\text{L}}$, is then defined at the height of maximum meteoric ionization, $\widehat{H}_I$, between 93 and 96 km above the earth's surface. The LINK grid is depicted in Fig. 3.47, where the potential trail-scatter points, $\mathscr{P}_s^{\text{L}}$, are located at the intersections of the downrange and cross-range earth-concentric arcs. At this point, the diversity algorithm diverges from the link modeling approach described in Section 3.2.2. Instead of specifying antenna locations and modeling trails determined to meet the resulting observability criteria ($\mathscr{P}_s^{\text{L}}, \hat{\mathbf{v}}_m^{\text{L}}$), METEORDIV models a large number of meteor trails and then determines the MR contribution to each MB link $\mathscr{T}_0^{\text{L}} - \mathscr{R}_i^{\text{L}}$, $i = 1, 2, \ldots, M_r$. This approach assures that the MR contribution to the AMR value from each observable trail is counted exactly once, an essential feature of the METEORDIV model.

An AMR contribution is calculated for each LINK grid point $\mathscr{P}_s^{\text{L}} = (x_{\text{L}}, y_{\text{L}}, z_{\text{L}})$ in Fig. 3.47 and the resulting values integrated over the corresponding common area. The total AMR contribution at a point $\mathscr{P}_s^{\text{L}}$ is found by integrating the incremental AMR contributions from trail orientations $\hat{\mathbf{v}}_m^{\text{L}}$ passing through an observation hemisphere (celestial hemisphere). This hemisphere is centered at each point $\mathscr{P}_s^{\text{L}}$ as shown in Fig. 3.48 and is used to define the observability criteria ($\mathscr{P}_s^{\text{L}}, \hat{\mathbf{v}}_m^{\text{L}}$) for each modeled meteor trail. These criteria are determined by appropriate selection of the angles

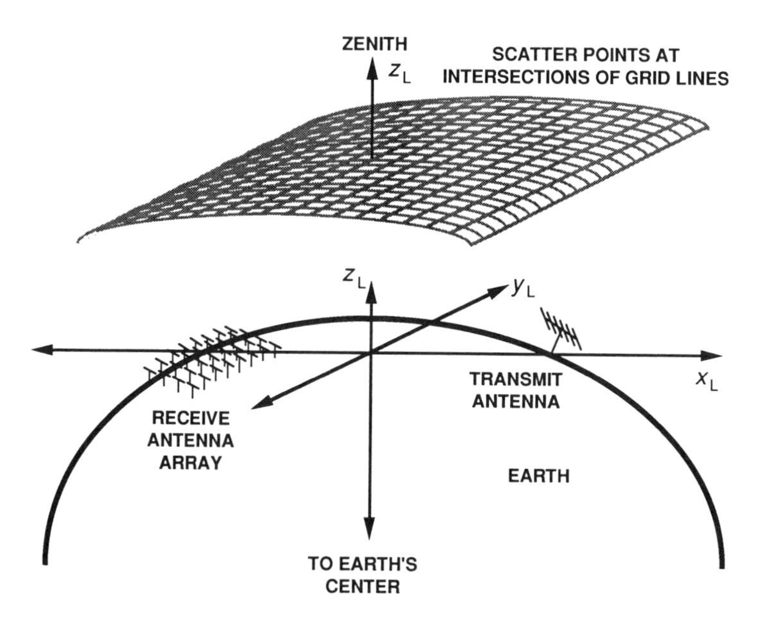

Figure 3.47. LINK grid for diversity calculation.

$\Delta\alpha_{sj}$ and $\Delta\beta_{sk}$ in Fig. 3.48 so that the corresponding solid angle $\Omega_s^{\alpha\beta} = \Delta\alpha_{sj}\,\Delta\beta_{sk}\sin(\beta_{sk})$ is approximately constant over the entire celestial hemisphere. A sufficient number of trail vectors $\hat{\mathbf{v}}_m^{\mathrm{L}}$ must then be chosen through each point $\mathscr{P}_s^{\mathrm{L}}$ to assure convergence of the quadrature technique employed.

Each trail orientation $\hat{\mathbf{v}}_m^{\mathrm{L}}$ through $\mathscr{P}_s^{\mathrm{L}}$ is evaluated to determine whether it

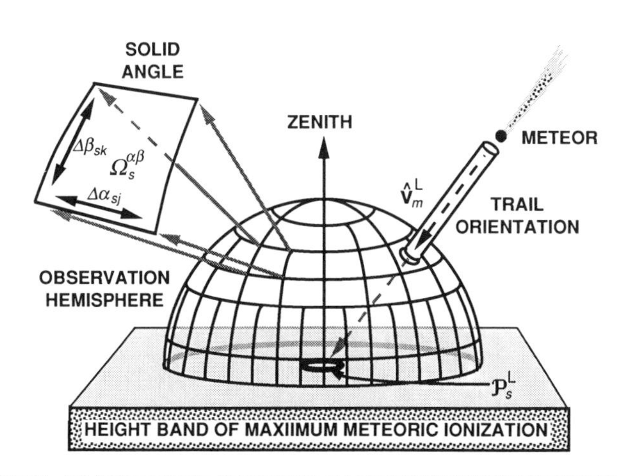

(NOTE: SCATTER POINT AT CENTER OF OBSERVATION HEMISPHERE)

Figure 3.48. Observation hemisphere for the determination of usable trail orientations.

would support a specular reflection from the transmitter at $\mathscr{T}_0^{\mathrm{L}}$ to each receiver at $\mathscr{R}_i^{\mathrm{L}}$, $i = 1, 2, \ldots, M_r$. This determination is based on the definition of an M plane, designated M_i, for each link $\mathscr{T}_0^{\mathrm{L}} - \mathscr{R}_i^{\mathrm{L}}$, $i = 1, 2, \ldots, M_r$, containing the point $\mathscr{P}_s^{\mathrm{L}}$. If the angle formed between the trail vector $\hat{\mathbf{v}}_m^{\mathrm{L}}$ and the M_{i_j} plane for some link $\mathscr{T}_0^{\mathrm{L}} - \mathscr{R}_{i_j}^{\mathrm{L}}$, $1 \le i_j \le M_r$, is less than the angle $\Delta\beta_{sk}$ (see Section 3.2.2.3.3), then $\hat{\mathbf{v}}_m^{\mathrm{L}}$ defines a specular trail orientation for the $\mathscr{T}_0^{\mathrm{L}} - \mathscr{R}_{i_j}^{\mathrm{L}}$ link. Assume that only M_r^* of the M_r receivers can employ meteor trails satisfying the observability criteria $(\mathscr{P}_s^{\mathrm{L}}, \hat{\mathbf{v}}_m^{\mathrm{L}})$, that is, only the links $\mathscr{T}_0^{\mathrm{L}} - \mathscr{R}_{i_j}^{\mathrm{L}}$, $j = 1, 2, \ldots, M_r^*$, can employ trails passing through $\mathscr{P}_s^{\mathrm{L}}$ with direction vector $\hat{\mathbf{v}}_m^{\mathrm{L}}$. After imposing the usability criteria $(L_{i_j}^*, \tau_{i_j}^*)$, the corresponding incremental MR values, $d^4\dot{N}_{0i_j}$, are computed for each MB link $\mathscr{T}_0^{\mathrm{L}} - \mathscr{R}_{i_j}^{\mathrm{L}}$, $j = 1, 2, \ldots, M_r^*$. The maximum of these $d^4\dot{N}_{0i_j}$ values, designated $d^4\dot{N}_1^*$, is then defined to be the AMR contribution satisfying the observability criteria $(\mathscr{P}_s^{\mathrm{L}}, \hat{\mathbf{v}}_m^{\mathrm{L}})$. This AMR contribution $d^4\dot{N}_1^*$ is the rate at which $k_r = 1$ or more receivers detect a usable trail-scattered signal. The second largest AMR contribution, $d^4\dot{N}_2^*$, corresponds to the rate at which $k_r = 2$ or more receivers detect the same trail, and so on. Integrating these AMR contributions independently for each value of k_r, $k_r = 1, 2, \ldots, M_r$, over the observation hemisphere gives the total AMR contribution, $d^2\dot{N}_{k_r}^*$, from the specified LINK grid point $\mathscr{P}_s^{\mathrm{L}}$ for a minimum of receivers $k_r = 1, 2, \ldots, M_r$.

Finally, the pointwise AMR contributions are integrated over the LINK grid for each number of receivers k_r yielding the AMR values $\dot{N}_{k_r}$, $k_r = 1, 2, \ldots, M_r$. Dividing the AMR value for $k_r = 1$ by the average link MR value, $\bar{\dot{N}}_L$, determined over all M_r receivers, gives the desired DIF value $F_{M_r} = \dot{N}_1 / \bar{\dot{N}}_L$. Subtracting the AMR value corresponding to $(k_r + 1)$ or more receivers, $\dot{N}_{k_r+1}$, from the AMR value derived for k_r or more receivers, $\dot{N}_{k_r}$, yields the MR value corresponding to *exactly* k_r receivers using trail-scattered signals produced by the same meteor. In fact, since all AMR contributions are determined for the same trail-scatter points, instead of anywhere along the trail as in the calculation of trail footprint, common use of a trail implies simultaneous reception. Dividing this result for each k_r value by the AMR value $\dot{N}_1$ yields an estimate of the probability of simultaneous reception

$$p_{k_r} = \Pr\{\text{simultaneous trail – scattered signal reception by exactly } k_r \text{ receivers}\}$$

This calculation, however, ignores finite meteor velocity effects which would produce simultaneous receptions in the receiver array both before and after the meteor passes through the point $\mathscr{P}_s^{\mathrm{L}}$ and suggests that p_{k_r} is a lower bound on the actual probability. The estimate of mutual reception probability improves in accuracy as the required burst duration τ^* increases, forcing link reliance on long-lasting, overdense trails. These trails are predominately found at the height of maximum meteoric ionization, $\widehat{H}_I$.

3.2.4.3 *Sensitivity Analysis.* A sensitivity analysis of MB AMR and DIF values versus receiver spacing and link range was performed using the METEORDIV computer model. A fixed array of 41 receivers was configured as shown in Fig. 3.49, designed to provide equal spacing between adjacent receivers. AMR values $\dot{N}_1$, DIF values F_{M_r}, average link MR values $\bar{\dot{N}}_L$, and the cumulative probability distribution of k_r receivers

$$P_{k_r} = \Pr\{\text{simultaneous trail} - \text{scattered signal reception by } k_r \textit{ or more } \text{receivers}\}$$

were computed using METEORDIV for east-west link ranges of 300, 1000, and 1600 km and receiver spacings of 8, 16, 24, 32, 48, 64, 80, 96, 120, and 160 km. Link range is measured from the transmitter, $\mathcal{T}_0^{\mathrm{L}}$, to the center of the receiver array shown in the figure. A transmitter power (P_t) of 1 kW was assumed with no other system losses. The predicted MR values and AMR values were determined for a required received power (P_r^*) of -110 dBm and a minimum burst duration (τ^*) of 0 s. All link predictions were performed for a frequency (f_L) of 40 MHz at local noon for a typical day in May.

The transmit and receive antennas used in this analysis were selected for best results at each range. At 300 km, the transmit antenna and all 41 receive antennas were folded crossed dipoles mounted at 1/2-wavelength height above ground. Use of omniazimuthal antennas at the short range of 300 km was necessary to provide links to receivers behind the transmitter for the larger values of receiver spacing. At the 1000-km range, the transmit

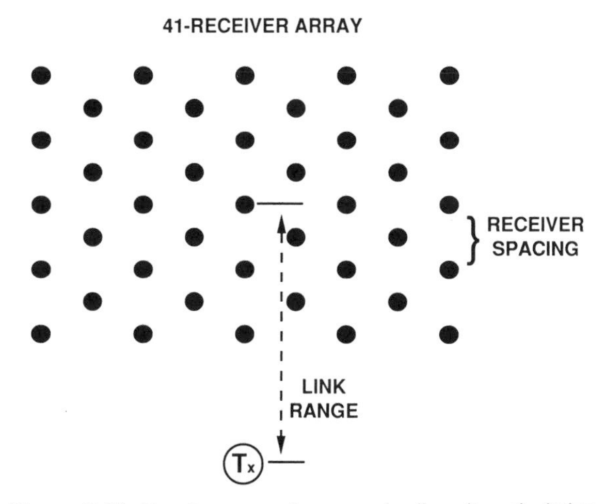

Figure 3.49. Receiver array for example diversity calculation.

antenna was changed to a five-element Yagi (1-wavelength height) to provide more efficient illumination of the common volume on the $\mathcal{T}_0^L - \Re_{eff}^L$ link. Finally, the transmit antenna and all 41 receive antennas were changed to five-element Yagis for the 1600-km range predictions. The Yagis at each of the 41 receivers were pointed along the GCP to $\mathcal{T}_0^L$. The transmit Yagi was pointed at the center of the $\Re_{eff}^L$ receiver array at the point $\mathcal{R}_0^L$ for both the 1000-and 1600-km range predictions.

Figure 3.50*a* is a plot of the AMR value, the average receiver MR value, and the corresponding DIF value for the 300-km link range versus receiver spacing. The figure shows that the AMR and DIF values increased steadily as the receiver spacing was increased to approximately 80 km. The AMR value decreased for larger values of receiver spacing. The average receiver MR value decreases as the distances from $\mathcal{T}_0^L$ to the more remote receivers in the array was increased. However, the AMR value was decreased more gradually than the average receiver MR value, so the corresponding DIF value continued to increase until receiver spacing was about 160 km. This result is indicative of the ambiguity in the DIF measure described in Section 3.2.4.1. Fig. 3.50*b* shows that increased receiver separation decreased the likelihood of multiple simultaneous receivers, as expected intuitively. The apparent "knee" in these curves becomes sharper for increasing values of receiver spacing as fewer and fewer receivers lie within the same trail footprint.

Figure 3.51*a* shows the AMR value, average receiver MR value, and corresponding DIF value versus receiver spacing for the 1000-km link range. The maximum AMR value was achieved for receiver spacing greater than about 110 km as compared to the spacing of 80 km observed for the 300-km link range. This result was due to the greater "stand-off" range of the transmitter at $\mathcal{T}_0^L$, which enlarges the effective common volume on the 1000-km $\mathcal{T}_0^L - \Re_{eff}^L$ link. The enlarged common volume combined with increased antenna gain from the Yagi transmit antenna yielded larger trail footprints. These larger footprints produced greater AMR values at 1000-km range than at the 300 km range. Ultimately, however, the outlying receivers contributed less and less to the AMR value as the receiver spacing was increased, despite the low probability of lying within the same trail footprint (see Fig. 3.51*b*). As observed at the 300-km range, the DIF value continued to increase despite the decreased AMR values.

The 1600-km $\mathcal{T}_0^L - \Re_{eff}^L$ link predictions are shown in Figs. 3.52*a* and *b*, respectively. Fig. 3.52*a* shows that the AMR values steadily increased for receiver spacings up to at least 160 km, although these values are below the corresponding values found for the 1000-km $\mathcal{T}_0^L - \Re_{eff}^L$ link range. Increased AMR values were caused by receivers moved closer to the transmitter at $\mathcal{T}_0^L$ as receiver spacing was increased. Note that the cumulative reception probabilities (Fig. 3.52*b*) have also decreased from the corresponding 1000-km values. This reduction was caused by a decrease in the common volume associated with the increased $\mathcal{T}_0^L - \Re_{eff}^L$ link range from 1000 to 1600 km.

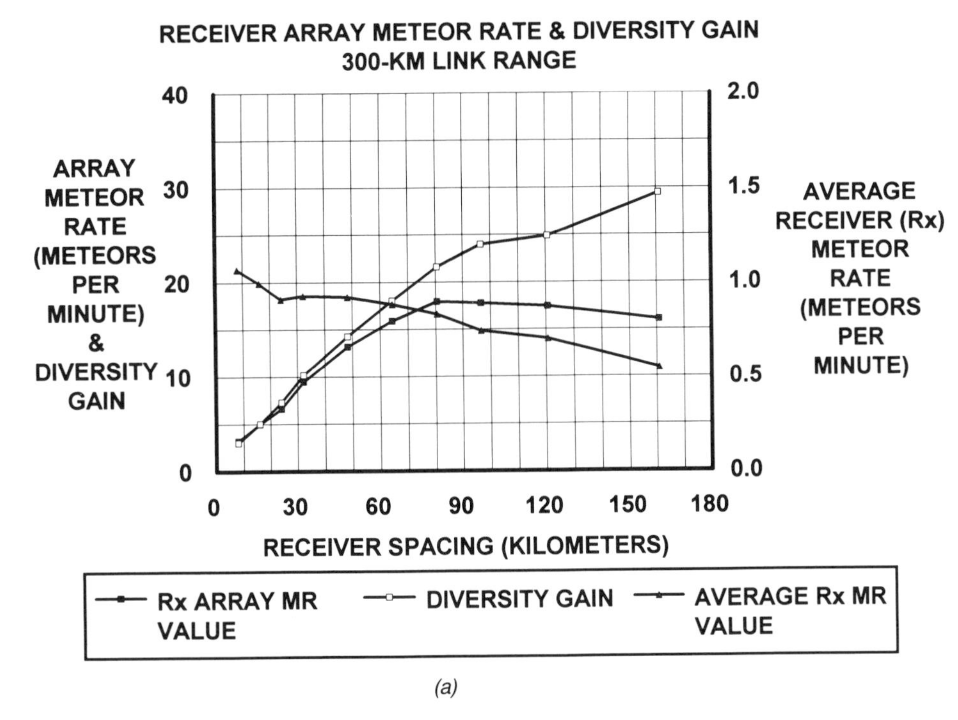

(a)

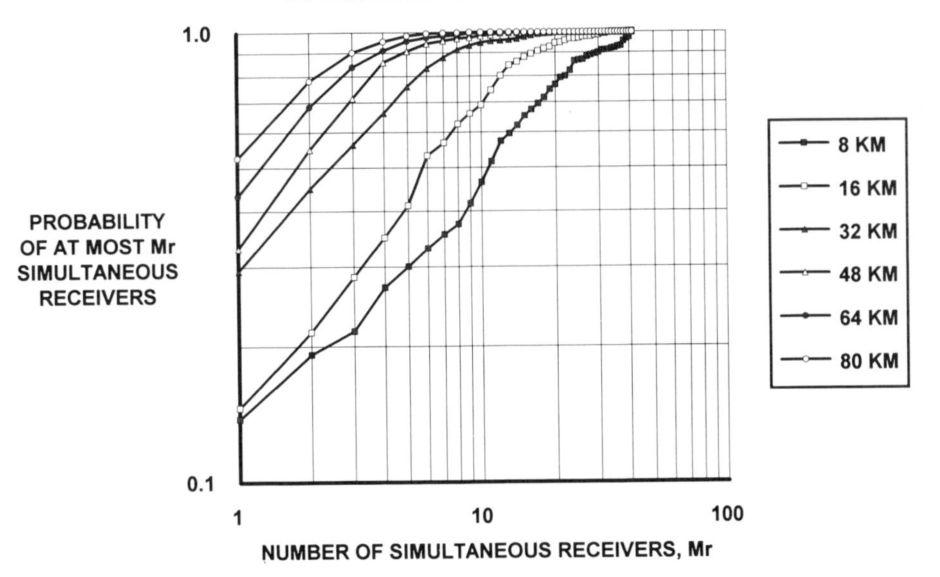

(b)

Figure 3.50. Diversity calculation for a 300-km effective link range: (*a*) AMR values; (*b*) simultaneous reception CDF.

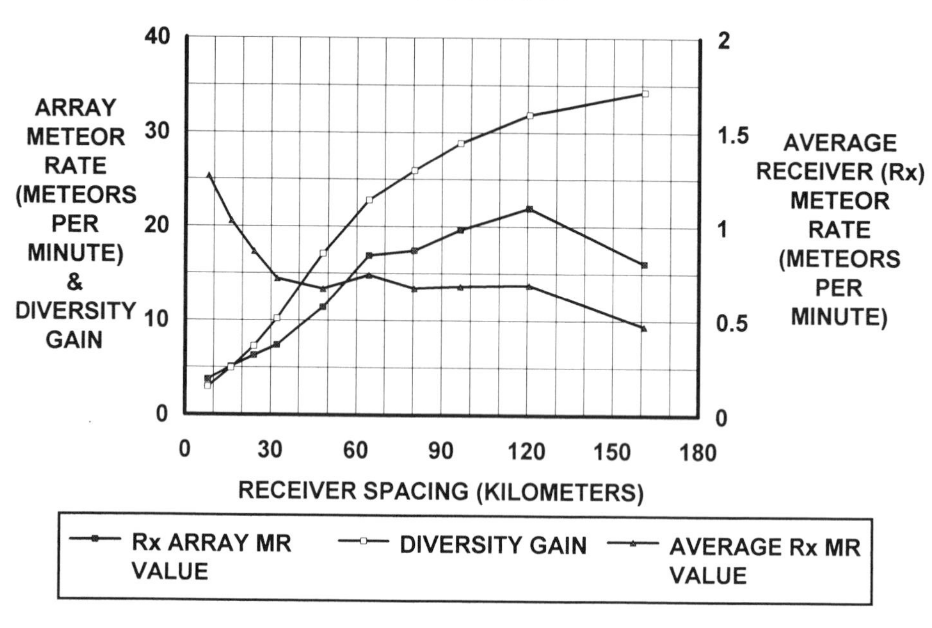

(a)

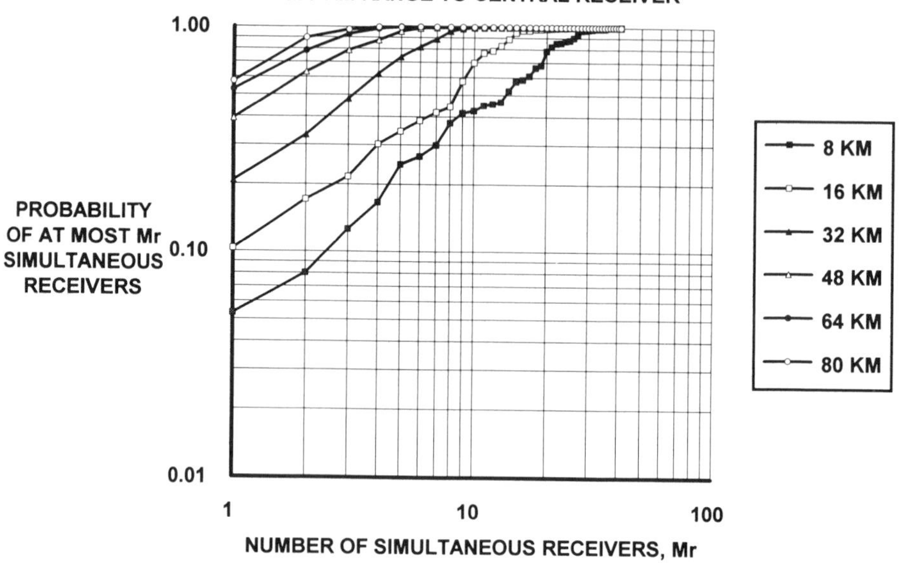

(b)

Figure 3.51. Diversity calculation for a 1000-km effective link range: (*a*) AMR values; (*b*) simultaneous reception CDF.

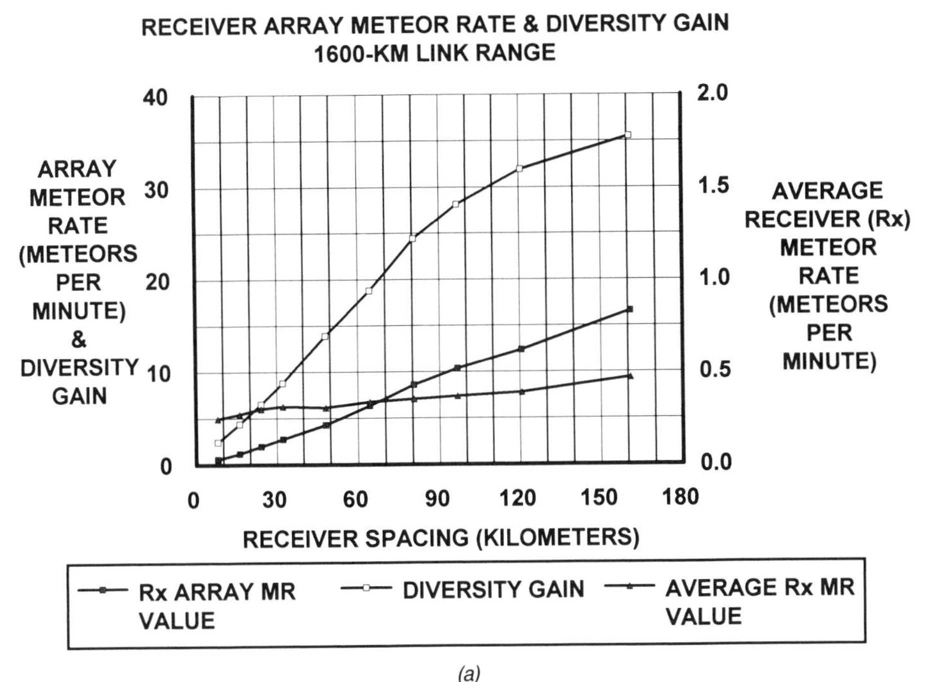

(a)

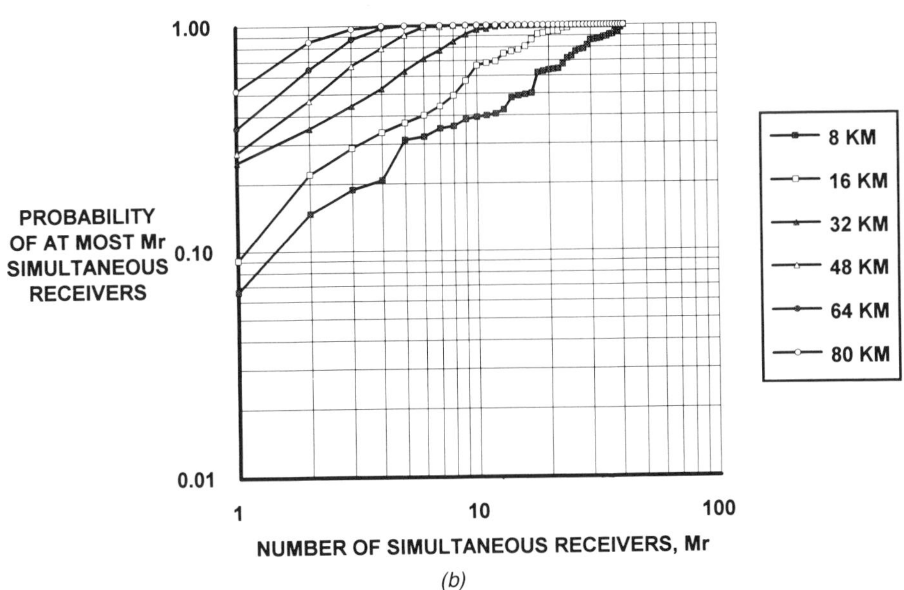

(b)

Figure 3.52. Diversity calculation for a 1600-km effective link range: (*a*) AMR values; (*b*) simultaneous reception CDF.

3.3 SIMULATION

3.3.1 Overview

The METEORLINK computer model predicts the mean and standard deviation of the link MR and DC values. Although these values have significance for the meteor scientist or MB communications system designer, they have little value for the communicator. A more useful measure of MB system performance from a communications perspective is the time between receipt of fixed length messages, MWT, typically measured in minutes. For example, a MR value of 0.5 meteors per minute and a corresponding DC value of 0.25% yields an average burst duration of 300 ms. In other words, the link will support a transmission of 300-ms duration once every two minutes. In this case, the MWT concept accurately conveys the intermittent, or bursty, nature of the MB channel. Messages exceeding the nominal single-burst duration may be transmitted during several independent bursts and reassembled at the receiver (message piecing). In this context, MWT includes only delays inherent in message transmission, such as the characteristic intermittent MB channel behavior or the limitations imposed by the terminal equipment, rather than transmit queue delay or priority message preemption.

A second measure used to characterize MB system performance is throughput, which implies continuous channel availability. The previous example of one 300-ms burst every 2 min, however, cannot be characterized as continuous availability. If link performance is significantly improved, say an MR value of 80 meteors per minute with a DC value of 40%, there are eighty trail-scattered signals per minute with an average 300-ms bursts duration. Although not continuous, this channel performance is better characterized by a throughput measure. Assuming these bursts support a 10-kbps channel rate, the first example yields a throughput of 38 words per minute (wpm) while the second example yields 6000 wpm. Of course, these example computations ignore the effect of protocol on link performance, which would reduce these numbers significantly in practice.

The conversion of link performance predictions of MR and DC into MWT or throughput can be performed analytically if certain simplifying assumptions are made regarding link protocol [160]. In general, however, half- and full-duplex protocols, fixed data block length, asymmetrical links, random burst start times, fading outages [161], variable data rates [162], and so on, complicate the analytical approach. These complications are further amplified by their consideration in a MB network communications environment. An alternative modeling technique capable of addressing these complications is variance reduction, also called Monte-Carlo simulation [163]. A simulation provides the capability to simply model complicated protocol and link operating procedures as well as provide a modeling tool that may be adapted to a wide variety of these procedures. To provide this capability, a computer Monte-Carlo simulation called METEORWAIT [164] was developed as a basis for the prediction of MB link and network performance.

METEORWAIT uses MR and DC values to generate meteor bursts with the occurrence rate and duration measured or predicted for a specified link or for several specified links in a network. Each modeled station or terminal (end user) employs both a transmit and receive queue, link operating procedures, and data exchange rules or protocol. Protocols governing the deterministic exchange of overhead and data frames between communicants are imposed on each simulated burst. Each burst may result in the modification of transmit and receive queues in the communicant's terminals. The simulation is completed when acceptable statistical confidence is achieved in the predicted results.

3.3.2 Link Simulation

3.3.2.1 Overall Approach. Many specific implementations of an MB link protocol are possible, including adaptations [165] of the amateur radio version of the CCITT X.25 data link protocol [166], called AX.25 [167]. These adaptations involve transmission of a stream of several data packets without requiring channel time for the exchange of "handshaking" frames. These packets may be received by the second terminal, which subsequently updates transmitted packet pointers to acknowledge correct reception. Standard link operating procedures, such as the probe-and-wait technique for MB communications, permit transmission from a remote transponder only in direct response to a channel probe signal from a master station. Half-duplex (HDX), single-frequency communications has also been demonstrated between two simultaneously probing master stations [168]. Given this diversity in MB communication techniques, a standard set of protocol and link operating procedures must be selected as a basis for simulation.

The Defense Communications Agency (DCA) has drafted a specification (MIL-STD-188-135C) [169] detailing an interoperability and performance standard for military MB communications. This standard provides the conventions required for the basis of a Monte Carlo link and network simulation. In this standard, master stations and remote terminals are arranged in a standard "star" [170] network configuration, as shown in Fig. 3.53. Remote terminals can establish communications with other remote terminals only through their respective master stations. Master stations serve as gateways between star networks, providing a trunking service across the network.

Master stations are the active control elements of the network, continuously transmitting a brief channel probe signal and listening for an acknowledgement. The probe identifies the master station and may interrogate one or more additional master stations or remote terminals. Remote terminals operate as passive elements, or transponders, activated by their respective

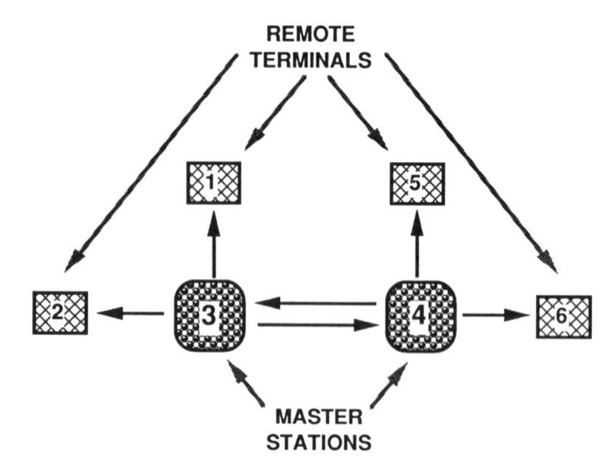

Figure 3.53. Example MB network configuration.

master stations. Remote terminals transmit only after correctly receiving the master stations probe signal and determining that they have been addressed to respond.

Communication between master stations or between master stations and remote terminals was defined in MIL-STD-188-135C as a three-stage process involving channel probing, link acquisition, and data exchange. Channel probing requires that a master station repeatedly transmit a brief signal to one or more masters and remote terminals, depending on the communications mission. This probe (PRB, 128 bits) signal contains a correlation pattern and master station identification. The link acquisition stage begins once a usable trail forms a link between the probing master station and a second master station or a remote terminal. The station or terminal receiving the PRB signal responds with the transmission of an acquire (ACQ, 152 bits) probe, which identifies the responding communicant and provides information necessary for data transmission. The master station sends an ACQ signal and, if correctly received by the second station or terminal, data transmission may commence. Message data exchange begins only after link acquisition probes have been exchanged.

During the data exchange stage, blocks (DTA, 544 bits maximum) consisting of a control frame and up to four message segments are transmitted. Messages segments consist of 14 seven-bit characters of data and a two-byte checksum. The control frame includes identification of the last correctly received message segment and serves as a "pointer" into the message currently being transmitted. If a master station or remote terminal does not have message data to transmit, a special control frame (SPC, 120 bits), notifying the communicant to "go-ahead" with its message segments, will be transmitted. The short SPC signal permits the data source to periodically determine the status of the MB link, that is, existence of a usable trail, as well as provide an updated message pointer.

The METEORWAIT simulation has been based on the MIL-STD-188-135C standard. Nevertheless, the number of bits in each protocol field may be varied to permit simulation of other link and network protocols or perform protocol sensitivity studies.

3.3.2.2 Burst Generation

3.3.2.2.1 Interarrival Time. The probability of k usable trail-scattered signals occurring within a specified time t is well approximated by the conditional Poisson distribution [171]:

$$\Pr\{k \mid \dot{N}_L\} = \frac{\exp(-\dot{N}_L t)(\dot{N}_L t)^k}{k!}$$

where $\dot{N}_L$ is the short-term average link MR value.

$\dot{N}_L$ may be calculated by a computer model, such as METEORLINK, or determined from channel measurements. If measured, $\dot{N}_L$ must be the rate of trail-scattered signals exceeding some fixed received power threshold P_r^*, not the rate of correct message receipt. Deviations from Poisson behavior have been observed for the time between the receipt of short messages when a message piecing protocol was employed [172]. This protocol combined interarrival times between independent trail occurrences into the interval between correct message receipt, thus distorting the Poisson meteor arrival statistics. Careful counting of meteor trail-scattered signals of at least $\tau^* = 20\,\text{ms}$ duration have shown excellent agreement with the Poisson distribution [173].

If the density of burst occurrences is Poisson distributed in the short-term, then the cumulative distribution function (CDF) of the random interarrival time, $\underset{\approx}{\tau}_a$, of usable meteors during this period is given by the exponential CDF

$$\Pr\{\underset{\approx}{\tau}_a = \tau_a \mid \dot{N}_L\} = 1 - \exp(-\dot{N}_L \tau_a) \tag{3.119}$$

where τ_a is a sample of the random variable $\underset{\approx}{\tau}_a$ and its mean value $\bar{\underset{\approx}{\tau}}_a = \dot{N}_L$. The CDF given by Eq. 3.119 is conditioned on the value of $\bar{\underset{\approx}{\tau}}_a = \dot{N}_L$. This distribution may be inverted to yield

$$\tau_a = -\frac{\ln(1 - \Pr\{\tau_a = \underset{\approx}{\tau}_a \mid \bar{\underset{\approx}{\tau}}_a\})}{\bar{\underset{\approx}{\tau}}_a} \tag{3.120}$$

Since $\Pr\{\underset{\approx}{\tau}_a = \tau_a \mid \bar{\underset{\approx}{\tau}}_a\}$ is a uniformly distributed random variable on the interval [0, 1] [174], the value

$$z_a = 1 - \Pr\{\underset{\approx}{\tau}_a = \tau_a \mid \bar{\underset{\approx}{\tau}}_a\} \tag{3.121}$$

is a sample of the random variable $\underset{\approx}{z}_a$ also uniformly distributed on [0, 1]. Generating a sample of the random variable $\underset{\approx}{z}_a$ according to the uniform density on [0,1], symbolized as $z_a \in \mu_{01}$, yields the exponentially-distributed burst interarrival time sample

$$\tau_a = -\frac{\ln(z_a)}{\bar{\underset{\approx}{\tau}}_a} \tag{3.122}$$

with the exponential CDF given by Eq. 3.119 and the short-term average MR value, $\dot{N}_L$, computed by METEORLINK.

In practice, a burst may be usable in one link direction but not in the reverse direction [175]. The principal causes of this phenomena are

- gyrotropic effects in the ionosphere and meteor-trail plasma
- differing transmit-receive antenna common volume for each link direction
- differing link power budgets (transmitter power, noise level, etc.) in each link direction.

Thus, a burst generated for one link direction may be unusable in the opposite direction. Of course, this consideration is irrelevant for simulation of a broadcast link.

Let the MR value for the MB link from site 1 to site 2 be $\dot{N}_{12}$, and the MR value in the opposite link direction be $\dot{N}_{21}$. One approach to including link asymmetry in the Monte-Carlo simulation is to compute the ratio

$$\xi_a(\dot{N}_{12}, \dot{N}_{21}) = \frac{\min\{\dot{N}_{12}, \dot{N}_{21}\}}{\max\{\dot{N}_{12}, \dot{N}_{21}\}} \tag{3.123}$$

If $\dot{N}_{12} > \dot{N}_{21}$, then ξ_a is an estimate of the probability that a trail meeting the observability $(\mathscr{P}_s^{\mathrm{L}}, \hat{\mathbf{v}}_m^{\mathrm{O}})$ and usability (L^*, τ^*) criteria on the site 1-site 2 link will meet the corresponding criteria on the site 2-site 1 link. Similarly, if $\dot{N}_{12} < \dot{N}_{21}$, then ξ_a estimates the probability that a burst generated for the site 2-site 1 link will also support the site 1-site 2 link. Obviously, any burst generated using the smaller of the two MR values, or for the case $\dot{N}_{12} = \dot{N}_{21}$ $(\xi_a = 1)$, would be useful in both link directions. The ratio ξ_a is used to determine meteor bursts that support simultaneous bidirectional communications between sites 1 and 2.

The generation of usable meteor bursts is most easily described using the timeline employed by the Monte-Carlo simulation. Let t_0 be the current absolute time corresponding to the end of a usable trail-scattered signal between sites 1 and 2 and define τ_w^{12} to be the time expired since the last C_{12}-character message was correctly received at site 2 from site 1 (site 1-site 2 MB link). Similarly, let τ_w^{21} be the time since the last correct C_{21}-character message was received at site 1 from site 2 (site 2-site 1 MB link). In this context, completed message reception occurs at the moment when all characters in the message have been received, either via single or multiple trail-scattered signals (message piecing). At this point, the simulation repeats this procedure to determine the occurrence time for the next bidirectional burst.

Without loss of generality, assume $\dot{N}_{12} > \dot{N}_{21}$, so the minimum burst interarrival times will be determined for the site 1-site 2 link. Generate $z_{12} \in \mu_{01}$ and the corresponding burst interarrival time $\tau_a^{12}(1) = \tau_a$ from Eq. 3.122, so the current time becomes $t_0 + \tau_a^{12}(1)$. Next, choose $z_{21} \in \mu_{01}$ and compare this value to ξ_a. If $z_{21} \leq \xi_a$, then the burst is usable in both link directions. Otherwise, either site 2 does not receive a correct PRB signal or site 1 cannot receive site 2's ACQ signal. In this case, a second site 1-site 2 burst occurrence time $\tau_a^{12}(1)$ is generated. This process is continued until $z_{21} \leq \xi_a$ for some $\tau_a^{12}(k)$ or until the simulation determines that the corresponding elapsed time $\tau_{\text{arv}}^{12} = \Sigma_{j=1}^{k} \tau_a^{12}(j)$ exceeds some preset limit. For this development, assume that bidirectional communications is possible during a burst starting at time $t_0 + \tau_{\text{arv}}^{12}$.

Although this simulation development is based on the establishment of bidirectional MB links, unidirectional communications is possible for some limited amount of transmitted data or for one-way MB broadcast communications. In the broadcast mode, the master station transmits the same message, or message segments, for an adequate period to achieve an acceptable probability of correct message receipt at the intended destination. The receiving station or terminal does not transmit, so the master station receives no acknowledgment of the success or failure of the attempted communication.

3.3.2.2.2 Burst Duration. Once the occurrence time $t_0 + \tau_{\text{arv}}^{12}$ for the current burst has been determined, a sample, τ_b, of the random burst duration, $\underset{\sim}{\tau}_b$, of a usable trail-scattered signal must be generated. Three different approaches are possible:

- the inverse exponential distribution conditioned on average burst duration $\dot{T}_L / \dot{N}_L$
- the inverse distribution of burst durations $\tau_q(q)$ generated by METEORLINK during its integration of link DC values over all trail orientations and line densities $q > q_{sm}^*$
- the inverse of an empirically-derived burst duration distribution.

Although the third alternative has the advantage of accuracy for the link whose duration distribution was employed, this accuracy cannot be easily extended to other hypothetical link configurations of interest.

The use of an exponential distribution for $\underset{\approx}{\tau}_b$ has been previously evaluated and found to adequately represent measured behavior [176]. Thus, let $\dot{N}_L$ and $\dot{T}_L$ be the MR and DC values measures or predicted for the link to be simulated. Compute the average burst duration

$$\tau_b = -\frac{\ln(z_b)}{\bar{\underset{\approx}{\tau}}_b} \tag{3.124}$$

where $\bar{\underset{\approx}{\tau}}_b = \dot{T}_L / \dot{N}_L$ and $z_b \in \mu_{01}$.

Although both underdense and overdense trails are modeled in METEORLINK, the distribution of trail-scattered signal durations, $\underset{\approx}{\tau}_b$, from the model may differ significantly from the exponential distribution for long duration bursts. From Eq. 3.88, $n_t(q)$ is the rate of usable bursts for line densities between q and $q + dq$ producing burst durations $\tau_q(q)$. The density of these τ values can be created by summing the corresponding $n_t(q)$ values into bins according to the rule

$$\tau_b \in [i\,\Delta\tau, (i+1)\,\Delta\tau) \Rightarrow n_\tau(i) \leftarrow n_\tau(i) + n_t(q)\,\Delta q$$

where the relation $\Rightarrow$ designates "implies," the "$\leftarrow$" designates "replaced by," Δq is the incremental q value used in the DC quadrature, and $n_t(q)$ is the number density of τ_b values between $i\,\Delta\tau$ and $(i+1)\,\Delta\tau$ for $i = 1, 2, \ldots, M_\tau + 1$ (M_τ bins, each $\Delta\tau$ length). The CDF of burst duration for the METEORLINK-modeled MB link becomes

$$\Pr\{\underset{\approx}{\tau}_b < \tau_k\} \cong \frac{\sum_{j=1}^{k} n_\tau(j)}{\sum_{i=1}^{M_\tau} n_\tau(i)} \tag{3.125}$$

Sample points of $\underset{\approx}{\tau}_b$ and $\Pr\{\underset{\approx}{\tau}_b < \tau_k\}$ are then used in a curve fit algorithm, with the $\Pr\{\underset{\approx}{\tau}_b < \tau_k\}$ values as the independent variable and the τ_b values as the dependent variable. Burst durations may be generated for each simulated burst occurrence by first creating the uniformly distributed random variable $z_b \in \mu_{01}$ and then using the curve fit to determine the corresponding value of τ_b.

RSL multipath fading, typically observed for trail-scattered signals lasting longer than about 400 ms [177], may cause modem "drop-outs" and necessitate resynchronization. The single contiguous burst duration, τ_b, is reduced by the fading outages and subdivided into multiple bursts. In fact, a single trail-scattered signal on a FDX link may produce different RSL fading behavior in opposite link directions. This phenomenon would occur if the electrical path lengths differed between the transmit and receive antennas used in opposite link directions. In this treatment, it will be assumed that the fades are simultaneous in both link directions. Thus, each nonfading interval may be treated as a new burst event, with the number and duration of these nonfading intervals determined from measured distributions.

In general, different burst durations τ_b^{12} and τ_b^{21} are generated for each link direction. If $\dot{N}_{12} > \dot{N}_{21}$, then the DC value $\dot{T}_{12} > \dot{T}_{21}$. Since $\dot{N}_{12}$ increases faster than $\dot{T}_{12}$, that is, the number of usable of underdense bursts increases faster than the corresponding link DC contribution from those bursts, the average burst duration $\bar{\tau}_b^{12} = \dot{T}_{12}/\dot{N}_{12} < \bar{\tau}_b^{21} = \dot{T}_{21}/\dot{N}_{21}$. Thus, given that a bidirectional burst has occurred, the simulation computes the ratio

$$\xi_d(\dot{T}_{12}, \dot{T}_{21}) = \frac{\min\{\dot{T}_{12}, \dot{T}_{21}\}}{\max\{\dot{T}_{12}, \dot{T}_{21}\}}$$

as a burst duration scale factor. Assuming $\dot{N}_{12} > \dot{N}_{21}$, the burst duration for the site 2-site 1 link may then approximated by $\tau_b^{21} = \xi_b \tau_b^{12}$, where the burst duration for the site 1-site 2 link, τ_b^{12}, is generated from Eq. 3.124. Thus, the next burst occurs at a time $t_0 + \tau_b^{12}$, producing τ_b^{12} and $\tau_b^{21} < \tau_b^{12}$ burst durations on the site 1-to-site 2 and site 2-to-site 1 MB links, respectively.

3.3.2.3 Burst Communications

3.3.2.3.1 Half-Duplex. The MIL-STD-188-135C MB protocol for HDX communications between a master station (M station) and a remote terminal (R terminal) is depicted in the timeline of Fig. 3.54. In this example, it was assumed that the burst duration from the M station to the R terminal was τ_b^{MR} and the corresponding duration from the R terminal to the M station was $\tau_b^{RM} < \tau_b^{MR}$. For each generated burst (or nonfading portion of a fading burst), the amount of time required for the R terminal to completely receive the M station's probe signal, PRB(M), is given by

$$\tau_1 = \tau_{off} + \tau_{PRB}$$

where

τ_{off} = random offset from the beginning of the burst to the start of the probe signal

τ_{PRB} = duration of the probe signal

The random offset τ_{off} occurs because the burst may start at any time after a PRB(M) signal has begun, so the PRB(M) signal is not correctly received at the R terminal and no ACQ(R) response is generated. Since HDX operation requires probe-and-wait, an interprobe delay, τ_{dly}, is fixed between PRB(M) transmissions to account for propagation delay and give the R

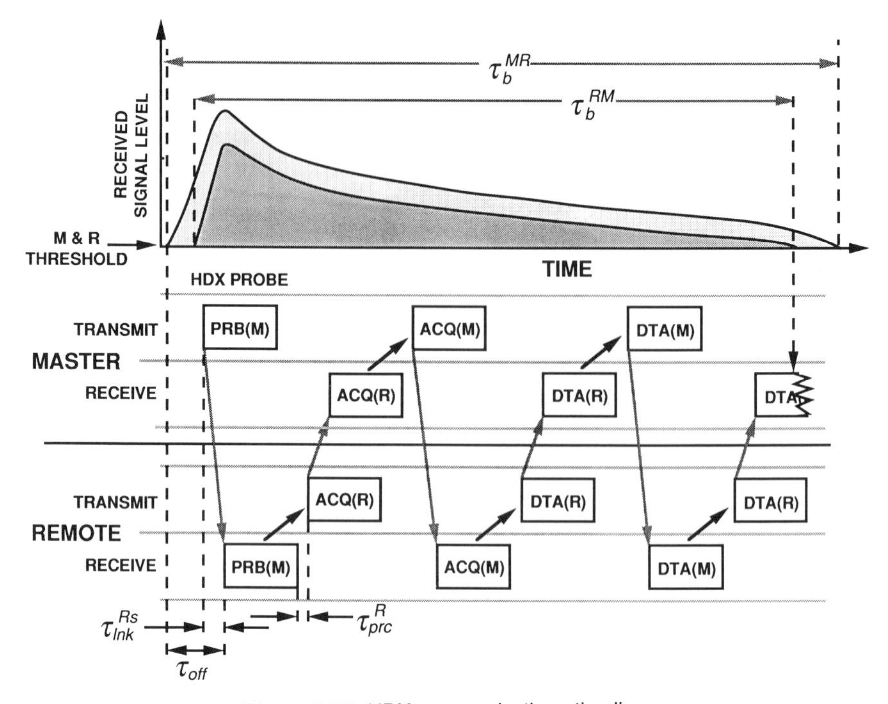

Figure 3.54. HDX communications timeline.

terminal adequate time to generate its ACQ(R) response. This delay includes time for the R terminal to receive the PRB(M) signal, generate the proper R terminal ACQ(R) response, switch-off the receiver, drive the transmitter to full output power, and transmit the ACQ(R) signal. The random offset τ_{off} can be generated by creating $z_{\text{off}} \in \mu_{01}$ and then computing

$$\tau_{\text{off}} = z_{\text{off}}(\tau_{\text{dly}} + \tau_{\text{PRB}})$$

where $\tau_{\text{dly}} + \tau_{\text{PRB}}$ is the maximum possible offset. If $\tau_1 > \tau_b^{\text{MR}}$, then the burst ends before the R terminal can receive the PRB(M) signal. If $\tau_b^{\text{MR}} > \tau_1 > \tau_b^{RM}$, the R terminal receives the PRB(M) signal, but the burst duration τ_b^{RM} is insufficient to support the R terminal's acknowledgment signal, ACQ(R). In either case, the M station continues to transmit the HDX PRB(M) signal and the remote continues channel monitoring. The simulated time since the last correct message receipt on the site 1-to-site 2, or M-to-R, link is increased to $\tau_w^{12} + \tau_{\text{arv}}^{12}$ and the corresponding time on the R-to-M link becomes $\tau_w^{21} + \tau_{\text{arv}}^{12}$. Finally, the absolute time is increased to $t_0 + \tau_{\text{arv}}^{12} + \tau_b^{\text{MR}}$.

If $\tau_b^{\text{MR}} > \tau_b^{\text{RM}} > \tau_1$, the remote correctly receives the PRB(M) signal, switches from receive to transmit, and sends its acquire signal, ACQ(R).

The ACQ(R) acknowledges that a channel exists, provides the R terminal's identification code (ID), and identifies the last correctly received message segment. In general, there is some PRB(M) signal processing delay before transmission of the acquire signal as well as propagation delay (negligible) from the R terminal to the trail-scatter point, $\mathcal{P}_s^{\mathrm{L}}$. Including these delays with the duration of the acquire signal, ACQ(R), yields

$$\tau_2 = \tau_1 + \tau_{\mathrm{prc}}^{\mathrm{R}} + \tau_{\mathrm{ACQ}}^{\mathrm{RM}} + \tau_{\mathrm{lnk}}^{\mathrm{Rs}}$$

where

τ_2 = minimum burst duration required for the M station to receive the ACQ(R) signal

τ_{prc}^{R} = processing time after PRB(M) reception prior to start of ACQ(R) transmission

$\tau_{\mathrm{ACQ}}^{\mathrm{RM}}$ = ACQ(R) signal duration

$\tau_{\mathrm{lnk}}^{\mathrm{R}s}$ = propagation delay

If $\tau_2 > \tau_b^{\mathrm{RM}}$, then the burst ends before the M station can completely receive the ACQ(R) signal. The M station will detect loss of signal and continue PRB(M) transmissions. Even if these PRB(M) transmissions begin before τ_b^{MR} has expired, no usable link exists for the master to receive the ACQ(R) signal. If $\tau_2 < \tau_b^{\mathrm{RM}}$, the M station will completely receive the ACQ(R), identity the R terminal ID, and determine the message pointer. If this pointer exceeds the message pointer retained by the M station, this reception would result in an update to the master's transmit queue. Assuming a usable trail exists, the master transmits its acquire signal, ACQ(M). If $\tau_2 > \tau_b^{\mathrm{MR}}$, however, the remote will not receive this ACQ(M) signal, nor will it receive any further R terminal transmissions during the current burst. After some minimal delay $\tau_{\mathrm{rtx}}^{\mathrm{M}}$, the M station will begin retransmitting its PRB(M) signal. Thus, if either $\tau_2 > \tau_b^{\mathrm{RM}}$ or $\tau_2 > \tau_b^{\mathrm{MR}}$, no message characters can be exchanged and a new burst must be generated.

Otherwise, the burst duration required for the R terminal to receive a correct ACQ(M) signal is given by

$$\tau_b^{\mathrm{MR}} > \tau_3 = \tau_2 + \tau_{\mathrm{prc}}^{\mathrm{M}} + \tau_{\mathrm{ACQ}}^{\mathrm{MR}} + \tau_{\mathrm{lnk}}^{\mathrm{Ms}}$$

where $\tau_{\mathrm{prc}}^{\mathrm{M}}$, $\tau_{\mathrm{ACQ}}^{\mathrm{MR}}$, and $\tau_{\mathrm{lnk}}^{\mathrm{Ms}}$ are the processing delay, ACQ(M) signal duration, and average propagation delay from site 1 (M) to the trail-scatter point, $\mathcal{P}_s^{\mathrm{L}}$, respectively.

In this HDX protocol scheme, the R terminal transmits message segments from its transmit queue in a DTA(R) block (see Fig. 3.54) immediately following receipt of a valid ACQ(M) signal. The DTA(R) block consists of a control frame followed by four message segments. Each message segment contains 14 characters (bytes) followed by two cyclic redundancy check (CRC) bytes. The four message segments must be received in the correct order, that is, there is no selective repeat feature in the version of MIL-STD-188-135 used as a guideline for simulation development. The control frame is used to update the message segment pointer at the M station. Thus, the duration of the DTA(R) block is

$$\tau_{\mathrm{DTA}}^{\mathrm{RM}} = \tau_{\mathrm{ctl}}^{\mathrm{RM}} + n_{\mathrm{seg}}^{\mathrm{RM}}(1)\tau_{\mathrm{seg}}^{\mathrm{RM}}$$

where

$n_{\mathrm{seg}}^{\mathrm{RM}}(1)$ = number of 14-character message segments in the first DTA(R) block received at the M station (1, 2, 3, or 4)

$\tau_{\mathrm{seg}}^{\mathrm{RM}}$ = duration of a single 14-character message segment

$\tau_{\mathrm{ctl}}^{\mathrm{RM}}$ = duration of the control frame

transmitted by the R terminal. Thus, if the trail duration

$$\tau_b^{\mathrm{RM}} > \tau_4 = \tau_3 + \tau_{\mathrm{prc}}^{\mathrm{R}} + \tau_{\mathrm{DTA}}^{\mathrm{RM}} + \tau_{\mathrm{lnk}}^{\mathrm{R}s}$$

for $n_{\mathrm{seg}}^{\mathrm{RM}}(1) = 1, 2, 3$, or 4, then the M station will receive $n_{\mathrm{seg}}^{\mathrm{RM}}(1) = 1, 2, 3$, or 4, respectively, segments from the R terminal's transmit queue. It has been assumed that the processing delay, $\tau_{\mathrm{prc}}^{\mathrm{R}}$, is also the delay from ACQ(M) reception prior to start of DTA(R) transmission. The M station will receive fewer than four segments if:

- fewer than four segments were in the R terminal's transmit queue for transmission
- one or more of the segments was received incorrectly due to trail-scattered RSL fading, noise, or interference
- the trail-scattered signal lifetime had expired.

The transmit queues at both the master and R terminal in the METEOR-WAIT simulation are effectively infinite, so link MWT values are determined from channel performance and protocol structure rather than message I/O delay. If $\tau_b^{\mathrm{RM}} < \tau_4$ for $n_{\mathrm{seg}}^{\mathrm{RM}}(1) = 1$, then the M station receives no segments from the R terminal, assumes the trail-scattered signal has become unusable, and resumes PRB(M) transmissions.

If message segments are received at the M station, that is, if $\tau_b^{\mathrm{RM}} > \tau_4$ for some $n_{\mathrm{seg}}^{\mathrm{RM}}(1)$ value ($n_{\mathrm{seg}}^{\mathrm{RM}}(1) = 4$ in Fig. 3.54), the M station will attempt to transmit its data block, DTA(M). This DTA(M) block contains the control frame (pointing at the last DTA(M) message segment correctly received from the R terminal) and up to four 14-character message segments for transmission to the R terminal. One or more of these DTA(M) blocks will be received by the R terminal if the burst duration

$$\tau_b^{MR} > \tau_5 = \tau_4 + \tau_{prc}^{M} + \tau_{DTA}^{MR} + \tau_{lnk}^{Ms}$$

where

$$\tau_{DTA}^{MR} = \tau_{ctl}^{MR} + n_{seg}^{MR}(1)\tau_{seg}^{MR}$$

with

$n_{seg}^{MR}(1)$ = number of 14-character message segments in the first DTA(R) block (maximum of four segments per block)

τ_{seg}^{MR} = duration of a single 14-character message segment

τ_{ctl}^{MR} = duration of the control frame

transmitted by the M station. If $\tau_b^{MR} \leq \tau_5$ for $n_{seg}^{MR}(1) = 1$, then the R terminal receives no message segments from the M station. This process continues with

$$\tau_b^{RM} > \tau_k = \tau_{k-1} + \tau_{prc}^{R} + \tau_{DTA}^{RM} + \tau_{\ln k}^{Rs}$$

$$\tau_b^{MR} > \tau_{k+1} = \tau_k + \tau_{prc}^{M} + \tau_{DTA}^{MR} + \tau_{\ln k}^{Ms}$$

as shown in the figure.

If $\tau_{k-1} < \tau_b^{RM} < \tau_k$ with $k \geq 4$ and $n_{seg}^{MR} = 1, 2$, or 3, the remote will have successfully transmitted

$$N_{seg}^{RM}(\tau_b^{RM}) = 4\,\mathrm{int}\left(\frac{k-3}{2}\right) + n_{seg}^{RM}(k)$$

14-character segments to the M station during the current burst, where int(u) is the greatest integer less than u. Let $\tau_k < \tau_b^{MR} < \tau_{k+1}$ $(\tau_b^{MR} > \tau_b^{RM})$ with $k \geq 4$ and $n_{seg}^{MR}(\mathrm{k}+1) = 1, 2$, or 3, then the M station will have successfully transmitted

$$N_{seg}^{MR}(\tau_b^{MR}) = 4\,\mathrm{int}\left(\frac{k-4}{2}\right) + n_{seg}^{MR}(k+1)$$

14-character message segments to the R terminal.

3.3.2.3.2 Full-Duplex. In FDX communications between two M stations, M_1 at site 1 and M_2 at site 2, at least one station must transmit channel probe signals. In Fig. 3.55, both M stations are shown transmitting their respective probe signals, PRB(M_1) and PRB(M_2), when a usable trail is

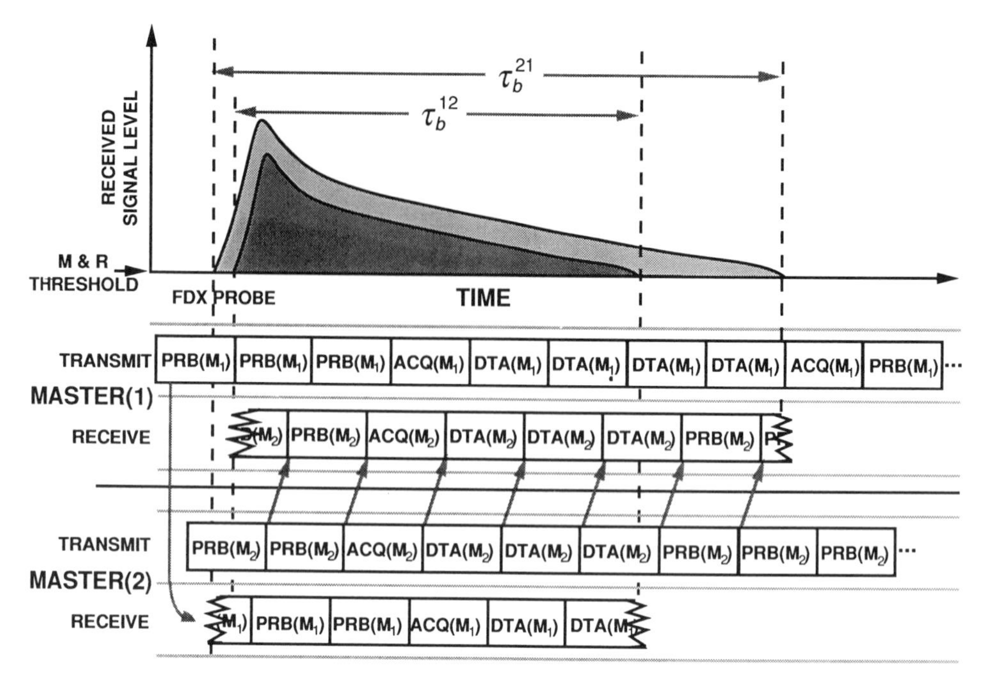

Figure 3.55. FDX communications timeline.

formed. Of course, these probe signals were not synchronized between stations M_1 and M_2, so PRB(M_1) and PRB(M_2) are offset in time as shown in the figure. It was assumed in the following development that a usable trail has formed simultaneously on the M_1-to-M_2 and M_2-to-M_1 MB links. The burst duration on the M_1-to-M_2 MB link was given by τ_b^{21} and the corresponding duration on the M_2-to-M_1 link was τ_b^{21}, where it was assumed that $\tau_b^{21} > \tau_b^{12}$. In addition, it will be assumed that both M_1 and M_2 stations operate at the same bit transmission, or burst, rate. This restriction may be necessary for the PRB(M_1) and PRB(M_2) signals in practice, but it is not valid in general for asymmetrical links employing trail-adaptive burst rates.

The simulation determines the absolute start time, t_0^{12}, of the M_1-to-M_2 burst from the start time, t_0^{21}, of the M_2-to-M_1 burst as

$$t_0^{12} = t_0^{21} + \tau_{\text{dif}}^{21} = t_0^{21} + z_0^{21}(\tau_b^{21} - \tau_b^{12})$$

where $z_0^{21} \in \mu_{01}$. The offset from the start of the M_1-to-M_2 burst until correct reception of the first PRB(M_1) signal is given by

$$\tau_{\text{off}}^{12} = z_{\text{off}}^{12} \tau_{\text{PRB}}^{12}$$

where $z_{\text{off}}^{12} \in \mu_{01}$ and τ_{PRB}^{12} is the duration of the PRB(M_1) signal. Similarly, τ_{off}^{21} can be generated from

$$\tau_{\text{off}}^{21} = z_{\text{off}}^{21} \tau_{\text{PRB}}^{21}$$

where $z_{\text{off}}^{21} \in \mu_{01}$ and τ_{PRB}^{21} is the duration of the PRB(M_2) signal. At this point, it will be assumed that $\tau_{\text{PRB}}^{21} = \tau_{\text{PRB}}^{12} = \tau_{\text{PRB}}$.

The timeline in Fig. 3.55 shows a scenario in which the M_2-to-M_1 burst begins prior to the start of the M_1-to-M_2 burst, but the PRB(M_1) signal is received at station M_2 before the PRB(M_2) signal is received at station M_1. This probe signal timing occurred because the PRB(M_2) signal was transmitted before the start of the M_2-to-M_1 burst, while the PRB(M_1) signal was transmitted after the start of the corresponding M_1-to-M_2 burst. In symbols, PRB(M_2) is correctly received at station M_1 beginning at time

$$t_{\text{PRB}}^{21} = t_0^{21} + \tau_{\text{off}}^{21}$$

while PRB(M_1) is received at station M_2 beginning at time

$$t_{\text{PRB}}^{12} = t_0^{12} + \tau_{\text{off}}^{12}$$

In this case, the difference in burst start times $\tau_{\text{dif}}^{21} \leq (\tau_{\text{off}}^{21} - \tau_{\text{off}}^{12})$. In general, if the PRB(M_1) signal duration $\tau_{\text{PRB}} < \tau_{\text{dif}}^{21}$, then the PRB($M_2$) signal will be received by station M_1 before the PRB(M_1) signal would be received by station M_2.

As shown in Fig. 3.55, both M stations transmitted their ACQ frames immediately after receiving a valid PRB signal and completing their current PRB signal transmissions. Since station M_2 correctly received the PRB(M_1) signal before station M_1 could receive the PRB(M_2) signal, station M_2 was the first to transmit its ACQ(M_2) frame. Station M_1 will receive the first valid ACQ frame from station M_2 if the burst duration

$$\tau_b^{21} > t_{\text{ACQ}}^{21} - t_0^{21} = (t_{\text{PRB}}^{21} + \tau_{\text{PRB}} + \tau_{\text{cmp}}^{2} + \tau_{\text{ACQ}}^{21}) - t_0^{21}$$

where

t_{ACQ}^{21} = absolute time at which station M_1 receives a valid ACQ(M_2) frame

τ_{cmp}^{2} = time from station M_2's receipt of a valid PRB(M_1) signal until station M_2 places an ACQ(M_2) signal next in the transmit queue, that is, completes the current PRB(M_2) signal

τ_{ACQ}^{21} = duration of the ACQ(M_2) signal

Otherwise, M_1 will not receive the ACQ(M_2) frame.

Shortly after station M_2 received the PRB(M_1) signal, station M_1 received the PRB(M_2) signal and transmitted its ACQ(M_1) signal. This ACQ(M_1) signal will be received at station M_2 if the burst duration

$$\tau_b^{12} > t_{\mathrm{ACQ}}^{12} - t_0^{12} = (t_{\mathrm{PRB}}^{12} + \tau_{\mathrm{PRB}} + \tau_{\mathrm{cmp}}^{1} + \tau_{\mathrm{ACQ}}^{12}) - t_0^{12}$$

where

t_{ACQ}^{12} = absolute time at which station M_2 receives a valid ACQ(M_1) frame

τ_{cmp}^{1} = time from station M_1's receipt of a valid PRB(M_2) signal until station M_1 places an ACQ(M_1) signal next in the transmit queue, that is, completes the current PRB(M_1) signal

τ_{ACQ}^{12} = duration of the ACQ(M_1) signal

If $\tau_b^{12} < t_{\mathrm{ACQ}}^{12} - t_0^{12}$, then station M_2 would not receive an ACQ(M_1) signal.

In the example shown in Fig. 3.55, station M_2 received the PRB(M_1) signal before station M_1 received the PRB(M_2) signal, so station M_2 was the first to begin transmission of its ACQ(M_2) signal and subsequent DTA(M_2) blocks. Station M_2 monitored its receiver for the ACQ(M_1) signal and DTA(M_1) blocks. It would maintain burst synchronization for the duration τ_b^{12} beginning at $t_0^{12} > t_0^{21}$ and receive data frames until

$$\tau_b^{12} < t_{\mathrm{ACQ}}^{12} - t_0^{12} + (N_{\mathrm{DTA}}^{12} + 1)\tau_{\mathrm{DTA}}^{12}$$

where

N_{DTA}^{12} = number of DTA(M_1) blocks correctly received at station M_2 during the current M_1-to-M_2 burst

and

$$\tau_{\mathrm{DTA}}^{12} = \tau_{\mathrm{ctl}}^{12} + 4\tau_{\mathrm{seg}}^{12}$$

with

τ_{seg}^{12} = duration of a single 14-character message segment transmitted by station M_1

τ_{ctl}^{12} = duration of the station M_1 control frame transmitted by station M_1

τ_{DTA}^{12} is the maximum duration of the station M_1 data block, DTA(M_1), transmitted to station M_2. The τ values τ_{DTA}^{21}, τ_{seg}^{21}, and τ_{ctl}^{21} defined for the M_2-to-M_1 link correspond to the values τ_{DTA}^{12}, τ_{seg}^{12}, and τ_{ctl}^{12} defined for the M_1-to-M_2 link.

In the example of Fig. 3.55, station M_2 was the first to lose bit synchronization. The number of segments received by station M_2 before synchronization was lost is given (in general) by

$$N_{\text{seg}}^{12}(\tau_b^{12}) = 4N_{\text{DTA}}^{12} + n_{\text{seg}}^{12}$$

where $n_{\text{seg}}^{12} < 4$ residue segments are correctly received from DTA(M_1) block number $N_{\text{DTA}}^{12} + 1$ before the M_1-to-M_2 burst ended. In other words, n_{seg}^{12} is the number of correctly received 14-character message segments from the last DTA(M_2) block received at the M_1 station (1, 2, or 3). Once synchronization was lost at station M_2, it completed transmission of the current data block, DTA(M_2), in the transmit queue and restarted transmission of its PRB(M_2) signals.

As shown in Fig. 3.55, station M_1 continued to receive DTA(M_2) blocks until station M_2 lost synchronization and restarted PRB(M_2) signal transmission. Station M_1 received these probe transmissions and determined that station M_2 had lost synchronization. At this point, station M_1 completed transmission of its current DTA(M_1) block and then transmitted an ACQ(M_1) signal in response to the PRB(M_2) signal. Finally, station M_1 detected the end of the usable trail duration, τ_b^{21}, and restarted PRB(M_1) signal transmissions. The number of segments received by station M_1 during this burst was

$$N_{\text{seg}}^{21}(\tau_b^{21}) = 4N_{\text{DTA}}^{21} + n_{\text{seg}}^{21}$$

where $n_{\text{seg}}^{12} < 4$ residue segments were correctly received from DTA(M_2) block number $N_{\text{DTA}}^{21} + 1$.

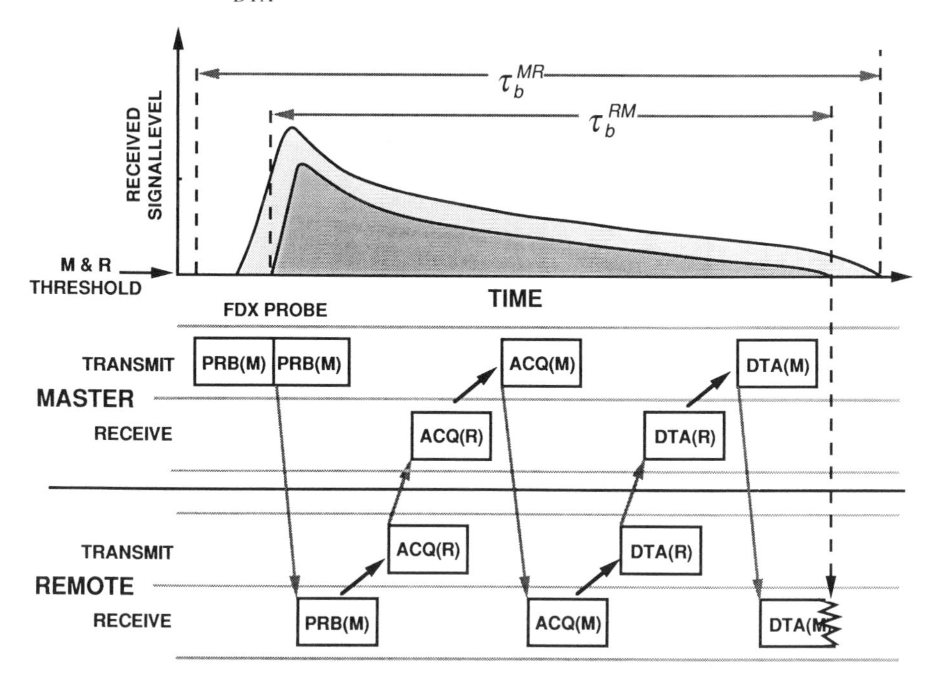

Figure 3.56. FDX probe communications timeline.

3.3.2.3.3 Full-Duplex Probe. A hybrid of the HDX and FDX protocols, called FDX probe, is shown in Fig. 3.56. In this configuration, the R terminal is capable of transmitting and receiving on two frequencies, albeit not simultaneously. The principal advantage of this technique over probe-and-wait is that the PRB(M) signals can be transmitted contiguously. The interprobe delay $\tau_{\text{dly}} = 0$, thus minimizing the potential usable burst duration lost in the classic HDX probe-and-wait protocol.

3.3.2.4 Status Update. Once the entire burst duration, τ_b, has been assigned to protocol overhead and message segments, the number of characters received at each M station or R terminal can be determined. The transmit queues at each station or terminal are refilled with more message segments from the current message or, if necessary, segments from the next message. A message may be considered successfully transmitted when all of its segments have been correctly received and reassembled at the intended destination. The generated MWT sample value, $\tau_w^{21} + \tau_{\text{arv}}^{12} + \tau_b^{21} (\tau_b^{21} > \tau_b^{12}$ in this example), is then calculated as the time expired since the immediately preceding message was completely received. For analysis purposes, it is assumed that all messages contain the same number of characters.

Alternative measures for MWT includes the

- time interval measured between valid message acknowledgments received by the message source
- time measured from the instant a message enters the transmit queue until it is correctly received at the intended destination
- time measured from the instant a message enters the transmit queue until a valid message acknowledgment is received by the message source

Each of these measures produces a different MWT sample and therefore different predictions of link performance. For example, Fig. 3.55 shows a DTA(M_2) block correctly received by station M_1, but the burst does not support further transmission to station M_2. Thus, the message pointer update in M_2's control signal (τ_{ctl}^{21}) is not received by station M_2 on the current burst. Assume this DTA(M_2) block completed the current message transmitted on the M_2-to-M_1 link. If the MWT sample is measured between correct message receipts, the resulting MWT sample is $\tau_w^{21} + \tau_{\text{arv}}^{12} + \tau_b^{21}$. If the MWT value is determined between receipt of message acknowledgments, the MWT sample has the minimum value $\tau_w^{21} + \tau_{\text{arv}}^{12} + \tau_b^{21} + \tau_a^{12}$, where τ_a^{12} is the arrival time of the next usable burst.

3.3.2.5 Link MWT Statistics. Each MWT sample, τ_{w_i}, generated from the application of the HDX, FDX, or FDX probe protocols to simulated bursts, is used to compute the sample mean, variance, and the distribution function

$P_{\tau_w}(\tau_{w_i})$. The sample mean MWT value is formed in the obvious way as

$$\bar{T}_w = \frac{\sum_{i=1}^{M_w} \tau_{w_i}}{M_w}$$

and the unbiased estimate of the variance becomes

$$s_w^2 = \frac{\sum_{i=1}^{M_w} (\tau_{w_i} - \bar{T}_w)^2}{M_w - 1}$$

$$= \frac{\sum_{i=1}^{M_w} (\tau_{w_i})^2 - (\bar{T}_w)^2}{M_w - 1}$$

The MWT value, $\bar{T}_w$, quantifies MB link communications performance for the simulated protocol, data rates, and message lengths as well as link MR and DC values. As M_w increases, the sample mean value, $\bar{T}_w$, approaches the actual mean MWT value, μ_w, and the variance estimate, s_w^2, approaches the actual MWT variance, σ_w^2.

The values μ_w and σ_w^2 describe moments of the MWT density function $p_{\tau_w}(\tau_w)$ whose distribution function, $P_{\tau_w}(\tau_w)$, may be approximated by $F_{\tau_w}(\tau_w)$. The function $F_{\tau_w}(\tau_w)$ can be determined by dividing the probable range of MWT sample values, τ_{w_i}, into accumulation bins. Each bin is assigned a counter value, n_{w_k}, corresponding to a τ_w interval $\tau_{w_k} \le \tau_w < \tau_{w_{k+1}}$. This counter value is incremented by 1, that is, $n_{w_k} \leftarrow n_{w_k} + 1$, for each interval $\tau_{w_k} \le \tau_w < \tau_{w_{k+1}}$ such that the τ_w sample value $\tau_{w_i} \ge \tau_{w_k}$. Dividing each bin counter value by M_w yields the approximate cumulative distribution function $F_{\tau_w}(\tau_w)$.

The MWT statistics μ_w, σ_w^2, and $p_{\tau_w}(\tau_w)$ quantify MB link performance given the Monte-Carlo simulation parameters and conditioned on the corresponding link MR and DC values. These MR and DC values $\dot{N}_L$ and $\dot{T}_L$, respectively, are actually short-term average values of random time functions $\underset{\approx}{\dot{\mathrm{N}}}_L(t)$ and $\underset{\approx}{\dot{\mathrm{T}}}_L(t)$, whose statistics are considered stationary for the short-term of 1 to 2 h. In symbols, $\dot{N}_L = \mathrm{E}\{\underset{\approx}{\dot{\mathrm{N}}}_L\}$ and $\dot{T}_L = \mathrm{E}\{\underset{\approx}{\dot{\mathrm{T}}}_L\}$. The remaining simulation parameters are essentially deterministic. Thus, the density function of MWT values is more accurately represented by the conditional density function $p_{\tau_w}(\tau_w | \underset{\approx}{\dot{\mathrm{N}}}_L = \dot{\mathrm{N}}_L, \underset{\approx}{\dot{\mathrm{T}}}_L = \dot{\mathrm{T}}_L)$, where $\dot{\mathrm{N}}_L$ and $\dot{T}_L$ are sample values of $\underset{\approx}{\dot{\mathrm{N}}}_L$ and $\underset{\approx}{\dot{\mathrm{T}}}_L$, respectively.

As shown in Section 3.2, the DC value $\dot{T}_L$ may be modeled as a deterministic function of MR value $\dot{N}_L$, or equivalently, $\dot{\mathrm{T}}_L$, is a deterministic function of $\dot{\mathrm{N}}_L$. Each $\dot{\mathrm{N}}_L$ value is ultimately the weighted sum of

SMRD-map contributions, where each contribution is a random variable Θ_n, with corresponding sample value Θ_n for $n = 1, 2, \ldots, M_\Theta$. The short-term link MWT statistics can then be related to the short-term statistics of the meteor arrival process. Integrating $p_{\tau_w}(\tau_w | \dot{N}_L = \dot{N}_L, \dot{T}_L = \dot{T}_L)$ over the joint density functions $p_{\tau_w}(\dot{T}_L | \dot{N}_L = \dot{N}_L)$, $p_{\dot{N}_L}(\dot{N}_L | \Theta_1, \Theta_2, \ldots, \Theta_{M_\Theta})$, and $p_{\Theta_n}(\Theta_1, \Theta_2, \ldots, \Theta_{M_\Theta})$ yields the ultimate distribution function for the short-term MWT value. In practice, $p_{\Theta_n}(\Theta_1, \Theta_2, \ldots, \Theta_{M_\Theta})$ is unknown, so MWT statistics must be generated for a range of predicted MR and DC values.

3.3.2.6 Simulation Control. The number of bursts needed to achieve the required precision for the estimates $\bar{T}_w$, s_w^2, and $F_{\tau_w}(\tau_w)$ determine the duration of the Monte-Carlo simulation. Once this precision has been achieved, the final MWT statistics have then determined and the simulation ends. An estimate of the number of samples, M_w^*, needed to state with confidence $100\,(1 - 2\,P_w)\%$ that $\bar{T}_w$ is within ϵ_w of μ_w can be given by [178] the expression

$$M_w^* = \left[\frac{z(P_w) s_w}{\epsilon_w} \right]^2 \qquad (3.126)$$

where

$z(P_w)$ = upper value of the random variable z_n with normal density n(0, 1) corresponding to a probability P_w, for example, 95% confidence implies $P_w = 0.025$ and $z(0.025) = 1.960$

ϵ_w = half-width of the interval surrounding $\bar{T}_w$ such that the probability $\Pr\{\bar{T}_w - \epsilon_w \le \mu_w \le \bar{T}_w + \epsilon_w\} = 1 - 2P_w$.

Since Eq. 3.126 is dependent on the estimate s_w^2 of σ_w^2, the required number of samples M_w^* changes as the simulation progresses.

The precision ϵ_w^* of the mean MWT estimate $\bar{T}_w$ may be computed for any number of τ_w samples $M_w > 30$, given the desired confidence level 100 $(1 - 2P_w)$ from [179]

$$\epsilon_w^* = \frac{z(P_w) s_w}{\sqrt{M_w^*}}$$

Similarly, the precision ϵ_w and number of τ_w samples, M_w, may be specified and the confidence level computed from

$$P_w^* = z_n^{-1}\left(\frac{\epsilon_w \sqrt{M_w}}{s_w} \right)$$

corresponding to $100(1 - 2P_w^*)\%$ confidence.

These calculations may be used to measure precision with confidence for other discrete points along the estimated τ_w sample distribution function $F_{\tau_w}(\tau_w)$ in addition to the mean value $\bar{T}_w$. Define the random variable

$$z_{ik}(\tau_{w_i}) = \begin{cases} 1 & \text{if } \tau_{w_i} \leq \tau_{w_k} \\ 0 & \text{if } \tau_{w_i} > \tau_{w_k} \end{cases}$$

for each τ_w bin $\tau_{w_k} \leq \tau_w < \tau_{w_{k+1}}$ with corresponding counter value, n_{w_k}, and each MWT sample τ_{w_i}, $i = 1, 2, \ldots, M_w$. Given $z_{ik}(\tau_{w_i})$,

$$\begin{aligned} E\{z_{ik}(\tau_{w_i})\} &= 1 \Pr\{z_{ik}(\tau_{w_i}) = 1\} + 0 \Pr\{z_{ik}(\tau_{w_i}) = 0\} = \Pr\{z_{ik}(\tau_{w_i}) = 1\} \\ &= F_{\tau_w}(\tau_w) \end{aligned}$$

so the estimate becomes

$$F_{\tau_w}(\tau_{w_i}) = \frac{\sum_k z_{ik}(\tau_{w_i})}{M_w}$$

and the techniques described relating P_w, ϵ_w, and s_w to the estimate $\bar{T}_w$ can be applied to the random variables $z_{ik}(\tau_{w_i})$ and therefore $F_{\tau_w}(\tau_w)$.

3.3.2.7 Sensitivity Analysis. The METEORWAIT simulation was used to predict MWT values for messages with fixed lengths of 10, 100, and 1000 characters using a simulated implementation of the draft protocol standard MIL-STD-188-135C. Results were provided for both HDX and FDX protocols assuming bidirectional link communication of equal-length messages. A fixed burst rate of 8 kbps was assumed with MR values of 0.5, 2.0, and 10 meteors per minute combined with a fixed DC value of 1.3%. These MR and DC values were chosen to provide average burst durations spanning typical performance for short-, medium-, and long-range MB links. In addition to the PRB, ACQ, and DTA bit lengths defined by the draft standard, a transmitter and receiver turn-on delay of 5 ms was employed and an inter-PRB spacing (response waiting time) of 30 ms was used in the HDX case. The results of this analysis are intended for illustrative purposes only and are not intended to characterize the performance of MIL-STD-188-135C in particular.

MWT values for the HDX link protocol are given in Table 3.4, where "M" designates master station and "R" designates the remote terminal with all tabulated MWT values provided in units of minutes. Clearly, the MWT values increase with increasing message length, from below 0.1 minutes for 10 character messages at 2 meteors per minute to as high as about 40 minutes for 1000-character messages at 10 meteors per minute. The higher MR values associated with fixed channel DC values force a shorter average burst duration $\bar{\tau}_b = \dot{T}_L / \dot{N}_L$, so fewer characters are exchanged per

TABLE 3.4 MWT Predictions for HDX Protocol Example

Message Length (characters)	DC: 1.3% MR[a]	HDX Protocol	
		R-TO-M	M-TO-R
10	0.5	0.098	0.110
	2.0	0.076	0.095
	10.0	0.181	0.629
100	0.5	0.528	0.572
	2.0	0.624	0.815
	10.0	1.380	4.770
1000	0.5	4.840	5.170
	2.0	5.290	6.840
	10.0	12.40	42.40

[a] MR value in usable meteors per minute.

burst and the ratio of overhead bytes to message characters necessarily increases.

The table shows that the R-to-M link MWT values are always smaller than the corresponding M-to-R MWT values. This result is due to the draft protocol standard implementation, which requires the remote to transmit its DTA(R) blocks earliest in the trail-scattered signal lifetime. This scheme biases HDX link protocol in favor of R terminal transmissions, because the M station cannot transmit its DTA(M) blocks until later in the signal's lifetime, after the R terminals first DTA(R) blocks (if any) have been received. Since burst waiting time increases as the minimum required signal duration increases, many signals last long enough to support the R-to-M link but degrade sufficiently to prevent some or all of the M station's DTA(M) blocks from being correctly received.

This approach is appropriate in practice, however, when the M station is much less physically and electrically constrained than the R terminal. In some MB applications, the R terminal must employ a relatively low gain antenna and low transmitter power. The M station, on the other hand, can employ increased receive antenna gain to compensate somewhat for the R terminal's lower effective radiated power (ERP). In fact, the M station may employ the adaptive antenna techniques described in Section 3.2.2.7.2. In addition, the M station may employ much higher transmitter powers than the R terminal, so the M-to-R MB link generally provides much higher MR and DC values than the corresponding R-to-M link. Typically, the R terminal must transmit a short message, usually telemetry, to the M station. Therefore, structuring the MB protocol to permit the R terminal to transmit its DTA(R) blocks early in the trail-scattered signal lifetime is consistent with this application.

Table 3.5 provides the FDX protocol MWT predictions for the 10-, 100-, and 1000-character message lengths between to M stations. The FDX MWT

TABLE 3.5 MWT Predictions for FDX Protocol Example

Message Length (characters)	DC: 1.3% MR[a]	FDX Protocol	
		M_1-to-M_2	M_2-to-M_1
10	0.5	0.072	0.066
	2.0	0.047	0.047
	10.0	0.174	0.180
100	0.5	0.228	0.228
	2.0	0.366	0.364
	10.0	1.130	1.140
1000	0.5	2.510	2.510
	2.0	3.000	3.000
	10.0	9.740	9.750

[a] MR value in usable meteors per minute.

values have decreased by less than a factor of 0.5 from the corresponding HDX values. This result is due to the more efficient use of the trail for simultaneous transmission and reception using separate transmit and receive frequencies, as compared to the alternating transmit/receive scheme characteristic of single frequency, HDX protocols. Furthermore, since there is no precedence for DTA block transmission, both M stations may transmit their corresponding DTA(M) blocks simultaneously. Thus, there is little difference between the MWT values predicted for opposite link directions.

3.3.3 Network Simulation

3.3.3.1 Overall Approach. The development of a Monte Carlo simulation for modeling either sensor telemetry systems, such as SNOTEL [180], or wide-area telecommunication systems, such as the Chinese network [181], was a direct outgrowth of the METEORWAIT Monte Carlo simulation. Since the METEORWAIT program was developed to predict the performance of general MB link protocols, the addition of multiple transmit/receive sites (network node), the concomitant network connectivity matrix (network links), and a burst-by-burst time base provided the fundamental components for a MB network simulation. In this context, the network nodes consist of M stations or R terminals that are geographically spaced to lie in independent trail footprints. In other words, each simulated trail-scatter event produces a usable channel between exactly two nodes in the network. If these node locations permitted a shared common volume to a third node, then the trail footprint (see Section 3.2.3) or diversity models (see Section 3.2.4) would provide the more appropriate modeling approaches.

The number of potential network performance measures includes the standard single MB link measures, but adds additional quantities endemic to the network environment. For example, probability density functions may be envisioned for

- MWT values for all MB links in the network (network-wide link performance)
- MWT values for message transfer from each node to all other nodes conditioned on the maximum number of relays allowed (network MB connectivity)
- MWT values for short message transfer from one or more source nodes to all other nodes (broadcast performance)
- node busy time, that is, percentage of time spent transmitting or receiving trail-scattered signals
- the number of relays required for message delivery
- transmit and receive queue delay

The use of MWT as a measurement objective may be replaced with network throughput, if the quantity of transmitted data, and not its timeliness, were of primary importance for a particular application. In general, the actual performance measures to be determined from the simulation are dependent on the communications objectives of the MB network to be modeled.

Assuming negligible contention for usable trail-scattered signals in a MB network with adequately spaced nodes (see Section 3.2.3 and 3.2.4) [182], the network simulation employs the following procedures:

1. Generate the time-sequenced occurrence of meteor bursts on each viable MB link in the network.
2. For the earliest burst occurrence anywhere in the network, determine the message traffic exchanged between the two nodes.
3. Update the appropriate transmit and receive queues at each of the two nodes linked by the burst.
4. Generate the next burst occurrence time for the link between the two nodes defining the previously simulated burst.
5. Repeat this procedure from step 2 until the specified performance estimates have been obtained with the desired precision and confidence.

The exchange of data between nodes as well as the update of node transmit and receive queues following each generated trail-scatter event is modeled exactly as described for the link simulation discussed in Section 3.3.2.

Both point-to-point and network messages may be modeled in the METEORWAIT network simulation. Point-to-point messages are designed for use between two nodes employing a single MB link, that is, they have a direct MB link within the network. An important use of point-to-point messages occurs in network consisting of multiple star subnetworks, in which each star configuration consists of an M station directly connected to each of several R terminals. In general, these R terminals cannot communicate with stations or terminals in the network other than the M station at the center of their star subnetwork. In this case, communication between the M station and the corresponding R terminals would exclusively employ point-to-point messages. Communication between star subnetworks is performed by the M stations, which serve as gateways between the network at large and their corresponding subnetworks. In general, these M stations may be separated by distances which cannot be spanned by a single MB link. In this example, network messages contain routing information to support MB relay route selection between star subnetworks. In this context, this information would consist of an ordered list of M stations describing an optimal connected path of MB links between subnetworks. Of course, a combination of point-to-point and network messages would be employed in a general, possibly time-varying, MB network environment. Although the point-to-point message transmission avoids multiple queue delays, these delays combined with intervening MB link MWT values may nevertheless yield shorter end-to-end MWT values than a poorly designed, or overutilized, direct link.

3.3.3.2 Generation of Network Burst History. Each of the $M_n^2 - M_n$ (one-way) links in an M_n-node MB network may be characterized in the short term (1 to 2 h) by the corresponding MR and DC values. Let $\mathbf{C}_{M_n} = [c_{ijk}^n]$ be the $M_n \times M_n \times 2$ connectivity matrix of an MB network of M_n nodes, where

c_{ij1}^n = the MR value for the MB link from node n_i to node n_j (n_i-to-n_j link)

c_{ij2}^n = the DC value for the n_i-to-n_j link MB link

Thus bidirectional MB communications between nodes n_i to node n_j is fully characterized in the short term by $c_{ij1}^n = \dot{N}_{ij}$, $c_{ji1}^n = \dot{N}_{ji}$, $c_{ij2}^n = \dot{T}_{ij}$, and $c_{ji2}^n = \dot{T}_{ji}$ (see Section 3.3.2.2). Of course, not all of the possible links implied by $\mathbf{C}_{M_n}$ may be possible given range and operating convention constraints (e.g., R terminals limited to single M-station access in a star subnetwork).

Given $\mathbf{C}_{M_n}$, a time history of trail-scattered signals occurring on any link in the network can be generated and subsequently updated throughout the simulation. Let this time history consist of a table of burst occurrence times and corresponding trail-scattered signal durations designated by the $M_n \times 3$ matrix $\mathbf{B}_t = [b_{lm}^t]$. The matrix $\mathbf{B}_t$ is generated using the connectivity matrix $\mathbf{C}_{M_n}$, with burst occurrence times generated from $\min\{c_{ijk}^n, c_{jik}^n\}$ for i,

$j = 1, 2, \ldots, M_n$. If a generated burst is usable in both link directions (see Section 3.3.2.2.1), the corresponding burst durations are also generated for each link direction. Otherwise, the burst generation procedure is repeated. Thus, $b^t_{l1} = \min\{c^n_{ij1}, c^n_{ji2}\}$, $b^t_{l2} = \tau^{ij}_b$, and $b^t_{l3} = \tau^{ji}_b$. This process is performed for every possible MB link described by the connectivity matrix, $\mathbf{C}_{M_n}$. Once burst occurrence times have been generated for each MB link in the network, the rows of the resulting burst matrix $\mathbf{B}_t$ matrix are sorted by the burst occurrence time values in the first column of $\mathbf{B}_t$ from minimum to maximum (i.e., earliest to latest occurrence time). The resulting matrix of time-sequenced bursts defines the burst history matrix $\mathbf{B}_t$, where the first burst corresponds to $l = 1$.

The burst history matrix, $\mathbf{B}_t$, contains the sequence of meteor bursts occurring throughout the network at the inception of the Monte Carlo simulation. At this point, METEORWAIT begins an iterative procedure to simulate the burst-by-burst transmissions occurring across the MB network. The first step in this procedure is to evaluate all PRB, ACQ, and DTA exchanges between the two network nodes n_i and n_j linked by the first trail-scatter event described in the $\mathbf{B}_t$ matrix, which corresponds to the first burst occurring anywhere in the network. This evaluation employs the principles and procedures associated with MB link simulation described in Section 3.3.2.

The network *state* will change as a result of this burst, where the network *state* is defined as the contents of the transmit and receive queues at each node. Once the new network state is determined, consisting of modifications to the transmit and receive queues at nodes n_i and n_j, a new burst event occurrence time is generated for the n_i-n_j link. This new occurrence time and the associated bidirectional burst durations are inserted into the burst history matrix, $\mathbf{B}_t$, and the rows of the matrix are then resorted based on the first column value, that is, the burst occurrence times. The simulation then evaluates the network state resulting from the new first burst in the $\mathbf{B}_t$ matrix, and so on. This procedure is iterated until one or more simulation prediction objectives have been achieved, such as some minimum confidence that specified MWT values have been estimated with the desired precision.

3.3.3.3 Network Communications. After the initial construction of the $\mathbf{B}_t$ matrix, the transmit queues for each of the M_n network nodes are created. Each node employs a transmit queue designated for each potential destination node in the MB network to which a direct MB link is possible. For example, node n_1 in Fig. 3.54 will have a three transmit queues, one for each of nodes n_2, n_3, and n_4. If node n_i must send a message to node n_j via a relay node, say n_r, then the message source node, n_i , places the message in the transmit queue designated for the relay mode, n_r. For example, if node n_1 intends to send a message to n_5, then node n_1 will place the message in its n_4 transmit queue. Once node n_4 receives some or all of the message segments destined for node n_5 from node n_1, node n_4 will then relay these segments to n_5. Messages are sorted in each node-designated transmit queue first on a priority basis, then on a first-in, first-out (FIFO) basis within each priority level. Messages are entered into the node transmit queues as specified by an input message rate parameter defined separately for each potential destination node. Thus, new messages are continuously generated throughout execution of the simulation.

Next, *optimum routes* are determined between each designated message source and destination node pair using the MR and DC values in the $\mathbf{C}_{M_n}$ matrix. Each optimum route consists of a sequence of network nodes and direct MB links that cumulatively minimize the end-to-end MWT values or maximize the end-to-end throughput, depending on network communication objectives. Each time a message moves to the head of the transmit queue, for transmission to a network node not directly MB linked to the source node, the corresponding optimum route is assigned to the message. During simulation execution, segments from this message will then be transmitted along the designated route.

3.3.3.4 Network State Update. Once the number of transmitted DTA blocks has been determined between two MB-linked nodes, their respective transmit and receive queues must be updated. For point-to-point messages, the source node's transmit queue message pointer is decreased by the number of transmitted segments acknowledged by the destination node until the entire message has been transferred. After a network message is fully assembled at an M station, it is placed in the transmit queue designated for the intended destination R terminal in the corresponding star subnetwork. In this development, only R terminals serve as source and destination nodes in an MB network, although the network simulation approach is easily extended to include M stations as message source and destination nodes. If the intended destination for the network message is not within the M station's subnetwork, the message is placed in the transmit queue of the next M station along the designated optimum route. The transmit queues are then resorted based on priority and FIFO in preparation for the next generated relevant trail-scatter event.

3.3.3.5 Performance Statistics. Once DTA blocks constituting a complete message have been exchanged between two nodes n_i and n_j, the corresponding distribution of MWT values is updated with the new MWT sample. These distributions may be accumulated for communications between two R terminals, R_1 and R_2, within the same subnetwork (i.e., the R_1-to-M-to-R_2 route) or between R terminals in different subnetworks through M_K intervening M stations (e.g., R_1-to-M_1-M_2-...-M_K-to-R_2). In the network context, performance measures are defined in terms of MWT or throughput values between two network nodes, whether or not they are directly MB linked. Thus, MWT or throughput distributions may be created to quantify the message transmission performance between any two network nodes n_i and n_j. In general, the distribution for n_i-to-n_j end-to-end communications may differ from the corresponding distribution for n_j-to-n_i communications. The MWT or throughput distributions may combine both MB link and transmit queue delays or different distributions may be generated independently for these delays. Separation of these distributions for MB link and transmit queue size provides diagnostic capability to identify network "bottlenecks," that is, network nodes whose message input rates exceed their message output rates. Of course, the sample mean and variance for each MWT or throughput measure are determined as well as the full sample distributions.

3.3.3.6 Simulation Control. Simulation control techniques for the network simulation are similar to those employed for the link simulation. In the network case, however, more than one performance measure may be generated and subsequently monitored for precision and confidence. For example, MWT distributions may be generated for multiple links in the network. The simulation would be continued until the desired precision had been achieved in estimating values for each distribution with the desired confidence.

3.3.3.7 Example Network Analysis. Figure 3.57 quantifies the connectivity between nodes for the modeled network shown in Fig. 3.53. This figure gives the MR and DC values for each link in the network, assuming identical values in both link directions. All links were operated at a burst transmission rate of 8 kbps using the HDX protocol described in Section 3.3.2.3.1 with fixed message lengths of 140 characters. Nodes n_3 and n_4 were

CONNECTIVITY MATRIX $C_{M_n} = [c^n_{ijk}]$ ENTRIES

LINK ID	COMMUNICANTS NODE ↔ NODE		METEORS PER MINUTE ($k = 1$)	DUTY CYCLE (%) ($k = 2$)
1	1	2	1.0	2.50
2	1	3	0.5	0.25
3	3	4	2.5	5.00
4	4	5	0.5	0.25
5	4	6	1.0	2.50

NOTE: $c^n_{jik} = c^n_{ijk}$

Figure 3.57. Network MWT analysis example: link MR and DC values.

M stations while the remaining nodes were R terminals. Fig. 3.58 contains the distribution of combined queue and channel MWT values generated by the METEORWAIT simulation. The figure shows the average MWT values between nodes with a 90% confidence and a 5% precision interval. For example, the average MWT value for messages transmitted from node n_1 to node n_3 was 25 s.

From the figure, it is apparent in this case that minimum MWT values were predicted for the point-to-point links, because these links did not involve message routing. Intuitively, if the individual MB link MWT values are not widely divergent, then multiple link routes would be expected to yield larger MWT values than point-to-point links. Note that the MWT values between node n_6 and the remaining nodes in the network were larger than the corresponding values for all other network nodes. This outcome was the result of message queue generation at each node, which used the node number as the message priority level. Thus, messages destined for node n_6 were placed at the end of each node's n_6 transmit queue. The additional queue delay accounts for lower MWT values for messages destined for node n_6.

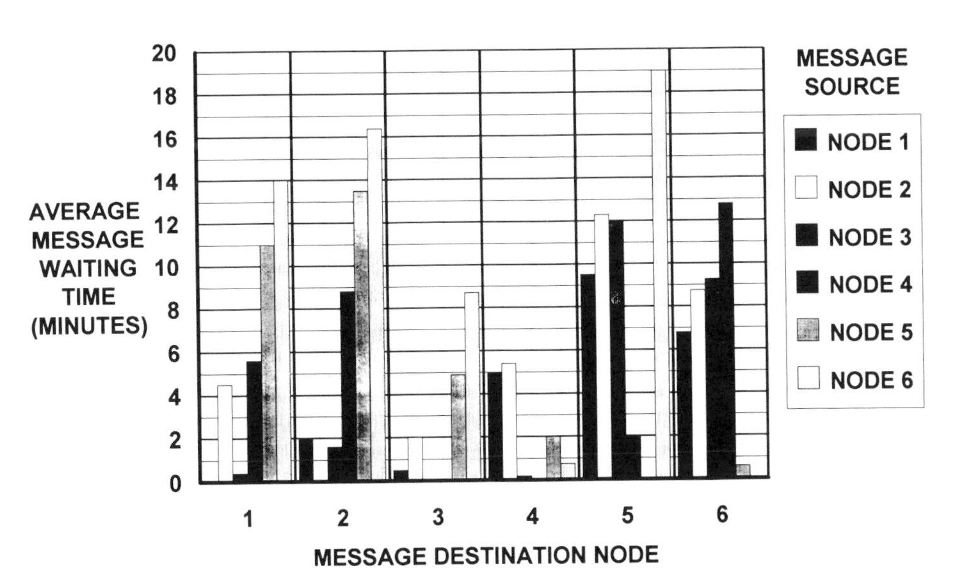

Figure 3.58. METEORWAIT sample network analysis results.

3.4 MODEL VALIDATION*

3.4.1 Purpose and Objectives

In general, computer model predictions of system performance must be compared with measurement to determine the accuracy with which they depict reality. Ideally, a computer model would reproduce empirical data derived from a diverse set of experimental conditions and thus enable confident extrapolation from these data to a wide variety of dissimilar system configurations. If these configurations include the system under study, the model may then be considered validated, for the intended modeling application. Each measurement–prediction comparison provides a sample of prediction error, so that a data base of these comparisons may be accumulated for use in quantifying prediction error.

This section presents measurement-prediction comparisons for the METEORLINK, METEORDIV, and METEORWAIT computer models. No METEORTRAK comparisons are provided, but the algorithm is nonetheless similar to that employed by the METEORDIV model. The purpose of these comparisons is to demonstrate the overall validation approach as much as to provide sample comparison results for these three models. In this context, the measurement-derived values used to perform the measurement-prediction comparison will be called *validation data*. Model comparisons were performed with three independent sources of validation data. Link MR, DC, and MWT comparisons were made using a small subset of the extensive, well-calibrated measurements performed on 700- and 1200-km MB test bed links operated by the Phillips Laboratory[1] [183] (PL) in Greenland. The 700-km link was planned, established, and operated in conjunction with the Naval Undersea Warfare Center (NUWC)[2].

A measurement-prediction comparison was performed using link MR and DIF values derived from measurements collected for a variety of range and antenna configurations in the northwest United States beginning in 1987. These measurements were performed by the Meteor Communications Corporation (MCC)[3], SRI International[4] and the Boeing Defense and Space Group (Boeing)[5] for the Ballistic Missile Organization (BMO) of the U.S. Air Force[6]. BMO undertook these measurements to provide a significant empirical basis for system design decisions regarding the use of MB radio to

* Cleared for public release, August 6, 1991, by OASD/PA. BMO Case File 91-069.

[1] Phillips Laboratory, Geophysics Directorate, Ionospheric Effects Division, Ionospheric Applications Branch, Hanscom AFB, Massachusetts.

[2] Naval Undersea Warfare Center, New London Laboratory, New London, Connecticut.

[3] Meteor Communications Corporation, Inc., 6020 South 190th Street, Kent, Washington.

[4] SRI International, 333 Ravenswood Avenue, Menlo Park, California.

[5] Boeing Defense and Space Group, Aerospace and Electronics Division, Seattle, Washington.

[6] Ballistic Missile Organization, U.S. Air Force Systems Command, Norton Air Force Base, California.

provide back-up communications between the Mobile Launch Control Center (MLCC) and the proposed Hard Mobile Launcher (HML) of the Small Intercontinental Ballistic Missile (SICBM) weapon system. An important objective of these measurements was to provide validation data for MB computer models.

Link MR predictions have also been compared with radar (backscatter) measurements compiled at Jodrell Bank, UK [184]. These radar measurements were performed to study the diurnal and seasonal variations of the sporadic meteor arrival rate. The backscatter geometry of radar detection provides a unique validation of the METEORLINK algorithm. In this case, the ellipsoidal scattering geometry (see Fig. 3.4) collapses to produces spheroidal geometry as the forward scattering angle (see Eq. 3.42) approaches zero. A measurement-prediction comparison using the Jodrell Bank results thus provides a demonstration of the versatility of the physical modeling approach.

3.4.2 Approach

The simplest measurement–prediction comparison is provided by a table or plot of corresponding measured and predicted values. This comparison permits simultaneous inspection of the absolute values and permits a "quick-look" at the prediction error. The magnitude of the prediction determines whether the results are within the precision of the computer model. For example, MR values on the order of several hours may extend beyond the applicability of the mass-rate exponent used in the model (see Eq. 3.25). In this case, both the MR values used to derive the model's SMRD map (e.g., see Fig. 3.19) and the MR values that constitute the validation data may have large statistical uncertainty.

Complete measurement-prediction comparisons require both absolute and trends analyses. Absolute analysis is based on some *validation measure*, $\mathcal{V}_\mu$, defined as a function of corresponding measured and predicted values. Let s_m be a sample from a measurement and s_p be the corresponding predicted value. One function of measurement and prediction yielding a useful validation measure is

$$\mathcal{V}_\mu = \log_{10}(s_{m/}s_p) \tag{3.127}$$

where the base ten logarithm permits equivalent $\mathcal{V}_\mu$ magnitudes for both *optimistic* ($s_p > s_m$) and *pessimistic* ($s_p < s_m$) predictions. Obviously, perfect agreement between prediction and measurement would yield $\mathcal{V}_\mu = 0$ and a ratio of 2.0 or 0.5 corresponds to $\mathcal{V}_\mu = \pm 0.3$, respectively. For example, the prediction is considered optimistic when predicted link MR (or DC) values exceed measured values, so that $\mathcal{V}_\mu < 0$. On the contrary, pessimistic predictions imply that measured MR (or DC) values exceed the corresponding predictions, yielding $\mathcal{V}_\mu > 0$.

Trend analysis is based on a comparison of the change in measured values and the corresponding change in predicted values due to some variation in system parameters. For example, if measured link MR values are given by $\dot{N}_L^{\mathrm{m}}(\tau_1^*)$ for a minimum required burst duration $\tau^* = \tau_1^*$ and $\dot{N}_L^{\mathrm{m}}(\tau_2^*)$ for $\tau^* = \tau_2^*$, then the ratio $\dot{N}_L^{\mathrm{m}}(\tau_2^*)/\dot{N}_L^{\mathrm{m}}(\tau_1^*)$ quantifies the burst duration trend from τ_1^* to τ_2^*. This value is then compared to the corresponding ratio of predicted values $\dot{N}_L^{\mathrm{p}}(\tau_2^*)/\dot{N}_L^{\mathrm{p}}(\tau_1^*)$ to evaluate the capability of the model to reproduce the measured burst duration trend. Additional link trends may be studied for range, transmit and receive antenna configuration, minimum required received power P_r^*, signal frequency, temporal parameters, and geographic location, among others.

Discrepancies between prediction and measurement may be due to one or more causes. First, input parameter errors, including inaccurate antenna patterns, must be discounted before considering model or measurement inadequacies as potential error sources. If model input parameters are verified, then both measurement statistical confidence and prediction accuracy must be considered in conclusions drawn from the comparison results. In addition, measured data must be reviewed for anomalistic behavior. For example, consider the MR enhancement for particular link geometries produced by the major meteor streams [185]. If stream radiants were not modeled in link MR predictions, then measured MR values would show this enhancement while the corresponding predictions would show no such enhancement. The resulting $\mathcal{V}_\mu$ values would show that the model was pessimistic, despite accurate predictions of the background sporadic meteor rate.

Once input parameters and measured data have been reviewed and verified, the accuracy of model physical parameters must be considered. In most cases, these parameters were derived indirectly from radio and photographic measurements, such as the initial trail radius r_0 and diffusion rate D, shown in Figs. 3.25*a* and *b*, respectively, of Section 3.2.2.3.2.1. If variation of the applicable model parameters within empirically derived limits fails to affect the measurement–prediction comparison, or if values of these parameters must change to achieve minimum discrepancy for different MB system measurements, then components of the model algorithm should be re-evaluated.

Any re-evaluation must consider improper computer algorithm implementation as well as inadequacies of the modeling algorithm. These inadequacies typically involve failure to incorporate important meteoric effects. For example, if the initial trail radius, r_0, is modeled as a function of trail-scatter point height (h_s, see Eqs. 3.47*a*, *b*) alone, then the actual dependence on meteor velocity (see Eq. 3.16) would force different r_0 values in the model at different times of day. In fact, the measurement–prediction comparison provides validation for the formulations used to compute model parameters as well as the model algorithm that employs these parameters.

3.4.3. Link Model Validation

This section gives a measurement–prediction comparison with data from the Greenland test bed, SICBM MB tests, and Jodrell Bank radar measurements. No subjective judgments of prediction accuracy are provided, although a measurement-to-prediction ratio between 2.0 and 0.5 (i.e., $\mathcal{V}_{\mu}$, = ±0.3) was chosen as an acceptability criterion during preliminary calibration of model parameters. The User's Guide for the Burst LINK (BLINK) computer program states that predictions within a factor of four of measured data are considered accurate [186]. This statement reflects the significant uncertainty in observed MB link performance due to variability in meteoric flux, trail-scattering phenomena, antenna patterns, and noise environments.

3.4.3.1 Greenland Meteor Burst Test Bed Measurements

3.4.3.1.1 Overview. The METEORLINK computer model was used to predict the MR and DC recorded on experimental MB links operated in Greenland by the Phillips Laboratory and established by the Rome Air Development Center (RADC) in 1984 [187]. The results of the absolute analysis and accompanying discussion presented in this section were derived from previous work [188]. These measurements were recorded on the Greenland Meteor Burst Test Bed with one-way MB links operating from Sondrestrom AFB to Thule AFB (1200 km) and separately from Sondrestrom AFB to Narsarsuaq, Greenland (700 km). METEORLINK predictions of MR, DC, underdense trail MR (UMR), and underdense trail DC (UDC) were compared with the corresponding measurements for 45, 65, 85, and 104 MHz; minimum RSL values of −120, −110, and −100 dBm; and 12 times per day for a total of five months. Several link antenna polarizations were employed, including horizontal, vertical, and cross-polarized transmit-receive combinations.

3.4.3.1.2 Measurements. MB measurements have been performed between the Sondrestrom AFB (SAFB) transmit site and the Thule AFB (TAFB) receive site since 1985. Recently, the Phillips Laboratory upgraded the original 1200-km SAFB–TAFB test bed link (ST link) to include high-quality, calibrated measurements for a variety of antenna configurations and frequencies. The PL test bed employs horizontal five-element Yagi antennas (HY5) at six frequencies (35, 45, 65, 85, 104, and 147 MHz) at the SAFB transmit site and both horizontal and vertical (VY5) five-element Yagis at the TAFB receive site. A nominal transmit power of 1 kW is used at each frequency. A second test bed was installed between SAFB and Narsarsuaq in southern Greenland (SN link). The 700-km SN link mimics the ST link but adds vertical whip (VWP) transmit antennas at 65 and 104 MHz. These links are depicted in Fig. 3.59 and summarized in Table 3.6.

The PL measurements include calibrated RSL samples recorded through-

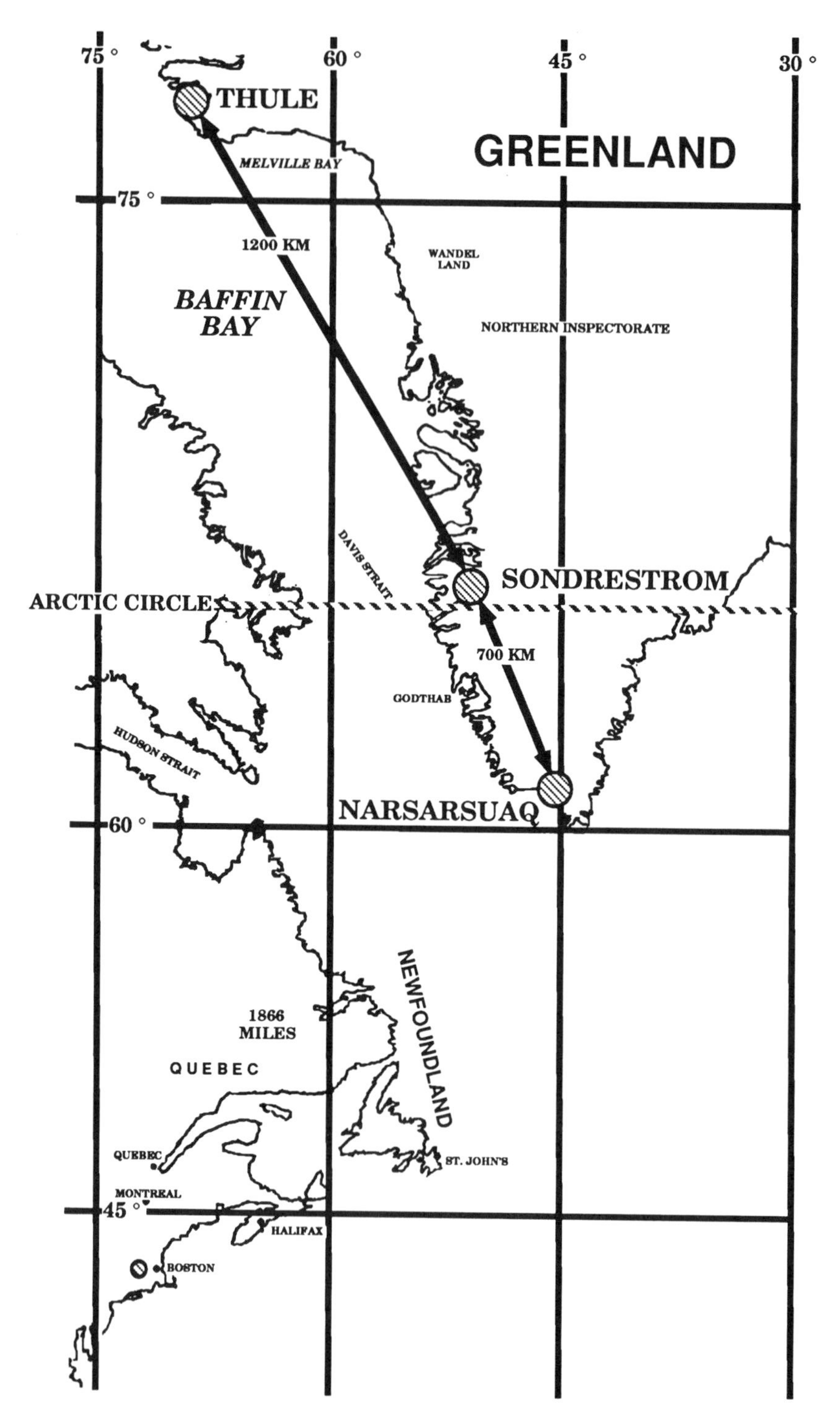

Figure 3.59. Greenland ST and SN MB links.

TABLE 3.6 ST- and SN-Link Parameters

	SAFB → TAFB, GREENLAND (1210 km)	
Transmit site (SAFB)	Latitude Longitude	66°59′ N 50°39′ W
Frequencies (f_L)		45, 65, 85, 104, & 147 MHz
Transmit antennas		Five-element Yagis @ all frequencies Horizontally polarized 1.5 wavelengths height 0° boom tilt
Transmit power (P_t)		1 kW (30 dBW) with ≅ 1 dB line loss
Receive site (TAFB)	Latitude Longitude	73°33′ N 67°51′ W
Receiver antennas		Five-element Yagis @ all frequencies Horizontally polarized 1.5 wavelengths height 0° boom tilt
	SAFB → NARSARSUAQ, GREENLAND (700 km)	
Transmit site (SAFB)	Latitude Longitude	66°59′ N 50°39′ W
Frequencies (f_L)		45, 65, 85, 104, & 147 MHz
Transmit antennas		Five-element Yagis @ all frequencies Horizontally polarized 1.5 wavelengths height 30° boom tilt Monopoles @ 65 & 104 MHz Vertically polarized 1.5 wavelengths height with radials
Transmit power (P_t)		1 kW (30 dBW) with ≅ 1 dB line loss
Receive site (Narsarsuaq)	Latitude Longitude	61°12′ N 45°26′ W
Receiver antennas) (all frequencies)		Five-element Yagis Horizontally & vertically polarized 1.5 wavelengths height 30° boom tilt
	COMMON LINK PARAMETERS	
Measurement bandwidth (BW_n)		100 Hz
Minimum required RSL (P_r^*)		Arbitrary above background noise (primarily Galactic limited)
Minimum burst duration (τ^*)		40-ms phase-lock loop acquisition

out all propagation events, typically trail-scatter or sporadic E, whenever the transmitted signal from Sondrestrom was identified. Trail-scattered signals are detected and recorded digitally for PL post-processing. Noise powers are periodically recorded and receiver measurements are monitored and periodically recalibrated. Measurement post-processing classifies sampled events as trail-scatter or sporadic-E, distinguishes between underdense and overdense trail RSL behavior [189], and yields the bihourly MR and DC for RSL values (P_r^*) exceeding specified thresholds. In addition, average noise measurements versus frequency and time are processed, typically showing the expected decrease in Galactic noise power with increasing frequency as well as its characteristic 2- to 3-dB diurnal variation. An important attribute of the PL measurement-derived MR and DC values is that noise uncertainty may be eliminated from the measurement–prediction comparisons.

3.4.3.1.3 Predictions. Input parameters for METEORLINK predictions corresponding to the Greenland measurements were derived directly from the ST- and SN-link configurations described in Table 3.6. A subset of these comparison results, spanning the full range of observed model accuracy, is summarized in this section. NEC-computed electric field patterns were generated by PL for the HY5, VY5, and VWP antennas at each relevant frequency. The antenna foregrounds at the TAFB and Narsarsuaq receive sites permitted pattern calculations including smooth-earth ground reflections. The SAFB patterns were generated using NEC-computed free space patterns modified to account for ground reflections produced by asymmetrical terrain features in the antenna foreground.

A small subset of measured data was used for limited calibration of model physical parameters. This calibration involved adjustment of these parameters within theoretical or empirically derived limits to achieve the best overall fit to the MR, DC, UMR, and UDC values obtained for the measurements recorded during March 1989 at 45 and 85 MHz on the ST link. For example, the plasma resonance factor, ρ, was varied from 1.0 (no resonance effect) to 2.0 [190]. The value $\rho = 2.0$ was found to yield predicted MR values closest to the corresponding measured values for short-duration underdense trail-scattered signals. Once this calibration was completed, the model was used to predict performance for the complete measurement set and an absolute measurement-prediction comparison was performed.

3.4.3.1.4 Comparisons. A measurement-prediction comparison of ST-link MR values, $\dot{N}_{\mathrm{ST}}(f_L)$, and DC values, $\dot{T}_{\mathrm{ST}}(f_L)$, is shown in Figs. 3.60*a*, *b* for $f_L = 45$ MHz and $f_L = 104$ MHz, respectively. The MR and DC values in these plots were computed for a -120-dBm minimum RSL measured on the ST link in March, 1989. Two MR and DC peak values are apparent at about 7 UT and 16 UT in both the Greenland measurements and the correspond-

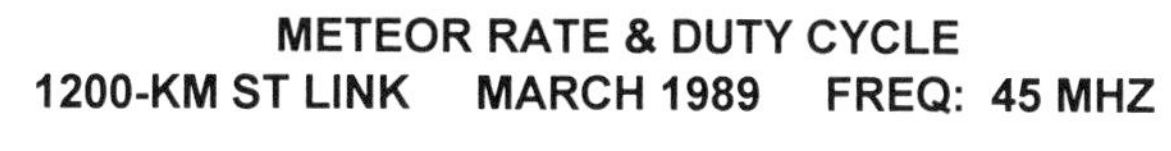

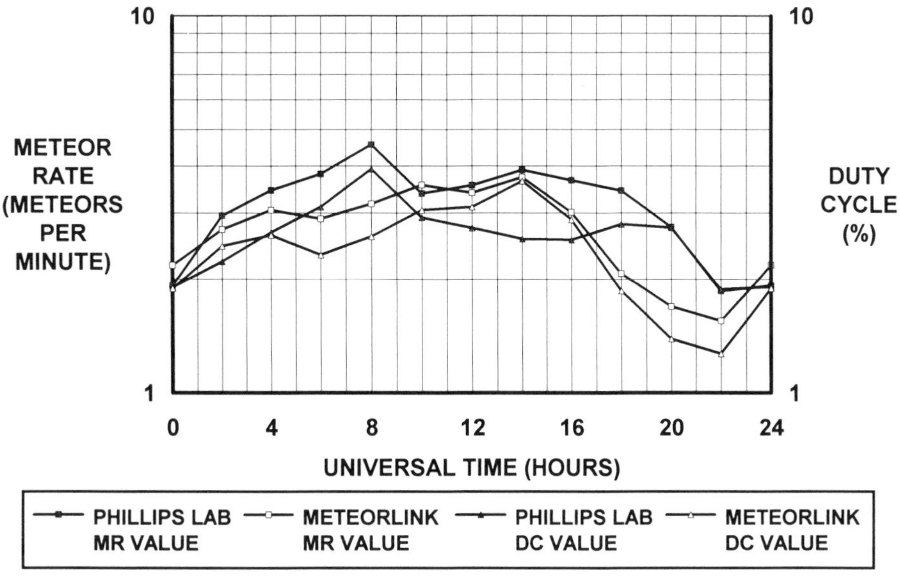

(a)

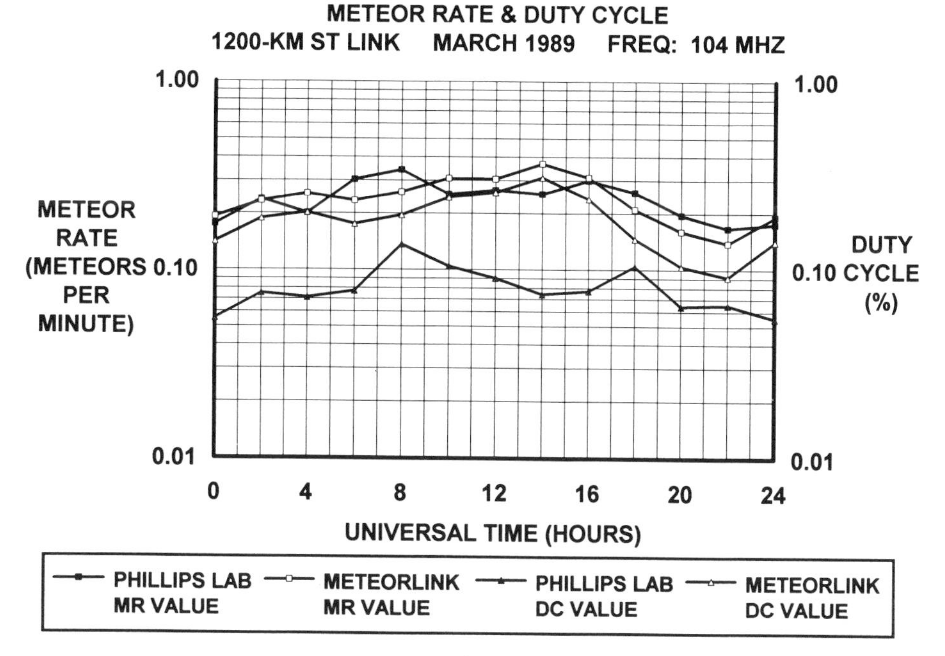

(b)

Figure 3.60. MR and DC comparison for the ST link, March, -120 dBm, HY5-to-HY5 link antennas: (*a*) 45 MHz; (*b*) 104 MHz.

ing METEORLINK predictions. Although the 7 UT peak values correspond to the classic morning maximum, the 16 UT peak is not observed in the classic mid-latitude diurnal variation on forward scatter links. This phenomenon was arguably caused by meteors passing over the north ecliptic pole (NEP, see Fig. 3.18) and producing link-observable trails in the early afternoon hours.

Comparing Fig. 3.60*a* with 3.60*b* shows that both measurement and prediction exhibited about 10 times greater daily average link MR (DAMR) values and 20 times greater daily average link DC (DADC) values at 45 MHz than at 104 MHz. Use of the DAMR value in these comparisons, rather than the bihourly MR and DC values, removed the complicating effects of diurnal variation from the measurement-prediction comparison. In symbols, $\bar{\dot{N}}_{\mathrm{ST}}$ (45 MHz) $\approx 10\bar{\dot{N}}_{\mathrm{ST}}$ (104 MHz) and $\bar{\dot{T}}_{\mathrm{ST}}$ (45 MHz) $\approx 20\bar{\dot{T}}_{\mathrm{ST}}$ (104 MHz). Therefore the ratio $\bar{\dot{T}}_{\mathrm{ST}}(f_L)/\bar{\dot{N}}_{\mathrm{ST}}(f_L)$, which is proportional to the average burst duration, $\bar{\tau}_b$ (see Eq. 3.124), decreases with increasing frequency. This decrease in $\bar{\tau}_b$ with increasing frequency occurred because both the burst duration, $\tau_q(q)$, and the corresponding incremental MR value, $m_t(q)$, decrease with increasing frequency. But link DC values are formed as the integrated product of $\tau_q(q)$ and $m_t(q)$ (see Eq. 3.89), so $\bar{\dot{T}}_{\mathrm{ST}}(f_L)$ decreased faster with increasing frequency than the corresponding MR value, $\bar{\dot{N}}_{\mathrm{ST}}(f_L)$. Thus, the average burst duration, $\bar{\tau}_b$, decreased with increasing frequency.

Both measured and predicted values for the SN link, July, -120 dBm, at 45 and 104 MHz are shown in Figs. 3.61*a* and *b*, respectively. The SN-link DAMR value is about three times the corresponding ST-link DAMR value at 45 MHz, that is, $\bar{\dot{N}}_{\mathrm{SN}}$ (45 MHz) $\approx 3\bar{\dot{N}}_{\mathrm{ST}}$ (45 MHz), while the SN-link DADC values are about twice the corresponding ST link values, or $\bar{\dot{T}}_{\mathrm{SN}}$ (45 MHz) $\approx 2\bar{\dot{N}}_{\mathrm{ST}}$ (45 MHz). These ratios reflect the greater observable sporadic meteor flux in July than in March and, to a lesser extent, an increase in link MR values from the addition of end-path to mid-path trail-scatter geometries [191]. The DADC ratio of two-to-one was less than the DAMR ratio of three-to-one because forward scattering angles, ϕ_s (see Eq. 3.42), on the SN link were smaller than the corresponding ϕ_s values on the longer-range ST link. The smaller ϕ_s values caused an effective reduction in usable trail-scattered signal lifetime on the SN link relative to the ST link.

The ratio of DAMR values at 45 and 104 MHz on the SN link (see Figs. 3.61*a* and *b*) was greater than the corresponding ratio on the ST link, that is,

$$\frac{\bar{\dot{N}}_{\mathrm{SN}}\ (45\ \mathrm{MHz})}{\bar{\dot{N}}_{\mathrm{SN}}\ (104\ \mathrm{MHz})} > \frac{\bar{\dot{N}}_{\mathrm{ST}}\ (45\ \mathrm{MHz})}{\bar{\dot{N}}_{\mathrm{ST}}\ (104\ \mathrm{MHz})}$$

This trend, apparent in both measurement and prediction, was due to the

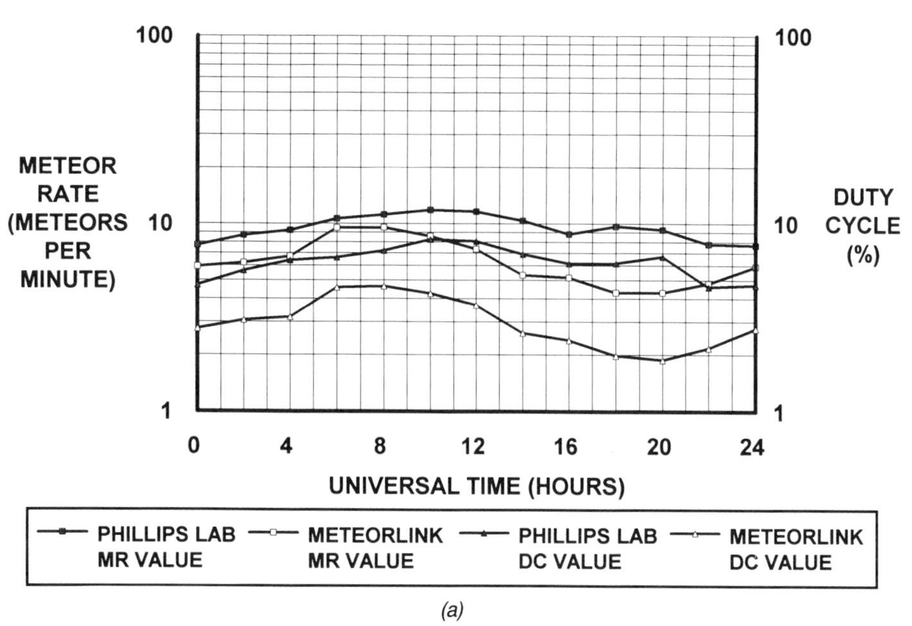

(a)

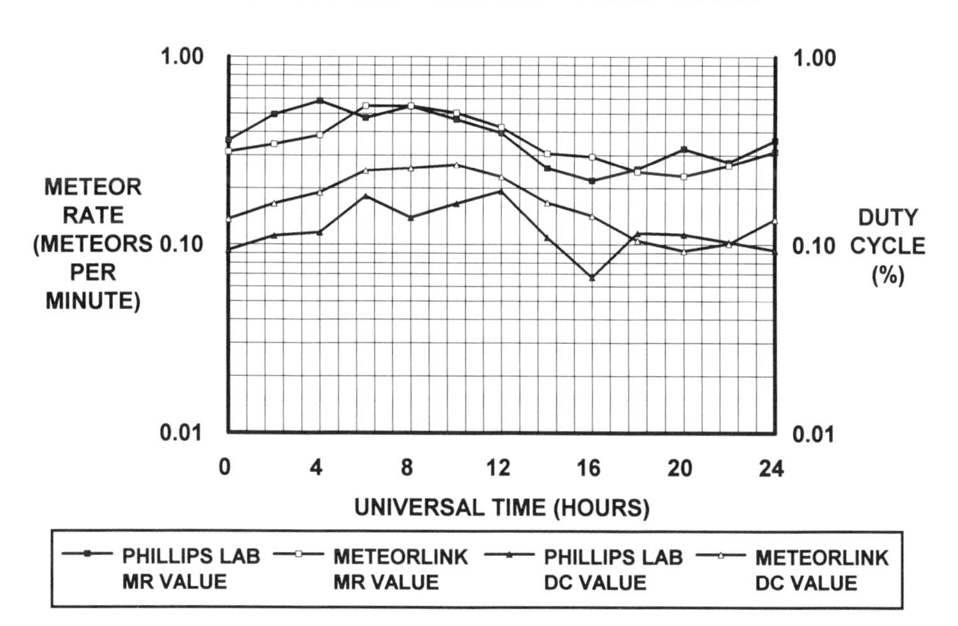

(b)

Figure 3.61. MR and DC comparison for the SN Link, July, -120 dBm, HY5-to-HY5 link antennas: (*a*) 45 MHz; (*b*) 104 MHz.

smaller ϕ_s values characteristic of the shorter-range SN link, exaggerated by the reduced signal wavelength at 104 MHz as compared to 45 MHz. In the case of underdense trail-scattered signals, the maximum RSL value is inversely proportional to $\exp[8\pi r_0^2 \cos(\phi_s)/\lambda_L^2]$. Increasing ϕ_s or decreasing wavelength λ_L reduces the maximum RSL value and thus reduces the link MR value, $\bar{\dot{N}}_{SN}(f_L)$.

Validation measure ($\mathcal{V}_\mu$, see Eq. 3.127) plots of the bihourly MR, UMR, DC, and UDC values for the ST link, March, and -120 dBm RSL are shown in Figs. 3.62*a*–*d*, respectively. The apparent temporal modulation of $\mathcal{V}_\mu$ values in Figs. 3.62 shows some trend of pessimistic prediction during the morning hours and optimistic prediction during the late afternoon hours. The figures show that this effect is most significant between 14 and 18 UT, where all $\mathcal{V}_\mu$ values become optimistic. This modulation of $\mathcal{V}_\mu$ values may result from differences between assumed and actual SMRD distributions, common volume illumination, and meteor velocities.

The ST- and SN-link measurements described in this study were performed in 1989 and 1990, and the Kazan–Mogadishu results were derived from measurements performed in the 1960s. Since SMRD values include sporadic meteors whose orbital period differs from the earth's orbital period [192], SMRD values may be expected to vary somewhat from year to year. This behavior has been observed in long-term radar measurements [193] of the sporadic meteor rate. In general, this year-to-year variation must be separated into orbital effects on the observed MR values and observed solar cycle effects [194]. This separation is beyond the scope of this work; nevertheless, the orbital component may be modeled as long term statistical fluctuation in SMRD values, while solar cycle effects may be treated through their impact on initial trail radius, r_0 (see Eqs. 3.16 and 3.47), diffusion rate, D (Eqs. 3.45 and 3.47*c*, *d*), and scale height, H_P (Eq. 3.12).

Note the relative pessimism exhibited by the DC $\mathcal{V}_\mu$ values at 45 MHz (see Fig. 3.62*a* and *c*) compared to the corresponding values at 65, 85, and 104 MHz. Fig. 3.62*b* and *d* show that the 45-MHz UMR and UDC $\mathcal{V}_\mu$ values suggest optimistic predictions, so the predicted DC contribution from overdense trails are pessimistic at 45 MHz. This behavior suggests sporadic-E spoofing of the trail classifier, more probable at 45 MHz than the higher frequencies. In this case, the number of sporadic-E events would be small relative to the 45 MHz MR values, while the DC contribution from sporadic-E would be far more significant. A second explanation is that the frequency-dependent underdense-overdense threshold, q_{sm}^T (see Eq. 3.57), forced more underdense trails to be modeled than were found by the trail classifier. Overdense trails produce a greater DC contribution per meteor than underdense trails with the same trajectory and link-relative location. As in the case of sporadic-E spoofing, the model would then exhibit greater pessimism in predicted DC values than the corresponding MR values. A third cause could be that the assumed mass-rate exponent s (see Section 3.2.2.2.2, Eq. 3.25, and Table 3.1) is too large, leading to pessimistic predictions for the link MR contribution from large meteors.

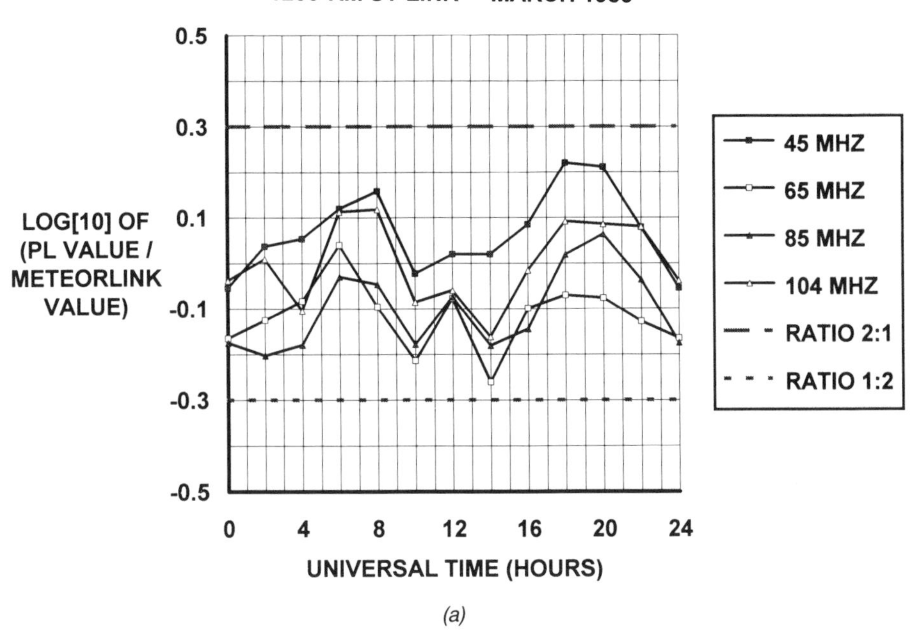

(a)

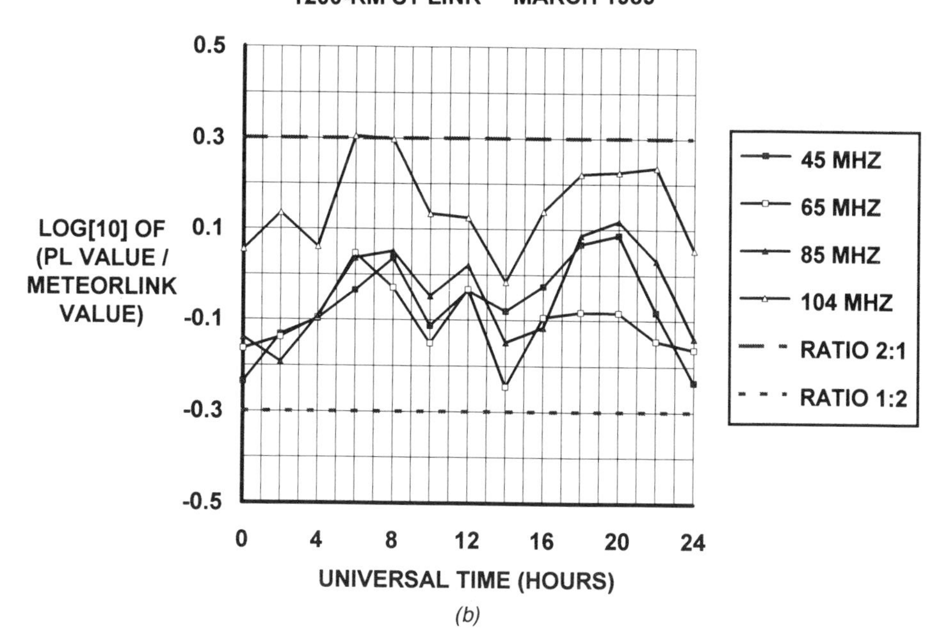

(b)

Figure 3.62. ST link: March; 45, 65, 85, 104 MHz; -120 dBm; HY5-to-HY5 link antennas: (*a*) MR $\mathcal{V}_\mu$ values; (*b*) UMR $\mathcal{V}_\mu$ values; (*c*) DC $\mathcal{V}_\mu$ values; (*d*) UDC $\mathcal{V}_\mu$ values.

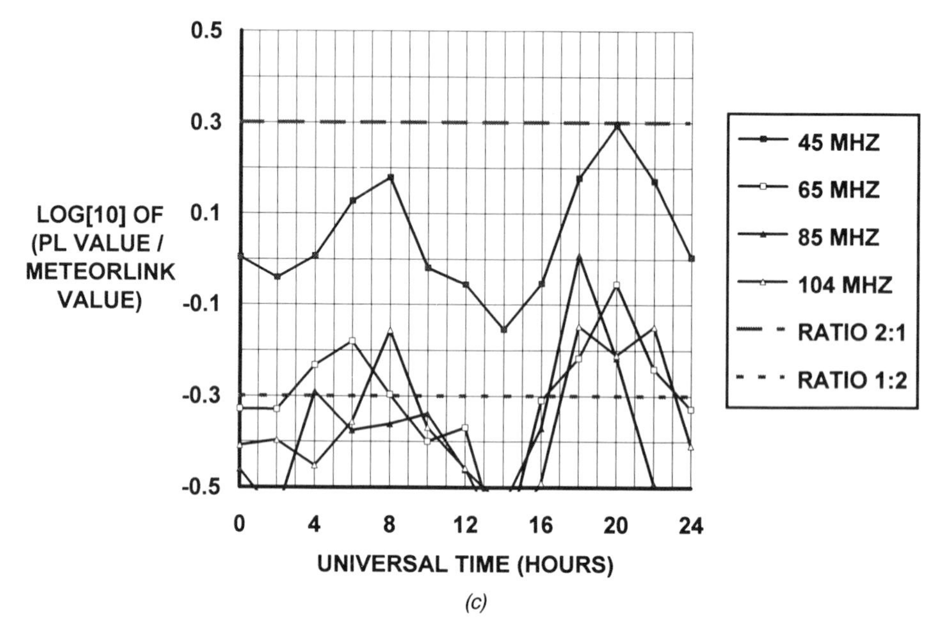

(c)

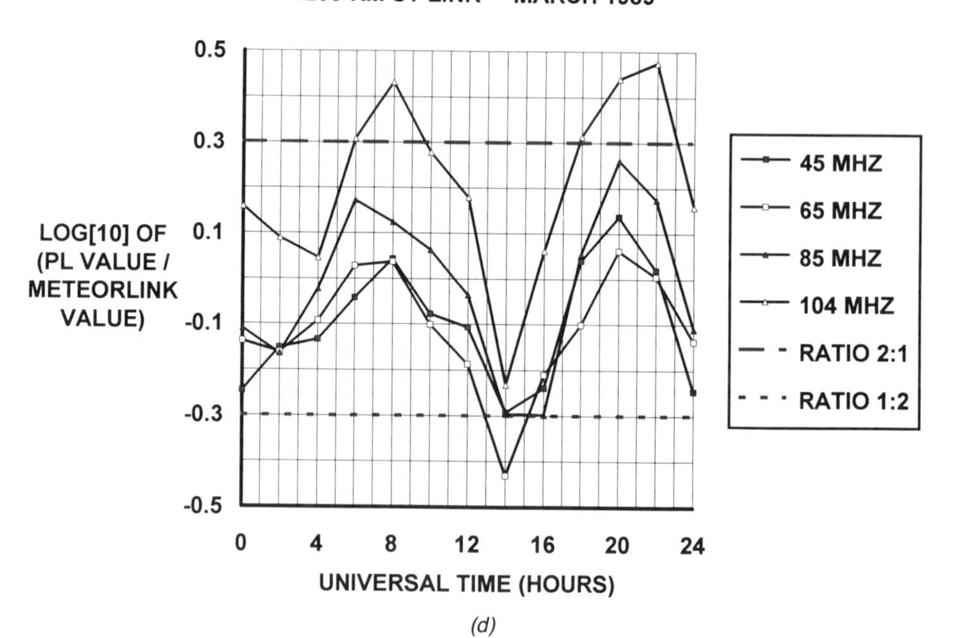

(d)

***Figure* 3.62.** (*Continued*)

Figures 3.62*b* and *d* show that the UMR and UDC exhibited increased optimism from 45 to 65 MHz, followed by increasing pessimism at 85 and 104 MHz, to the point that the 104-MHz UMR and UDC results showed the most pessimism. One explanation for this behavior involves the initial trail radius, r_0, and its dependence on meteor speed, v. These predictions assumed a speed-averaged r_0 value dependent only on trail height, h_s (see Eq. 3.47*b*), and elongation, ϵ (see Eq. 3.21). If this r_0 value is too small, optimistic predictions result from links employing a significant percentage of underdense trails. The effect diminishes as the frequency increases to the point at which MR values are dominated by overdense trails.

This effect may explain the UMR and UDC frequency trend observed from 45 to 104 MHz in Figs. 3.62*b* and *d*. Moreover, this effect may explain the increased prediction optimism apparent between 14 and 18 UT. If meteors are clearing the NEP and producing usable trail-scattered signals about 16 UT, then the speeds of these meteors would produce larger r_0 values (see Eq. 3.16). These larger r_0 values would reduce trail detectability, particularly as signal frequency is increased, until MR values become dominated by the overdense trail contribution.

Figures 3.63*a*–*d* plot $\mathcal{V}_\mu$ values for vertically polarized transmit and receive antennas (VWP-to-VY5) on the SN link at 65 and 104 MHz. The prediction error is greater for the vertically polarized SN link than for horizontally polarized ST and SN links, so the plot scales in Fig. 3.63 show the range $-1.0 \leq \mathcal{V}_\mu \leq 1.0$. Figure 3.63 shows that the 65-MHz SN-link predictions were consistently optimistic, with about twice the measured MR values and three times the measured DC values ($\mathcal{V}_\mu = \pm 0.5$). This result is consistent with the optimism found for the ST link at 65 and 85 MHz, exaggerated by the smaller ϕ_s values characteristic of the SN link.

At 104 MHz, the DC predictions are optimistic ($\mathcal{V}_\mu \cong -0.7$) but the corresponding UDC values are excessively pessimistic ($\mathcal{V}_\mu > +1.0$). Note, however, that the measured UDC values at 104 MHz on the VWP-to-VY5 SN link varied between 0.0 and 0.003%, whereas the corresponding predicted values were on the order of 0.0001%. This increased model pessimism on the VWP-to-VY5 SN link was caused by MR reliance on more overdense trail-scattered signals than necessary for the corresponding HY5-to-HY5 link, a result of the reduced power margin on the VWP-to-VY5 SN link versus the HY5-to-HY5 link. The VWP-to-VY5 SN link employed less transmit peak antenna gain than the HY5-to-HY5 link ($\cong$ 6 dBi for the VWP antenna versus $\cong$ 15 dBi for the HY5 antenna) while the corresponding receive antennas (VY5 and HY5) had comparable peak gain values. In addition, the decreased trail scattering efficiency at 65 and 104 MHz as compared to 45 MHz also contributed to the lower MR values. At these diminutive MR values, model inaccuracies (e.g., mass exponent s, initial radius r_0, etc.) combined with inadequate measurement confidence may have produced these significant $\mathcal{V}_\mu$ values.

Figure 3.63 also plots $\mathcal{V}_\mu$ values measured on the SN link, July, and

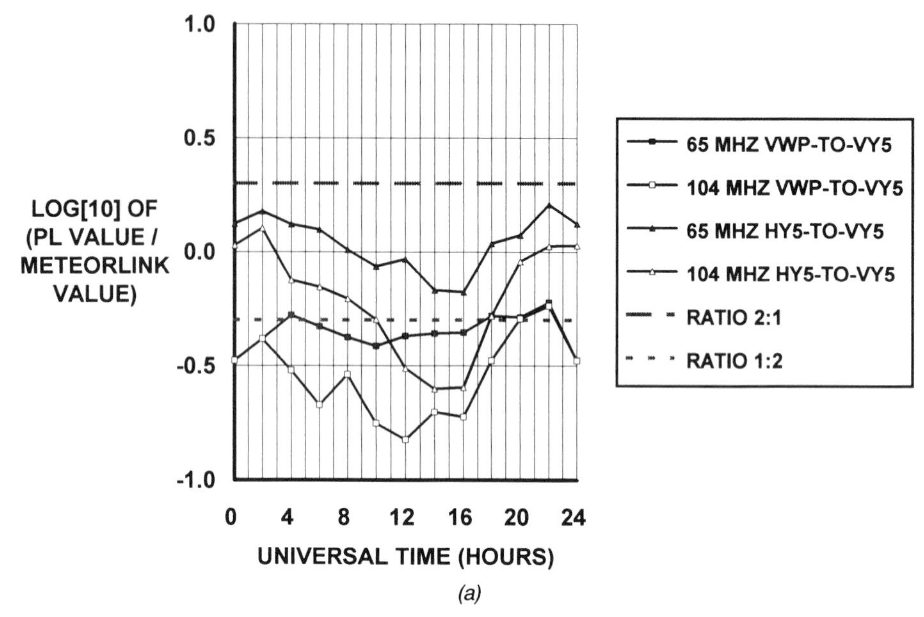

(a)

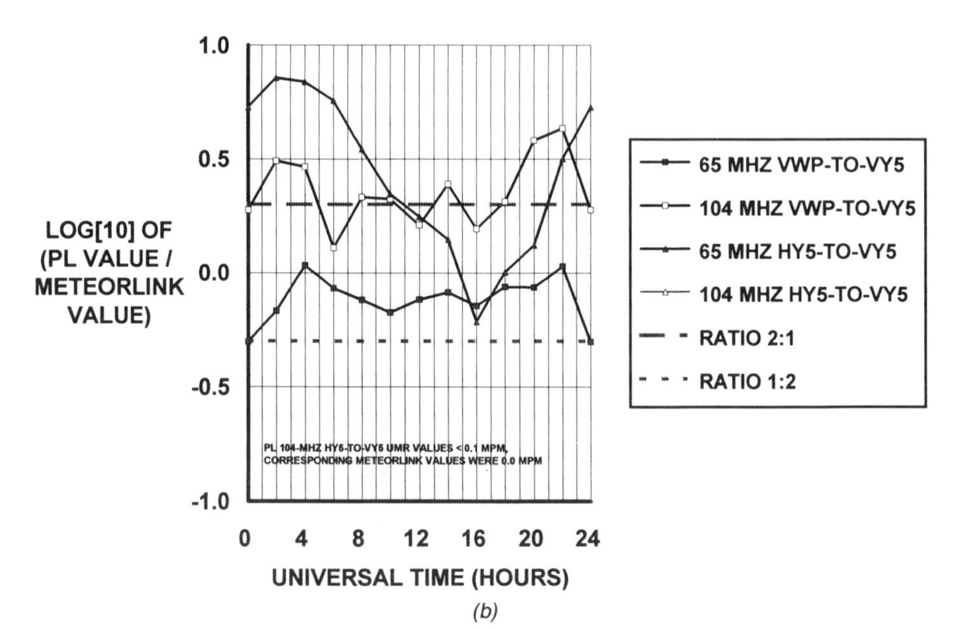

(b)

Figure 3.63. SN link: July; 65 & 104 MHz; -120 dBm; VWP-to-VY5 & HY5-to-VY5 link antenna combinations: (*a*) MR $\mathscr{V}_\mu$ values; (*b*) UMR $\mathscr{V}_\mu$ values; (*c*) DC $\mathscr{V}_\mu$ values; (*d*) UDC $\mathscr{V}_\mu$ values.

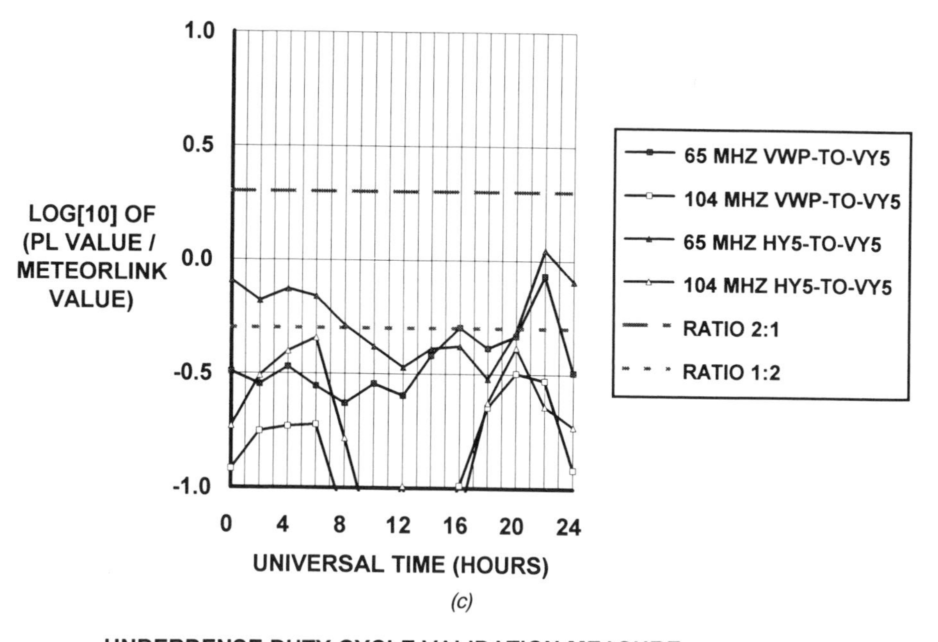

(c)

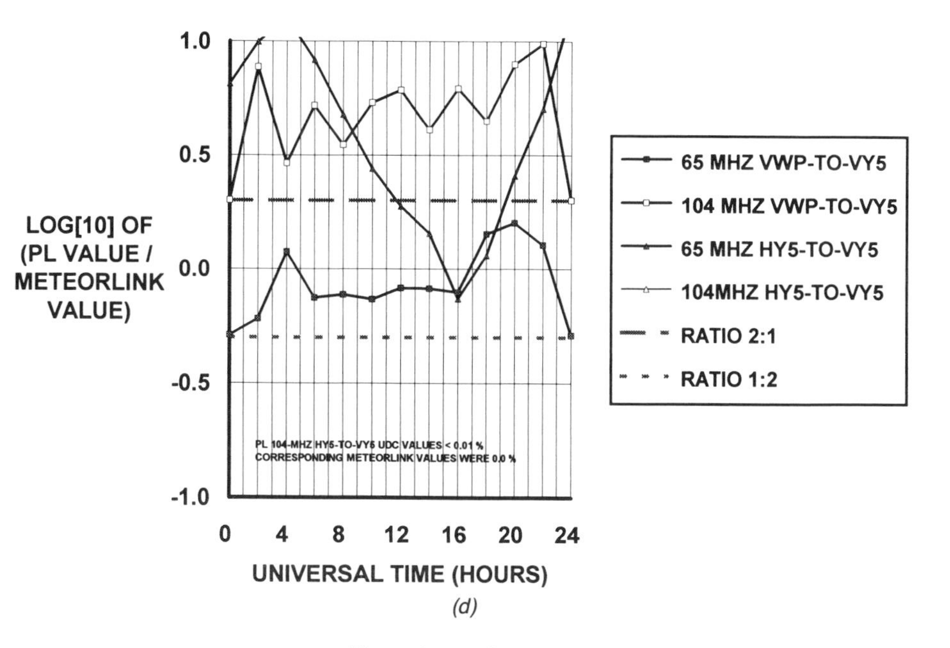

(d)

***Figure* 3.63.** (*Continued*)

-120 dBm, using horizontally polarized five-element Yagi transmit antennas and vertically polarized five-element Yagi receive antennas HY5-to-VY5, that is, a cross-polarized link, at both 65 and 104 MHz. At 65 MHz, the MR and UMR $\mathcal{V}_{\mu}$ values are pessimistic by about 0.2 and 0.4, respectively. The measured MR values for this case are similar to the VWP-to-VY5 case at 65 MHz, a minimum of 0.1 meteors per minute throughout the average day in July. Thus, both measurement and prediction demonstrate MB link operation using cross-polarized transmit and receive antennas. In this case, the negligible measured and predicted link MR and DC values preclude significant conclusions regarding model trends.

Since the link geometry and frequency suggest negligible Faraday rotation effects [195], this cross-polarization phenomenon may be rooted in the transmit and receive E-field phasor dot products in the link common volume. This hypothesis is supported by model link-diagnostic calculations that show significant E field dot products in the common volume above and to each side of the receive VY5 antenna (site 2) as shown in the LINK grid of Fig. 3.64. The E phasors accepted by the VY5 antenna pattern at high elevation angles have significant *horizontally* polarized components. These components are aligned with the horizontally polarized E-field phasors launched by the horizontally polarized HY5 transmit antenna (site 1) that illuminate the common volume above the receive VY5 antenna at site 2. The bilateral asymmetry along the x_{L} and y_{L} axes apparent in the figure is due to the SAFB HY5 antenna pattern, which included asymmetrical terrain blockage effects in the antenna foreground. These 65 MHz results are repeated at 104 MHz, but at lower MR and DC values and with excessive pessimism in the UMR and UDC $\mathcal{V}_{\mu}$ values.

3.4.3.2 Small ICBM MB Link Measurements

3.4.3.2.1 Overview. MCC performed a series of link measurements during the summer of 1988 designated Phase 2 Link Tests [196]. An important objective of these tests was to evaluate the effects of antenna polarization and directivity on single-burst MWT versus link range, time-of-day, and burst duration τ^*. In this regard, five combinations of transmit and receive antennas were selected for the antenna comparison tests. Each test was operated continuously for several days at each of seven east–west link ranges in Montana during July, August, and September of 1988. A control link, intended to provide a performance benchmark, was operated in a single, fixed configuration throughout the antenna comparison tests.

Science Applications International Corporation (SAIC) employed the METEORLINK program to predict the absolute and trend performance of the Phase 2 Link Tests [197]. In addition to test variables, these predictions required modeling of each link antenna as well as input parameters derived from measured noise values, system losses, and modem-required E_{b}/N_0 values. A second measurement–prediction comparison was performed in-

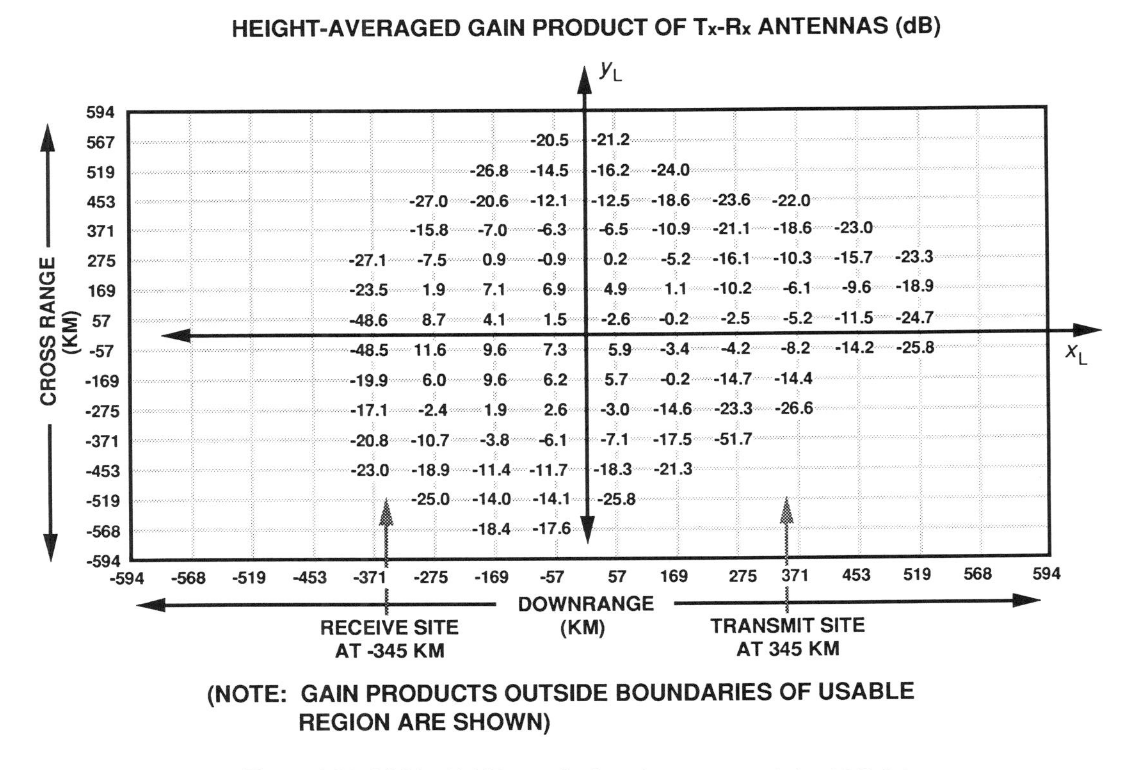

Figure 3.64. LINK grid MR contributions for a cross-polarized MB link.

cluding both an absolute and trend analysis of the daily average message waiting time (DAWT). The absolute analysis considered temporal and range variation, while the trend analysis focused on antenna polarization and single-burst message duration effects. The results of this DAWT analysis are presented in this section.

3.4.3.2.2 Measurements. Since test objectives included exploration of antenna polarization and gain pattern effects versus range, link antenna types were chosen to provide simple, easily realizable antennas with separable polarizations and patterns. Omniazimuthal antennas consisted of a horizontally polarized, folded crossed-dipole (HXD) and vertical whip (VWP) in a transportable receive-only (RO) test configuration. The HXD antennas were configured at both 1/4- and 1/2-wavelength heights above ground. Ground screens were used beneath each antenna to minimize the impact of varying ground electrical parameters (conductivity and permitivity) at different RO sites.

Fixed transmit-only (TO) sites, one located at Bozeman, Montana, and the other at Maple Valley, Washington, transmitted a fixed 8-ms channel probe signal identifying the source location. Simultaneous operation of dual MCC-6520 Master Station transmitters at each TO site, combined with the inherent space diversity in the MB channel, permitted simultaneous antenna configuration tests on closely spaced frequencies (49.18 and 49.93 MHz) at each link range. The Maple Valley facility also served as the receive site for the control link, with control link MCC-6520 probe transmission from Bozeman at 46.18 MHz. The geographic locations of the TO and RO sites and the seven test ranges employed are provided in Table 3.7.

The link transmit–receive antenna configurations used at each range are summarized in Table 3.8. Stacked horizontal and vertical log periodic antennas (HLPS and VLPS, respectively) provided directional gain at each transmit site, while the HXD and VWP antennas were used for omniazimuthal transmit antennas. These antennas were operated at both 4.5- and

TABLE 3.7 Phase 2 Link Sites

Transmit Sites			Latitude	Longitude
Boseman, Montana			*45°41′N*	*110°09′W*
Maple Valley, Washington			*47°27′N*	*122° 2′W*
			Range (km) from	
Receive Sites (Montana)	Latitude (North)	Longitude (West)	Bozeman	Maple Valley
Graycliff	*45°43′*	*111°31′*	—	*825*
Molt	*45°54′*	*109°4′*	*160*	*1000*
Colstrip	*45°45′*	*106°33′*	*360*	*1200*
Ekelaka	*45°47′*	*104°32′*	*510*	*1350*
Maple Valley (Control link)	*47°27′*	*122°2′*	*840*	—

TABLE 3.8*a* Phase 2 Antenna Configurations

	Transmit Antennas		Receive Antennas	
Test	Antenna Description	Phase Center Height	Antenna Description	Phase Center Height
1A	Stacked vertical log periodics, VLPS	$3/4\lambda$ 4.5 m	Vertical whip, VWP	Ground 0 m
1B	Stacked horizontal log periodics, HLPS	$3/4\lambda$ 4.5 m	Horizontal crossed dipoles, HXD	$1/4\lambda$ 1.5 m
2A	Stacked vertical log periodics, VLPS	$3/4\lambda$ 4.5 m	Vertical whip, VWP	Ground 0 m
2B	Stacked horizontal log periodics, HLPS	$3/4\lambda$ 4.5 m	Horizontal crossed dipoles, HXD	$1/2\lambda$ 3.0 m
3A	Stacked horizontal log periodics, HLPS	2λ 12.0 m	Horizontal crossed dipoles, HXD	$1/4\lambda$ 1.5 m
3B	Stacked horizontal log periodics, HLPS	$3/4\lambda$ 4.5 m	Horizontal crossed dipoles, HXD	$1/4\lambda$ 1.5 m
4A	Vertical whip, VWP	Ground 0 m	Vertical whip, VWP	Ground 0 m
4B	Horizontal crossed dipoles, HXD	$1/4\lambda$ 1.5 m	Horizontal crossed dipoles, HXD	$1/4\lambda$ 1.5 m
5A	Vertical whip, VWP	Ground 0 m	Vertical whip, VWP	Ground 0 m
5B	Horizontal crossed dipoles, HXD	$1/2\lambda$ 3.0 m	Horizontal crossed dipoles, HXD	$1/2\lambda$ 3.0 m
Control link	Single horizontal log periodic, HLP	$3/2\lambda$ 9.0 m	Single horizontal log periodic, HLP	$3/2\lambda$ 9.0 m

12.3-m heights with main boom tilt angles from the horizontal designed to improve midpath illumination. The control link employed a single horizontally-polarized LPA (HLP) at both transmit and receive sites, spanning a total path length (see Table 3.7) of about 840 km. The LPAs were pointed along the GCP between transmit and receive sites while the HXD antennas were oriented with one element pointed toward true north.

The occurrence times of 8-ms probe signals correctly received by dual MCC-6570 remote terminals were recorded during the several days of each link test. This data was processed by TRW [198] using a sliding time window to recover each burst occurrence and its proper duration despite possible RSL fading. Periods of sporadic-E and aircraft scatter behavior were excised by establishing rules governing its expected occurrence, for example, events of excessive duration. The number of bursts versus duration provided the daily average MR value versus τ^* for each test link configuration and the control link. The corresponding DAWT was then approximated as the

TABLE 3.8*b* Phase 2 Antenna Configurations (Continued)

Phase 2 Transmit Antenna Configurations		Phase 2 Test Receive Sites			
		Graycliff	Molt	Colstrip	Ekelaka
Bozeman	Range (km)	—	160	360	510
Stacked vertical log periodics @ 4.5 m	Main boom tilt angles (°)	—	48	30	20
Stacked horizontal log periodics @ 4.5 m	Main boom tilt angles (°)	—	48	30	30
Stacked horizontal log periodics @ 12.3 m	Main boom tilt angles (°)	—	7.5	7.5	7.5
Maple Valley	Range (km)	825	1000	1200	1350
Stacked vertical log periodics @ 4.5 m	Main boom tilt angles (°)	10	10	10	10
Stacked horizontal log periodics @ 4.5 m	Main boom tilt angels (°)	0	0	0	0

Control Link Antenna Configuration		Transmit Antenna at Bozeman	Receive Antenna at Maple Valley
Single horizontal log Periodic @ 9.0 m	Main boom tilt angle (°)	15	15

inverse of each DAMR value. Test link τ^* values of 20, 100, 200, and 400 ms produced increasing DAWT values and provided the basis for the measurement–prediction comparison with the Phase 2 Link Tests. TRW post-processing of the control link data was performed for burst durations of 200 ms.

MCC also recorded noise values (P_n) at the beginning of each test hour. These measurements showed about 3-dB higher noise value on the HXD

TABLE 3.9 Phase 2 Link Power Budget Parameters

Test	Dates in 1988	Range(s) (km)	Transmit Power (dBW)	Minimum Required RSL (P_r^*) Versus Time of Day (Pacific Time)					
				00	04	08	12	16	20
VLPS-VWP (2A)	8/10-13	160, 1000	26.0	−114.4	−114.8	−115.0	−115.0	−115.0	−114.3
	8/29-31	360, 1200	25.7	−114.6	−115.2	−115.1	−115.2	−114.3	−114.0
	9/11-13	510	25.7	−114.7	−115.0	−115.1	−115.1	−115.2	−114.2
	7/26-28	825	27.3	−114.5	−115.0	−115.2	−115.2	−114.7	−114.0
HLPS-HXD (2B)	8/10-13	160, 1000	25.7	−111.7	−112.0	−113.2	−113.3	−112.9	−111.7
	8/29-31	360, 1200	25.7	−111.9	−113.4	−113.5	−113.4	−111.8	−111.7
	9/11-13	510	25.7	−111.9	−112.5	−112.7	−112.2	−111.7	−112.1
	7/26-28	825	26.0	−112.0	−112.2	−113.2	−113.4	−113.1	−110.5
VWP-VWP (5A)	8/13-17	160, 1000	26.0	−114.4	−114.8	−115.0	−115.0	−115.0	−114.3
	9/2-4	360	25.7	−114.0	−114.8	−115.2	−114.9	−114.3	−113.5
HXD-HXD (5B)	8/13-17	160, 1000	25.7	−111.8	−112.4	−113.4	−113.2	−112.1	−111.6
	9/2-4	360	25.7	−112.1	−113.0	−113.4	−113.4	−113.3	−111.8
	9/5-7	510	25.7	−114.8	−114.1	−115.8	−115.5	−114.7	−114.5

antenna, P_n(HXD), than measured for the VWP antenna, P_n(VWP), a consequence of the greater average gain of the HXD as compared to the VWP antenna. These values also showed a diurnal variation characteristic of the changing orientation of the receive antenna relative to the principal Galactic noise sources [199]. The values $P_n(\text{HXD}) \cong -111 \pm 1.5\,\text{dBm}$ and $P_n(\text{VWP}) \cong -113 \pm 1.5\,\text{dBm}$) provided a critical input parameter for the METEORLINK predictions when combined with the additional link budget parameters, including line losses ($L_t \cong L_r \cong 2\,\text{dB}$), transmit power levels ($P_t \cong 500\,\text{W}$), receiver noise figures ($N_f \cong 5\,\text{dB}$), and modem error rate performance (P_e versus $E_b/N_0 \cong 7\,\text{dB}$) to compute the minimum required RSL values, P_r^*.

3.4.3.2.3 Predictions. METEORLINK predictions were performed for the test and control link ranges and the corresponding antenna configurations in italics in Tables 3.7 and 3.8. Predictions were produced for single-burst durations (τ^*) of 20, 100, 200, and 400 ms. MR values computed for six 4-h predictions spanning a day in the middle of each link antenna configuration test period were found adequate to determine DAMR values and, ultimately, the DAWT values. The 4-h average power budget and temporal parameters employed for each test and control link prediction are provided in Table 3.9. NEC-modeled antenna patterns were provided by Boeing for each link transmit and receive antenna. These antennas were modeled using the appropriate ground screens and compared with low elevation angle gain measurements performed on-site by MCC [200]. Predictions reflecting uncertainty in power budget parameters were also performed; however, only the results corresponding to the average power budget parameters are presented in this summary.

3.4.3.2.4 Comparisons

3.4.3.2.4.1 Absolute Comparison. Figure 3.65 is a plot of DAWT $\mathcal{V}_\mu$ values for the VLPS-to-VWP link (Test 2A: directional, vertically polarized transmit antenna and omniazimuthal, vertically polarized receive antenna) versus burst duration $\tau^* = 20$, 100, 200, and 400 ms for several test dates and their corresponding ranges. Since the validation data consists of DAWT values, the validation measure has been redefined for this comparison as

$$\mathcal{V}_\mu = \log_{10}(s_p/s_m)\,, \tag{3.128}$$

where s_p and s_m are the predicted and measured values, respectively. The horizontal lines in the plot correspond to $\mathcal{V}_\mu = \pm 0.3$ for prediction / measurement ratios of 2.0 and 0.5, respectively. Missing $\mathcal{V}_\mu$ values are due to the lack of the corresponding measured or post-processed data points. It is important to note that variation of $\mathcal{V}_\mu$ values in these plots may be due to the effects of *test date* as much range.

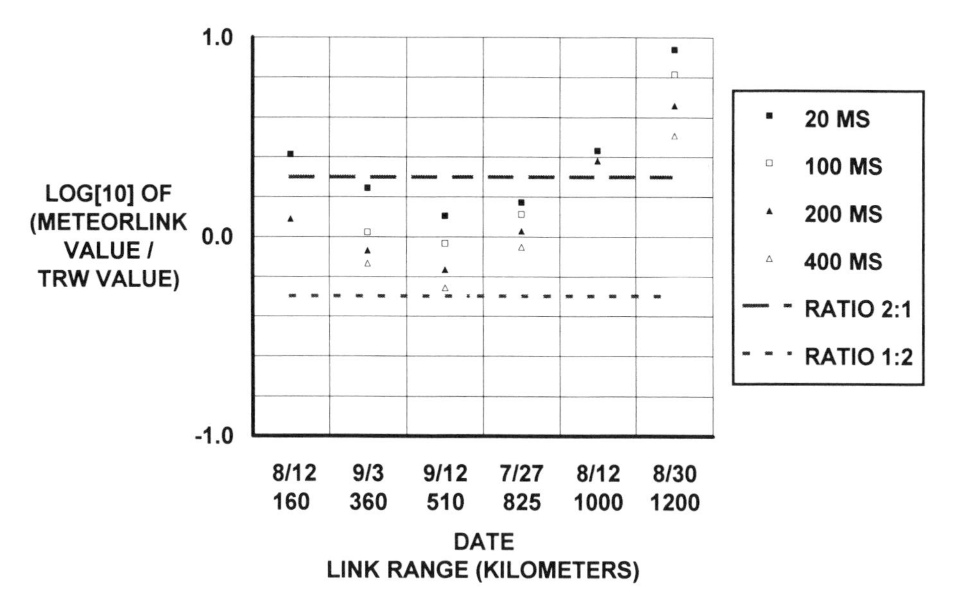

Figure 3.65. VLPS-to-VLP (Test 2A) DAWT $\mathscr{V}_\mu$ values versus date, range, and burst duration.

The VLPS-to-VWP (Test 2A) and HLPS-to-HXD (Test 2B: directional, horizontally polarized transmit antenna and omniazimuthal, horizontally polarized receive antenna) measurements for 160-and 1000-km link ranges were performed during the Perseids meteor shower, which achieves its peak MR value on the 12–13 of August [201]. This peak value can be observed in the control link $\mathscr{V}_\mu$ values plotted in Fig. 3.66. The annotated point in Fig. 3.67 corresponds to the shower-induced decrease observed in the measured control link DAWT value (12 August) relative to the surrounding non-shower DAWT values. Since the effects of the Perseid shower were not modeled, the measured decrease in DAWT values would produce pessimistic Phase 2 test link predictions. Thus, a TRW-derived factor of 1.37 was used to scale the measured DAWT values before calculation of the corresponding $\mathscr{V}_\mu$ values plotted in Fig. 3.65.

The increased optimism in the predictions for 9/12 (510 km) may indicate sporadic meteor flux values (SMRD values, see Section 3.2.2.2.2) in excess of observed values, a change in model accuracy with range, or disagreement between actual and modeled antenna patterns. Since this behavior is also observed in the control link results (Fig. 3.66), the effect is probably due to excessive assumed SMRD values for September rather than a numerical coincidence between the VLPS-to-VWP (Test 2A) and HLPS-to-HXD (Test 2B) test results and control link DAWT predictions. This example illustrates

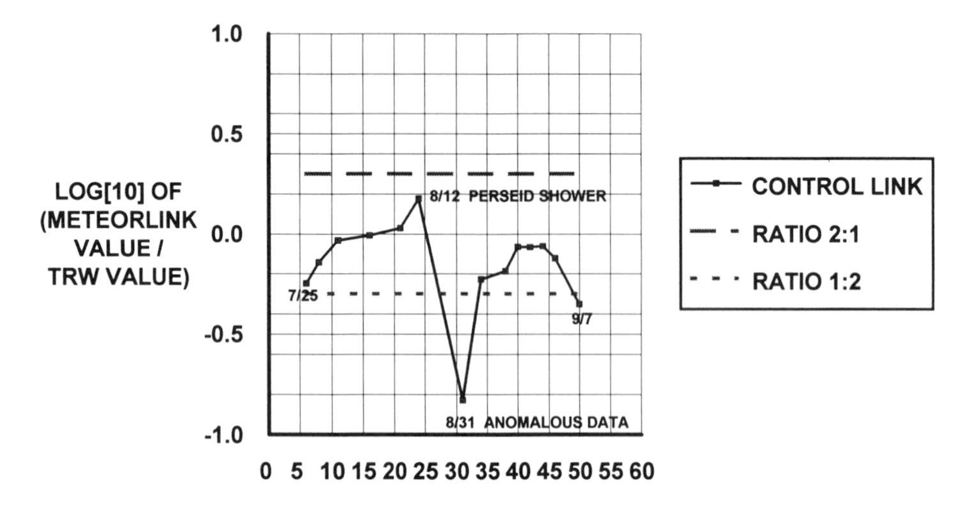

Figure 3.66. Control link (HLP-to HLP) DAWT $\mathscr{V}_\mu$ values versus test day.

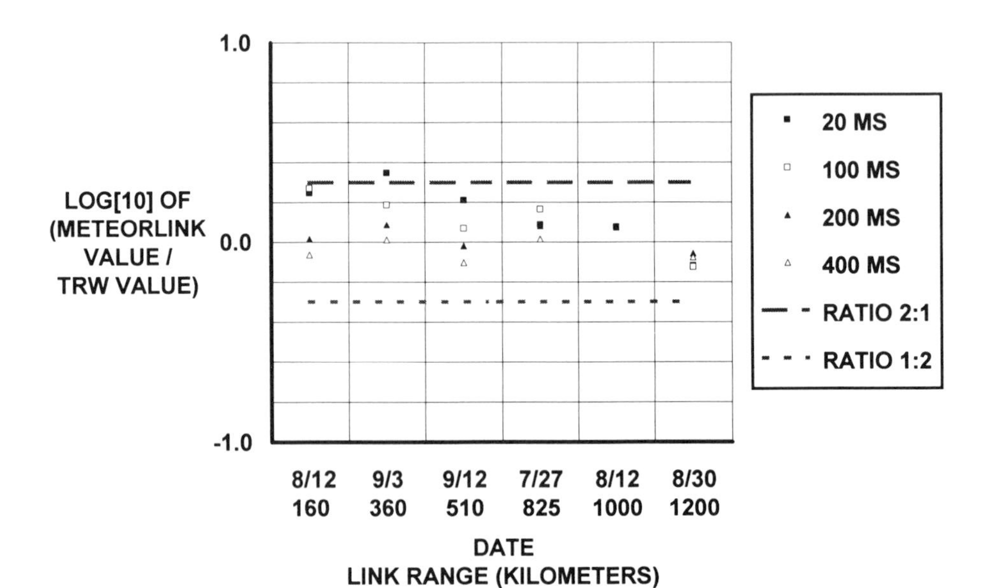

Figure 3.67. HLPS-to-HXD (Test 2B) DAWT $\mathscr{V}_\mu$ values versus date, range, and burst duration.

the importance of a control link in the execution of MB link tests, or radio tests in general, in which long-term variations in channel propagation characteristics may influence the performance measurements.

Figure 3.67 is a plot of HLPS-to-HXD (Test 2B) $\mathcal{V}_{\mu}$ values versus test date and range for burst durations $\tau^* = 20$, 100, 200, and 400 ms. Since these tests were performed simultaneously with VLPS-to-VWP (Test 2A, see Fig. 3.65) and control link test day 24, the 160-and 1000-km link range tests were performed during the Perseids shower. The shower radiants enhanced the normal SMRD distribution but were not included in the METEORLINK prediction, thus resulting in the expected control link prediction pessimism. The TRW-derived DAWT-multiplier of 1.37 was applied to the measured HLPS-to-HXD DAWT to extract the effect of the Perseids shower from the corresponding measurement-prediction comparison. Again, predictions made for September were optimistic, tracking both the HLPS-to-HXD and control link results. Nevertheless, most test link $\mathcal{V}_{\mu}$ values are within ± 0.3 for both $\tau^* = 20$- and 200-ms burst durations.

$\mathcal{V}_{\mu}$ values for the VWP-to-VWP (Test 5A: omniazimuthal, vertically polarized transmit and receive antennas) and HXD-to-HXD (Test 5B: omniazimuthal, horizontally polarized transmit and receive antennas) links are plotted in Figs. 3.68 and 3.69, respectively. Both measured and predicted values were available only for the VWP-to-VWP 160-and 360-km ranges and the HXD-to-HXD 160-, 360-, and 510-km link ranges. These results show increasing optimism with burst duration, τ^*, beginning at pessimistic values for the 20-ms results in both antenna tests.

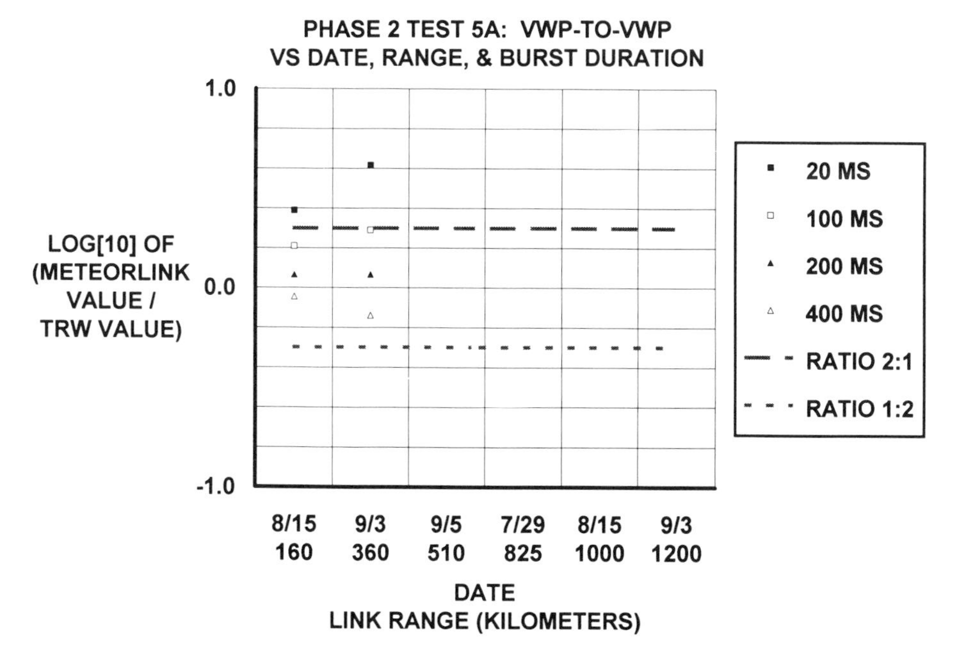

Figure 3.68. VWP-to-VWP (Test 5A) DAWT $\mathcal{V}_{\mu}$ values versus date, range, and burst duration.

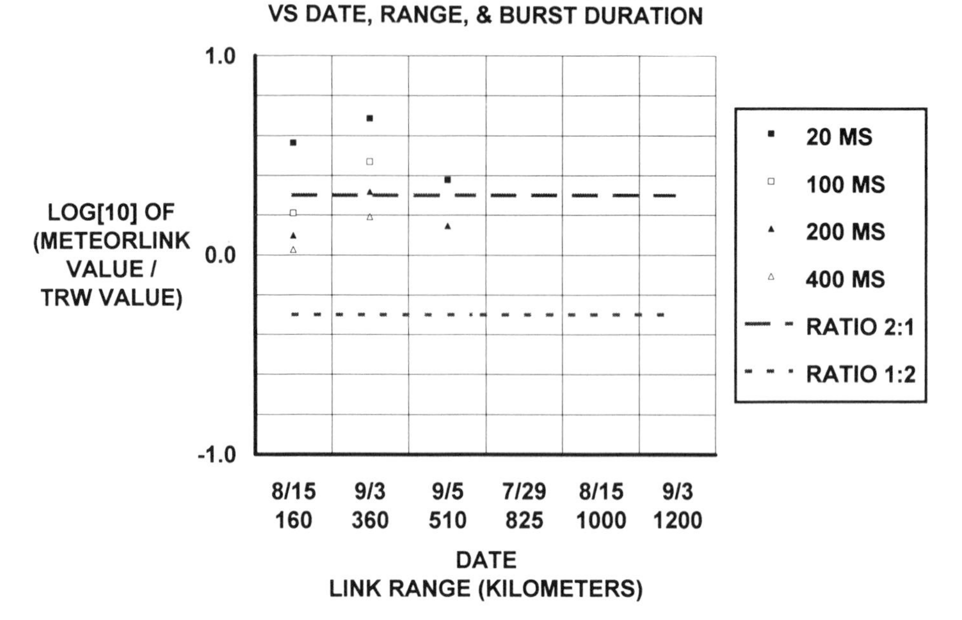

Figure 3.69. HXD-to-HXD (Test 5B) DAWT $\mathcal{V}_\mu$ values versus date, range, and burst duration.

3.4.3.2.4.2 Trend Analysis

Polarization Comparisons.

High-Gain Transmit to Low-Gain Receive Antennas. Figures 3.70*a*, *b* are plots of the ratio of HLPS-to-HXD (Test 2B) DAWT values to VLPS-to-VWP (Test 2A) DAWT values for burst durations of 20 and 400 ms, respectively. Each figure plots both the ratio of measured values and the corresponding ratio of predicted values. Since the VLPS-to-VWP (Test 2A) link employed vertically polarized antennas and the HLPS-to-HXD (Test 2B) link tests were performed simultaneously, these ratios provide a relative measure of link antenna configuration performance. More importantly, they provide a means to compare predicted trends with the corresponding measurement-derived trends. If both ratios are above unity or below unity at a given range, then the predicted trend match the measured trend at that range inasmuch as relative performance is concerned. For example, if both the predicted and measured ratios were below unity, then both the predictions and the measurements would determine that the VLPS-to-VWP link produced higher DAWT values (i.e., lower DAMR values) than the HLPS-to-HXD link.

In Fig. 3.70*a*, the polarization trends match at all ranges with the exception of 1200 km, showing that the HLPS-to-HXD (Test 2B) link out-performed the VLPS-to-VWP (Test 2A) link. At 1200 km and 20-ms

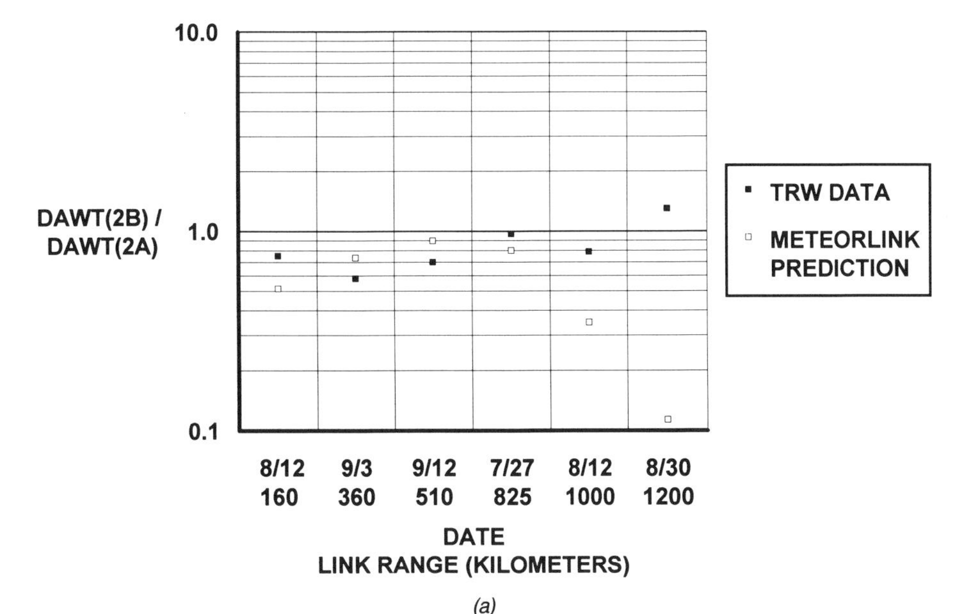

(a)

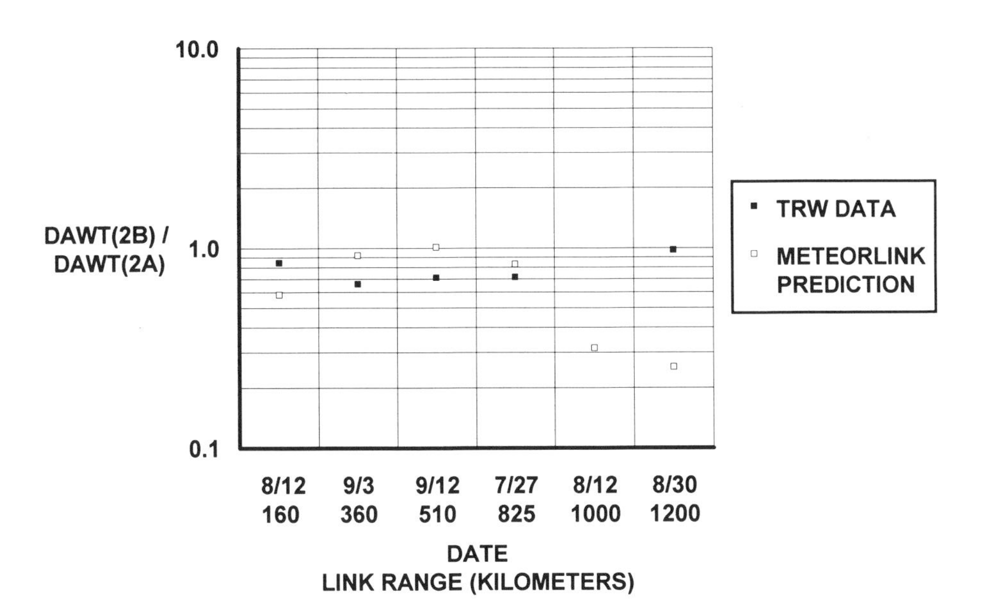

(b)

Figure 3.70. Directional-to-omni (Test 2) measurement-prediction polarization trends: (*a*) 20-ms burst duration; (*b*) 400-ms burst duration.

burst duration, predictions showed that the HLPS-to-HXD link produced higher DAWT values than the VLPS-to-VWP link, in contrast to the corresponding measurements. At 400-ms burst duration, Fig. 3.70*b* shows that the trends agreed at all ranges except 510 km, where the predictions incorrectly determined the VLPS-to-VWP link to yield the lower DAWT values. The predicted and measured trend ratios showed the greatest difference for 400 ms at 1200 km; the predicted DAWT values are less than the corresponding measurements on the HLPS-to-HXD link by about a factor of five. No measured data were available for the 1000-km test link in the 400-ms case.

These plotted ratios provide a relative measure of horizontal versus vertical polarization. It is important to note, however, that the average power gain of the HXD antenna is greater than the average power gain of the VWP antenna by about 3 dB [202]. Adding this gain difference to the VLPS-to-VWP link transmitter would reduce the apparent polarization trend, suggesting that antenna directivity and gain, not polarization, determine link performance (see Table 3.3). Apparently, antenna polarization would therefore be chosen to provide the maximum gain within the usable antenna azimuth and elevation angles (see Section 3.2.2.7.2).

Low-Gain Transmit and Receive Antennas. Figures 3.71*a*, *b* plot ratios for low gain transmit and receive antennas, that is, the ratio of DAWTs from HXD-to-HXD (Test 5B) results to those obtained for VWP-to-VWP (Test 5A). Since no VWP-to-VWP measurements were available at 510 km, no measurement-derived ratio is shown. Similarly, no measured ratio could be formed for the 1000-km, 400-ms case. The predicted trends match the measured trends for the available data. The greatest difference in the ratios, more than a factor of ten, occurs at 1000 km for the 20-ms burst duration.

Burst Duration Trends. Figure 3.72*a* is a plot of the VLPS-to-VWP (Test 2A) predicted and measured burst duration (τ^*) trends for 100 ms, that is, ratios of DAWT for a 100-ms burst to the DAWT for a 20-ms burst. No measurements were available for 160 or 1000 km, so no measured ratios are shown. All burst duration trend plots show the expected result that MWT values increase with increasing burst duration; that is, all burst duration trend ratios exceed unity.

The VLPS-to-VWP 200-ms burst duration trend, shown in Fig. 3.72*b*, differs from measurement ranges. This divergence is greatest for the VLPS-to-VWP 400-ms duration (see Fig. 3.72*c*), reaching a peak error of about a factor of three at the 360-km range. At this range, the VWP-to-VWP link MR values are dominated by overdense-trail scattered signals. As this dominance gives way to more underdense trail MR contributions at the longer ranges (e.g., 825 km), the measured and predicted ratios converge. The overdense dominance begins to recur at 1200 km and the ratios begin to diverge again. These results show that the overdense trail model is less sensitive to required burst duration than is apparent from the Phase 2 measured data.

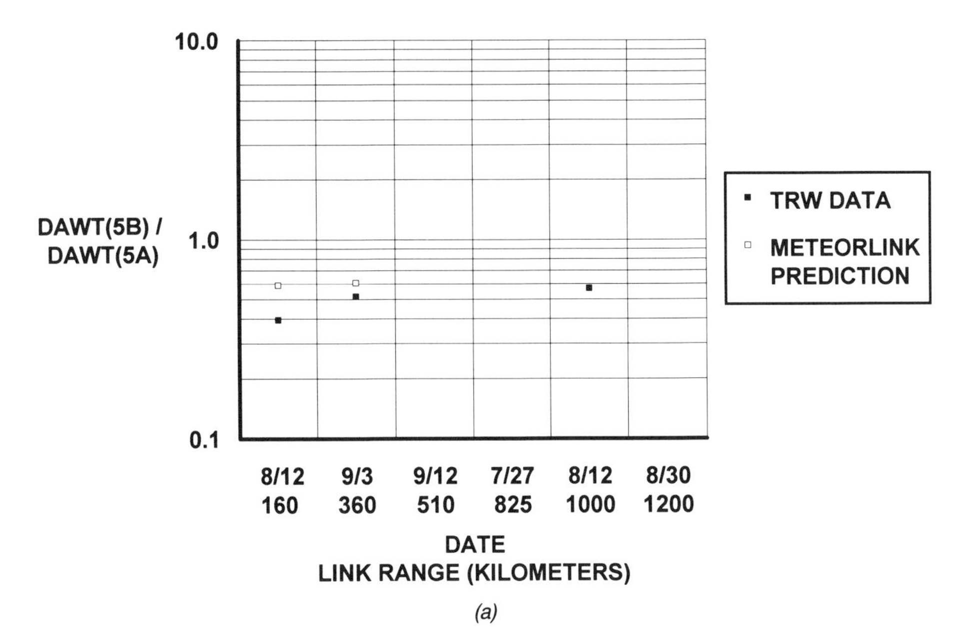

(a)

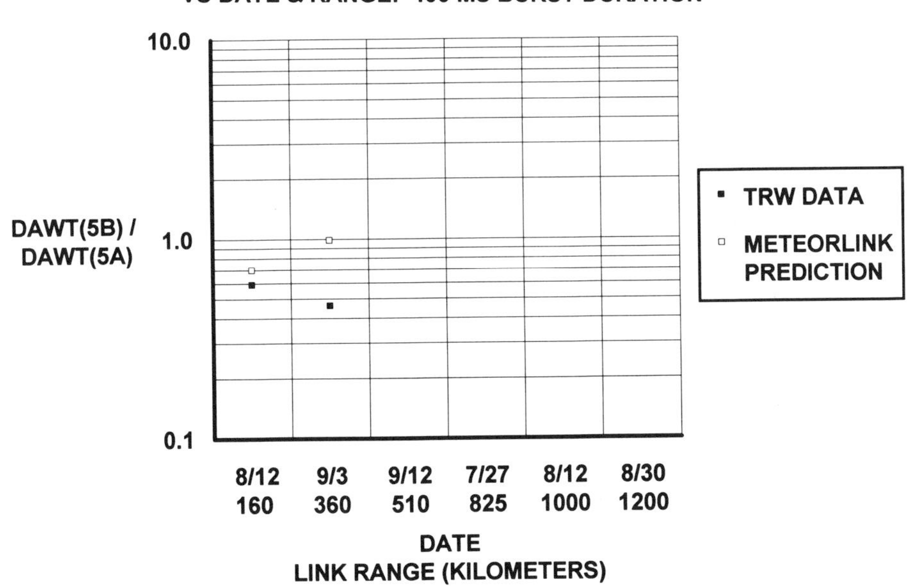

(b)

Figure 3.71. Omni-to-omni (Test 5) measurement-prediction polarization trends: (*a*) 20-ms burst duration; (*b*) 400-ms burst duration.

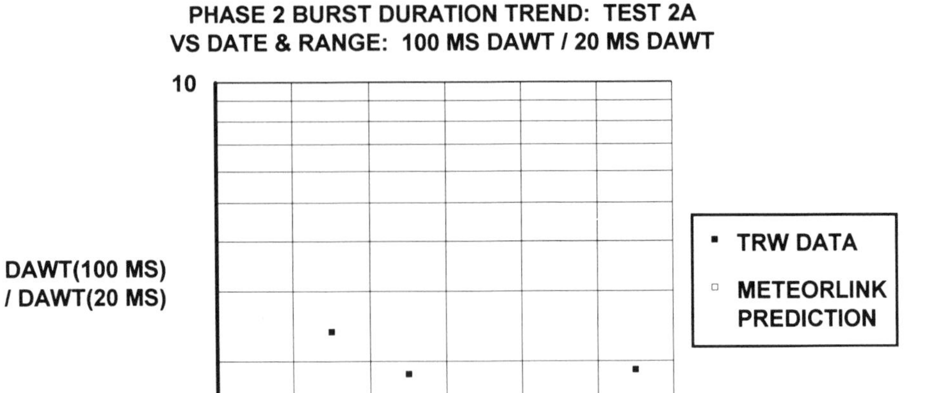

(a)

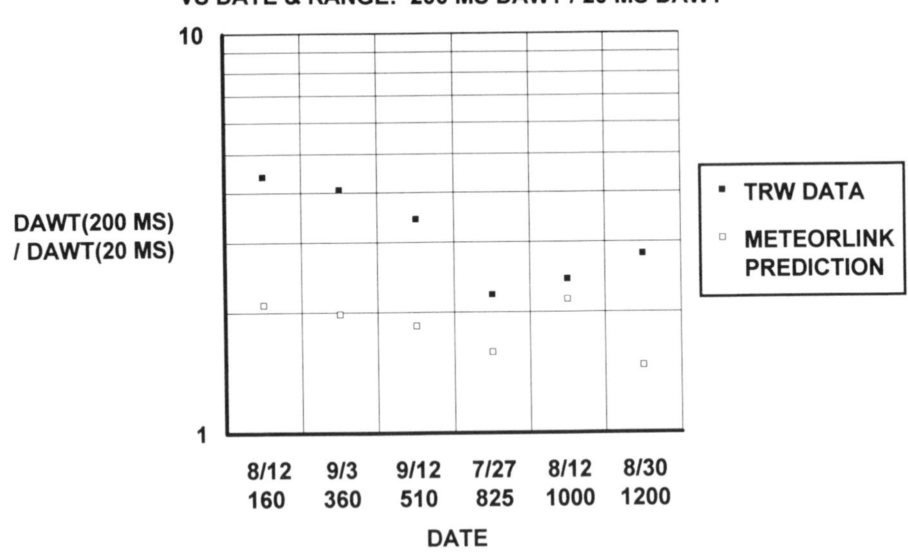

(b)

Figure 3.72. VLPS-to-VWP (Test 2A) measurement-prediction burst duration trend: (*a*) DAWT (100 ms) DAWT (20 ms); (*b*) DAWT (200 ms)/DAWT (20 ms); (*c*) DAWT (400 ms)/DAWT (20 ms).

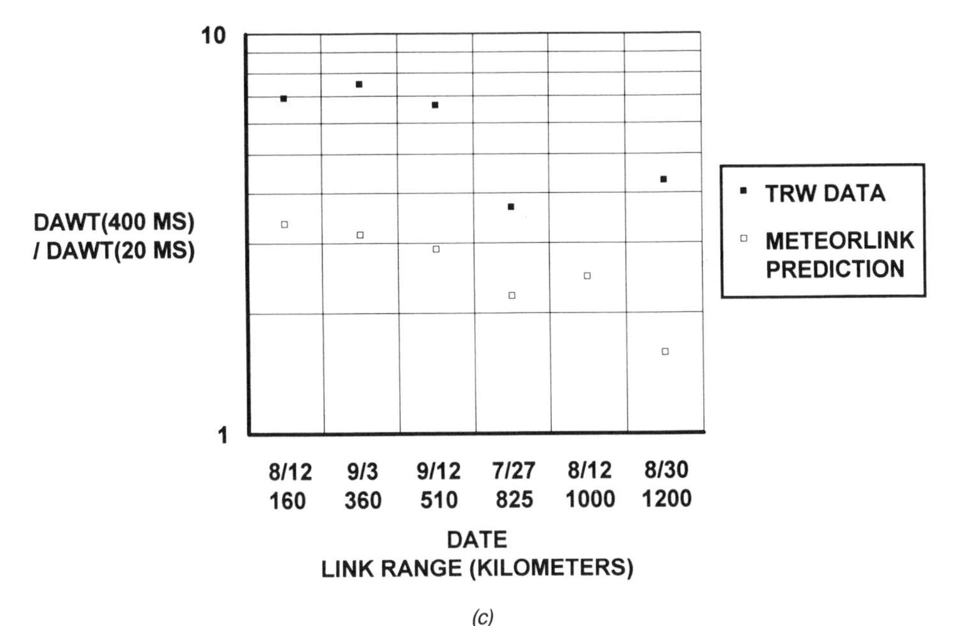

(c)

***Figure* 3.72.** (*Continued*)

3.4.3.3 Jodrell Bank Meteor Radar Measurements

3.4.3.3.1 Overview. A meteor radar was constructed and operated at the Jodrell Bank Experimental Station of the University of Manchester, United Kingdom, beginning in 1949 [203]. The radar's mission was to measure the hourly rate of sporadic meteors as well as to determine the radiants of both major and minor meteor streams [204]. In the radar case, the scattering ellipsoid (see Fig. 3.4) becomes a sphere and the potential common volume is maximized for omnidirectional antenna patterns. The Jodrell Bank results, however, consist of MR values measured with highly directional antenna beams. Thus, MR predictions of the Jodrell Bank measurements must track the interaction between observable SMRD values, narrow-beamed antennas, and spherical scattering geometry (backscatter), to reproduce the observed diurnal variation.

3.4.3.3.2 Measurements. The Jodrell Bank (2°18′W, 53°14′N) measurements were performed at a frequency of 72 MHz using two independent antenna arrays, each illuminating different portions of the sky. Each array reportedly [205] consisted of six horizontally polarized six-element Yagi antennas with 6-dB beamwidths in the horizontal and vertical planes of 10° and a corresponding peak gain of approximately 24 dBi at an elevation angle of 8.5°. The two arrays were placed at equal distances from either side of the building containing the transmitter and receiver. The antenna beams were fixed to point along different bearings, 242° (southwest beam) and 292° (northwest beam) clockwise from true north, respectively. It has been separately stated that the antenna bearings were ±25° of true west [206] for the sporadic meteor survey, although the former values were used in this measurement–prediction comparison.

The radar's peak transmitter power was 5 kW (37 dBW) and its minimum detectable power P_r^* was 7×10^{-14} W ($\cong -102$ dBm). A burst duration (radar pulse) of 8 μs duration was employed with a repetition rate of 150 pulses per second. Dual pulses spaced 300 μs apart were used to minimize noise spoofing of the scrolled photographic recording of received pulses. Transmissions were made on both antennas simultaneously, but receptions were recorded through separate receive chains from each antenna. Thus, the transmitted power into each antenna was about 2.5 kW. Plots of received pulses during the Geminid (December) meteor shower of 1949 clearly show the stream radiant passing through the southwest beam and then, about one hour later, passing through the northwest beam, as the earth rotates in the stream [206].

Average hourly MR values were derived from continuous radar measurements that spanned contiguous months from October 1949 through September 1951 [207]. In other words, the averaged MR values for each hour represent the average of all MR values collected for that hour in each day of the specified month. MR values attributed to known meteor showers were extracted. Since the two antenna beams pointed to different bearings, two independent sets of averaged MR values were obtained simultaneously. No statistical fluctuation (error bars) were provided for these measurements. Otherwise, these results provide an excellent source of validation data for measurement–prediction comparisons of diurnal and seasonal variation.

3.4.3.3.3 Predictions. METEORLINK predictions were performed using the documented radar parameters, although no NEC-generated pattern was available to model the Yagi arrays. Instead, an antenna pattern for two vertically stacked (half-wavelength separation) horizontally polarized 12-element Yagis at five wavelengths height was used to approximate the Jodrell Bank array. The 3-dB beamwidths were increased to about 35° in the horizontal plane and 40° in the vertical plane with a corresponding reduction in peak gain to about 18 dBi at an elevation angle of 8°. These antenna patterns were oriented at 242 and 292° clockwise from true north, thus approximating the southwest and northwest antenna beams, respectively, of the Jodrell Bank radar.

A 5-kW transmitter output power, P_t, was assumed into *each* antenna, unlike the 2.5 kW used by the Jodrell Bank radar. The additional 3 dB of transmit power into each antenna somewhat offset the difference in actual versus modeled peak antenna gains. Nevertheless, the assumed antenna patterns can only be considered an approximation to the actual patterns. Since the modeled antenna illuminates a greater sky volume than the Jodrell radar antenna (wider main beam, albeit with reduced power gain), the absolute MR values would be expected to vary somewhat between measurement and prediction. As in the measurements, a radar frequency (f_R) of 72 MHz was assumed with a required received power P_r^* of -102 dBm. A required burst duration of 10 ms was assumed to account for the radar repetition rate (150 per second) as an *average* offset from the start of a burst until the arrival of a radar pulse. The measured data used in this comparison was collected for the period from October 1950 through September 1951.

3.4.3.3.4 Comparison. A measurement–prediction comparison of the Jodrell Bank results from March 1951, June 1951, September 1951, and December 1950 is shown in Figs. 3.73–3.76. These figures plot the monthly-averaged hourly MR values from both northwest and southwest radar beams [208]. Absolute MR values, rather than $\mathcal{V}_\mu$ values, were plotted to illustrate the diurnal variation experienced by the northwest and southwest radar beams as well as to report the results of the measurement-prediction comparison. Note that the MR values are expressed in meteors per hour. Plots of the measured and METEORLINK predicted values for all 12 months, as well as the monthly averages, are provided in Appendix C.

Figures 3.73*a*, *b* show plots of the measured and predicted MR values for March using the northwest and southwest antenna beams, respectively. Both measurement and prediction show modulation throughout the day as the antenna beams illuminate varying SMRD values. Since Jodrell Bank is in the Greenwich Time Zone, local time corresponds to Universal time. Thus, the minimum measured and predicted MR values occur from 14 to 20 h UT, when the apex is blocked by the earth. The northwest beam measurements show a peak at 4 UT, while the corresponding prediction peaks at 6 UT. Peak MR values occurs at 7 UT in the Jodrell Bank southwest beam, while the corresponding prediction achieves a peak at about 4 UT.

Measurement and prediction-based relative MR values for both northwest and southwest beams in June are plotted in Figs. 3.74*a*, *b*. Peak values for both beams occur between 8 and 11 UT with minimum values between 14 and 20 UT. Measured and predicted occurrence times for the peak values of the southwest beam MR values differ significantly. In addition, the swing in measured MR values from a maximum of about 35 per hour at 10 UT to a minimum of about 2.5 meteors per hour 4 h later at 2 UT was not predicted. This behavior may be due to the narrower antenna beam used by the radar than employed in the corresponding predictions. A similar maximum-to-minimum swing is also evident in both measurement and prediction for the northwest beam, although the predicted values lag the corresponding measurements by about 2 h.

The measurement–prediction comparison for September plotted in Figs. 3.75*a*, *b* shows a significant difference in the absolute MR values for the

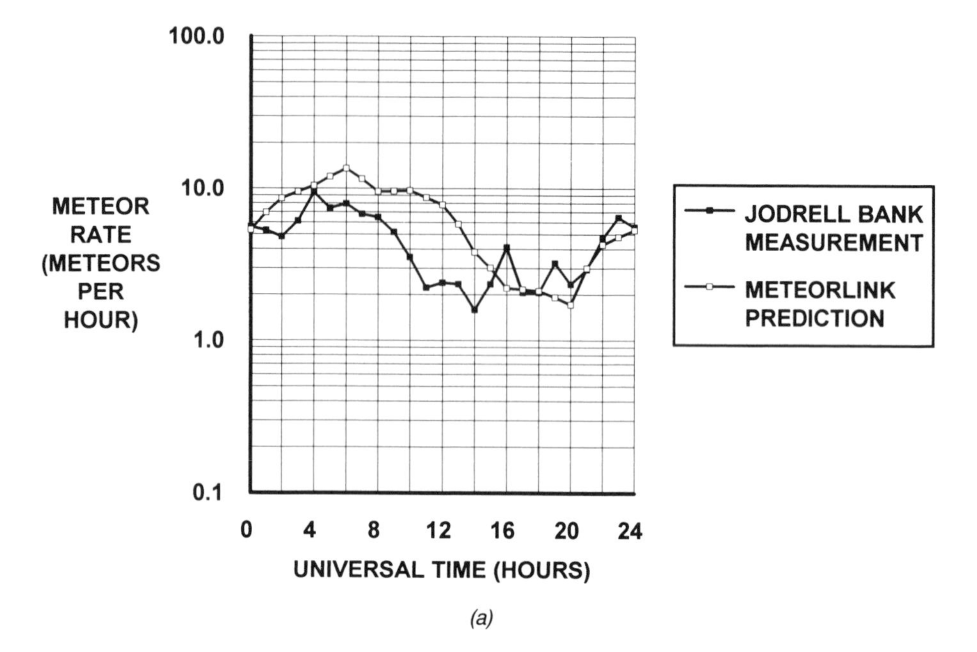

(a)

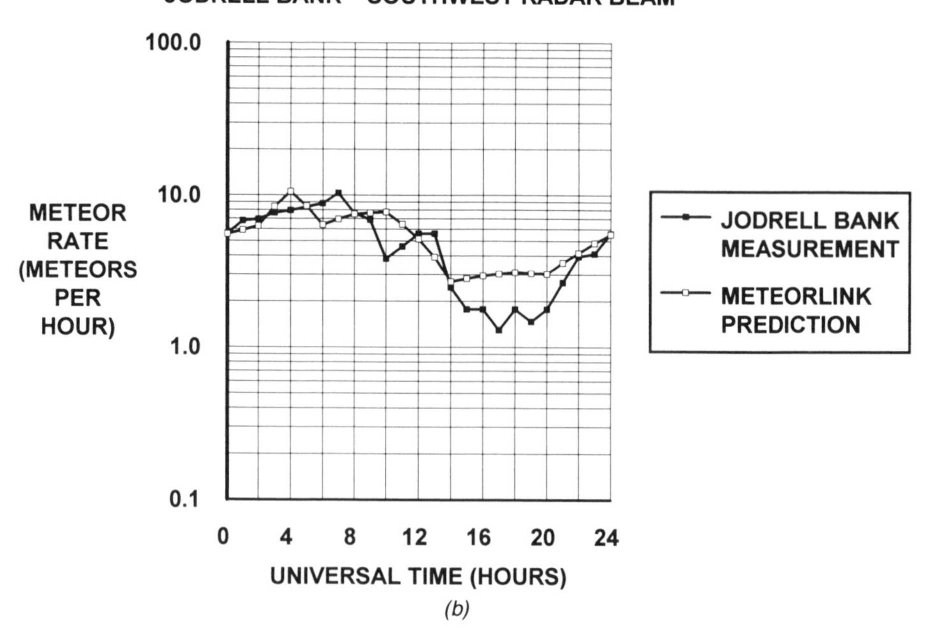

(b)

Figure 3.73. Jodrell Bank MR comparison in March: (*a*) northwest radar beam; (*b*) southwest radar beam.

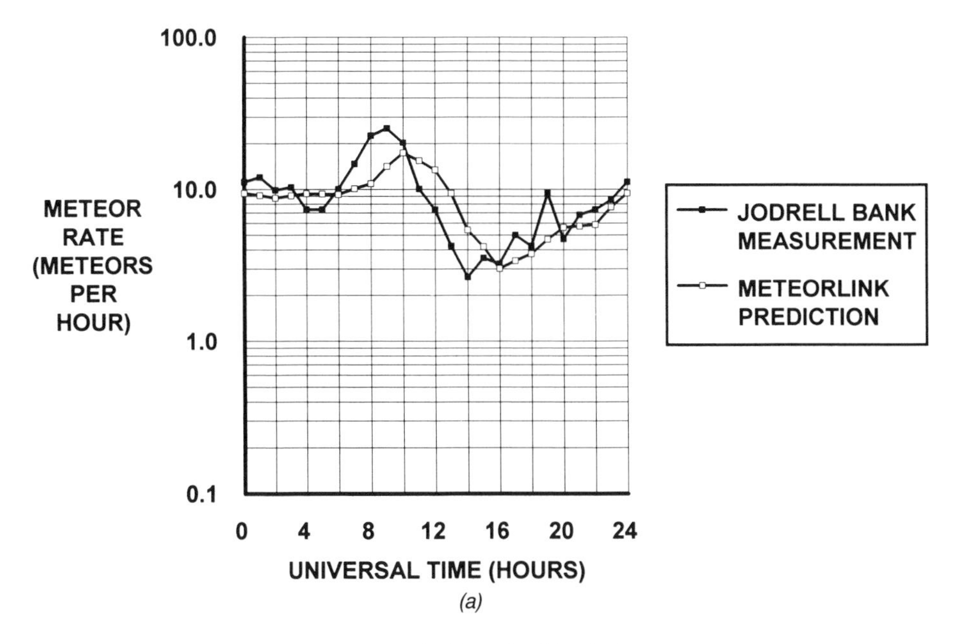

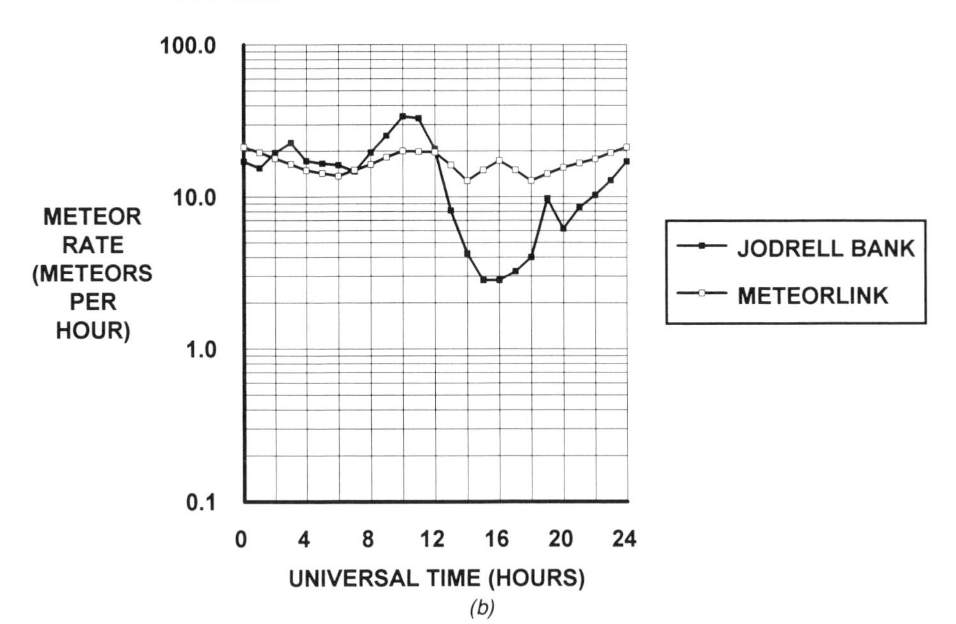

Figure 3.74. Jodrell Bank MR comparison in June: (*a*) northwest radar beam; (*b*) southwest radar beam.

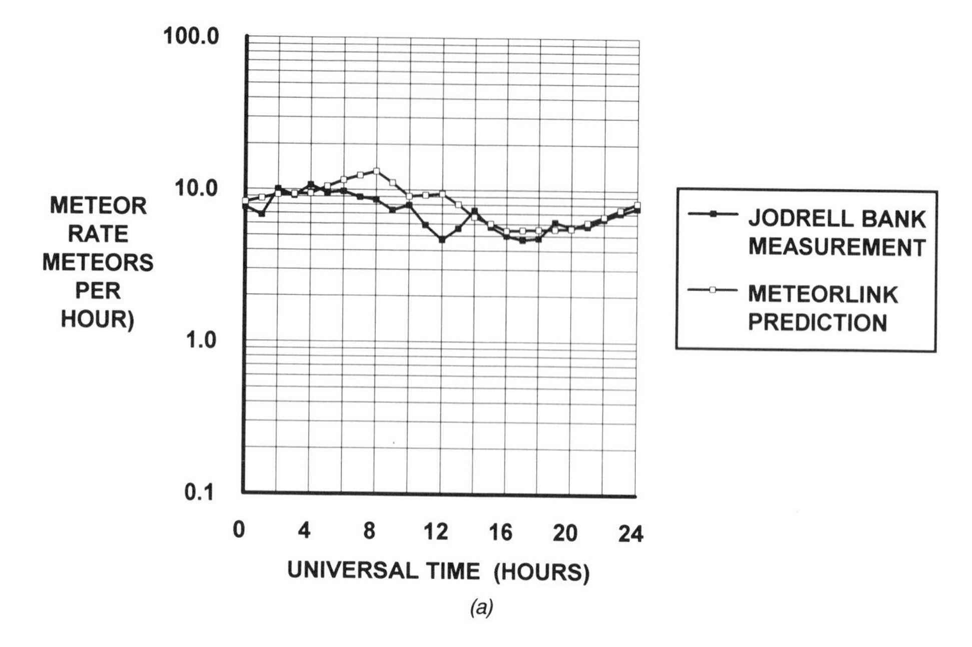

(a)

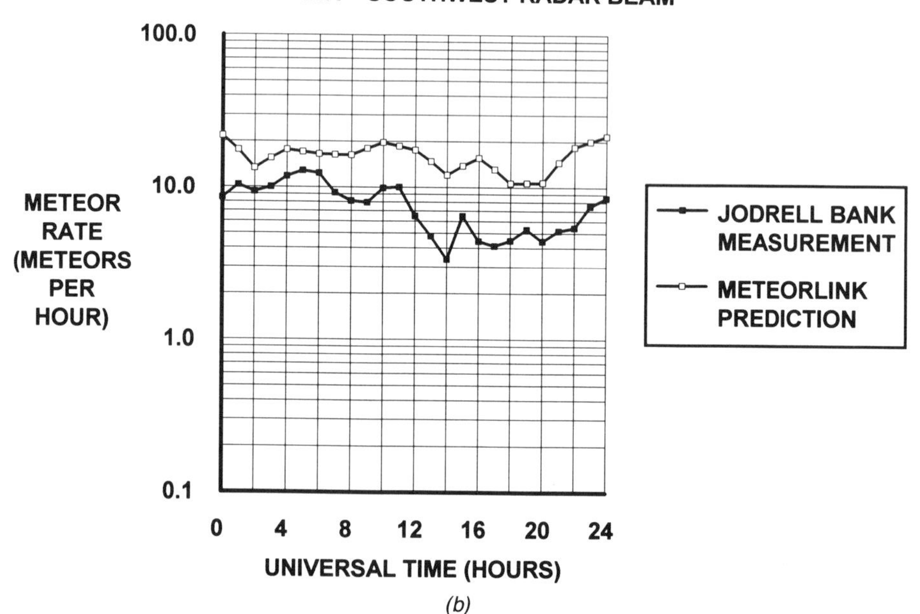

(b)

Figure 3.75. Jodrell Bank MR comparison in September: (*a*) northwest radar beam; (*b*) southwest radar beam.

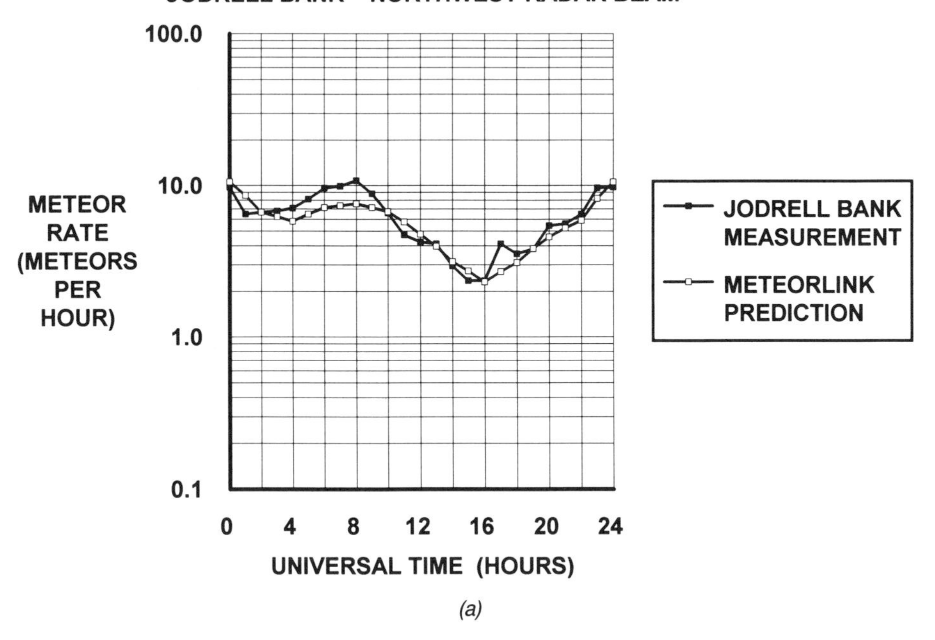

(a)

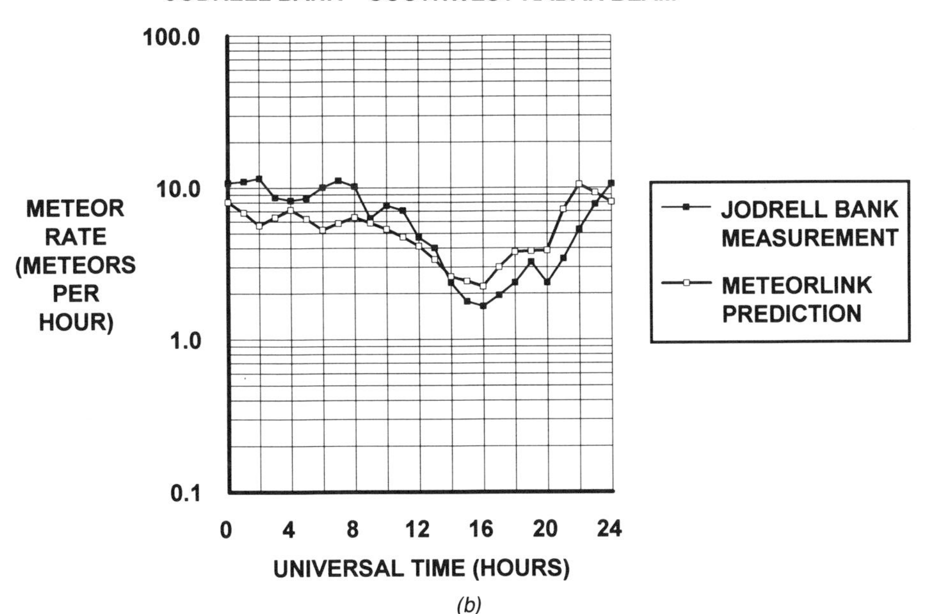

(b)

Figure 3.76. Jodrell Bank MR comparison in December: (*a*) northwest radar beam; (*b*) southwest radar beam.

southwest beam, particularly during the afternoon hours, and in the amount of diurnal MR maximum–minimum swing. Corresponding measured and predicted values for the northwest beam match quite closely. The measurement–prediction comparison for December is shown in Figs. 3.76*a*, *b*. As in the March and June results, the prediction and measurement show similar diurnal modulation, achieving peak values between 6 and 8 UT and minimum MR values at about 16 UT. The significant prediction optimism apparent the southwest results in September has disappeared and the predictions track the measured diurnal variation.

Several possible causes for the significant prediction optimism in September, despite more accurate predictions in March, June, and December, may be considered. First, the higher incidence of sporadic-E during the summer months may have spoofed the Kazan–Mogadishu radars used to generate the SMRD values used in the METEORLINK program. These radars operated at 8 m, which is more susceptible to backscatter from pockets of intense E-layer ionization than was likely for the Jodrell Bank radar operating at 4.2 m. No description of sporadic-E spoofing was available for the Kazan–Mogadishu SMRD measurements. It is doubtful, however, that diurnal variation in the occurrence of sporadic-E would correspond to so closely to the diurnal variation in predicted MR values, or that the northwest beam results would have been unaffected.

A second source of measurement–prediction discrepancy is variation between radar-observed MR values recorded in different years. The Jodrell Bank survey was conducted from 1949 to 1951, while the Kazan–Mogadishu survey spanned the mid to late 1960s. Radar measurements at Kharkov [209] conducted from 1972 through 1976 showed different average hourly MR values in the same month of consecutive years, sometimes by as much as a factor of two. This effect is described in Section 3.4.3.1.4 and suggests that year-to-year variation in the arriving sporadic meteor flux may explain the apparent prediction error.

A third source of measurement–prediction discrepancy may be due to the Jodrell Bank radar system. The detection equipment was modified to include a discriminator circuit and the transmitter was configured to transmit a two-pulse (300-μs separation) signal. The discriminator reduced detector spoofing due to noise and interference, while the dual-pulse provided means of radar signal identification [210]. This modification considerably reduced spoofing, although no quantitative description of the effects on detection of short-duration events was provided. If this measurement-based discrepancy had occurred, however, it would have affected the March, June, and December results as well as those found in September.

3.4.3.4 Summary of Link Comparison Results. The Jodrell Bank measurement–prediction comparison provides an excellent starting point for the synthesis of link validation results into recommendations for model improvement. The significant prediction optimism found for September is also apparent in the measurement–prediction comparison results for August through November presented in Appendix C. This consistent prediction optimism during the late summer and early autumn months coincides with the portion of earth's orbit for which the northern latitudes are exposed to the greatest sporadic meteor speeds. The northern latitudes are tilted forward during these months due to the obliquity of the ecliptic ($\cong 23.45°$, see Appendix A), that is, toward the apex direction. Since average geocentric meteor speeds increase with decreasing elongation angle, (see Eq. 3.21), the radar detects scattered signals from trails formed by faster meteors during these months than at other times of year.

Given these considerations, *observed* meteor speeds in the northern latitudes may be expected to achieve their peak values on 21 September, the autumnal equinox, when the north celestial pole achieves its greatest forward tilt. After this date, the observed meteor speeds would decrease. This effect is best shown by the plot of month-by-month DAMR values for each beam shown in Figs. 3.77*a*, *b*. Prediction optimism achieves peak values from July through October, with a maximum in July on the northwest beam and in September on the southwest beam. These results, particularly for the southwest beam, correlate well with the period of maximum sporadic meteor speeds. In fact, the optimism in the September Jodrell Bank predictions corresponds to the optimism apparent in the Phase 2 test link and control link validation (see Section 3.4.3.2.4.1 and Figs. 3.65-3.67) performed using data from early September, 1988.

During the late summer and early autumn months, the Jodrell Bank radar and Phase 2 link common volume would have been exposed to higher meteor speeds than during other months. The resulting velocity-selective effects on meteor detectability (see Section 3.2.2.2.2) provides one possible explanation for the apparent prediction optimism. In fact, an annual peak of MR values in September producing short-duration trail-scattered signals has been observed in practice [211]. Short-burst durations are characteristic of higher meteor velocities, because the initial trail radius, r_0, increases with increasing meteor speed (see Eq. 3.16) and yields a more diffuse ionization. In other words, more underdense trail-scattered signals are detected in the morning hours than in the late afternoon, but with decreasing peak trail-scattered power. In effect, this velocity selectivity reduces the *apparent* diurnal variation in the meteor arrival rate observed by radar relative to the actual arrival rate.

Assuming the Jodrell Bank validation data is accurate, systematic prediction error may explain the velocity-selective effects apparent in the link measurement-prediction comparisons. The METEORLINK development in Section 3.2.2.2.2 assumed that the meteor velocity distribution function, $p_{\mathbf{V}_a}(v, \hat{\mathbf{v}}_m^{\mathrm{O}})$, was adequately approximated by a delta function at the average atmospheric speed for each elongation angle (see Eq. 3.23), that is, $p_{\mathbf{V}_a}(v, \hat{\mathbf{v}}_m^{\mathrm{O}}) \cong c_a n(\epsilon)\, \delta_D[v - \bar{v}_\epsilon]$. This assumption yields the classic result that the difference in radar sensitivity between the Jodrell Bank and Kazan-

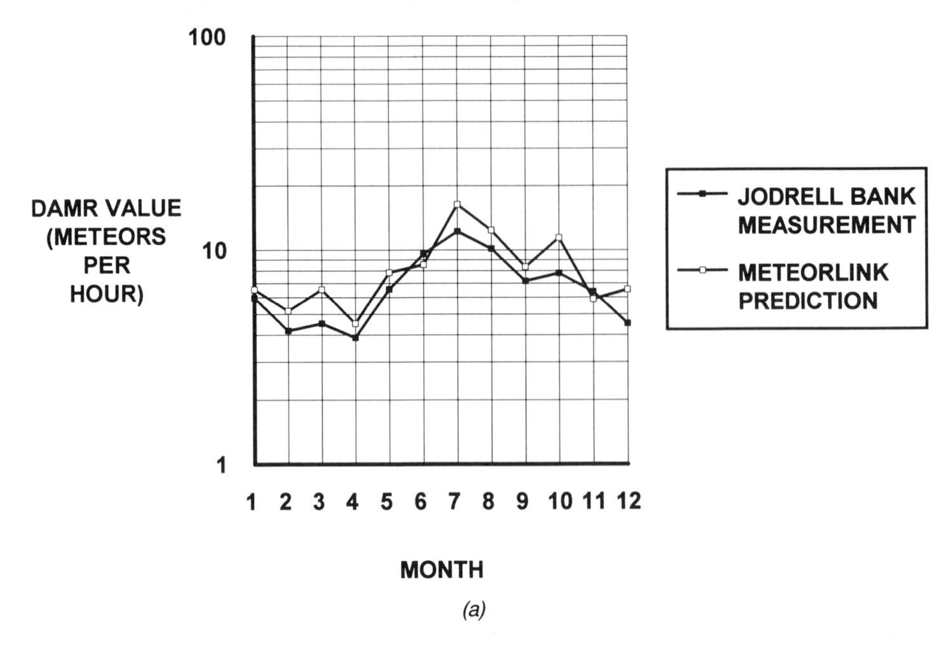

(a)

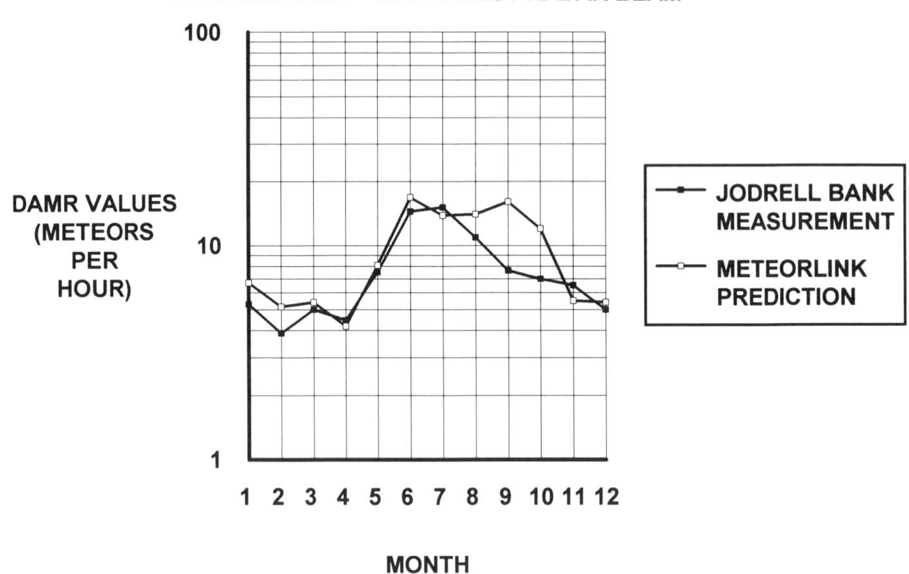

(b)

Figure 3.77. Monthly average Jodrell Bank MR comparison: (*a*) northwest radar beam; (*b*) southwest radar beam.

Mogadishu radars can be approximated by the ratio of their respective minimum radar-detectable mass values at the height of maximum meteoric ionization, $\widehat{H}_I$. The critical assumption necessary for this result was that the minimum detectable mass value for both the link and radar corresponded to the *same* optimum meteor speed $v^* = \bar{v}_\epsilon$. This assumption is valid when (1) the *effective* link wavelength, $\lambda_L \sec(\phi_s)$, and the radar signal wavelength, λ_R, are similar, *or*, (2) the density function for meteor velocity $p_{\mathbf{V}_a}(v, \hat{\mathbf{v}}_m^{\mathrm{O}})$ is concentrated near the mean value, $\bar{v}_\epsilon$. Of course, the assumption is most appropriate when both conditions (1) and (2) are true. If neither of these conditions are met, then the assumption is invalid and the result

$$d^3N_L = X_m(\hat{\mathbf{v}}_m^{\mathrm{O}})\left[\frac{\widehat{m}_R(\bar{v}_\epsilon, \hat{\mathbf{v}}_m^{\mathrm{O}})}{\widehat{m}^*_{sm}(\bar{v}_\epsilon, \hat{\mathbf{v}}_m^{\mathrm{O}})}\right]^{(s-1)} dS\, d\Omega\, dt \tag{3.128}$$

given by Eq. 3.32 would produce systematic prediction error. This error may have been responsible for the prediction optimism not only observed in the Jodrell Bank measurement-prediction comparisons, but also apparent in the Phase 2 and PL comparisons as well.

Consider removing the assumption that the meteor velocity density function is well-approximated by $p_{\mathbf{V}_a}(v, \hat{\mathbf{v}}_m^{\mathrm{O}}) = c_a n(\epsilon)\delta_D[v - \bar{v}_\varepsilon]$. The expression for the link MR contribution, d^3N_L, meeting the observability criteria $(\mathscr{P}_s^{\mathrm{L}}, \hat{\mathbf{v}}_m^{\mathrm{L}})$ and usability criteria (L^*, τ^*) given by Eq. 3.32 may be combined with Eq. 3.11 to yield, after some manipulation,

$$d^3N_L = X_m(\hat{\mathbf{v}}_m^{\mathrm{O}}) \frac{\displaystyle\int_{v_l}^{v_u} p_{\mathbf{V}_a}(v, \hat{\mathbf{v}}_m^{\mathrm{O}})[\widehat{m}^*_{sm}(v, \hat{\mathbf{v}}_m^{\mathrm{O}})]^{(1-s)}\, dv}{\displaystyle\int_{v_l}^{v_u} p_{\mathbf{V}_a}(v, \hat{\mathbf{v}}_m^{\mathrm{O}})[\widehat{m}_R(v, \hat{\mathbf{v}}_m^{\mathrm{O}})]^{(1-s)}\, dv}\, dS\, d\Omega\, dt$$

The minimum link and radar detectable meteor mass values, $\widehat{m}^*_{sm}(v, \hat{\mathbf{v}}_m^{\mathrm{O}})$ and $\widehat{m}_R(v, \hat{\mathbf{v}}_m^{\mathrm{O}})$, respectively, may be replaced by the corresponding expressions for trail line density (from Eqs. 3.15 and 3.31), to yield

$$d^3N_L = X_m(\hat{\mathbf{v}}_m^{\mathrm{O}}) \frac{\displaystyle\int_{v_l}^{v_u} p_{\mathbf{V}_a}(v, \hat{\mathbf{v}}_m^{\mathrm{O}})[\beta_I(v)]^{(s-1)}[\widehat{q}^*_{sm}(v, \hat{\mathbf{v}}_m^{\mathrm{O}})]^{(1-s)}\, dv}{\displaystyle\int_{v_l}^{v_u} p_{\mathbf{V}_a}(v, \hat{\mathbf{v}}_m^{\mathrm{O}})[\beta_I(v)]^{(s-1)}[\widehat{q}_R(v, \hat{\mathbf{v}}_m^{\mathrm{O}})]^{(1-s)}\, dv}\, dS\, d\Omega\, dt \tag{3.129}$$

The q values in the numerator and denominator of Eq. 3.128 are determined from the initial trail radius by the factors

$$\widehat{q}^{*}_{sm}(v, \hat{\mathbf{v}}^{\mathrm{O}}_{m}) \propto \exp\left\{\frac{8\pi[r_0(h_s, v)]^2}{\lambda_L^2 \sec^2(\phi)}\right\} \quad \text{and} \quad \widehat{q}_R(v, \hat{\mathbf{v}}^{\mathrm{O}}_{m}) \propto \exp\left\{\frac{8\pi[r_0(h_s, v)]^2}{\lambda_R^2}\right\} \tag{3.130}$$

for long wavelength underdense trail-scattered signals detected by the MB link and meteor radar, respectively. For similar *effective* wavelengths

$$\lambda_L \sec(\phi_s) \approx \lambda_R$$

or a velocity density function of the form $p_{\mathbf{v}_a}(v, \hat{\mathbf{v}}^{\mathrm{O}}_{m}) \cong c_a n(\epsilon)\delta_D[v - \bar{v}_\epsilon]$, the *ratio* of numerator and denominator in Eq. 3.129 depends primarily on the relative power budgets of the link and radar and not on meteor speed v. In other words, the ratio of the v integrals in Eq. 3.128 will approach unity and the approximation given by Eq. 3.128 is appropriate.

In the Jodrell Bank measurements, $\sec(\phi_s) = 1$ and the signal wavelength $\lambda_L = 4.2$ m while $\lambda_R = 8$ m for the Kazan-Mogadishu radars used to derive the METEORLINK SMRD maps. Since the wavelengths differ by a factor of two, the arguments of the exponentials in Eqs. 130 differ by a factor of four. This difference will be significant for many underdense trail-scattered signals, so condition (2) above is not met. If condition (1) were met, then the apparent velocity selective effects would not have been observed. This result suggests that the meteor velocity distribution is not concentrated near the mean value, v_ϵ, but is more broadly distributed about this value. Similarly, it can be shown that the optimum meteor speed, v^*, that produces the minimum-detectable meteor mass, is also a function of signal wavelength. This result may explain the Jodrell Bank prediction error determined for those months with the highest observed meteor velocities and suggests that more accurate predictions require a more realistic meteor velocity distribution function, $p_{\mathbf{v}_a}(v, \hat{\mathbf{v}}^{\mathrm{O}}_{m})$. The cost of this improved accuracy would be the addition of two numerical integrations over meteor speed as shown in Eq. 3.129.

These velocity-selective effects may explain the model trends observed in the PL measurement–prediction comparison (see Section 3.4.3.1). In particular, the 65- and 85-MHz UMR and UDC results were optimistic for both ST- and SN-link predictions. This optimism was arguably caused by the velocity-selective effects relating the Kazan–Mogadishu SMRD measurements to the Greenland MB Test Bed predictions. These effects, however, were not observed at 104 MHz, possibly due to the negligible link MR dependence on underdense trails. The 45 MHz link frequency was sufficiently close to the Kazan-Mogadishu frequency of 37.5 MHz that the effects of velocity selectivity were minimized. Furthermore, the SN-link predictions showed greater optimism than the corresponding ST-link predictions. This trend mimics the increased prediction optimism evident with the decreased forward scattering angle, ϕ_s, associated with the 700-km SN link versus the

1200-km ST link. These ϕ_s values can significantly effect the magnitude of velocity selective effects (see Eqs. 3.130 and 3.129).

3.4.4 Diversity Model Validation

3.4.4.1 Overview. The METEORDIV MB space diversity model was used to predict the DIF values (see Section 3.2.4.1) derived from a series of landmark measurements performed by SRI International [212]. These measurements, designated the Phase 3 Diversity Tests, were performed for Boeing and BMO to provide empirical verification of trail footprints and space diversity improvement. This verification provided a data base for footprint and diversity model validation versus range, antenna polarization/directivity, and link-geographic bearing. In addition, two short tests were conducted to ascertain the effect of a 3-dB reduction in link power budget *and* offset in the pointing of directional transmit antennas from the $\mathscr{T}_0^L - \mathscr{R}_0^L$ link GCP (see Section 3.2.4.1).

3.4.4.2 Measurements. SRI International deployed an array of 33 receive sites in the configuration shown in Fig. 3.78 located in central Montana and positioned two transportable transmit sites at ranges of 280, 640, and 1000 km. These sites were operated simultaneously to create east-to-west

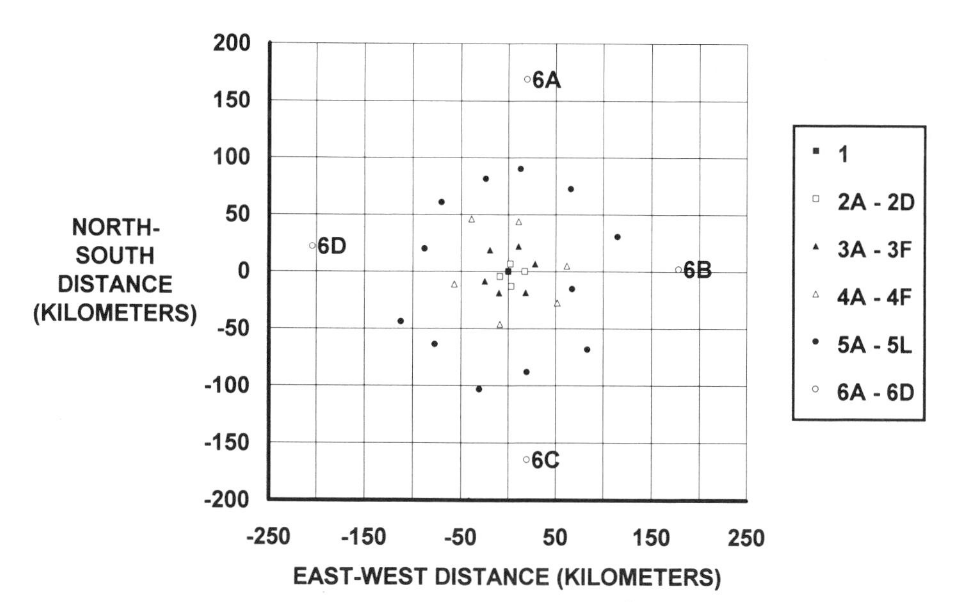

Figure 3.78. Phase 3 diversity measurement receiver-array geometry.

and south-to-north links along the GCP to the center of the array at Lewistown, Montana, as shown in Fig. 3.79. In addition, a single west-to-east test link was operated at a range of 640 km totaling the seven transmit sites labeled in Fig. 3.79. The tests were begun in early December 1988 and ended in early April 1989. In addition, the control link operated during the Phase 2 MB link tests (see Section 3.4.3.2.2) was operated throughout the diversity measurements.

Each receive site consisted of four independent receive channels permitting simultaneous measurements on both horizontally and vertically polarized antennas for both south-to-north and east-to-west link bearings. The frequencies of each channel are shown in Fig. 3.79. A single antenna structure was employed at each range, containing both horizontal crossed dipole (HXD) and vertical whip (VWP) antennas. At 280 and 640 km, the "small" [213] antenna structure was mounted over a 24-radial ground screen with the VWP antenna at ground plane level and the HXD antenna at a 3-m height. At 1000-km range, the "large" structure [214] was employed to achieve additional antenna height. The large structure placed the crossed dipole at about a 10-m height with the VWP antenna mounted above the HXD antenna at 12-m height.

Each channel employed a 5-dB noise figure (N_f) receiver, non-coherent FSK demodulator operating at a channel rate (R_c) of 2.5 kbps in an 8 kHz demodulator bandwidth (BW_n). A 40-ms minimum burst duration (τ^*) was realized by a 100-bit message consisting of 60 bits for detection (pseudo-noise, or PN, sequence pattern correlation), 32 bits for message number identification, and a corresponding cyclic redundancy check (CRC). Each receive system recorded a description of each burst, including receiver ID, frequency, occurrence time, burst duration, message number received (if any), RSL and noise measurements [215]. A required E_b/N_0 of about 6 dB provided the required bit error probability (P_e). Receive cable losses (L_r) were 1.5 dB for the small structure HXD antenna and 1 dB for the small structure VWP antenna. For the large antenna structure, the corresponding cable losses were 2.5 dB and 1.0 dB, respectively.

Six transmitters were employed for the space diversity tests. Two transmitters were deployed with each transportable transmit site to provide dual frequency transmissions using both horizontally and vertically polarized antennas. One transmitter at Lewistown provided timing signals to maintain synchronization at each receiver in the array. Both directional and omniazimuthal transmit antennas were used at the transportable transmit sites during the diversity measurements. The directional antenna was a five-element Yagi with 0° tilt at all ranges. At 280 and 640-km ranges, both vertically and horizontally polarized Yagis (designated VY5 and HY5, respectively) were erected to a height of 6 m. At 1000-km range, the Yagis were fixed at a height of 13.5 m. Omniazimuthal HXD and VWP antennas as well as directional antennas were used on the 280-km link. The HXD antenna was fixed at a 3-m height and the VWP antenna was mounted at the

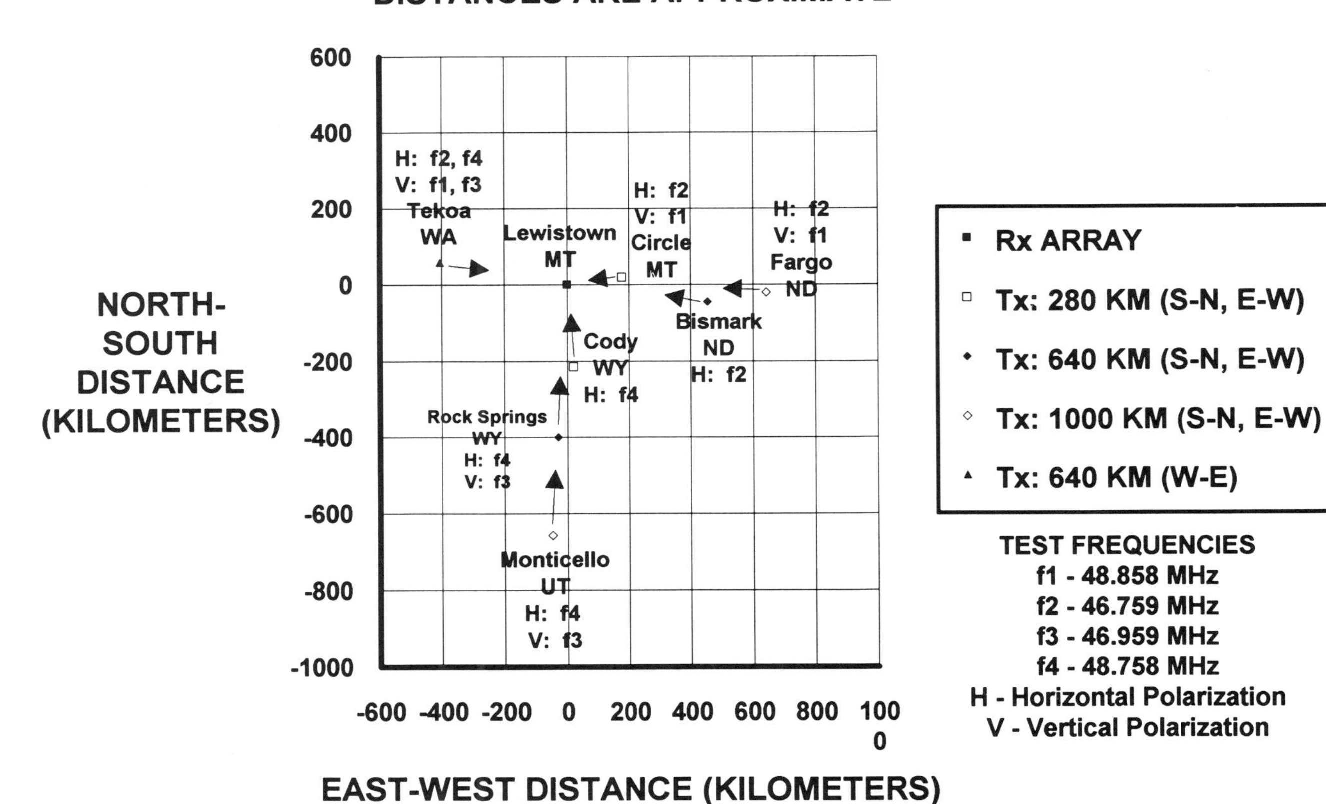

Figure 3.79. Phase 3 transmitter-to-receiver array link geometry.

TABLE 3.10 Phase 3 Link Test Configurations

			Transmit Antennas (horizontal and vertical)		Receive Antennas (horizontal and vertical)	
Test Number	Range (km)	Link Bearing	Antenna Type	Height (m)/ Tilt Angle (°)	Antenna Type	Height (m)
1.1A, B	640	W-to-E	HY5, VY5	6/0, 6/0	HXD, VWP	3, 0
2.1A	280	E-to-W	HY5, VY5	6/0, 6/0	HXD, VWP	3, 0
2.1B	280	S-to-N	HY5, VY5	6/0, 6/0	HXD, VWP	3, 0
2.3A	280	E-to-W	HXD, VWP	3/0, 0/0	HXD, VWP	3, 0
2.3B	280	S-to-N	HXD, VWP	3/0, 0/0	HXD, VWP	3, 0
3.1A	640	E-to-W	HY5, VY5	6/0, 6/0	HXD, VWP	3, 0
3.2A[a]	640	E-to-W	HY5, VY5	6/0, 6/0	HXD, VWP	3, 0
3.1B	640	S-to-N	HY5, VY5	6/0, 6/0	HXD, VWP	3, 0
3.2B[a]	640	S-to-N	HY5, VY5	6/0, 6/0	HXD, VWP	3, 0
4.1A	640	E-to-W	HY5, VY5	14/0, 15/0	HXD, VWP	9, 11
4.2A[b]	640	E-to-W	HY5, VY5	14/0, 15/0	HXD, VWP	9, 11
4.1B	640	S-to-N	HY5, VY5	14/0, 15/0	HXD, VWP	9, 11
4.2B[b]	640	S-to-N	HY5, VY5	14/0, 15/0	HXD, VWP	9, 11

[a] 3-dB reduced transmit power.
[b] 15° counterclockwise azimuthal offset in transmit antenna bearing.

center of the ground screen. Both omniazimuthal antennas were mounted above 24-radial wire ground screens, where each radial wire was 7 m in length.

Raw data was processed to yield a file containing the time history of received trail-scattered signals with periods of sporadic E events excised. This time history was further processed to determine both the daily average AMR value (see Section 3.2.4.1) and the daily average receiver MR value. The ratio of these values provides a measurement of the daily average DIF value. The AMR value and average receiver MR value were computed over all days providing usable data from each test configuration summarized in Table 3.10. Measurement-derived DIF values for this measurement-prediction comparison were extracted from an extensive data post-processing effort performed by TRW [216]. The daily average DIF value was the only validation measure considered for the space diversity model in this study, although the AMR value nevertheless provided the ultimate performance measure for the receiver array.

3.4.4.3 Predictions. METEORDIV predictions were performed for the ten transmitter configurations and three receiver array configurations shown in Table 3.11, creating a total of 30 transmitter-to-receiver array links. Both directional and omniazimuthal transmit antennas in both polarizations were used for the measurement-prediction comparison of the 280-km east-to-west link results. For the south-to-north link at 280-km range, predictions were generated only for the horizontally polarized omniazimuthal antennas link

TABLE 3.11 Phase 3 Link Test Configurations for Measurement-Prediction Comparison

Test Number	Range (km)	Link Bearing	Link Antenna Configurations	Number of Usable Receivers/Receiver Spacing (km)			Link Test Configuration Summary
				7/50	7/60	19/100	
2.3A	280	E-to-W	HXD-to-HXD	6	6	13	Short range,
		E-to-W	VWP-to-VWP	6	6	13	omni-to-omni
2.3B	280	S-to-N	HXD-to-HXD	6	6	13	
2.1A	280	E-to-W	HY5-to-HXD	6	6	12	Short range,
		E-to-W	VY5-to-VWP	6	6	13	directional-to-omni
3.1A	640	E-to-W	HY5-to-HXD	6	7	14	Medium range,
	640	E-to-W	VY5-to-VWP	6	6	14	directional-
3.1B	640	S-to-N	HY5-to-HXD	6	6	12	to-omni
4.1A	640	E-to-W	HY5-to-HXD	6	7	14	Long range,
	640	E-to-W	VY5-to-VWP	6	6	14	directional-to-omni

configuration. At the 640-km range, both horizontally and vertically polarized link antenna configurations were employed in the predictions of east-to-west DIF values while only horizontal polarization was used for the corresponding south-to-north DIF prediction. Finally, both polarizations were used on the 1000-km east-to-west link. Not all of the south-to-north link configurations were predicted because the east-to-west versus south-to-north link orientation did not have a significant impact on predicted DIF values.

Three different receiver array configurations were employed: a maximum of seven receivers spaced 50 km apart, a maximum of seven receivers spaced 100 km apart, and a maximum of 19 receivers spaced 50 km apart. In general, fewer receivers than the stated maximum number were used in the predictions because not all of the measurement receive sites produced usable data. Thus, Table 3.11 gives the actual number of receivers used to compute both measured and predicted DIF values. The specific latitude and longitude of each usable receive site was employed in each DIF prediction, so the modeled geometry of the transmit-to-receive array link was accurate.

Each transmit and receive antenna was modeled by the NEC program, including radial ground screens. All link budget parameters were taken from the final report [217] except measured noise values, which were deemed unreliable for prediction purposes. Instead, Galactic noise values measured on the HXD and VWP antennas used during the SICBM Phase 2 link measurements were employed. To determine the potential effects of error in the assumed noise values, a set of predictions was performed to determine the effect of link budget error on DIF predictions.

TABLE 3.12 Phase 3 DIF Measurement–Prediction Comparison

Test Number	Range (km)	Link Bearing	Link Antenna Configurations	Measurement/Prediction		
				7/50	7/60	19/100
2.3A	280	E-to-W	HXD-to-HXD	4.5/4.4	5.3/5.2	8.8/ 8.7
		E-to-W	VWP-to-VWP	4.3/4.1	4.8/4.8	8.4/ 7.7
2.3B	280	S-to-N	HXD-to-HXD	3.7/4.1	5.2/5.4	8.1/ 9.5
2.1A	280	E-to-W	HY5-to-HXD	3.5/4.2	4.9/5.3	6.9/ 8.9
		E-to-W	VY5-to-VWP	4.1/4.6	5.4/5.2	8.7/ 8.6
3.1A	640	E-to-W	HY5-to-HXD	3.7/4.8	5.3/6.6	6.7/10.8
	640	E-to-W	VY5-to-VWP	3.8/4.6	5.0/5.1	8.1/ 9.5
3.1B	640	S-to-N	HY5-to-HXD	3.5/4.9	4.7/5.6	6.6/10.7
4.1A	640	E-to-W	HY5-toHXD	3.9/4.0	4.9/5.2	7.0/ 7.8
	640	E-to-W	VY5-to-VWP	3.9/3.9	5.1/4.7	8.3/ 7.2

3.4.4.4 Comparisons. Table 3.12 gives the measured and predicted diversity for each transmitter-to-receive array configuration used in this study. Neither measurement-derived nor predicted DIF values show a significant variation with link range, a result consistent with the DIF sensitivity analysis presented in Section 3.2.4.3. The greatest prediction error at each range is produced for the link configurations employing directional, horizontally polarized antennas. In fact, the prediction error with directional antennas at short range exceeds the error achieved for the same range using the omniazimuthal antennas. The prediction error is greatest for the 19 MAX @ 50 km case, as would be expected given the increased DIF values associated with more receivers. The error is at minimum for the 7 MAX @ 50 km case at long range, where the relative common volume is smaller than for the other two cases.

The anticipated trend of increased DIF values with increased receiver spacing (50-100 km) for a fixed number of receivers, or increased number of receivers (from seven maximum to 19 maximum) at a constant spacing, is apparent in both measurement and prediction. Increased spacing moves receivers to lower probability contours on the trail footprint (see Section 2.3.3); therefore, the AMR value increases relative to the average link MR value, so the corresponding DIF value increases. Similarly, adding additional receivers extends the geographic extent of the receiver array as well as the number of potential receivers, thus increasing the DIF value.

3.4.5 Waiting Time Model Validation

3.4.5.1 Overview. Trail-scattered RSL recordings from the Greenland ST link in 1985 were used by the Rome Laboratory (RL)[1] to simulate the data

[1]Rome Laboratory, Griffis AFB, Rome, New York, formerly the Rome Air Development Center (RADC). The Greenland ST link is currently operated and maintained by the Phillips Laboratory (see Section 3.4.3.1).

exchange on a HDX MB link and estimate link MWT values [218]. METEORWAIT was used to simulate the protocol employed in the RL work, so the resulting MWT predictions could be compared with the RL-derived MWT values. This comparison was first performed for the METEORLINK predecessor called METWAIT [219]. This comparison provides validation data which may be used to evaluate the METEORWAIT approach, such as the use of exponentially-distributed burst interarrival times and burst duration.

3.4.5.2 Measurements. The RL-measured RSL time histories provided an excellent source of channel measurements for the simulation of a link communications protocol. For each burst, that is, RSL excursion above P_r^* for the specified E_b/N_0 values, the protocol-controlled exchange of bits can be apportioned the available burst duration for each measured trail-scatter event. Once the burst ends, MWT statistics are updated and the next recorded burst is treated. This process continues until the desired confidence is achieved in the measurement-derived results. Although this technique guarantees accurate trail occurrence and duration statistics, results are necessarily limited to link antenna configurations and ranges similar to the Greenland ST link.

In particular, a HDX protocol was employed to compute bihourly 400-bit MWT values using the recorded RSL values collected for May, 1985. Ideal coherent BPSK was assumed with a bit error rate $P_e = 10^{-4}$ and a transmission rate of 15 kbps. Total link initialization time, including time to complete probe reception, was 50 ms. The remote response consisted of 16 overhead bits followed by message bits transmitted in 100-bit packets constituting a 50-character message. One-way message transmission was assumed from the R terminal to the M station. In this simulation, at most one message was queued for transmission during each recorded trail-scatter event.

3.4.5.3 Predictions. METEORWAIT was used to predict the bihourly MWT values for comparison with RL ST link results at 45, 65, and 104 MHz for May, 1985. The bihourly average MR and DC values ($\dot{N}_{\mathrm{ST}}$ and $\dot{T}_{\mathrm{ST}}$, respectively) used by METEORWAIT for each MWT prediction were extracted from the corresponding May 1985 measured values. These values are plotted along with the corresponding METLINK (METEORLINK predecessor) predictions in Figs. 3.80*a* and *b*. As described in Section 3.3.2.2, these average MR and DC values were used to generate burst interarrival times and durations from inverse exponential distributions. Bits were assigned for each generated burst according to the HDX protocol conventions employed in the RL simulation. The transmit queue length was limited to a single 50-character message such that at most one 50-character message could be transmitted per trail-scatter event. A fixed burst transmission rate of 15 kbps was assumed.

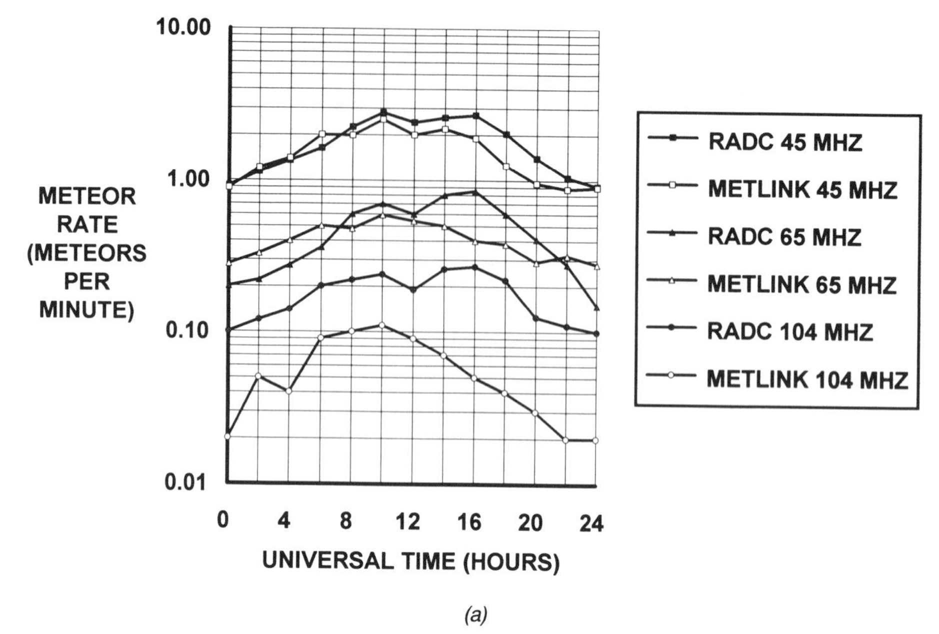

(a)

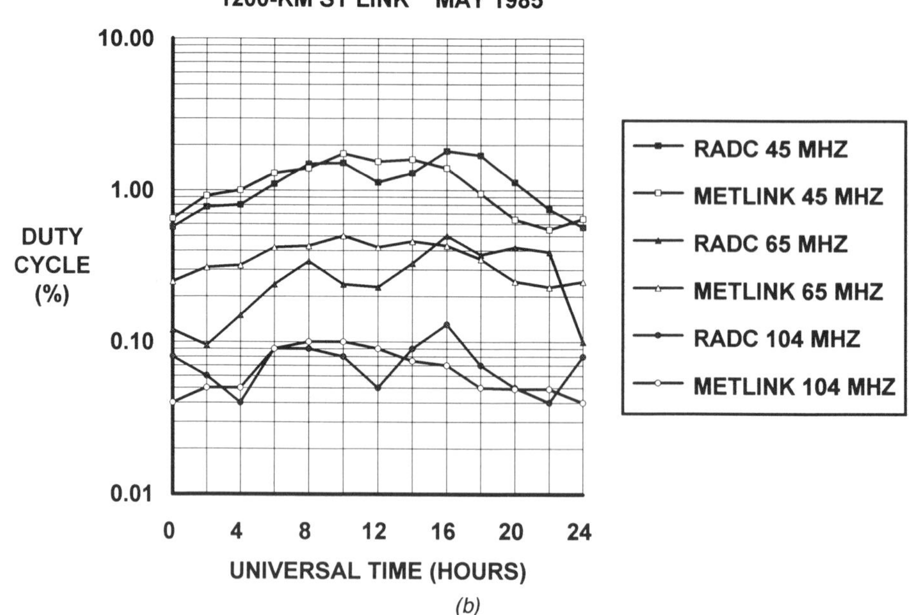

(b)

Figure 3.80. RADC-measured and METLINK-predicted ST-link performance for May 1985: (*a*) MR values; (*b*) DC values. (Approved for public release by Air Force Electronic Systems Division, Public Affairs Office, Case File 86-997.)

3.4.5.4 Comparisons. Measurement-derived 50-character MWT values and the corresponding METEORWAIT-predicted values are plotted in Fig. 3.81 for 45, 65, and 104 MHz. Both measurement and prediction show minimum MWT values occurred twice a day, from 6-10 UT and 14 UT-18 UT. These two periods of minimum MWT values correspond to the peak MR and DC values occurring at the corresponding times as shown in Fig. 3.80. It was suggested in Sections 3.2.2.7.1.1 and 3.4.3.1.4 that the dual MR peaks resulted from meteors passing above the NEP (see Fig. 3.A.1) and creating usable trails on the high-latitude Greenland ST link in the afternoon hours.

The $\mathscr{V}_\mu$ values corresponding to the results shown in Fig. 3.81 are plotted in Fig. 3.82, which shows that all $\mathscr{V}_\mu$ values are bounded by ± 0.3. These results showed increasing pessimism with increasing signal frequency. This pessimism corresponds to a dominance of overdense trail-scattered signals in the predicted ST-link MR values. Since all METEORWAIT burst durations were generated from an exponential distribution, this pessimism suggests that this distribution underestimates the number of long duration trail-scattered signals. This hypothesis may explain the increased pessimism of the METEORWAIT predictions with increasing frequency, despite the use of measured MR and DC values in the simulation.

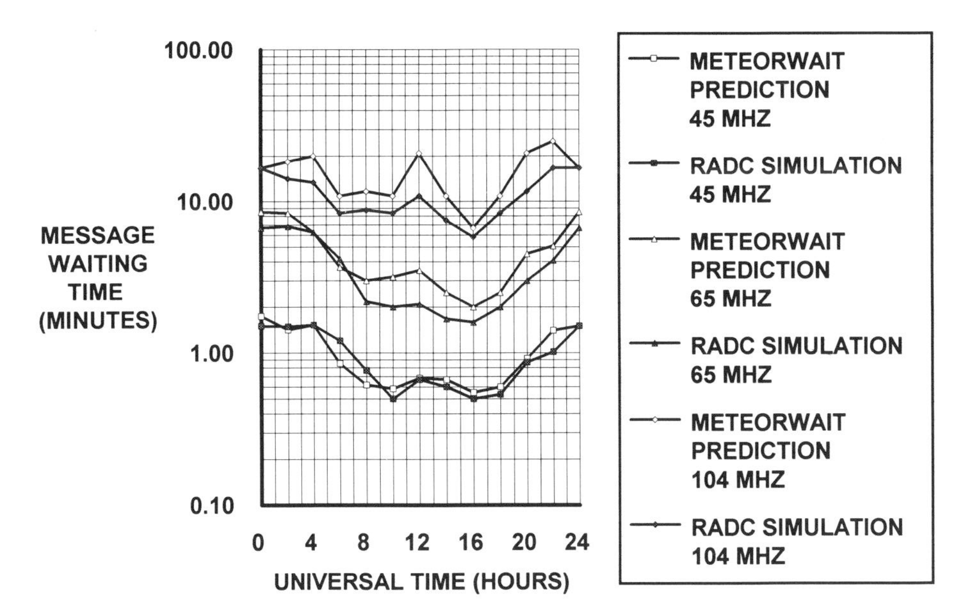

Figure 3.81. Measurement-derived and METEORWAIT-predicted 400-bit MWT values at 45, 65, and 104 MHz. (Approved for public release by Air Force Electronic Systems Division, Public Affairs Office, Case File 86-997.)

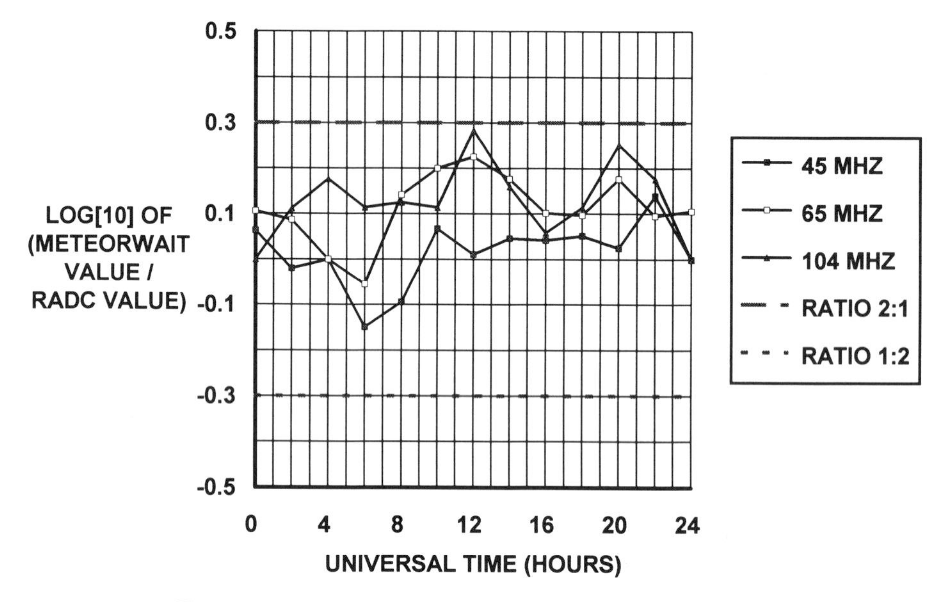

Figure 3.82. 400-bit MWT $\mathcal{V}_{\mu}$ values at 45, 65, and 104 MHz.

A significant result of this prediction-measurement comparison was apparent from the $\mathcal{V}_{\mu} < \pm 0.3$ accuracy achieved by the METEORWAIT simulation. The measurement-derived results employed the actual RSL samples recorded on the Greenland ST link. Thus, the RSL samples represented actual RSL time behavior, interarrival times and burst durations, compared to the idealized model behavior (see Section 3.3.2.2) inherent in the use of METEORWAIT simulation. Nevertheless, a prediction accuracy of $\mathcal{V}_{\mu} < \pm 0.3$ was achieved using only two numbers, the link MR value ($\dot{N}_L$) and DC value ($\dot{T}_L$), to completely characterize MB channel behavior. In other words, the exponential distributions for burst interarrival times and burst durations yielded MWT predictions with $\mathcal{V}_{\mu} < \pm 0.3$ accuracy.

3.5 SUMMARY

3.5.1 Meteor Burst Modeling and Analysis

The METEORLINK physical model for predicting the availability of meteoric trail-scattered signals on an arbitrary MB communication link has been presented. Model development began with a description of the incident SMRD distribution used to estimate the total available sporadic meteor flux producing link-observable meteor trail orientations. These observable trails are then filtered by usability criteria derived from the link power budget and trail-scatter phenomena to predict link MR and DC values, the standard performance measures for link design and analysis. Sensitivity analyses were performed to demonstrate the effects of antenna pattern, power margin, range, geographic location, and temporal variations on these predictions. This link model was used as a foundation for the development of trail footprint and spatial diversity models, METEORTRAK and METEORDIV, respectively.

A Monte Carlo simulation of MB communications, called METEORWAIT, was described as a useful approach for the prediction of link MWT values for a wide variety of data exchange protocols. This simulation was designed to generate usable MB propagation events from MR and DC values computed by the METEORLINK model or measured on an actual MB link. The link simulation was shown to be a special case of the more general network simulation, which used a table of burst occurrence times to sequence network traffic according to the performance of the individual MB links. The Monte Carlo format provided the versatility to model any protocol or combination of protocols employed in a MB network.

Several measurement-prediction comparisons were described for the link, diversity, and protocol simulation models to stress the importance of validation in the development of MB modeling and analysis capability as well as to quantify achievable prediction accuracy. A wide variety of antenna types, frequencies, ranges, burst durations, power budgets, temporal, and geographic factors were employed for these comparisons. The results of these validation efforts suggested the need for link model improvements as well as renewed MB measurement efforts.

3.5.2 Model Improvements

3.5.2.1 Validation Results. The validation results presented in Section 3.4 showed that a physical model can achieve prediction accuracy within a factor of 0.5 to 2.0 across a diverse set of MB link parameter values. Greatest prediction error was observed when

- the MB link MR was determined exclusively by overdense trail-scattered signals
- the link signaling frequency was varied significantly from the Kazan-Mogadishu radar frequency of 37.5 MHz.

Several model improvements may be considered with the simultaneous objectives of minimizing prediction error without

- significantly affecting those cases in which the predictions may be considered accurate
- remaining within the bounds of acceptable meteor scientific theory and engineering practice

Additional meteor scatter research, measurements, and analysis are required to achieve these objectives and to develop and validate model improvements.

3.5.2.2 Meteor Arrival Model

3.5.2.2.1 SMRD Map. The SMRD map is a fundamental component of any physical MB link performance model. Although the SMRD map employed by METEORLINK provides adequate granularity covering most of the celestial sphere, it includes measurements recorded in different years (Kazan, mid-1960's, and Mogadishu, 1968-1970). The arriving sporadic meteor flux may vary in the same month in subsequent years [220], however, and solar atmospheric effects are different in the same month in different years. Thus, the combination of the Kazan and Mogadishu measurements from different years may have produced deficiencies in the resultant SMRD map, although similar MR trends were observed between the two measurement sets [221].

Simultaneous flux measurements from multiple radar sites or forward scatter MB links at significantly different latitudes could be combined using techniques similar to those employed for the Kazan-Mogadishu measurements [222]. In addition, solar flux data collected during these measurements could be correlated with both short [223] and long-term [224] variation in the observed MR values. Since computer capabilities have advanced significantly since the 1960's, a much more extensive and carefully processed SMRD data base could be established. The radio measurement, data collection, and post-processing expertise needed to perform this task, as well as the compilation of an extensive data base of trail-scattered signals, has been proven in Greenland [225] and in the former Soviet Union [226]. Of course, the mass-rate exponent, s, and its variation over the celestial sphere, should be derived in conjunction with the improved SMRD map [227]. The results of this data collection and processing effort would provide not only a valuable data base for MB link modeling, but also advance meteor science generally.

3.5.2.2.2 Meteor Velocity Distribution. The results of the Jodrell Bank measurement-prediction comparison suggested that a more accurate calculation of the radar detection factor (see Eq. 3.9) was necessary. This calculation requires an accurate distribution of sporadic meteor atmospheric velocities over the celestial sphere. It also requires an integration over meteor speed for each link-observable trail modeled by METEORLINK. At a minimum, theoretical velocity distributions derived from measurement should be considered [228, 229] for this purpose.

3.5.2.3 Trail-Scatter Models. The METEORLINK model, used for the measurement-prediction comparison results presented in Section 4, employed the classic overdense trail model (see Eq. 3.59). The most significant prediction error occurred when the modeled link parameters forced prediction dependence on the long wavelength overdense trail-scatter signal model. Conclusions must be drawn carefully from this improvement, however, because more accurate modeling of the mass-rate exponent (see Eq. 3.25 and Table 3.1) for large meteors may equally account for the prediction pessimism observed when the link model is overdense-reliant.

Improved trail-scatter models may be derived from exact solutions of Maxwell's equations for a cylindrical distribution of electron scatterers [230]. These models should *continuously* span the full range of MB link-detectable electron line densities, from about 10^{11} epm to 10^{18} epm, rather than delineating separate underdense and overdense regimes. In addition, model results should be polarization sensitive, so that an accurate representation of plasma resonance effects can be achieved [231]. These new models should be capable of direct solution for the required q value, because computer processing speeds are still too slow to perform iterative series solutions in search of the minimum required q value for each modeled trail. A useful alternative approach has been developed that combines exact trail-scatter solutions with minimal CPU time requirements [232]. This approach employs a table look-up of solutions to the general oblique trail-scatter problem given the incident and trail-scattered E field vectors and relative trail orientations.

3.5.2.4 Meteor Fragmentation Effects. The effective lengths of meteor trail-scatterers are shorter and the maximum trail line density is greater for fragmenting meteors than for the single solid-body meteor. Measured results [233] have shown that the solid body assumption is inaccurate for many meteors. Thus, the single body equations used in this development (see Eqs. 3.12 and 3.18) must be altered for meteors that fragment during their descent. Since METEORLINK is not a Monte Carlo simulation, the effect of fragmentation must be addressed in an average sense.

3.5.2.5 Initial Radius, Diffusion, Attachment, and Turbulence. This model development stressed the importance of the initial trail radius, r_0, and diffusion constant, D, for accurate predictions of underdense trail-scattered signal amplitudes and both underdense and overdense signal durations. These parameters vary not only with scatter point height, h_s, and meteor atmospheric velocity, v, but also with atmospheric density in the meteoric ionization layer (see Eqs. 3.16 and 3.45, respectively). Currently,

METEORLINK employs a static density-height profile [234]. Use of a latitude and time-dependent density profile, however, may provide further improvement of model prediction accuracy. In this context, available empirical data should be used to validate model predictions of enduring trail-scattered signals (>10 s). If necessary, additional data should be collected and model adjustments performed to improve the prediction accuracy of the link DC contribution from long duration trail-scatter events.

3.5.2.6 Quantifying Model Uncertainty. MB performance predictions are unusable in practice unless the accuracy of these predictions can be quantified and made available to potential model users. This verification is impractical if based solely on theoretical uncertainty estimates. A better approach to quantify uncertainty may be derived from an extensive set of measurement-prediction comparisons like those described in Section 3.4. For example, consider the cumulative distribution function (CDF) of $\mathcal{V}_\mu$ values generated from MR values collected from a wide variety of well instrumented MB test bed links. Thus, MB links similar in design to the test bed link may be expected to exhibit similar prediction errors.

In this approach, link model uncertainty would be determined by an extensive data base of measurement-prediction comparisons. Each comparison yields a logarithmic $\mathcal{V}_\mu$ value which augments an existing CDF for the appropriate category of link parameters. Given an average link bihourly MR prediction, $\dot{N}_L$, and a desired confidence $P_c(\%)$, the corresponding $\mathcal{V}_\mu$ value, $\mathcal{V}^c_\mu$, may be used to determine the modified prediction

$$\dot{N}^c_L = 10^{\mathcal{V}^c_\mu} \dot{N}_L$$

where the expected link MR value is less than $\dot{N}^c_L$ with confidence P_c. Of course, multiple CDFs are required because prediction uncertainty varies with respect to link geometry, required link power budget, and burst duration, among others. In practice, the prediction uncertainty data base would be accumulated over long-term application of the MB model to predict performance of known forward scatter links or meteor radars.

APPENDIX A: DERIVATION OF COORDINATE TRANSFORMATIONS

Several coordinate systems can be defined to simplify development of the MB link mathematical model as well as to clarify link geometrical relationships. These coordinate systems relate the transmit and receive antenna patterns to the orientation of observable meteor trails. This relationship determines the performance of MB communication links.

3.A.1 The TERRESTRIAL and ORBITAL Systems

The locations of the transmit and receive antennas on a MB link determine the orientations of observable meteor trails. Normally, the link antenna coordinates are specified in the TERRESTRIAL coordinate system (T system) by

l_{ϕ_j} = longitude measured east of the Greenwich meridian ($-\pi \le l_{\phi_j} \le \pi$)

l_{θ_j} = latitude measured north of the terrestrial equator ($-\pi/2 \le l_{\phi_j} \le \pi/2$)

h_{a_j} = height of the antenna at site j above sea level measured along the zenith above (l_{ϕ_j}, l_{θ_j})

where $j = 1, 2$, for each of the two link antenna sites. Without loss of generality, assume that the transmit antenna is located at site 1 and the receive antenna is located at site 2. The antenna coordinates (l_{ϕ_j}, l_{θ_j}, h_{a_j}), $j = 1, 2$, are used to specify vectors that locate these antennas in the T system as well as the projection of these locations on the earth's surface. The T-system coordinates of the transmit and receive antenna, measured relative to the Greenwich meridian, are given by the vectors $\mathbf{a}_j^{\mathrm{T}} = \langle a_{j1}^{\mathrm{T}}, a_{j2}^{\mathrm{T}}, a_{j3}^{\mathrm{T}} \rangle$, $j = 1, 2$, with components

$$a_{j1}^{\mathrm{T}} = R_{a_j} C_{\theta_j} C_{\phi_j} \tag{3.A.1a}$$

$$a_{j2}^{\mathrm{T}} = R_{a_j} C_{\theta_j} S_{\phi_j} \tag{3.A.1b}$$

$$a_{j3}^{\mathrm{T}} = R_{a_j} S_{\theta_j} \tag{3.A.1c}$$

where

$$R_{a_j} = R_E + h_{a_j}$$

R_E = earth radius ($\cong$ 6370 km)

and

$$C_{\phi_j} = \cos(l_{\phi_j}) \quad C_{\theta_j} = \cos(l_{\theta_j}) \tag{3.A.2a}$$

$$S_{\phi_j} = \sin(l_{\phi_j}) \quad S_{\theta_j} = \sin(l_{\theta_j}) \tag{3.A.2b}$$

the corresponding vectors $\mathbf{p}_j^{\mathrm{T}} = \langle \mathrm{p}_{j1}^{\mathrm{T}}, \mathrm{p}_{j2}^{\mathrm{T}}, \mathrm{p}_{j3}^{\mathrm{T}} \rangle$, $j = 1, 2$, pointing to the site geographic locations at sea level are given by their components

$$p_{j1}^{\mathrm{T}} = R_E C_{\theta_j} C_{\phi_j} \tag{3.A.3a}$$

$$p_{j2}^{\mathrm{T}} = R_E C_{\theta_j} S_{\phi_j} \tag{3.A.3b}$$

$$p_{j3}^{\mathrm{T}} = R_E S_{\theta_j} \tag{3.A.3c}$$

for $j = 1, 2$.

The local time at link site 1 is given by

$$t_y = \text{year (for solar cycle effects)}$$

$$t_m = \text{month of year } t_y$$

$$t_d = \text{day of month } \mathrm{t}_m$$

$$t_{h1} = \text{hour in local time at site 1}$$

$$t_{GW_1} = \text{hours behind Greenwich at the site 1 meridian.}$$

These temporal parameters, combined with the link site terrestrial coordinates and the sporadic meteor radiant density (SMRD), determine the link-observable meteor arrival rate.

SMRD values are specified in the ecliptic ORBITAL (O) coordinate system, or O system, shown in Fig. 3.A.1. The $-y_O$ axis in the O system

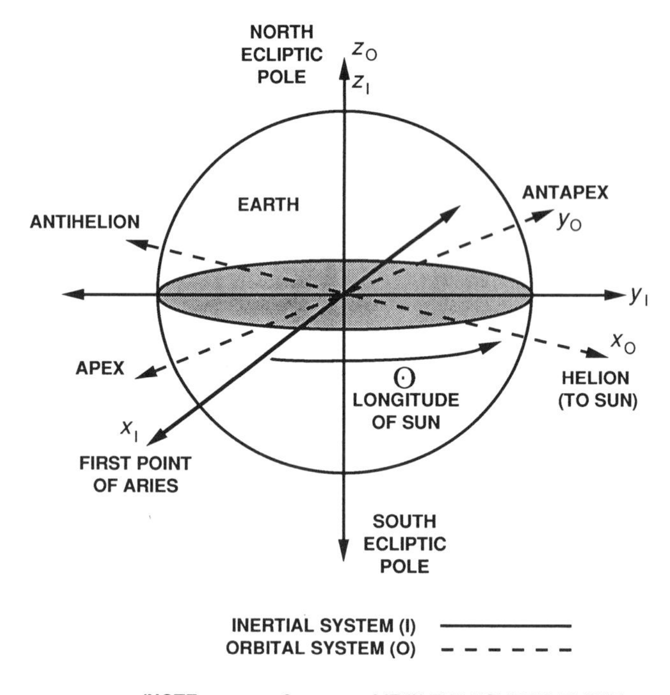

Figure 3.A.1. The INERTIAL and ORBITAL coordinate systems.

points in the direction of earth's motion (apex), the $+x_O$ axis points toward the sun (helion), and the $+z_O$ axis is normal to the plane of the earth's orbit. SMRD maps have been derived that cover most of the celestial sphere in both geocentric [235] and heliocentric [236] reference frames.

The first objective is to calculate the O-system coordinates of the link sites $(l_{\phi j}, l_{\theta j})$ for $j = 1, 2$ at the desired link prediction time. This calculation requires the establishment of a reference earth position and time. Consider the moment of vernal equinox (VE) at Greenwich. At that time, the sun is at 0° ecliptic longitude (solar longitude $\odot$) measured relative to the INERTIAL (I) coordinate system shown in Fig. 3.A.1. The earth-centered I system remains in a fixed orientation relative to the First Point of Aries, while the O system rotates to keep the x_O axis along the helion at the desired link calculation time. The solar longitude $\odot$ needed to provide the desired rotation is given by

$$\odot = \tau_l + 1.915 \sin(\tau_g) + 0.02 \sin(2\tau_g)^\circ$$

where

$$\tau_l = 280.46 + 0.9856474\tau_n$$

$$\tau_g = JD_{GW} - 2451545$$

and

$$JD_{GW} = \text{Julian Day number at the prediction time}$$

JD_{GW} can be computed from standard algorithms given time factors t_y, t_m, t_d, t_{h_1}, and t_{GW_1}. The transformation matrix from the I system to the O system at the calculation time JD_{GW} is given by

$$\mathbf{T}_{IO} = [t_{ik}^{IO}] = \begin{bmatrix} \cos(\odot) & \sin(\odot) & 0 \\ -\sin(\odot) & \cos(\odot) & 0 \\ 0 & 0 & 1 \end{bmatrix}$$

The $\mathbf{a}_j^T$ $j = 1, 2$, vectors in the T system can be transformed into the I system by

- Rotation of the Greenwich meridian from the calculation time into the reference time at VE
- Rotation by the obliquity of the ecliptic $\epsilon_E \cong 23.45^\circ$

These rotations are illustrated in Fig. 3.A.2. The first rotation takes the T system at the prediction time, designated T_t, into the T system at VE, designated T_{VE}. The rotation angle θ_{GW}, called Greenwich sidereal time, is

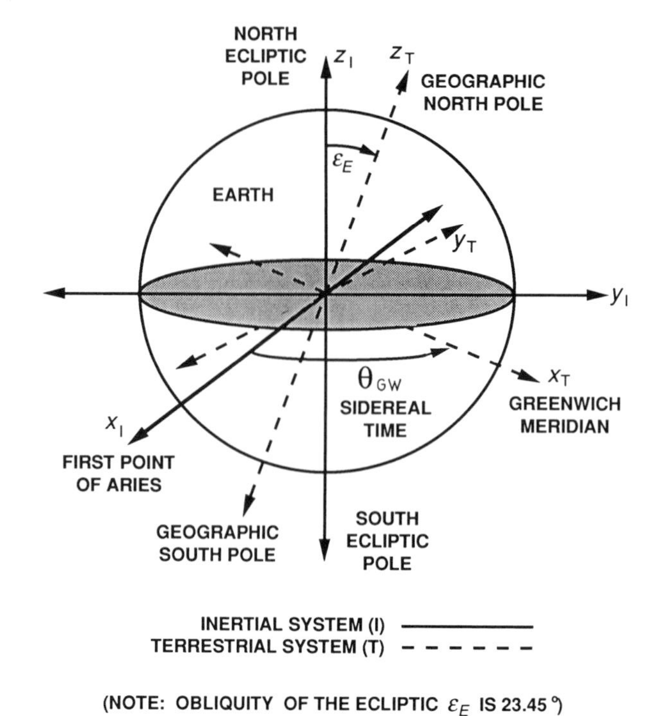

Figure 3.A.2. The INERTIAL and TERRESTRIAL coordinate systems.

measured from the First Point of Aries to the Greenwich meridian along the earth's celestial equator. The corresponding transformation matrix is given by

$$\mathbf{T}_{\mathrm{T}_t\mathrm{T}_{VE}} = [t_{ik}^{\mathrm{T}_t\mathrm{T}_{VE}}] = \begin{bmatrix} \cos(\theta_{GW}) & -\sin(\theta_{GW}) & 0 \\ \sin(\theta_{GW}) & \cos(\theta_{GW}) & 0 \\ 0 & 0 & 1 \end{bmatrix}$$

where

$$\theta_{GW} = 99.690983292 + 36000.768903\tau_u + 0.0038707997\tau_u^2 \,{}^\circ$$

and

$$\tau_u = \frac{JD_{GW} - 2415020.0}{36525.0}$$

The second rotation transforms the T_{VE} system into the I system via the transformation

$$\mathbf{T}_{\mathrm{T}_{VE}\mathrm{I}} = [t_{ik}^{\mathrm{T}_{VE}\mathrm{I}}] = \begin{bmatrix} 1 & 0 & 0 \\ 0 & \cos(\epsilon_E) & \sin(\epsilon_E) \\ 0 & -\sin(\epsilon_E) & \cos(\epsilon_E) \end{bmatrix}$$

Combining these transformation matrices yields

$$\mathbf{T}_{\mathrm{TO}} = \mathbf{T}_{\mathrm{IO}} \cdot \mathbf{T}_{\mathrm{T}_{VE}\mathrm{I}} \cdot \mathbf{T}_{\mathrm{T}_{\mathrm{T}'}\mathrm{T}_{VE}} \tag{3.A.4}$$

where $(\cdot)$ designates matrix-matrix or matrix-vector multiplication.

3.A.2 The LINK and SCATTER Systems

The antenna position vectors $\mathbf{p}_j^{\mathrm{T}}$ and $\mathbf{a}_j^{\mathrm{T}}$, converted to the O system at the calculation time, are given by

$$\mathbf{p}_j^{\mathrm{O}} = \mathbf{T}_{\mathrm{TO}} \cdot \mathbf{p}_j^{\mathrm{T}} \quad \text{and} \quad \mathbf{a}_j^{\mathrm{O}} = \mathbf{T}_{\mathrm{TO}} \cdot \mathbf{a}_j^{\mathrm{T}} \tag{3.A.5}$$

for $j = 1, 2$. Define the LINK (L) coordinate system such that the x_{L} axis passes through the earth between the two link antenna locations as shown in Fig. 3.A.3. Thus, the direction numbers defining the x_{L} axis in the O system are given by

$$\mathbf{x}_{\mathrm{L}}^{\mathrm{O}} = \mathbf{p}_2^{\mathrm{O}} - \mathbf{p}_1^{\mathrm{O}} \tag{3.A.6}$$

where the positive x_{L} axis points along the line from transmit $(j = 1)$ to receive antenna $(j = 2)$. The corresponding unit vector is given by

$$\hat{\mathbf{x}}_{\mathrm{L}}^{\mathrm{O}} = \frac{\mathbf{x}_{\mathrm{L}}^{\mathrm{O}}}{|\mathbf{x}_{\mathrm{L}}^{\mathrm{O}}|}$$

where $(|\ |)$ designates vector magnitude. The z_{L} axis bisects the angle formed by the $\mathbf{p}_1^{\mathrm{O}}$ and $\mathbf{p}_2^{\mathrm{O}}$ vectors, giving the direction numbers

$$\mathbf{z}_{\mathrm{L}}^{\mathrm{O}} = \mathbf{p}_1^{\mathrm{O}} + \mathbf{p}_2^{\mathrm{O}} \tag{3.A.7}$$

and the corresponding unit vector

$$\hat{\mathbf{z}}_{\mathrm{L}}^{\mathrm{O}} = \frac{\mathbf{z}_{\mathrm{L}}^{\mathrm{O}}}{|\mathbf{z}_{\mathrm{L}}^{\mathrm{O}}|}$$

The y_{L} axis is defined in the normal right-handed sense, so the corresponding direction numbers are given by the unit vector

$$\hat{\mathbf{y}}_{\mathrm{L}}^{\mathrm{O}} = \frac{\hat{\mathbf{z}}_{\mathrm{L}}^{\mathrm{O}} \times \hat{\mathbf{x}}_{\mathrm{L}}^{\mathrm{O}}}{|\hat{\mathbf{z}}_{\mathrm{L}}^{\mathrm{O}} \times \hat{\mathbf{x}}_{\mathrm{L}}^{\mathrm{O}}|} \tag{3.A.8}$$

where ($\times$) designates vector cross product. These unit vectors can be used to specify the transformation matrix from the O system to the L system, namely

$$\mathbf{T}_{\mathrm{LO}} = [t_{ik}^{\mathrm{LO}}] = [\hat{\mathbf{x}}_{\mathrm{L}}^{\mathrm{O}} | \hat{\mathbf{y}}_{\mathrm{L}}^{\mathrm{O}} | \hat{\mathbf{z}}_{\mathrm{L}}^{\mathrm{O}}] \tag{3.A.9}$$

Since $\mathbf{T}_{\mathrm{LO}}$ is formed from basis vectors, the inverse transformation matrix

$$\mathbf{T}_{\mathrm{OL}} = [t_{ik}^{\mathrm{OL}}] = (\mathbf{T}_{\mathrm{LO}})^{-1} = (\mathbf{T}_{\mathrm{LO}})^{\mathrm{t}} = [t_{ki}^{\mathrm{LO}}]$$

where (^{t}) denotes matrix transpose.

Replacing the vector $\mathbf{p}_j^{\mathrm{T}}$ with $\mathbf{a}_j^{\mathrm{T}}$, $j = 1, 2$, in Eqs. 3.A.6–3.A.8 yields the unit vectors defining the SCATTER (S) coordinate system (S system) $\hat{\mathbf{x}}_{\mathrm{S}}^{\mathrm{O}}$, $\hat{\mathbf{y}}_{\mathrm{S}}^{\mathrm{O}}$, and $\hat{\mathbf{z}}_{\mathrm{S}}^{\mathrm{O}}$ as shown in Fig. 3.A.3. Thus, the transformation matrix from the S system to the O system is given by

$$\mathbf{T}_{\mathrm{SO}} = [t_{ik}^{\mathrm{SO}}] = [\hat{\mathbf{x}}_{\mathrm{S}}^{\mathrm{O}} | \hat{\mathbf{y}}_{\mathrm{S}}^{\mathrm{O}} | \hat{\mathbf{z}}_{\mathrm{S}}^{\mathrm{O}}] \tag{3.A.10}$$

and the inverse transformation given by

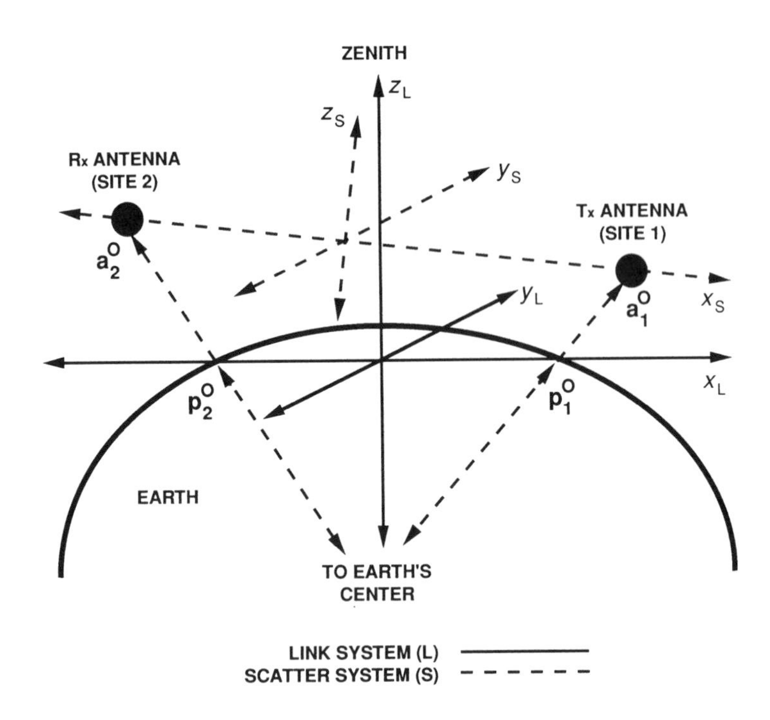

Figure 3.A.3. The LINK and SCATTER coordinate systems.

$$\mathbf{T}_{\mathrm{OS}} = [t_{ik}^{\mathrm{SO}}] = (\mathbf{T}_{\mathrm{SO}})^{\mathrm{t}} = (\hat{\mathbf{x}}_{\mathrm{S}}^{\mathrm{O}} | \hat{\mathbf{y}}_{\mathrm{S}}^{\mathrm{O}} | \hat{\mathbf{z}}_{\mathrm{S}}^{\mathrm{O}}]^{\mathrm{t}} = \begin{bmatrix} \hat{\mathbf{x}}_{\mathrm{S}}^{\mathrm{O}} \\ \hline \hat{\mathbf{y}}_{\mathrm{S}}^{\mathrm{O}} \\ \hline \hat{\mathbf{z}}_{\mathrm{S}}^{\mathrm{O}} \end{bmatrix} \tag{3.A.11}$$

The S system is used to determine the required trail orientations for coherent trail-scatter propagation. Obviously, a vector in the L system can be related to the corresponding vector in the S system by

$$\mathbf{T}_{\mathrm{LS}} = (\mathbf{T}_{\mathrm{SL}})^{\mathrm{t}} = \mathbf{T}_{\mathrm{OS}} \cdot \mathbf{T}_{\mathrm{LO}} \tag{3.A.12}$$

where $\mathbf{T}_{\mathrm{SL}}$ is the transformation from the S system to the L system.

The S system can be related to the T system since the direction numbers for the x_{S} axis are given by

$$\mathbf{x}_{\mathrm{S}}^{\mathrm{T}} = \mathbf{a}_2^{\mathrm{T}} - \mathbf{a}_1^{\mathrm{T}}$$

with the corresponding unit vector

$$\hat{\mathbf{x}}_{\mathrm{S}}^{\mathrm{T}} = \frac{\mathbf{x}_{\mathrm{S}}^{\mathrm{T}}}{|\mathbf{x}_{\mathrm{S}}^{\mathrm{T}}|}$$

Similarly, the z_{S} axis is given by

$$\mathbf{z}_{\mathrm{S}}^{\mathrm{T}} = \mathbf{a}_1^{\mathrm{T}} + \mathbf{a}_2^{\mathrm{T}}$$

with the unit vector

$$\hat{\mathbf{z}}_{\mathrm{S}}^{\mathrm{T}} = \frac{\mathbf{z}_{\mathrm{S}}^{\mathrm{T}}}{|\mathbf{z}_{\mathrm{S}}^{\mathrm{T}}|}$$

The unit direction vector for the y_{S} axis is then defined by

$$\hat{\mathbf{y}}_{\mathrm{S}}^{\mathrm{T}} = \frac{\hat{\mathbf{z}}_{\mathrm{S}}^{\mathrm{T}} \times \hat{\mathbf{x}}_{\mathrm{S}}^{\mathrm{T}}}{|\hat{\mathbf{z}}_{\mathrm{S}}^{\mathrm{T}} \times \hat{\mathbf{x}}_{\mathrm{S}}^{\mathrm{T}}|}$$

Thus, the transformation matrix from the S system to the T system is given by

$$\mathbf{T}_{\mathrm{ST}} = [t_{ik}^{\mathrm{ST}}] = [\hat{\mathbf{x}}_{\mathrm{S}}^{\mathrm{T}} | \hat{\mathbf{y}}_{\mathrm{S}}^{\mathrm{T}} | \hat{\mathbf{z}}_{\mathrm{S}}^{\mathrm{T}}] \tag{3.A.13}$$

Again, the inverse transformation matrix is simply the transpose

$$\mathbf{T}_{\mathrm{TS}} = (\mathbf{T}_{\mathrm{ST}})^{\mathrm{t}}$$

The $\mathbf{T}_{\mathrm{ST}}$ matrix is needed to relate the antenna electric field (E field) pattern at each link site to the observable meteor trail orientations.

3.A.3 The METEOR and TRAIL Systems

The link MR and DC values are computed from point-wise contributions integrated over the observable meteor region between link sites. The L system is used to define the integration variables, the corresponding limits, and the number of integration points. The link transmit and receive antennas are located along the zenith vectors through the x_{L}-axis points $(x_1^{\mathrm{L}}, 0, 0)$ and $(x_2^{\mathrm{L}}, 0, 0)$ in the L system, where

$$x_1^{\mathrm{L}} = -x_2^{\mathrm{L}} = \frac{\sqrt{\Sigma_{j=1}^{3}\,(p_{j2}^{\mathrm{O}} - p_{j1}^{\mathrm{O}})}}{2} \tag{3.A.14a}$$

The corresponding antenna locations $(x_1^{\mathrm{S}}, 0, 0)$ and $(x_2^{\mathrm{S}}, 0, 0)$ in the S system are determined from vectors

$$x_1^{\mathrm{S}} = -x_2^{\mathrm{S}} = \frac{\sqrt{\Sigma_{j=1}^{3}\,(a_{j2}^{\mathrm{O}} - a_{j1}^{\mathrm{O}})}}{2} \tag{3.A.14b}$$

Let $\mathbf{v}_{\mathrm{S}}^{\mathrm{L}} = \langle x_s^{\mathrm{L}}, y_s^{\mathrm{L}}, z_s^{\mathrm{L}} \rangle$ be a vector in the L system pointing from the origin to a potential scatter point, $\mathcal{P}_s^{\mathrm{L}} = (x_s^{\mathrm{L}}, y_s^{\mathrm{L}}, z_s^{\mathrm{L}})$, in the meteoric ionization layer (80- to 120-km altitude). Converting $\mathbf{v}_s^{\mathrm{L}}$ into the S system vector $\mathbf{v}_s^{\mathrm{S}}$ yields

$$\mathbf{v}_s^{\mathrm{S}} = \langle x_s^{\mathrm{S}}, y_s^{\mathrm{S}}, z_s^{\mathrm{S}} \rangle = \mathbf{T}_{\mathrm{LS}} \cdot \left[\mathbf{v}_s^{\mathrm{L}} - \mathbf{T}_{\mathrm{OL}} \cdot \left(\frac{\mathbf{z}_{\mathrm{S}}^{\mathrm{O}} - \mathbf{z}_{\mathrm{L}}^{\mathrm{O}}}{2} \right) \right] \tag{3.A.15}$$

If the antenna heights $h_{a_j} = 0$, for $j = 1, 2$, then the L system and S system are coincident so $\mathbf{v}_s^{\mathrm{S}} = \mathbf{v}_s^{\mathrm{L}}$.

Define the R vectors $\mathbf{r}_{sj}^{\mathrm{S}}$, pointing from each link site to the trail-scatter point, $\mathcal{P}_s^{\mathrm{L}}$, using the vector $\mathbf{v}_s^{\mathrm{S}}$ as:

$$\mathbf{r}_{sj}^{\mathrm{S}} = \mathbf{v}_s^{\mathrm{S}} - \langle x_j^{\mathrm{S}}, 0, 0 \rangle \tag{3.A.16a}$$

for $j = 1, 2$. The corresponding unit vectors are given by

$$\hat{\mathbf{r}}_{sj}^{\mathrm{S}} = \frac{\mathbf{r}_{sj}^{\mathrm{S}}}{|\mathbf{r}_{sj}^{\mathrm{S}}|} \tag{3.A.16b}$$

for $j = 1, 2$.

A meteor trail passing through $\mathcal{P}_s^{\mathrm{S}} = (x_s^{\mathrm{S}}, y_s^{\mathrm{S}}, z_s^{\mathrm{S}})$ provides the maximum scattering cross section when it is tangent to an ellipsoid whose foci are $(x_1^{\mathrm{S}}, 0, 0)$ and $(x_2^{S}, 0, 0)$. Meteor trails with this orientation are observable on the MB link at $\mathscr{P}_s^{\mathrm{S}}$, which corresponds to the center of the first, or principal, Fresnel zone. This criterion is equivalent to requiring that observable trails are perpendicular to the unit normal vector $\hat{\mathbf{n}}_s^{\mathrm{S}}$ defining the trail-scatter ellipsoid at $\mathscr{P}_s^{\mathrm{S}}$. The unit normal vector is given by

$$\hat{\mathbf{n}}_s^{\mathrm{S}} = \frac{\hat{\mathbf{r}}_{s1}^{\mathrm{S}} + \hat{\mathbf{r}}_{s2}^{\mathrm{S}}}{|\hat{\mathbf{r}}_{s1}^{\mathrm{S}} + \hat{\mathbf{r}}_{s2}^{\mathrm{S}}|} \tag{3.A.17}$$

The $\mathbf{r}_{sj}^{\mathrm{S}}$ and $\hat{\mathbf{n}}_s^{\mathrm{S}}$ vectors determine half of the the forward scattering angle

$$\phi_s = \cos^{-1}(\hat{\mathbf{r}}_{s1}^{\mathrm{S}} \cdot \hat{\mathbf{n}}_s^{\mathrm{S}}) \tag{3.A.18}$$

The unit normal vector $\hat{\mathbf{n}}_s^{\mathrm{S}}$ defines the z_{M} axis in the METEOR (M) or M system; thus

$$\hat{\mathbf{z}}_{\mathrm{M}}^{\mathrm{S}} = \hat{\mathbf{n}}_s^{\mathrm{S}}$$

Meteors are directed toward the earth's surface both by gravity (zenith attraction) such that the maximum zenith angle, that is, the acute angle formed between the trail axis and the zenith at $\mathscr{P}_s^{\mathrm{S}}$, is 90°. The zenith vector at $\mathscr{P}_s^{\mathrm{S}}$ is given by

$$\boldsymbol{\zeta}_s^{\mathrm{S}} = \mathbf{T}_{\mathrm{OS}} \cdot \left(\frac{\mathbf{z}_{\mathrm{S}}^{\mathrm{O}}}{2} + \mathbf{T}_{\mathrm{SO}} \cdot \mathbf{v}_s^{\mathrm{S}} \right) \tag{3.A.19a}$$

and the corresponding unit vector

$$\hat{\boldsymbol{\zeta}}_s^{\mathrm{S}} = \frac{\boldsymbol{\zeta}_s^{\mathrm{S}}}{|\boldsymbol{\zeta}_s^{\mathrm{S}}|} \tag{3.A.19b}$$

The unit normal vector $\hat{\mathbf{n}}_s^{\mathrm{S}}$ and zenith vector $\hat{\boldsymbol{\zeta}}_s^{\mathrm{S}}$ determine the x_{M} axis, defining the limiting orientations of observable trails at $\mathscr{P}_s^{\mathrm{S}}$, as the normalized cross product:

$$\hat{\mathbf{x}}_{\mathrm{M}}^{\mathrm{S}} = \frac{\hat{\boldsymbol{\zeta}}_s^{\mathrm{S}} \times \hat{\mathbf{n}}_s^{\mathrm{S}}}{|\hat{\boldsymbol{\zeta}}_s^{\mathrm{S}} \times \mathbf{n}_s^{\mathrm{S}}|}$$

The y_{M} axis is defined in the normal right-handed sense, so that

$$\hat{\mathbf{y}}_{\mathrm{M}}^{\mathrm{S}} = \frac{\hat{\mathbf{z}}_{\mathrm{M}}^{\mathrm{S}} \times \hat{\mathbf{x}}_{\mathrm{M}}^{\mathrm{S}}}{|\hat{\mathbf{z}}_{\mathrm{M}}^{\mathrm{S}} \times \hat{\mathbf{x}}_{\mathrm{M}}^{\mathrm{S}}|}$$

These unit vectors define the transformation matrix from the M system to the S system as

$$\mathbf{T}_{\mathrm{MS}} = [t_{ik}^{\mathrm{MS}}] = [\hat{\mathbf{x}}_{\mathrm{M}}^{\mathrm{S}} | \hat{\mathbf{y}}_{\mathrm{M}}^{\mathrm{S}} | \hat{\mathbf{z}}_{\mathrm{M}}^{\mathrm{S}}] \tag{3.A.20a}$$

and the inverse transformation matrix given by

$$\mathbf{T}_{\mathrm{SM}} = (\mathbf{T}_{\mathrm{MS}})^{\mathrm{t}} \tag{3.A.20b}$$

The M system is illustrated at the potential scatter point, $\mathscr{P}_s^{\mathrm{S}}$, in Fig. 3.A.4. The range of observable meteor trail orientation angles is shown shaded in the figure. Let α_m be the angle measured counterclockwise from the $+x_{\mathrm{M}}$ axis to the vector coincident with the meteor velocity unit vector $\hat{\mathbf{v}}_m^{\mathrm{M}}$ shown in the figure. The $\hat{\mathbf{v}}_m^{\mathrm{M}}$ vector has components:

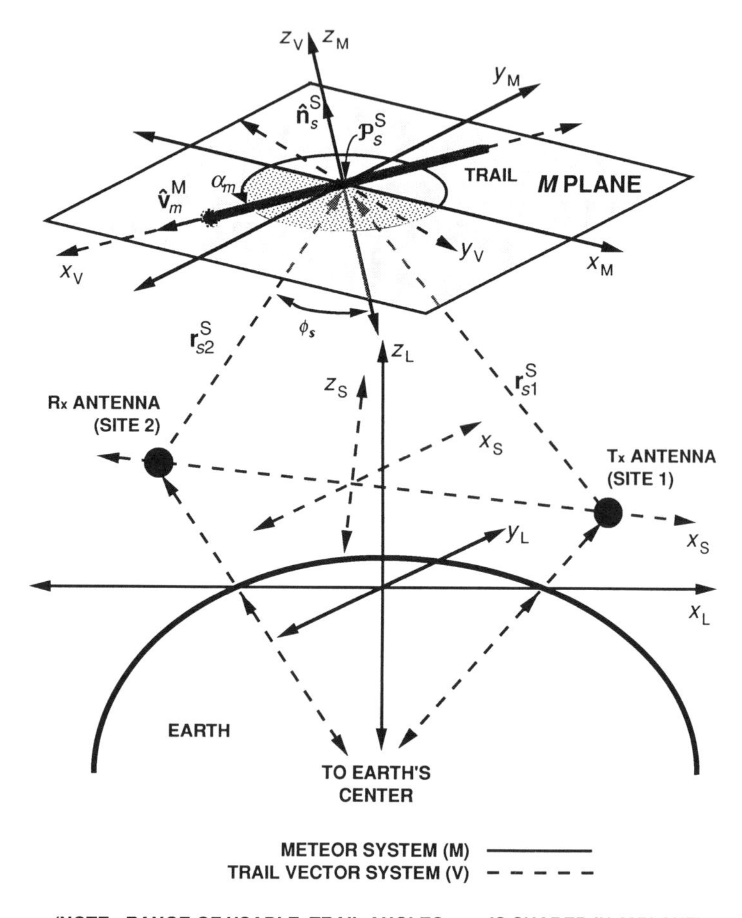

Figure 3.A.4. The METEOR and TRAIL VECTOR coordinate systems.

$$\hat{v}_{m1}^{\mathrm{M}} = \cos(\alpha_m)$$
$$\hat{v}_{m2}^{\mathrm{M}} = \sin(\alpha_m) \tag{3.A.21}$$
$$\hat{v}_{m3}^{\mathrm{M}} = 0$$

The meteor trail unit vector, $\hat{\mathbf{v}}_m^{\mathrm{M}}$, can be transformed to the O system using Eqs. 3.A.10 and 3.A.20*a* as follows:

$$\hat{\mathbf{v}}_m^{\mathrm{O}} = \mathbf{T}_{\mathrm{MO}} \cdot \hat{\mathbf{v}}_m^{\mathrm{M}} = (\mathbf{T}_{\mathrm{SO}} \cdot \mathbf{T}_{\mathrm{MS}}) \cdot \hat{\mathbf{v}}_m^{\mathrm{M}} \tag{3.A.22}$$

where $\mathbf{T}_{\mathrm{MO}}$ is the METEOR-to-ORBITAL transformation matrix. This matrix permits determination of the SMRD value corresponding to any chosen link-observable meteor trail. Similarly, the trail vector $\hat{\mathbf{v}}_m^{\mathrm{O}}$ can be converted to the S and L systems by

$$\hat{\mathbf{v}}_m^{\mathrm{S}} = \mathbf{T}_{\mathrm{MS}} \cdot \hat{\mathbf{v}}_m^{\mathrm{M}} \tag{3.A.23a}$$

$$\hat{\mathbf{v}}_m^{\mathrm{L}} = \mathbf{T}_{\mathrm{ML}} \cdot \hat{\mathbf{v}}_m^{\mathrm{M}} = (\mathbf{T}_{\mathrm{SL}} \cdot \mathbf{T}_{\mathrm{MS}}) \cdot \hat{\mathbf{v}}_m^{\mathrm{M}} \tag{3.A.23b}$$

respectively, where $\mathbf{T}_{\mathrm{ML}}$ is the METEOR-to-LINK transformation matrix. The trail vector $\hat{\mathbf{v}}_m^{\mathrm{S}}$ and the *R* vector $\mathbf{r}_{s1}^{\mathrm{S}}$ pointing from the transmit site ($j = 1$) to the trail-scatter point, $\mathscr{P}_s^{\mathrm{S}}$, determine the incidence angle of the transmitted signal

$$\gamma_{sm} = \cos^{-1}(\hat{\mathbf{r}}_{s1}^{\mathrm{S}} \cdot \hat{\mathbf{v}}_m^{\mathrm{S}}) \tag{3.A.24}$$

The transformation matrix from the M system to the TRAIL VECTOR (V) system, or V system, is determined for each value of α_m from

$$\mathbf{T}_{\mathrm{MV}} = [t_{\mathrm{ik}}^{\mathrm{MV}}] = \begin{bmatrix} \cos(\alpha_m) & \sin(\alpha_m) & 0 \\ -\sin(\alpha_m) & \cos(\alpha_m) & 0 \\ 0 & 0 & 1 \end{bmatrix} \tag{3.A.25}$$

The transformation matrix from the S system to the V system (see Fig. 3.A.4) is then given by

$$\mathbf{T}_{\mathrm{SV}} = (\mathbf{T}_{\mathrm{VS}})^{\mathrm{t}} = (\mathbf{T}_{\mathrm{MV}} \cdot \mathbf{T}_{\mathrm{SM}}) \tag{3.A.26}$$

which is used to transform electric vectors in the S system to the V system.

3.A.4 The GEOGRAPHIC, ANTENNA, and PATTERN Systems

The complex, time-varying electric vectors, or phasors, radiated by the transmit antenna (site 1) *toward* the trail-scatter point $\mathscr{P}_s^{\mathrm{S}}$ and accepted by the receive antenna (site 2) *from* the trail-scatter point $\mathcal{P}_s^{\mathrm{S}}$ are required for the calculation of the antenna gain product, G_{sm}^{12} (see Eq. 3.38) at $\mathcal{P}_s^{\mathrm{S}}$. These E-field phasors are determined from the unit R vectors, $\hat{\mathbf{r}}_{sj}^{\mathrm{S}}$, $j = 1, 2$, which locate the point $\mathcal{P}_s^{\mathrm{S}}$ relative to each link antenna site. Before their transformation into the antenna coordinate systems, the unit R vectors may first be modified to account for the effect of atmospheric refraction or ray bending. Refraction bends the ray towards the denser atmosphere, which extends the radio line-of-sight (see Section 3.2.2.5.2) from each antenna to $\mathcal{P}_s^{\mathrm{S}}$ and increases the effective elevation angle of the ray. This increased elevation angle could significantly effect the E-field phasor associated with the antenna in the direction of $\mathcal{P}_s^{\mathrm{S}}$. Finally, the unit R vectors are transformed into the coordinate systems of the associated antennas to index the appropriate E-field phasors from the corresponding antenna patterns.

First, determine the zenith vectors in the S system at the antenna sites $\mathcal{P}_j^{\mathrm{S}}$ from

$$\hat{\boldsymbol{\zeta}}_j^{\mathrm{S}} = \frac{\mathbf{T}_{\mathrm{TS}} \cdot \mathbf{a}_j^{\mathrm{T}}}{|\mathbf{T}_{\mathrm{TS}} \cdot \mathbf{a}_j^{\mathrm{T}}|}$$

for $j = 1, 2$. The zenith angle of each R vector is then measured between the site zenith vector, $\hat{\boldsymbol{\zeta}}_j^{\mathrm{S}}$, and the corresponding $\hat{\mathbf{r}}_{sj}^{\mathrm{S}}$ vector, yielding

$$\theta_{sj}^{\mathrm{S}} = \cos^{-1}(\hat{\boldsymbol{\zeta}}_j^{\mathrm{S}} \cdot \hat{\mathbf{r}}_{sj}^{\mathrm{S}})$$

for $j = 1, 2$. A rough approximation for the elevation angle effect of refraction assuming 4/3 earth radius (see Section 3.2.2.5.2) for zenith angles θ_{sj}^{S} above 80° is given by the expression

$$(\theta_{sj}^{\mathrm{S}})_{\mathrm{eff}} \cong 84.7912 - 0.9292 \cos^{-1}(\hat{r}_{sj3}^{\mathrm{S}}) \tag{3.A.27}$$

for $\cos^{-1}(\hat{r}_{sj3}^{\mathrm{S}})$ measured in degrees and $\cos^{-1}(\hat{r}_{sj3}^{\mathrm{S}}) > 80°$, $j = 1, 2$. The modified elevation angle $90° - \theta_{sj}^{\mathrm{S}}$ from Eq. 3.A.27 is plotted versus the corresponding non-refracted elevation angle in Fig. 3.A.5. The effective unit R vectors in the S system then become

$$(\hat{\mathbf{r}}_{sj}^{\mathrm{S}})_{\mathrm{eff}} = \langle \hat{r}_{sj1}^{\mathrm{S}}, \hat{r}_{sj2}^{\mathrm{S}}, (\hat{r}_{sj3}^{\mathrm{S}})_{\mathrm{eff}} \rangle \tag{3.A.28a}$$

where

$$(\hat{r}_{sj3}^{\mathrm{S}})_{\mathrm{eff}} = \cos\left[\frac{84.8 - 0.9 \cos^{-1}(\hat{r}_{sj3}^{\mathrm{S}})}{0.9}\right] \tag{3.A.28b}$$

The difference between $(\hat{r}_{sj3}^{\mathrm{S}})_{\mathrm{eff}}$ and $\hat{r}_{sj3}^{\mathrm{S}}$ is small, from 1 to 2° maximum. The only significant effect on MR and DC predictions occurs for modeled MB links approaching the natural range limit of about 2000 km, and then only

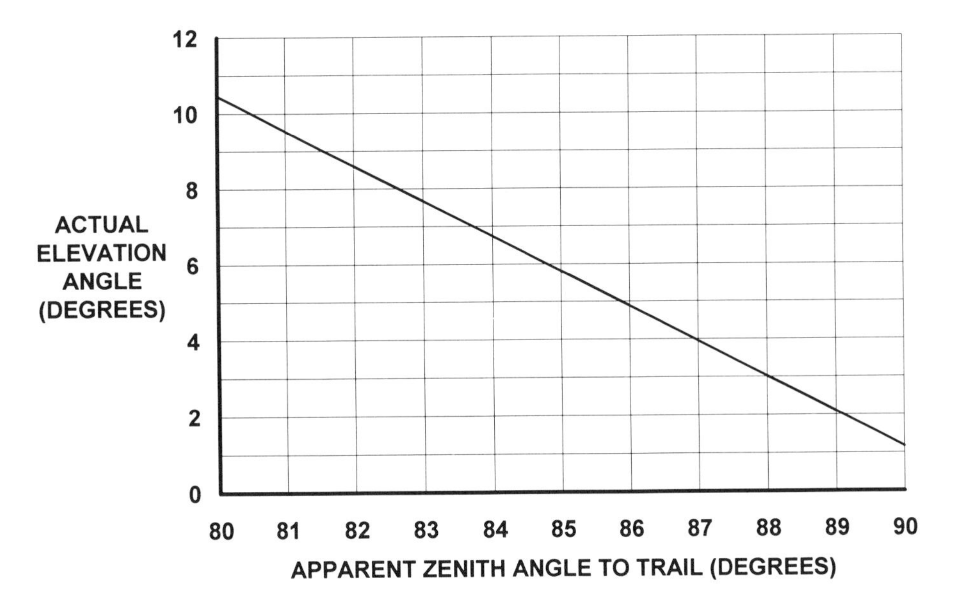

Figure 3.A.5. Atmospheric refraction effects on elevation angle.

for antenna patterns with significant gain at the horizon. Even at these ranges, the uncertainties of antenna pattern, terrain blockage, and changing atmospheric conditions would produce significant differences between measurement and prediction.

Define the GEOGRAPHIC coordinate system (G_j system) with its origin at each antenna site $\mathcal{P}_j^{S}$, such that the $+x_{G_j}$ axis points toward geographic north along the longitude meridian passing through $\mathcal{P}_j^{S}$ and the $+z_{G_j}$ axis is coincident with the zenith vector $\hat{\boldsymbol{\zeta}}_j^{S}$ at $\mathcal{P}_j^{S}$ for $j = 1, 2$. The S-system R vector, $(\hat{r}_{sj3}^{S})_{eff}$, may be transformed into the T system using Eq. 3.A.13, yielding

$$\hat{\mathbf{r}}_{sj}^{T} = \mathbf{T}_{ST} \cdot (\hat{r}_{sj3}^{S})_{\text{eff}} \tag{3.A.29}$$

for $j = 1, 2$. The $\hat{\mathbf{r}}_{sj}^{T}$ vector can then be transformed into the G_j system for each link antenna site j, $j = 1, 2$, using the transformation:

$$\mathbf{T}_{TG_j} = [t_{ik}^{TG_j}] = \begin{bmatrix} -S_{\phi_j} & C_{\phi_j} & 0 \\ -S_{\theta_j} C_{\phi_j} & S_{\theta_j} S_{\phi_j} & C_{\theta_j} \\ C_{\theta_j} C_{\phi_j} & C_{\theta_j} S_{\phi_j} & S_{\theta_j} \end{bmatrix} \tag{3.A.30}$$

In general, the antenna pattern at site j, $j=1,2$, is rotated at some bearing angle, θ_{B_j}, relative to geographic north as shown in Fig. 3.A.6, that is, the $+x_{G_j}$ axis. If the antenna height above ground, h_{a_j}, is many wavelengths (an aircraft antenna, for example), the relevant free space antenna pattern may be pitched by an angle θ_{P_j} and rolled by an angle θ_{R_j}. In this regard, define the ANTENNA system for site j (A_j system) for $j=1, 2$, to be derived from the G_j system by the ordered rotations θ_{B_j}, θ_{P_j}, and θ_{R_j}. Thus, a vector in the G_j system may be represented in the A_j system by the transformation

$$\mathbf{T}_{G_jA_j} = \mathbf{T}_{R_j} \cdot \mathbf{T}_{P_j} \cdot \mathbf{T}_{B_j} \tag{3.A.31}$$

where

$$\mathbf{T}_{R_j} = \begin{bmatrix} 1 & 0 & 0 \\ 0 & \cos(\theta_{R_j}) & \sin(\theta_{R_j}) \\ 0 & -\sin(\theta_{R_j}) & \cos(\theta_{R_j}) \end{bmatrix}$$

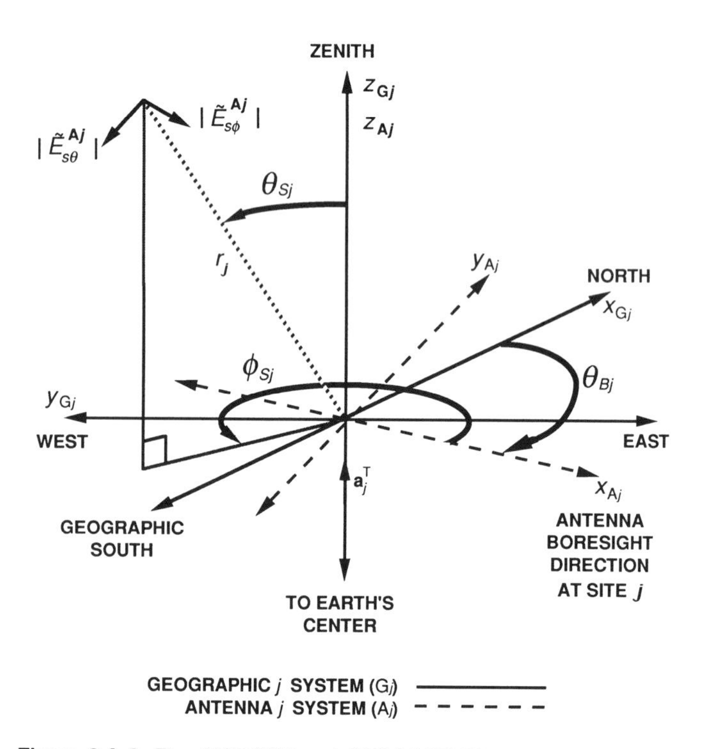

Figure 3.A 6. The ANTENNA and GEOGRAPHIC coordinate systems.

$$\mathbf{T}_{P_j} = \begin{bmatrix} \cos(\theta_{P_j}) & 0 & \sin(\theta_{P_j}) \\ 0 & 1 & 0 \\ -\sin(\theta_{P_j}) & 0 & \cos(\theta_{P_j}) \end{bmatrix}$$

$$\mathbf{T}_{B_j} = \begin{bmatrix} \sin(\theta_{B_j}) & \cos(\theta_{B_j}) & 0 \\ -\cos(\theta_{B_j}) & \sin(\theta_{Bj}) & 0 \\ 0 & 0 & 1 \end{bmatrix}$$

and the order of matrix multiplication is significant.

The R vectors $\hat{\mathbf{r}}_{sj}^{\mathrm{T}}$, $j = 1, 2$, in the T system may be converted to the A_j system by the transformation

$$\hat{\mathbf{r}}_{sj}^{\mathrm{A}_j} = \mathbf{T}_{\mathrm{G}_j\mathrm{A}_j} \cdot \mathbf{T}_{\mathrm{TG}_j} \cdot \hat{\mathbf{r}}_{sj}^{\mathrm{T}} \tag{3.A.32}$$

The azimuth and zenith angles locating the point $\mathscr{P}_s^{\mathrm{S}}$ in the A_j system are then computed from

$$\phi_s^{\mathrm{A}_j} = \tan^{-1}\left(\frac{\hat{r}_{sj2}^{\mathrm{A}_j}}{\hat{r}_{sj1}^{\mathrm{A}_j}}\right) \tag{3.A.33a}$$

$$\theta_s^{\mathrm{A}_j} = \cos^{-1}(\hat{r}_{sj3}^{\mathrm{A}_j}) \tag{3.A.33b}$$

These angles index E-field vectors in the antenna pattern.

Each pattern is specified in the standard spherical PATTERN (P_j) coordinate system at antenna site j, $j = 1, 2$, shown in Fig. 3.A.6. The R vectors $\hat{\mathbf{r}}_{sj}^{\mathrm{T}}$, $j = 1, 2$, in the T system may be converted to the P_j system by the transformation

$$\hat{\mathbf{r}}_{sj}^{\mathrm{P}_j} = \mathbf{T}_{\mathrm{A}_j\mathrm{P}_j} \cdot \mathbf{T}_{\mathrm{G}_j\mathrm{A}_j} \cdot \mathbf{T}_{\mathrm{TG}_j} \cdot \hat{\mathbf{r}}_{sj}^{\mathrm{T}} = \mathbf{T}_{\mathrm{TP}_j} \cdot \hat{\mathbf{r}}_{sj}^{\mathrm{T}} \tag{3.A.34}$$

where $\mathbf{T}_{\mathrm{TG}_j}$ is given by Eq. 3.A.30 and $\mathbf{T}_{\mathrm{G}_j\mathrm{A}_j}$ is given by Eq. 3.A.31. The transformation from the Cartesian A_j system to the spherical P_j system is given by

$$\mathbf{T}_{\mathrm{A}_j\mathrm{P}_j} = [t_{ik}^{\mathrm{A}_j\mathrm{P}_j}] = \begin{bmatrix} \sin(\theta_s^{\mathrm{A}_j})\cos(\phi_s^{\mathrm{A}_j}) & \sin(\theta_s^{\mathrm{A}_j})\sin(\phi_s^{\mathrm{A}_j}) & \cos(\theta_s^{\mathrm{A}_j}) \\ \cos(\theta_s^{\mathrm{A}_j})\cos(\phi_s^{\mathrm{A}_j}) & \sin(\theta_s^{\mathrm{A}_j})\cos(\phi_s^{\mathrm{A}_j}) & -\sin(\theta_s^{\mathrm{A}_j}) \\ -\sin(\phi_s^{\mathrm{A}_j}) & \cos(\phi_s^{\mathrm{A}_j}) & 0 \end{bmatrix}$$

with $\mathbf{T}_{\mathrm{P}_j\mathrm{A}_j} = (\mathbf{T}_{\mathrm{A}_j\mathrm{P}_j})'$, for $j = 1, 2$.

Finally, the complex electric phasor vector determined at the trail-scatter point, $\mathscr{P}_s^{\mathrm{S}}$, at a distance $r_j = |\mathbf{r}_{sj}^{S}|$ from the antenna at site j, $j = 1, 2$, is defined by

$$\begin{aligned}\tilde{\mathbf{E}}_s^{\mathrm{P}_j} &= \tilde{E}_{sr}^{\mathrm{P}_j}\hat{\mathbf{r}}_{\mathrm{P}_j} + \tilde{E}_{s\phi}^{\mathrm{P}_j}\hat{\boldsymbol{\phi}}_{\mathrm{P}_j} + \tilde{E}_{s\theta}^{\mathrm{P}_j}\hat{\boldsymbol{\theta}}_{\mathrm{P}_j} \\ &= 0\hat{\mathbf{r}}_{\mathrm{P}_j} + \|\tilde{E}_{s\phi}^{\mathrm{P}_j}\| \exp(-\mathrm{j}\,\Phi_{s\phi}^{\mathrm{P}_j})\hat{\boldsymbol{\phi}}_{\mathrm{P}_j} + \|\tilde{E}_{s\theta}^{\mathrm{P}_j}\| \exp(-\mathrm{j}\Phi_{s\theta}^{\mathrm{P}_j})\hat{\boldsymbol{\theta}}_{\mathrm{P}_j} \\ &= E_{s\phi}^{\mathrm{P}_j} \exp(-\mathrm{j}\Phi_{s\phi}^{\mathrm{P}_j})\hat{\boldsymbol{\phi}}_{\mathrm{P}_j} + E_{s\theta}^{\mathrm{P}_j} \exp(-\mathrm{j}\Phi_{s\theta}^{\mathrm{P}_j})\hat{\boldsymbol{\theta}}_{\mathrm{P}_j}\end{aligned} \tag{3.A.35}$$

where

$$E_{s\phi}^{\mathrm{P}_j} = \|\tilde{E}_{s\phi}^{\mathrm{P}_j}\|, \quad E_{s\theta}^{\mathrm{P}_j} = \|\tilde{E}_{s\theta}^{\mathrm{P}_j}\|, \quad \text{and}$$

$\hat{\mathbf{r}}_{\mathrm{P}_j}$ = radial unit vector in the P_j system

$\hat{\boldsymbol{\phi}}_{\mathrm{P}_j}$ = tangential unit vector normal to $\hat{\mathbf{r}}_{\mathrm{P}_j}$ in the azimuth plane of the P_j system

$\hat{\boldsymbol{\theta}}_{\mathrm{P}_j}$ = tangential unit vector normal to $\hat{\mathbf{r}}_{\mathrm{R}_j}$ and $\hat{\boldsymbol{\phi}}_{\mathrm{P}_j}$ in a vertical plane in the P_j system

$\tilde{E}_{sr}^{\mathrm{P}_j}$ = complex magnitude of the $\hat{\mathbf{r}}_{\mathrm{P}_j}$ component of the radiated electric field phasor, $\tilde{\mathbf{E}}_s^{\mathrm{P}_j}$ at $\mathscr{P}_s^{\mathrm{S}}$, $\tilde{E}_{sr}^{\mathrm{P}_j} = 0$

$\tilde{E}_{s\phi}^{\mathrm{P}_j}$ = complex magnitude of the $\hat{\boldsymbol{\phi}}_{\mathrm{P}_j}$ component of the radiated electric field phasor, $\tilde{\mathbf{E}}_s^{\mathrm{P}_j}$, at $\mathscr{P}_s^{\mathrm{S}}$

$\Phi_{s\phi}^{\mathrm{P}_j}$ = argument of the $\hat{\boldsymbol{\phi}}_{\mathrm{P}_j}$ component of the radiated electric field phasor, $\tilde{\mathbf{E}}_s^{\mathrm{P}_j}$, at $\mathscr{P}_s^{\mathrm{S}}$

$\tilde{E}_{s\theta}^{\mathrm{P}_j}$ = complex magnitude of the $\hat{\boldsymbol{\theta}}_{\mathrm{P}_j}$ component of the radiated electric field phasor, $\tilde{\mathbf{E}}_s^{\mathrm{P}_j}$, at $\mathscr{P}_s^{\mathrm{S}}$

$\Phi_{s\theta}^{\mathrm{P}_j}$ = argument of the $\hat{\boldsymbol{\theta}}_{\mathrm{P}_j}$ component of the radiated electric field phasor, $\tilde{\mathbf{E}}_s^{\mathrm{P}_j}$, at $\mathscr{P}_s^{\mathrm{S}}$

and $\mathrm{j} = \sqrt{-1}$. The embellishment (~) in Eq. 3.A.35 designates a complex number and the symbol (‖ ‖) indicates complex absolute value. In general, these phasor vector components may be supplied by the Numerical Electromagnetics Code [237] or other numerical or closed form calculation of antenna performance. Equation 3.A.35 may be rewritten as

$$\tilde{\mathbf{E}}_s^{\mathrm{P}_j} = E_{s\phi}^{\mathrm{P}_j}\hat{\boldsymbol{\phi}}_{\mathrm{P}_j} + E_{s\theta}^{\mathrm{P}_j} \exp(-\mathrm{j}\Phi_s^{\mathrm{P}_j})\hat{\boldsymbol{\theta}}_{\mathrm{P}_j} \tag{3.A.36}$$

where $\Phi_s^{\mathrm{P}_j} = \Phi_{s\theta}^{\mathrm{P}_j} - \Phi_{s\phi}^{\mathrm{P}_j}$ is the relative phase angle between the $\hat{\boldsymbol{\phi}}_{\mathrm{P}_j}$ and $\hat{\boldsymbol{\theta}}_{\mathrm{P}_j}$ phasor-vector components and the superfluous phase factor $\exp(-\mathrm{j}\Phi_{s\phi}^{\mathrm{P}_j})$ has been suppressed.

The phasor-vector $\tilde{\mathbf{E}}_s^{\mathrm{P}_1}$ launched by the transmit antenna ($j = 1$) or accepted by the receive antenna ($j = 2$) may be modified by passage through the lower ionosphere due to gyrotropic effects induced by the earth's magnetic field. Let ψ_{sj}^{F} be the Faraday rotation angle imparted to the $\tilde{\mathbf{E}}_s^{\mathrm{P}_j}(r_j)$

vector when propagating from site j to the trail-scatter point $\mathscr{P}_s^{\mathrm{S}}$. The rotated phasor vector is then given by

$$\tilde{\mathbf{E}}_{s_\psi}^{\mathrm{P}_j} = \tilde{E}_{s_\psi\phi}^{\mathrm{P}_j}\hat{\mathbf{\Phi}}_{\mathrm{P}_j} + \tilde{E}_{s_\psi\theta}^{\mathrm{P}_j}\hat{\mathbf{\theta}}_{\mathrm{P}j} \tag{3.A.37}$$

where the component phasors are given by

$$\tilde{E}_{s_\psi\phi}^{\mathrm{P}_j} = E_{s\phi}^{\mathrm{P}_j}\cos(\psi_{sj}^{\mathrm{F}}) + E_{s\theta}^{\mathrm{P}_j}\sin(\psi_{sj}^{\mathrm{F}})\exp(-\mathrm{j}\Phi_s^{\mathrm{P}_j})$$

$$\tilde{E}_{s_\psi\theta}^{\mathrm{P}_j} = -E_{s\phi}^{\mathrm{P}_j}\sin(\psi_{sj}^{\mathrm{F}}) + E_{s\theta}^{\mathrm{P}_j}\cos(\psi_{sj}^{\mathrm{F}})\exp(-\mathrm{j}\Phi_s^{\mathrm{P}_j})$$

Ignoring a constant phase offset, Eq. 3.37 can be put into the form of Eq. 3.36 as follows:

$$\tilde{\mathbf{E}}_{s_\psi}^{\mathrm{P}_j} = E_{s_\psi\phi}^{\mathrm{P}_j}\hat{\mathbf{\Phi}}_{\mathrm{P}_j} + E_{s_\psi 0}^{\mathrm{P}_j}\exp(-\mathrm{j}\Phi_{s_\psi}^{\mathrm{P}_j})\hat{\mathbf{\theta}}_{\mathrm{P}_j} \tag{3.A.38}$$

where

$$E_{s_\psi\phi}^{\mathrm{P}_j} = \|\tilde{E}_{s_\psi\phi}^{\mathrm{P}_j}\|$$

$$E_{s_\psi 0}^{\mathrm{P}_j} = \|\tilde{E}_{s_\psi 0}^{\mathrm{P}_j}\| = \frac{\tilde{E}_{s_\psi\theta}^{\mathrm{P}_j}\cdot(\tilde{E}_{s_\psi\phi}^{\mathrm{P}_j})^{\mathrm{c}}}{\|\tilde{E}_{s_\psi\phi}^{\mathrm{P}_j}\|}$$

The new relative phase angle between the two phasor-vector components is given by

$$\Phi_{s_\psi}^{\mathrm{P}_j} = \tan^{-1}\left(\frac{\mathrm{Im}\{\tilde{E}_{s_\psi 0}^{\mathrm{P}_j}\}}{\mathrm{Re}\{\tilde{E}_{s_\psi 0}^{\mathrm{P}_j}\}}\right)$$

The E field phasor vector corrected for Faraday rotation may be converted to the A_j system by

$$\tilde{\mathbf{E}}_{s_\psi}^{\mathrm{A}_j} = \mathbf{T}_{\mathrm{P}_j\mathrm{A}_j}\cdot\tilde{\mathbf{E}}_{s_\psi}^{\mathrm{P}_j} \tag{3.A.39a}$$

and ultimately into the S system by

$$\tilde{\mathbf{E}}_{s_\psi j}^{\mathrm{S}}(\mathbf{T}_{\mathrm{TS}}\cdot\mathbf{T}_{\mathrm{G}_j\mathrm{T}}\cdot\mathbf{T}_{\mathrm{A}_j\mathrm{G}_j})\cdot\tilde{\mathbf{E}}_{s_\psi}^{\mathrm{A}_j} = \mathbf{T}_{\mathrm{A}_j\mathrm{S}}\cdot\tilde{\mathbf{E}}_{s_\psi}^{\mathrm{A}_j} \tag{3.A.39b}$$

for calculation of the antenna gain product G_{sm}^{12} at $\mathscr{P}_s^{\mathrm{S}}$ from Eq. 3.38.

3.A.5 Summary

Given earth coordinates for the transmit and receive sites of a MB link and the relevant temporal parameters, the following coordinate systems have been determined:

- INERTIAL (I system) to provide a fixed reference system to relate the earth's orbital position and the MB links location on the earth relative to the incident SMRD map
- ORBITAL (O system) used to index appropriate values in the SMRD map determined to meet the link observability criteria $(\mathscr{P}_s^{\mathrm{L}}, \hat{\mathbf{v}}_m^{\mathrm{O}})$
- TERRESTRIAL (T system) to provide earth coordinates of the transmit and receive sites
- LINK (L system) used to integrate the MR and DC contributions computed throughout the meteoric ionization region
- SCATTER (S system) used to specify meteor trail orientations meeting the link observability criteria $(\mathscr{P}_s^{\mathrm{L}}, \hat{\mathbf{v}}_m^{\mathrm{O}})$
- METEOR (M system) to specify usable trail orientation at each trail-scatter point $\mathscr{P}_s^{\mathrm{S}}$
- TRAIL VECTOR (V system) for resolution of the parallel and perpendicular components of the incident and scattered E-field phasor vector at the meteor trail
- GEOGRAPHIC system at antenna site j, $j = 1, 2$, (G_j system) to provide a reference coordinate system for the specification of antenna pattern orientation relative to the MB link
- ANTENNA (A_j system) used to locate each trail-scatter point $\mathscr{P}_s^{\mathrm{S}}$ relative to the antenna at site j, $j = 1, 2$
- PATTERN (P_j system, $j = 1, 2$) used to determine the appropriate E-field phasor-vector components from the antenna pattern in the direction of each scatter point $\mathscr{P}_s^{\mathrm{S}}$

For each calculation point $\mathscr{P}_s^{\mathrm{L}}$ in the L system, the S system is employed to determine link-observable trail orientations $\hat{\mathbf{v}}_m^{\mathrm{S}}$ at $\mathscr{P}_s^{\mathrm{S}}$ ($\mathscr{P}_s^{\mathrm{L}}$ relative to the S system). Each trail orientation, $\hat{\mathbf{v}}_m^{\mathrm{S}}$, is transformed into the O system $(\hat{\mathbf{v}}_m^{\mathrm{O}})$ for determination of the incident SMRD value. Next, the path loss value L^* at $\mathscr{P}_s^{\mathrm{L}}$ is determined from a link power budget calculation using the antenna gain product G_{sm}^{12} at $\mathscr{P}_s^{\mathrm{S}}$. This gain product is the complex dot product of the normalized (see Eqs. 3.36-3.38) transmit and receive E-field phasor vectors. These vectors are determined from the NEC-generated antenna pattern at both ($j = 1, 2$) link sites using the T-, G_j-, A_j-, and P_j coordinate systems. Finally, the link-usable portion of the arriving meteor flux with direction vector $\hat{\mathbf{v}}_m^{\mathrm{S}}$ at $\mathscr{P}_s^{\mathrm{S}}$ is determined from the maximum allowable path loss L^* and minimum required burst duration τ^*. These usability criteria, (L^*, τ^*), determine the total MR and DC contribution from each modeled meteor trail with direction vector $\hat{\mathbf{v}}_m^{\mathrm{O}}$ at the trail-scatter point $\mathscr{P}_s^{\mathrm{L}}$.

Integrating these MR and DC contributions over

- trail orientation (α_m) at each modeled scatter point
- scatter point height above ground (z_{L}, or h_s)
- link cross-range location (y_{L})

- link downrange location (x_L)

yields the total link MR and DC values. The coordinate systems introduced in this appendix enable direct selection of meteor trails that satisfy both the observability criteria ($\mathscr{P}_s^L, \hat{\mathbf{v}}_m^O$) and usability criteria (L^*, τ^*). The representation of each planar rotation by a single transformation matrix simplified debugging of the METEORLINK, METEORTRAK, and METEORDIV computer programs, and clarified the geometric and spatial relationships that are fundamental in the understanding of meteor trail-scatter phenomena.

APPENDIX B: LINK DIAGNOSTIC GRIDS

This appendix contains link diagnostic grids generated by the METEORLINK computer program. These grids were computed for the Sondrestrom-Thule MB Test Bed link (1200-km, ST link) operated by the Phillips Laboratory in Greenland [238]. Predicted values were computed for the ST link at 6 UT in March, 1989. The geometry of these grids is determined by the LINK coordinate system shown in Fig. 3.B.1. Link grids are provided as follows:

Figure 3.B.2: MR contributions to the link MR value in meteors per minute

Figure 3.B.3: DC contributions to the link DC value in percent time

Figure 3.B.4: percentage of underdense trails constituting each MR contribution (UMR)

Figure 3.B.5: percentage of underdense trails constituting each DC contribution (UDC)

Figure 3.B.6: antenna gain product measured in dB

Figure 3.B.7: maximum allowable path loss in dB to achieve the required received power P_r^*

Figure 3.B.8: transmit antenna elevation angles whose corresponding MR contributions exceed 1% of the net link MR value

Figure 3.B.9: transmit antenna azimuth angles whose corresponding MR contributions exceed 1% of the link MR-value

Figure 3.B.10: Faraday rotation angles determined from the transmit antenna to each modeled trail-scatter point, $\mathscr{P}_s^L$

In general, the link grids for the receive antenna (site 2) differ from the corresponding Figs. 3.B.8, 3.B.9, and 3.B.10 for the transmit antenna. The uneven spacing of grid calculation points apparent in these figures is due to the Gauss quadrature technique employed.

Figure 3.B.11 shows observable trail orientations emanating from each modeled trail-scatter point above the link as well as the orientation of a

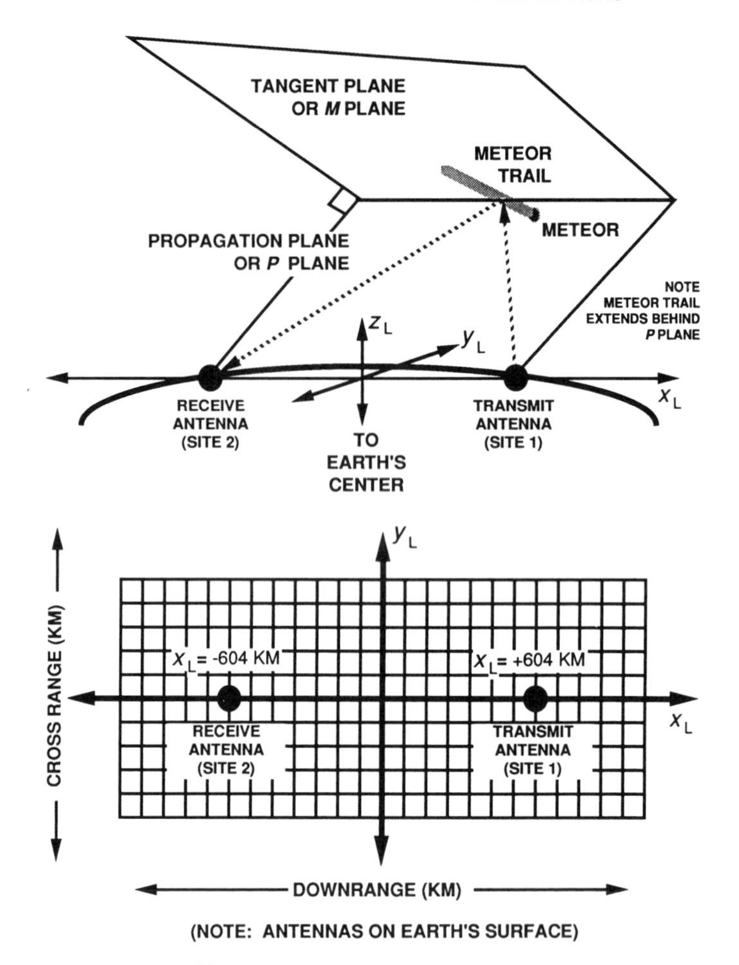

Figure 3.B.1. LINK grid conventions.

phasor vector launched from the transmit site (T_x) by a perfect horizontally-polarized antenna (no θ component). This figure demonstrates the trail-scatter and E phasor vector geometry determined by METEORLINK. In addition, the figure shows the orientation of E phasor vectors relative to observable trails necessary for the macroscopic study of plasma resonance effects on MB links. Similarly, Fig. 3.B.12 shows observable trails and a vertically-polarized electric phasor vector (no ϕ component) launched from the transmit site.

Finally, Fig. 3.B.13 gives the relative contribution to the link MR value from each cell of the Kazan-Mogadishu SMRD map. The value 100 in the 0 to 15° ecliptic latitude interval and the 216° to 240° ecliptic longitude indicates the strongest contributing radiant. Since the apex direction corresponds to 270° ecliptic longitude in the O system, the radiants contributing to the 0 to 15° source are approaching *from* the apex direction. This result corresponds to the ST link position on the apex side of the earth at 6 UT.

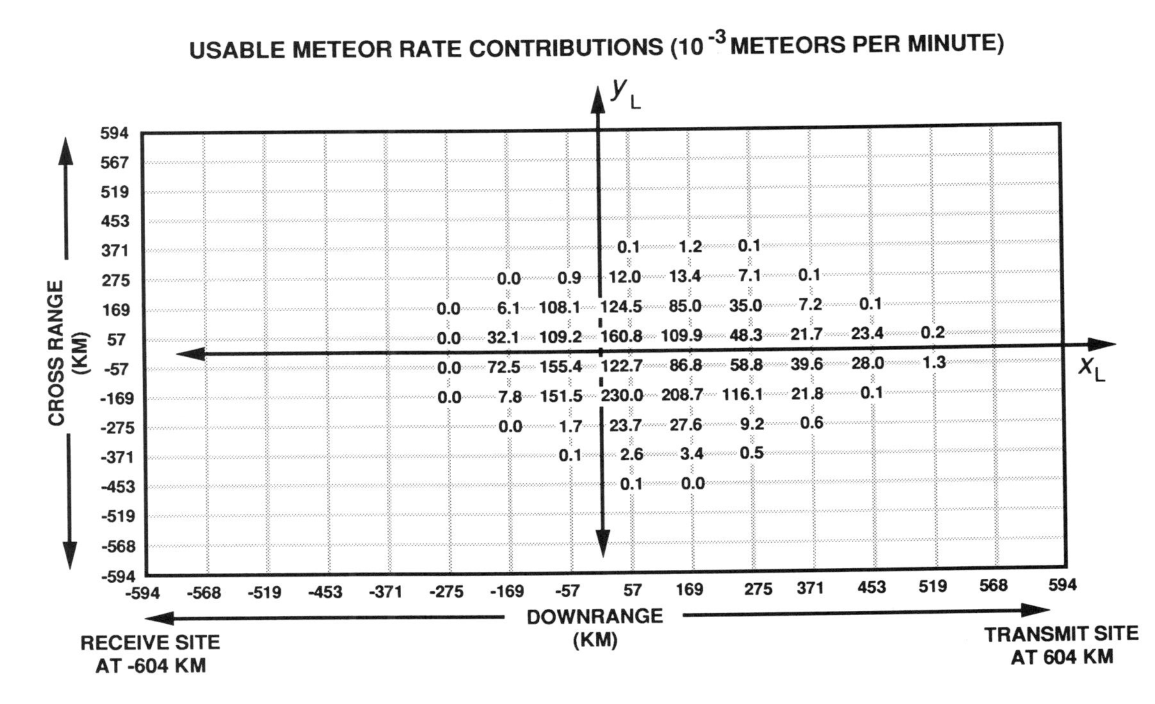

Figure 3.B.2. ST-link MR contributions.

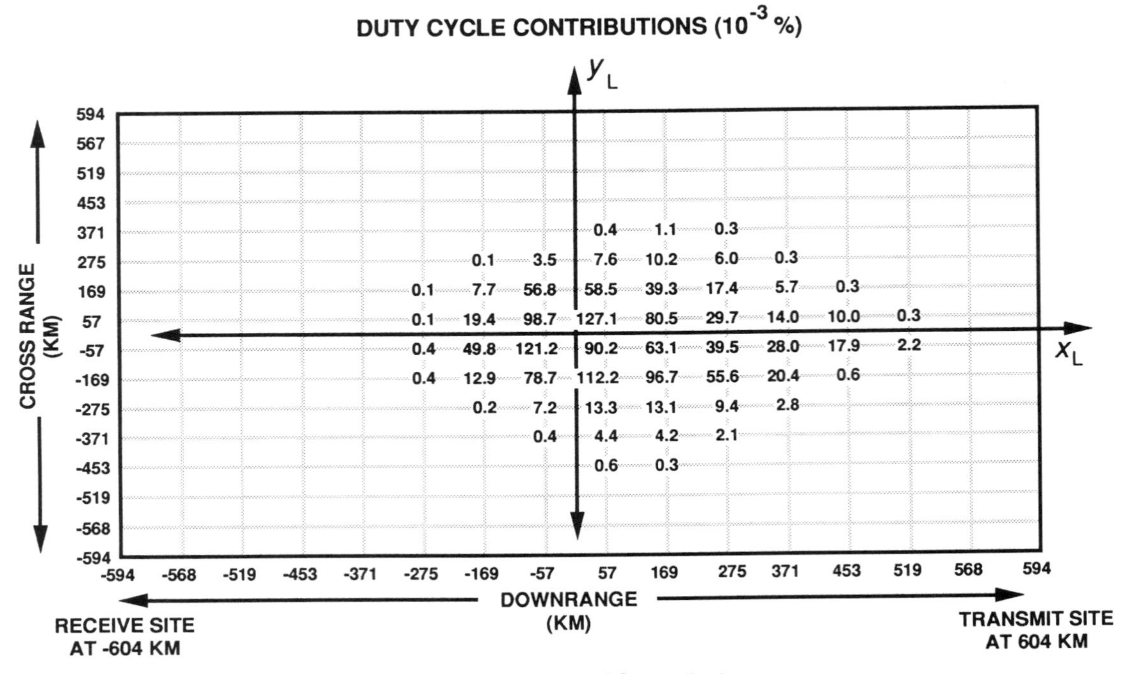

Figure 3.B.3. ST-link DC contributions.

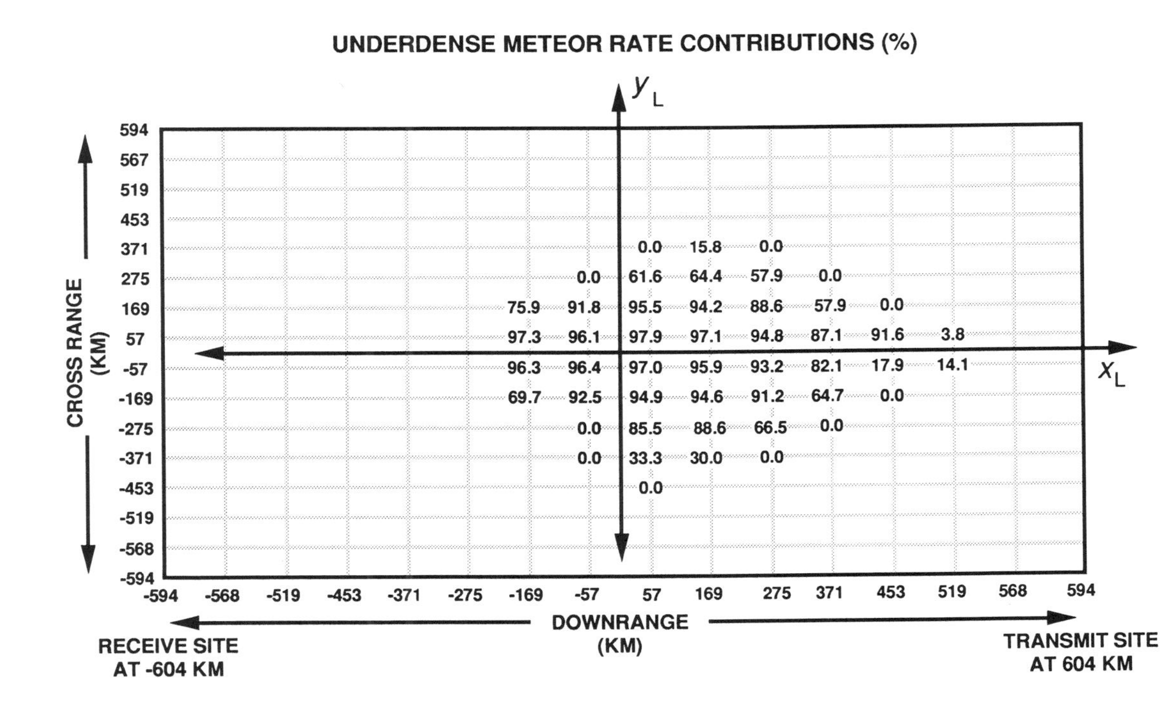

(NOTE: EACH VALUE IS THE PERCENTAGE OF THE CORRESPONDING USABLE METEOR RATE CONTRIBUTION PROVIDED BY UNDERDENSE METEOR TRAILS)

Figure 3.B.4. Percentage of underdense trails for each MR contribution.

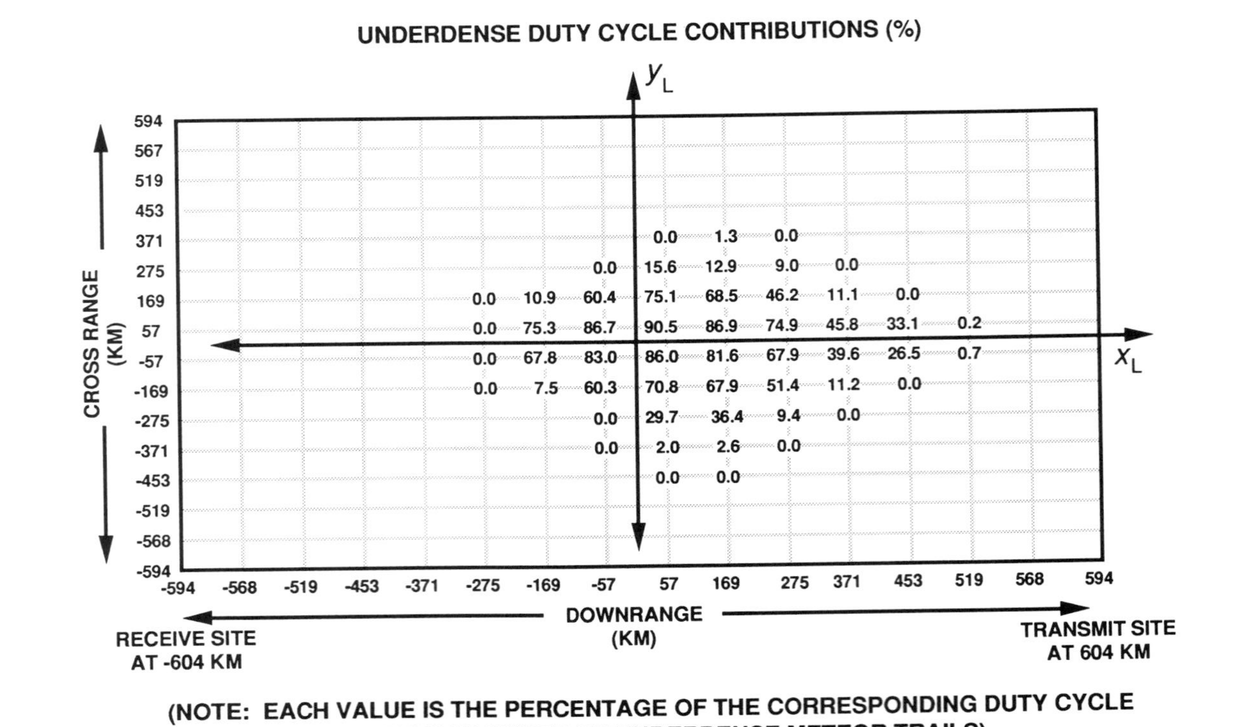

Figure 3.B.5. Percentage of underdense trails for each DC contribution.

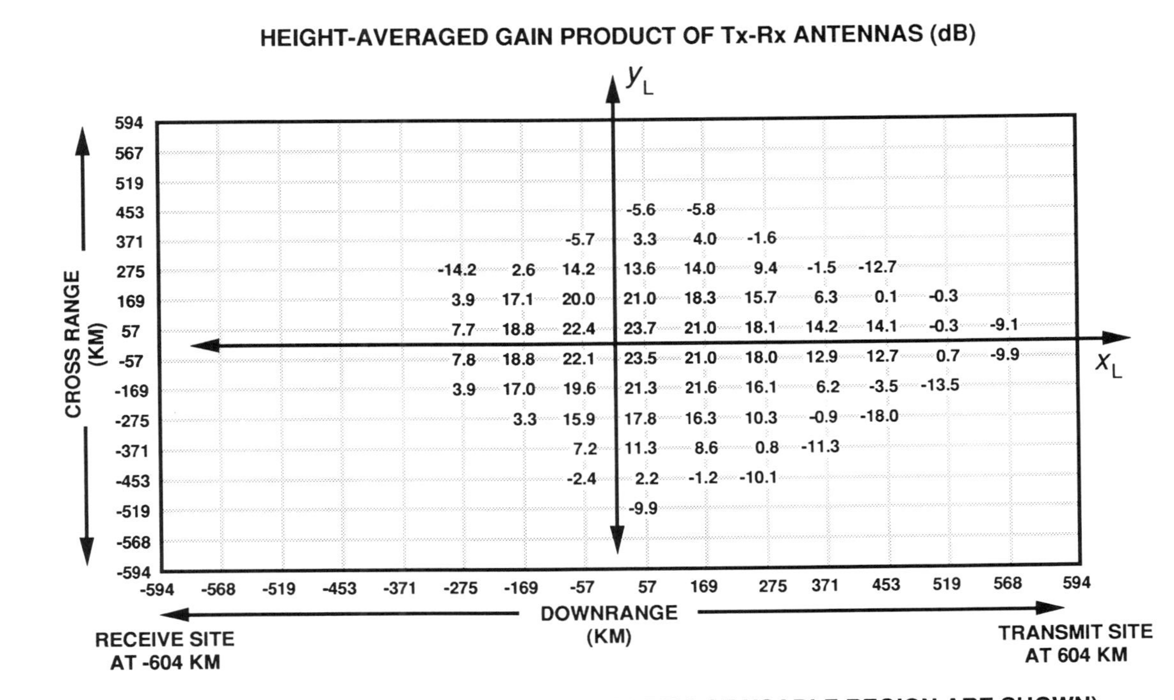

(NOTE: GAIN PRODUCTS OUTSIDE BOUNDARIES OF USABLE REGION ARE SHOWN)

Figure 3.B 6. Antenna gain products.

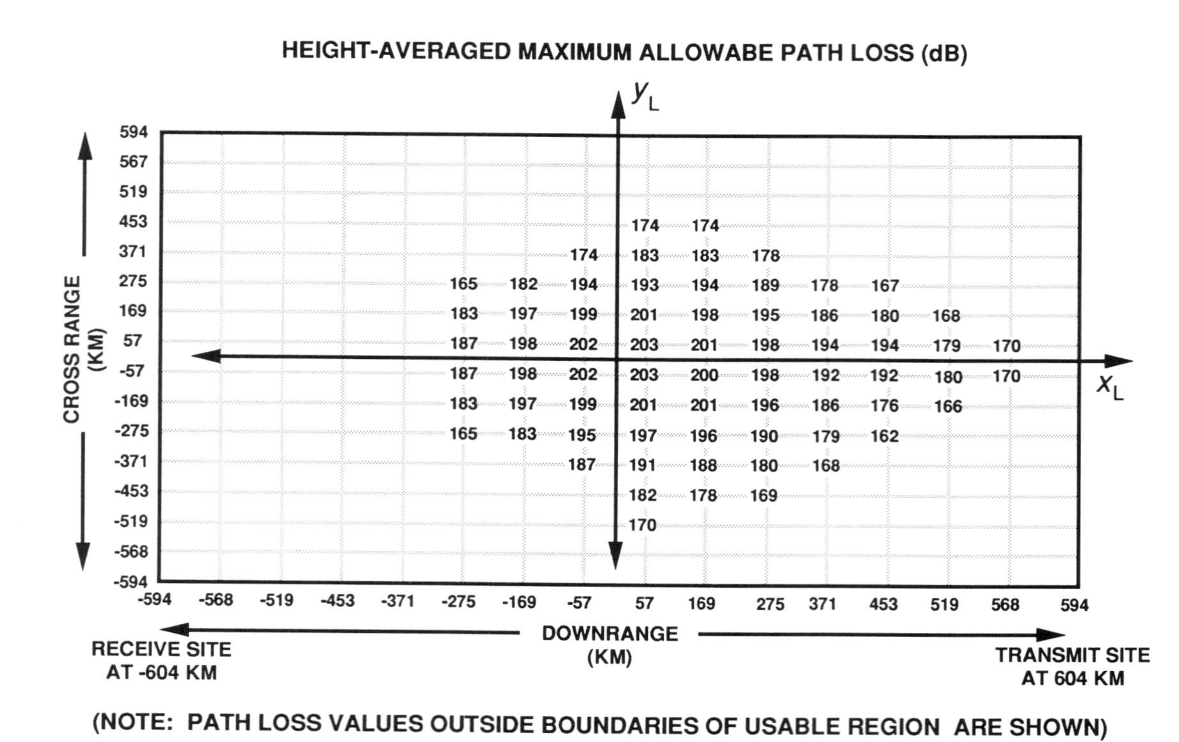

Figure 3.B.7. Maximum allowable path loss values.

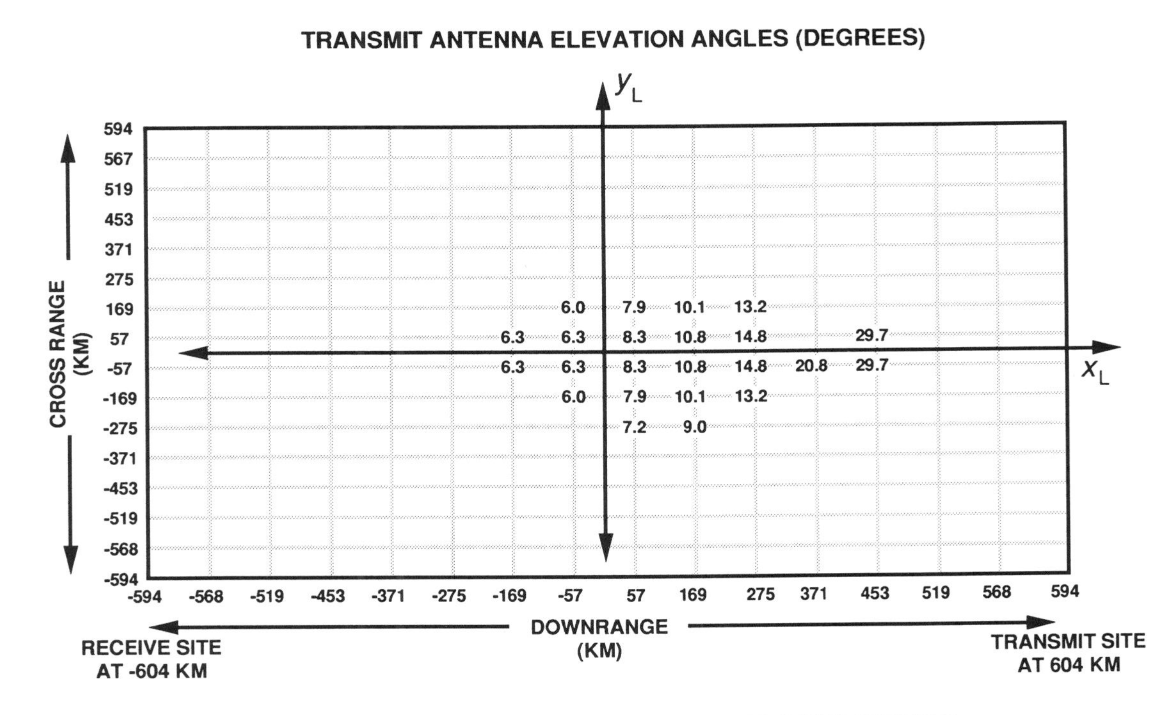

Figure 3.B.8. Usable transmit antenna elevation angles.

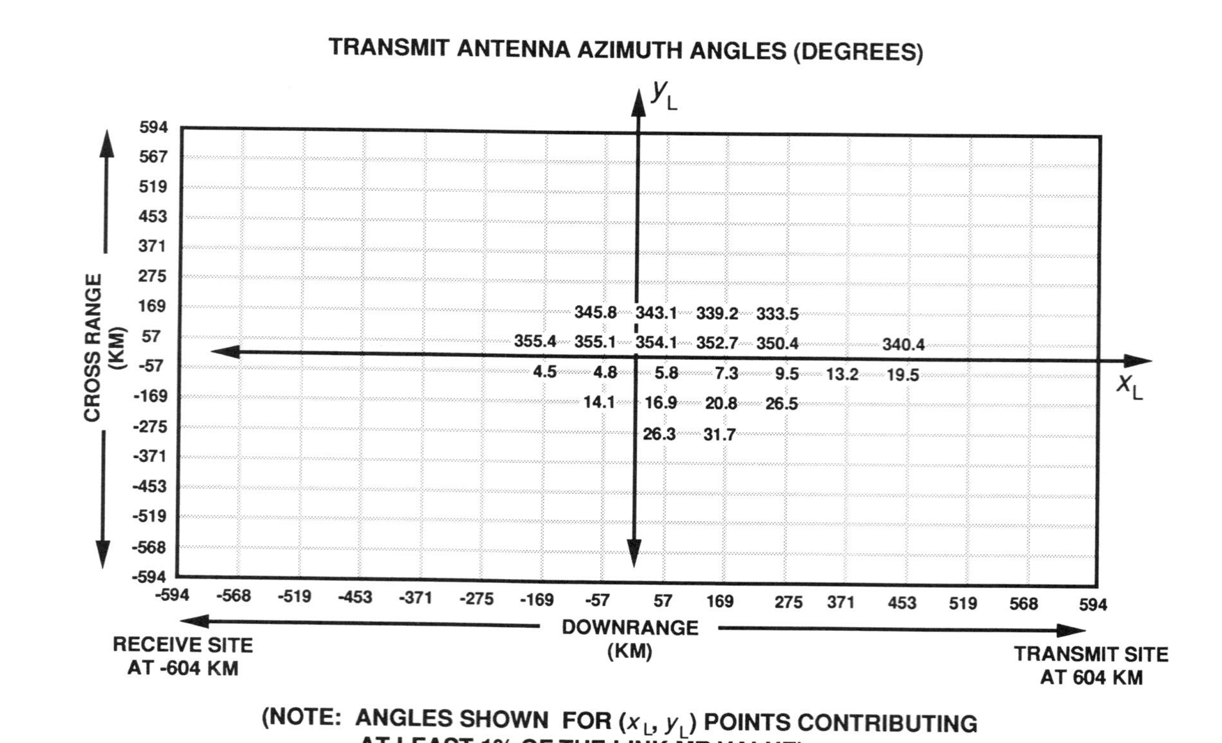

Figure 3.B.9. Usable transmit antenna azimuth angles.

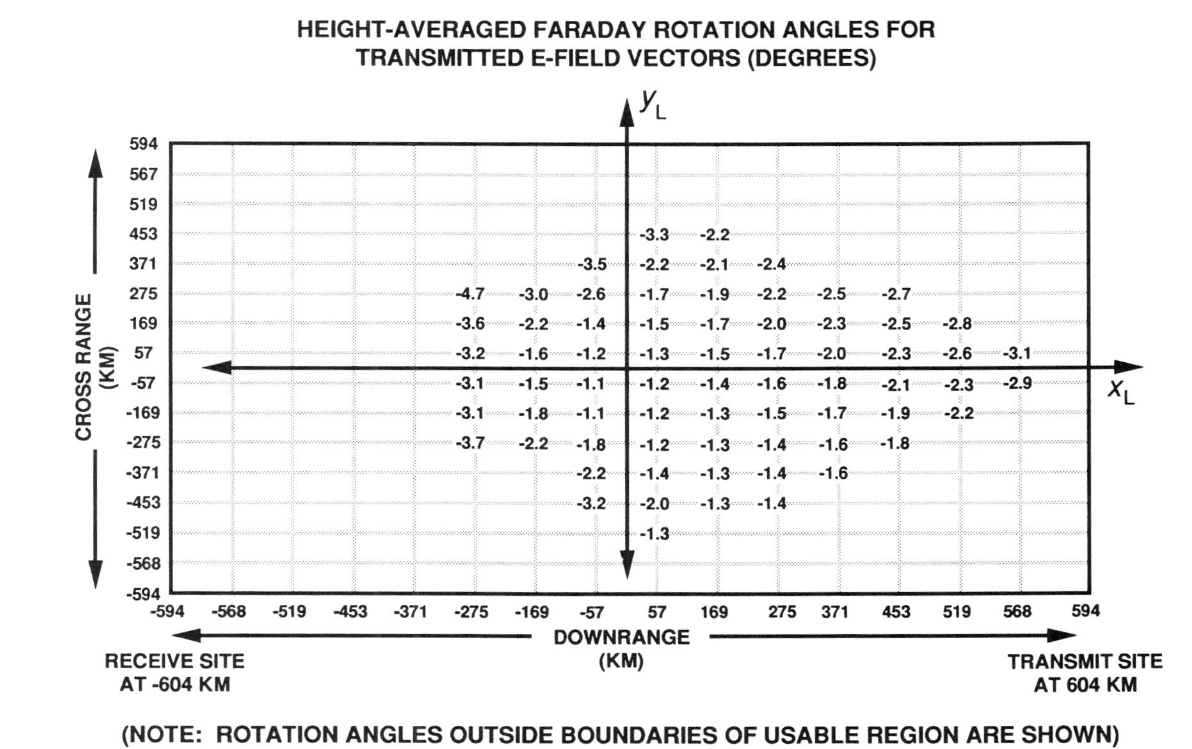

Figure 3.B.10. Faraday rotation angles for transmitted E-field phasors.

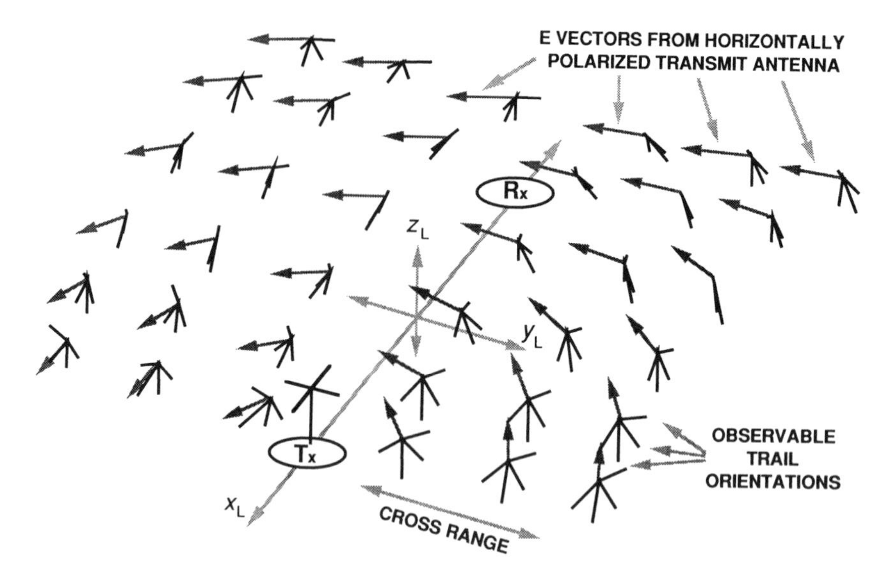

Figure 3.B.11. Observable trails and corresponding E-field phasors relative to a horizontally polarized antenna (four trails shown at each scatter point).

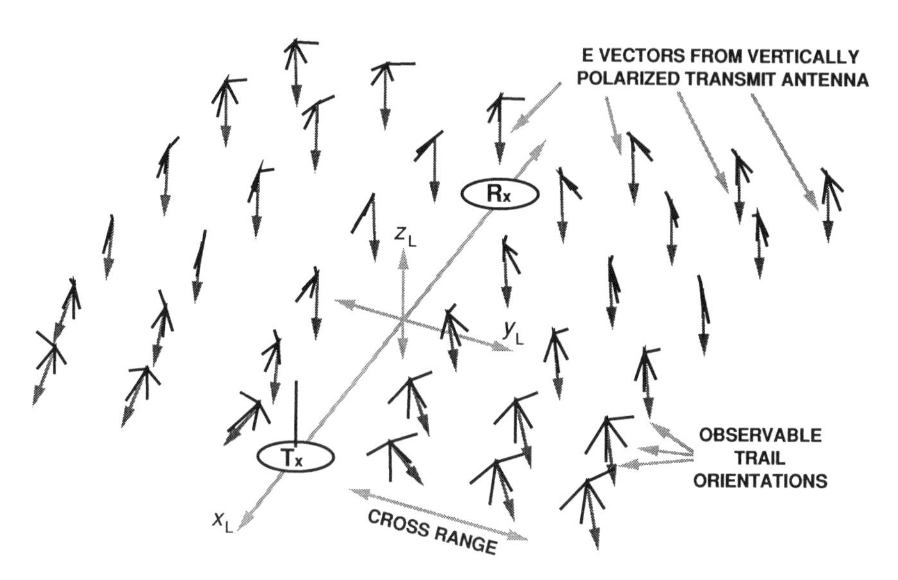

Figure 3.B.12. Observable trails and corresponding E field phasors relative to a vertically polarized antenna (four trails shown at each scatter point).

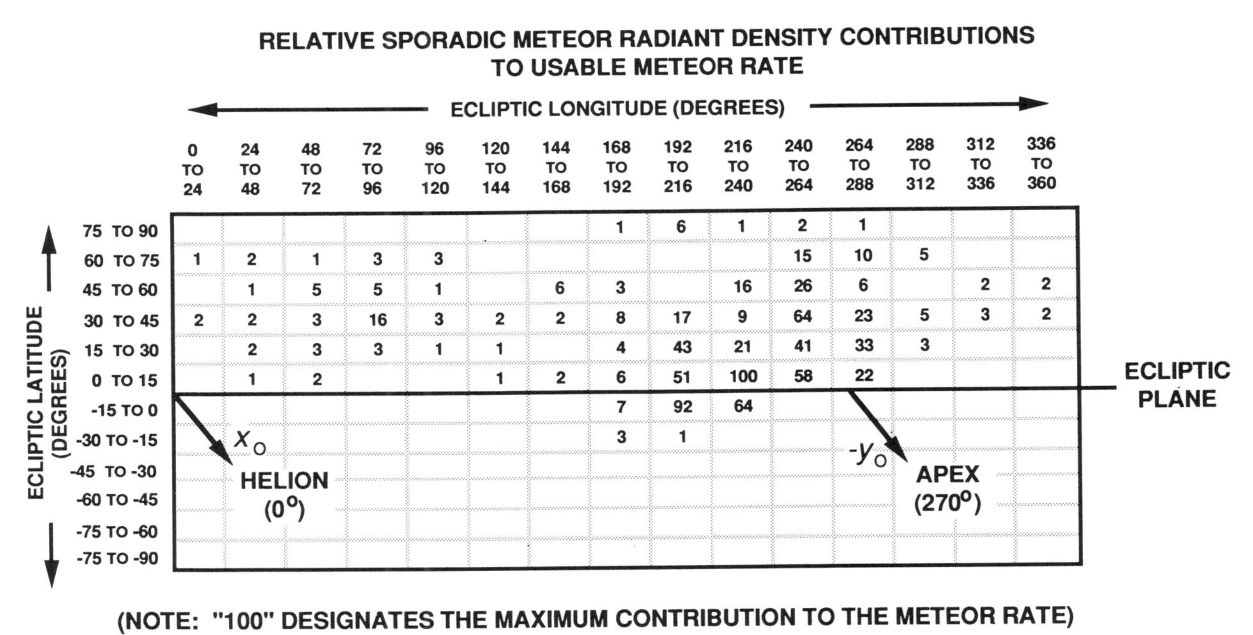

RELATIVE SPORADIC METEOR RADIANT DENSITY CONTRIBUTIONS TO USABLE METEOR RATE

ECLIPTIC LATITUDE (DEGREES) \ ECLIPTIC LONGITUDE (DEGREES)	0 TO 24	24 TO 48	48 TO 72	72 TO 96	96 TO 120	120 TO 144	144 TO 168	168 TO 192	192 TO 216	216 TO 240	240 TO 264	264 TO 288	288 TO 312	312 TO 336	336 TO 360
75 TO 90								1	6	1	2	1			
60 TO 75	1	2	1	3	3						15	10	5		
45 TO 60		1	5	5	1		6	3		16	26	6		2	2
30 TO 45	2	2	3	16	3	2	2	8	17	9	64	23	5	3	2
15 TO 30		2	3	3	1	1		4	43	21	41	33	3		
0 TO 15		1	2			1	2	6	51	100	58	22			
-15 TO 0								7	92	64					
-30 TO -15								3	1						
-45 TO -30															
-60 TO -45															
-75 TO -60															
-75 TO -90															

(NOTE: "100" DESIGNATES THE MAXIMUM CONTRIBUTION TO THE METEOR RATE)

Figure 3.B.13. SMRD contributions to link MR value.

APPENDIX C: COMPARISON OF METEORLINK PREDICTIONS WITH JODRELL BANK RADAR MEASUREMENTS

This appendix contains the results of a measurement-prediction comparison performed using 72-MHz radar-detected sporadic meteors collected at the Jodrell Bank Experimental Station, UK, in 1950 and 1951 [239] (Figs. 3.C.1–3.C.13). Plots of measured and METEORLINK-predicted MR values (meteors per hour) have been provided for both the "northwest" and "southwest" radar beams (see Section 3.4.3.3). The comparison is shown for each month from the period from January 1951 through December 1951 followed by the period from October 1950 through November 1950. In addition, both measured and predicted daily average MR values (DAMR) values for both radar beam azimuths angles are plotted versus month. The measured data presented in these plots were extracted from published results [240].

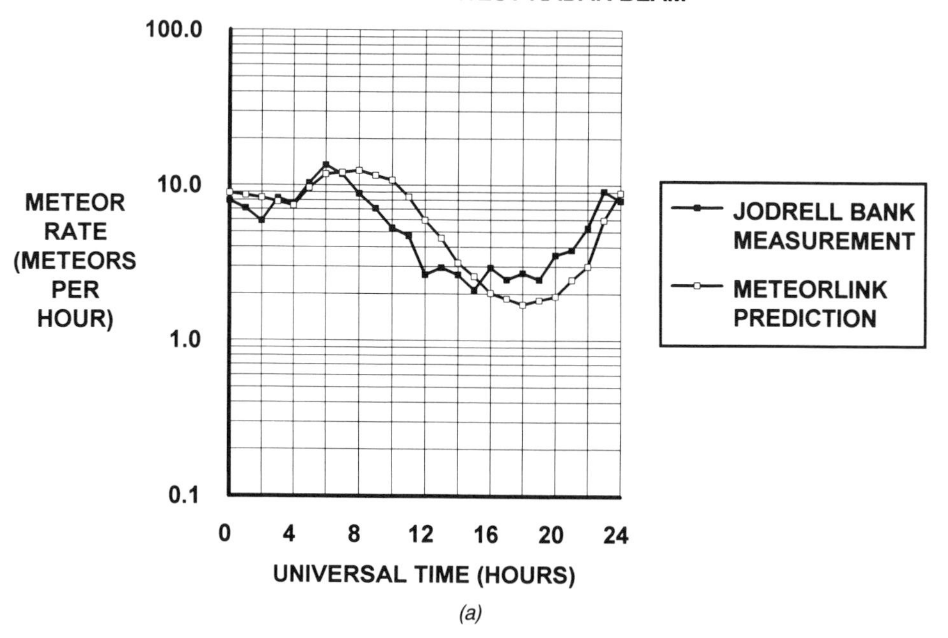

(a)

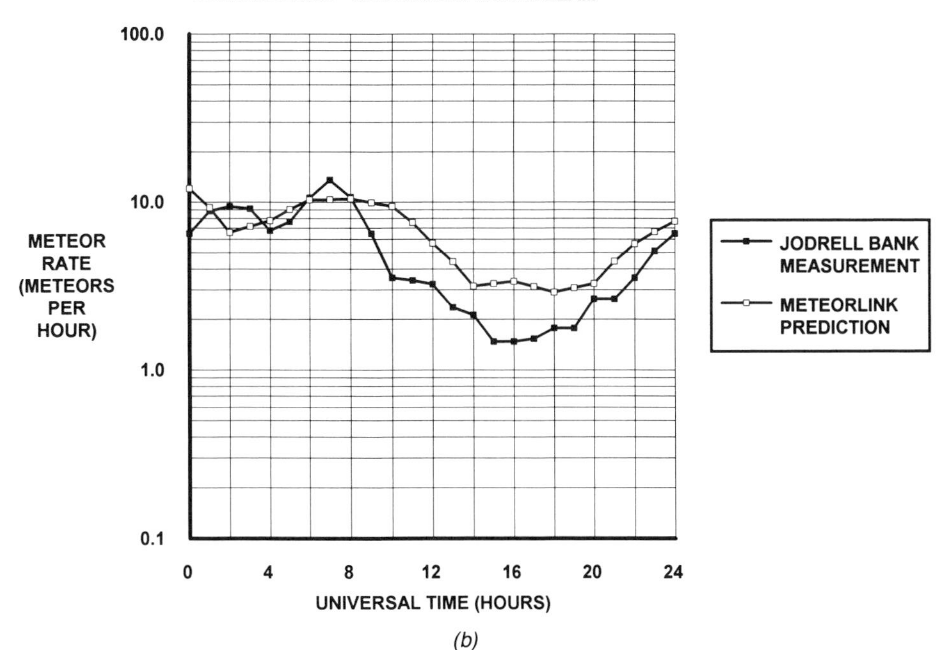

(b)

Figure 3.C.1. Jodrell Bank MR measurement-prediction comparison for January 1951 (*a*) northwest radar beam; (*b*) southwest radar beam.

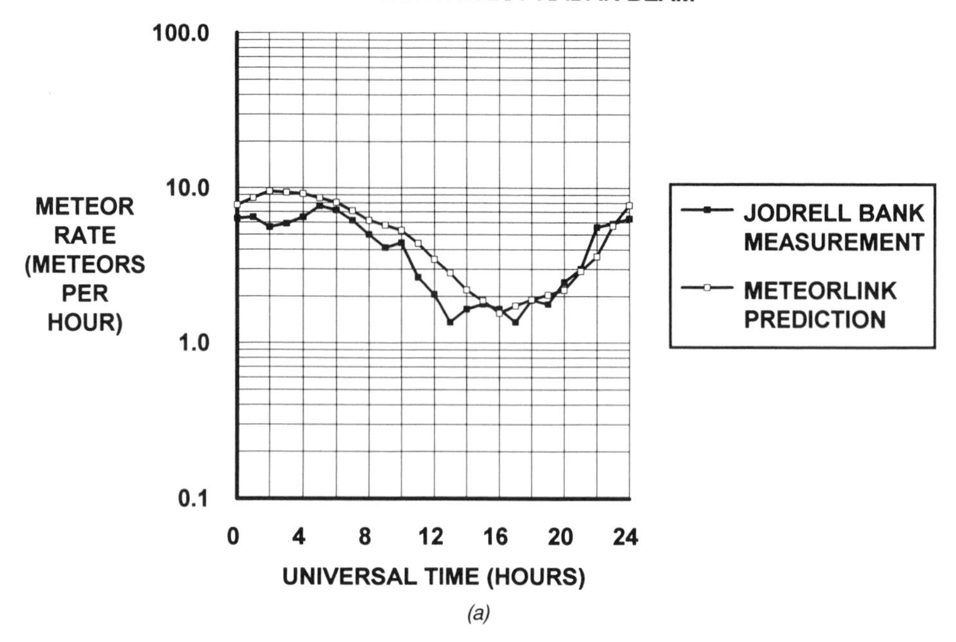

(a)

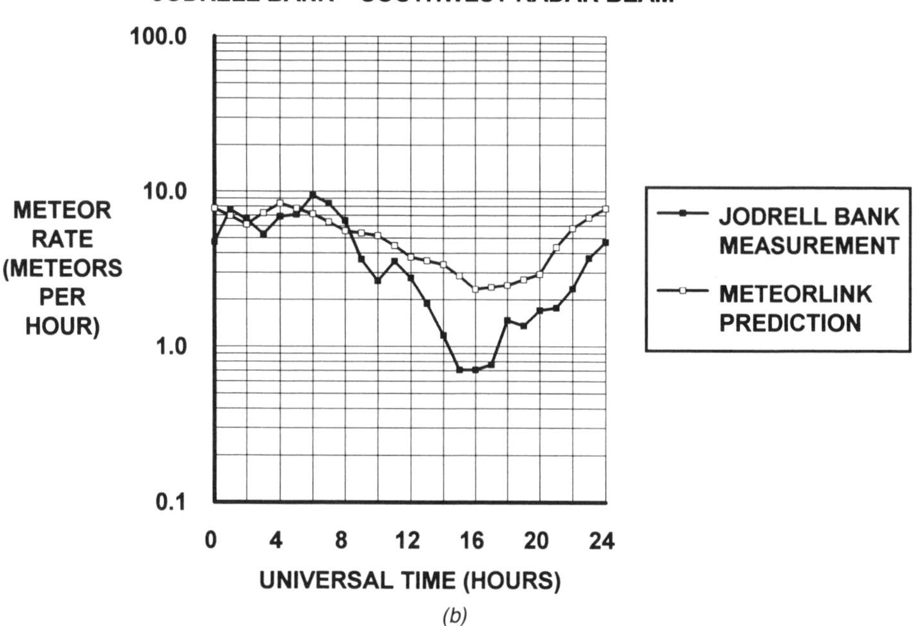

(b)

Figure 3.C.2. Jodrell Bank MR measurement-prediction comparison for February 1951 (*a*) northwest radar beam; (*b*) southwest radar beam.

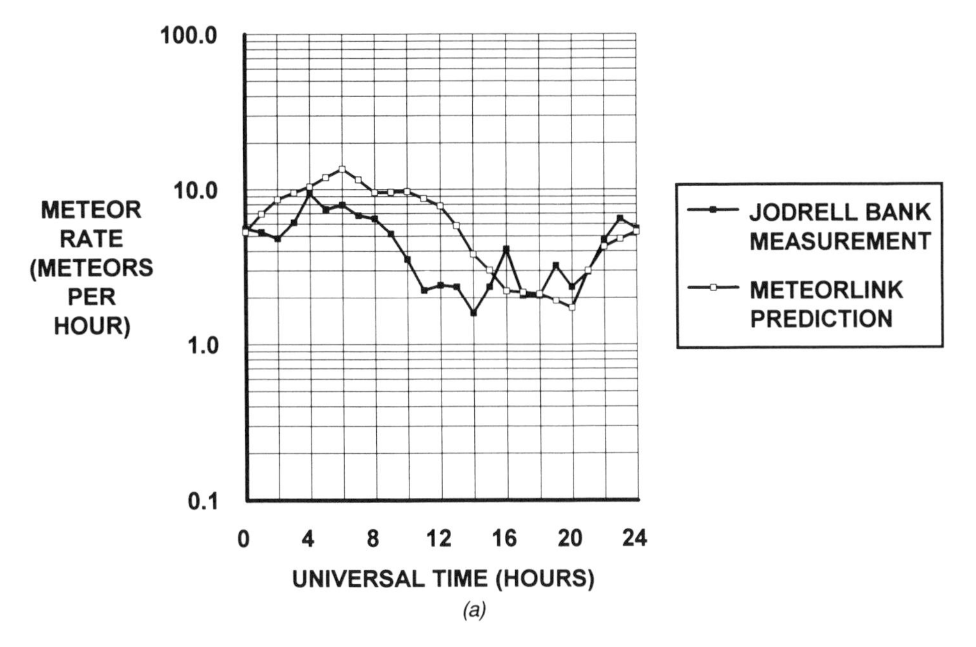

(a)

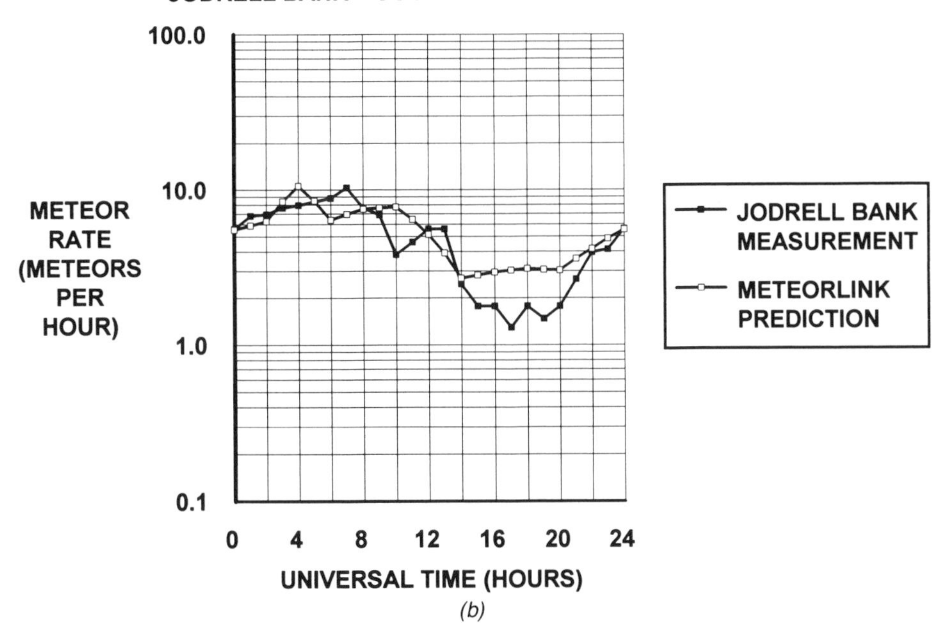

(b)

Figure 3.C.3. Jodrell Bank MR measurement-prediction comparison for March 1951 (*a*) northwest radar beam; (*b*) southwest radar beam.

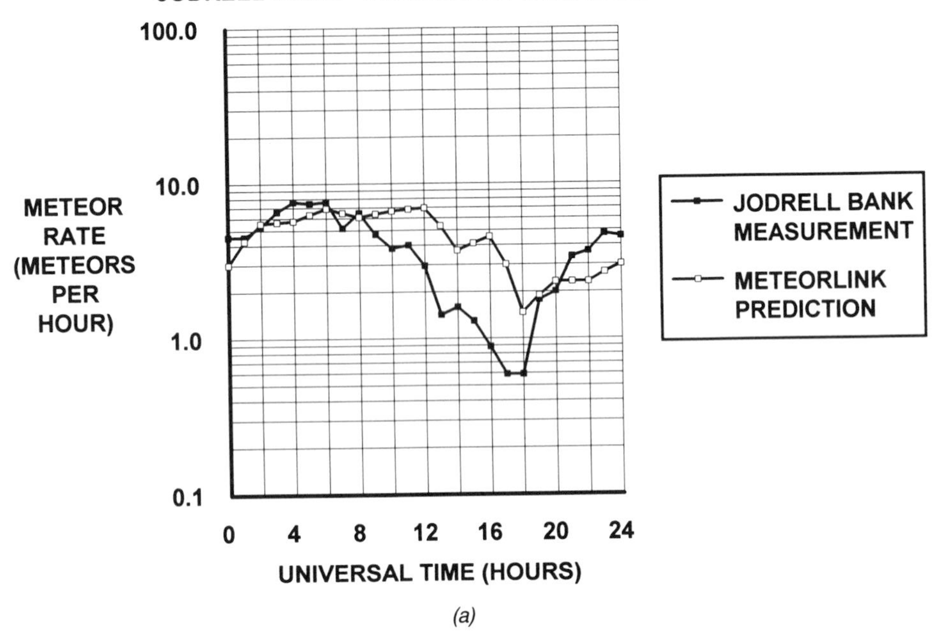

(a)

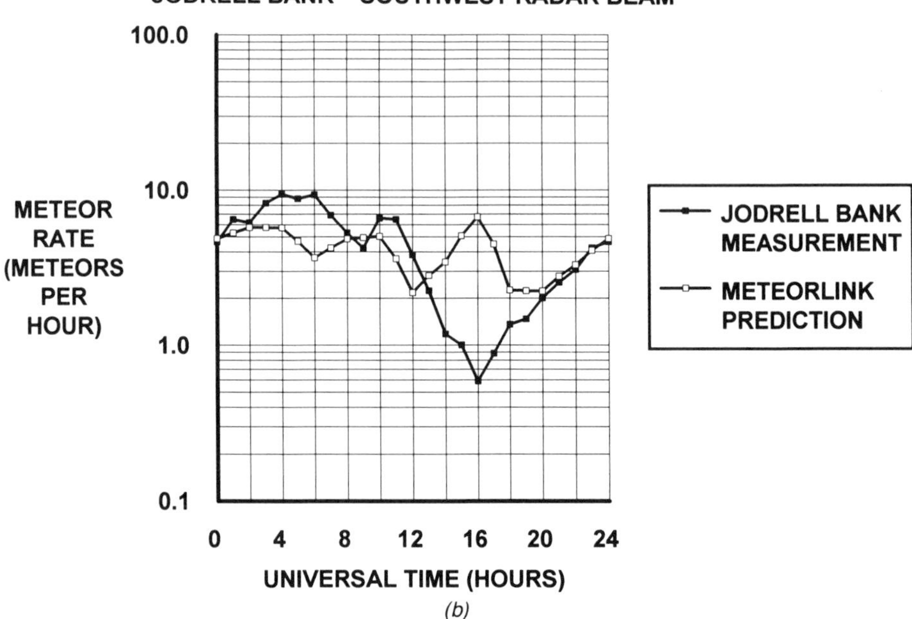

(b)

Figure 3.C.4. Jodrell Bank MR measurement-prediction comparison for April 1951 (*a*) northwest radar beam; (*b*) southwest radar beam.

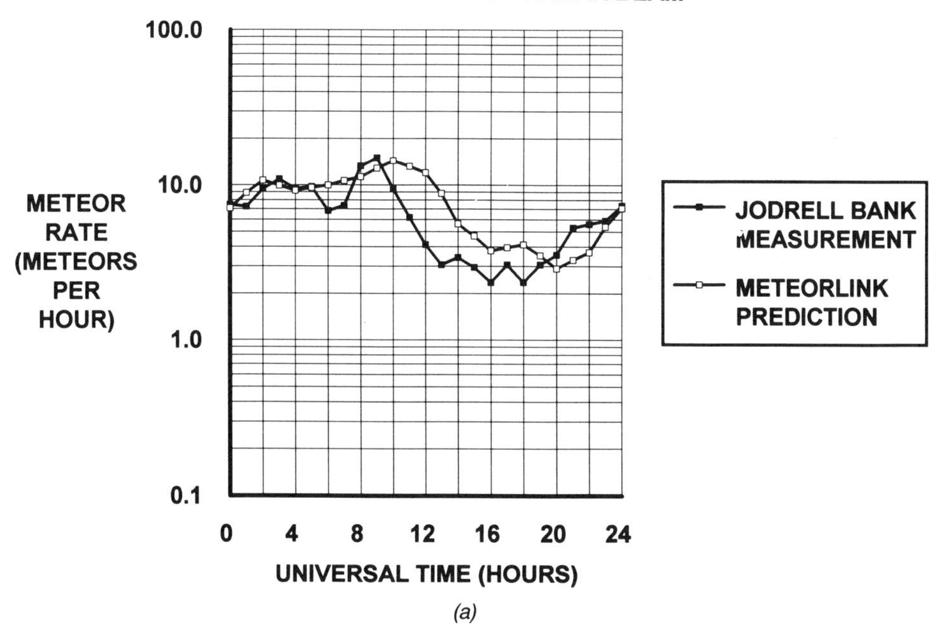

(a)

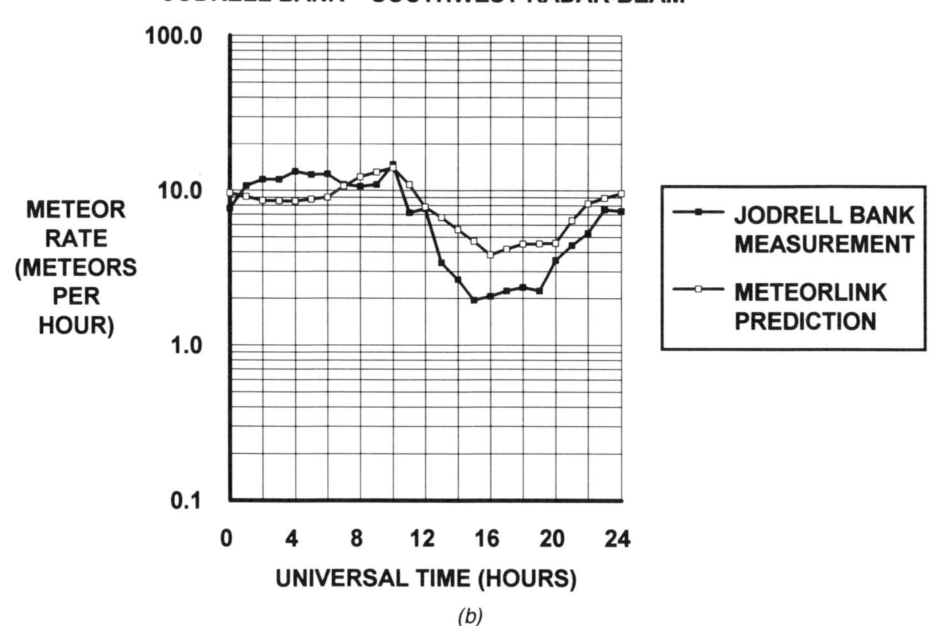

(b)

Figure 3.C.5. Jodrell Bank MR measurement-prediction comparison for May 1951 (*a*) northwest radar beam; (*b*) southwest radar beam.

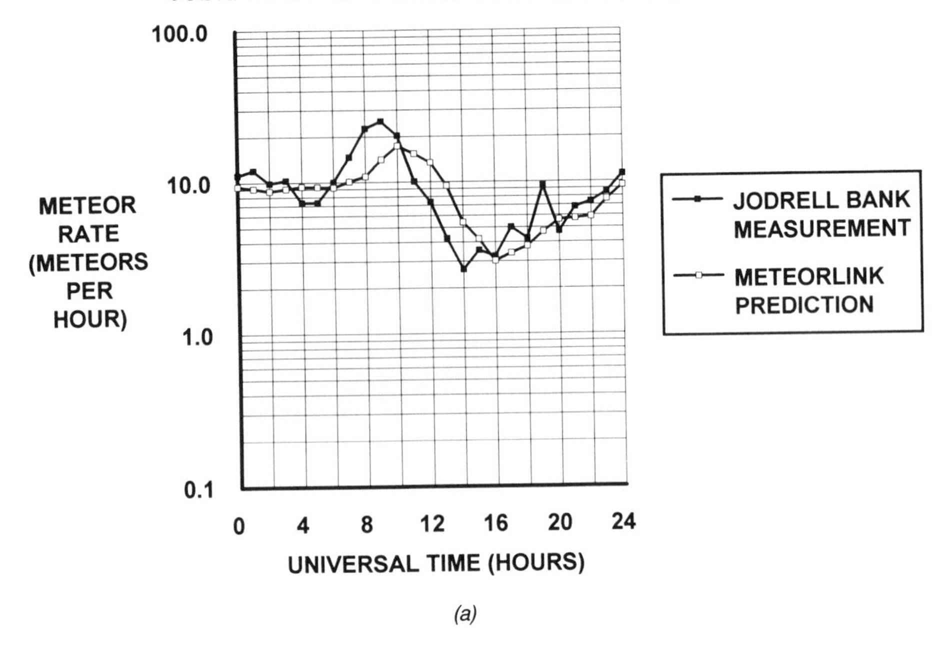

(a)

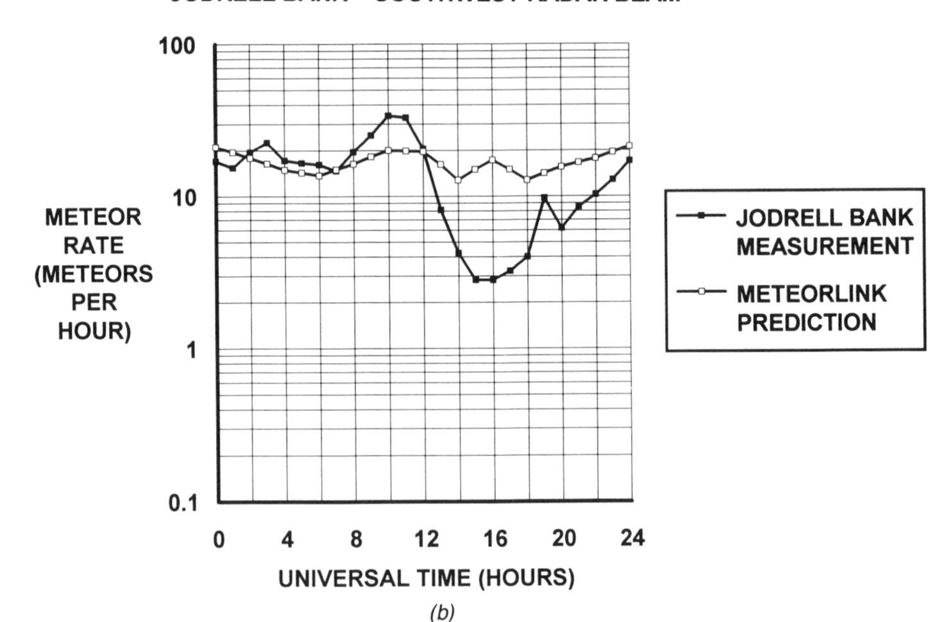

(b)

Figure 3.C.6. Jodrell Bank MR measurement-prediction comparison for June 1951 (*a*) northwest radar beam; (*b*) southwest radar beam.

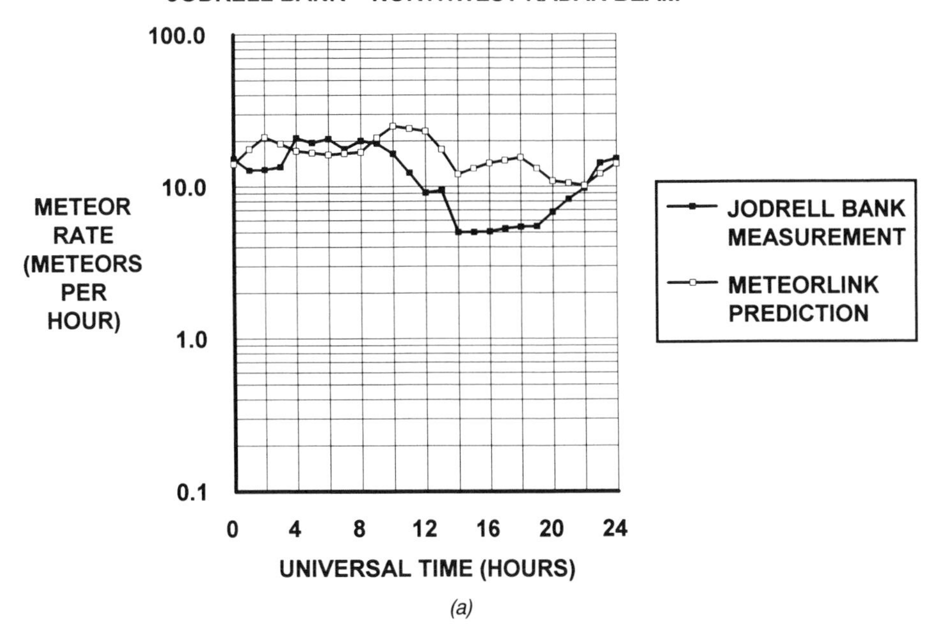

(a)

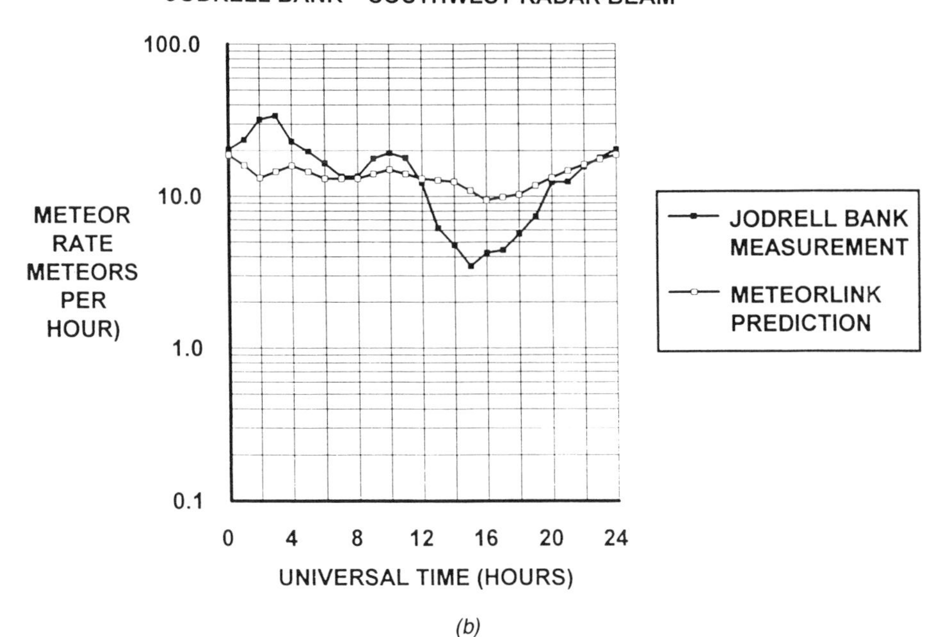

(b)

Figure 3.C.7. Jodrell Bank MR measurement-prediction comparison for July 1951 (*a*) northwest radar beam; (*b*) southwest radar beam.

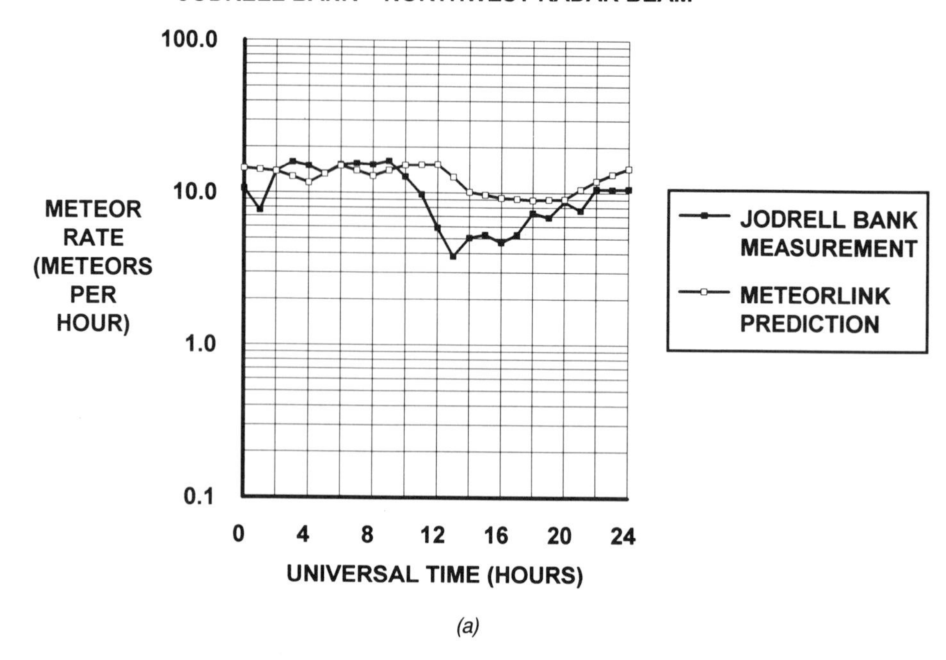

(a)

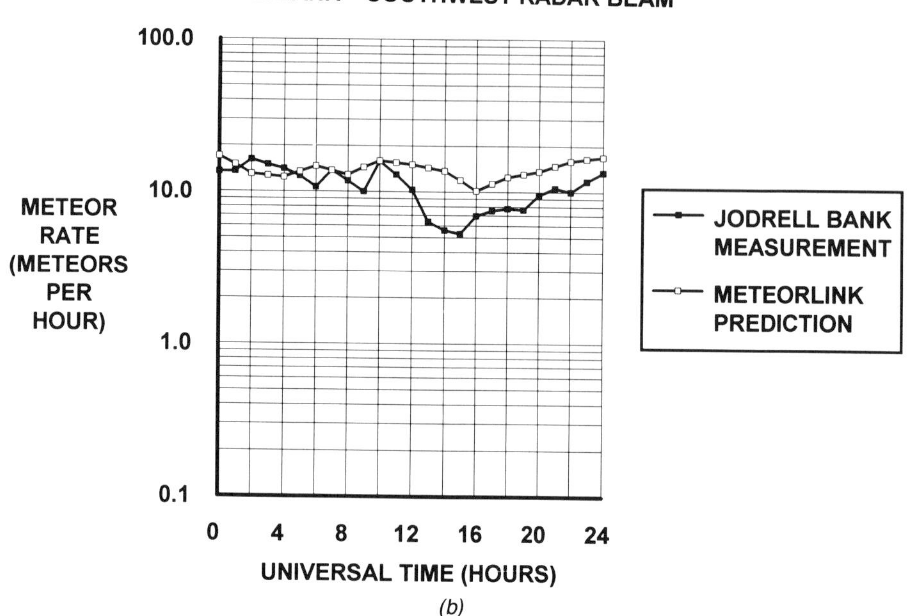

(b)

Figure 3.C.8. Jodrell Bank MR measurement-prediction comparison for August 1951 (*a*) northwest radar beam; (*b*) southwest radar beam.

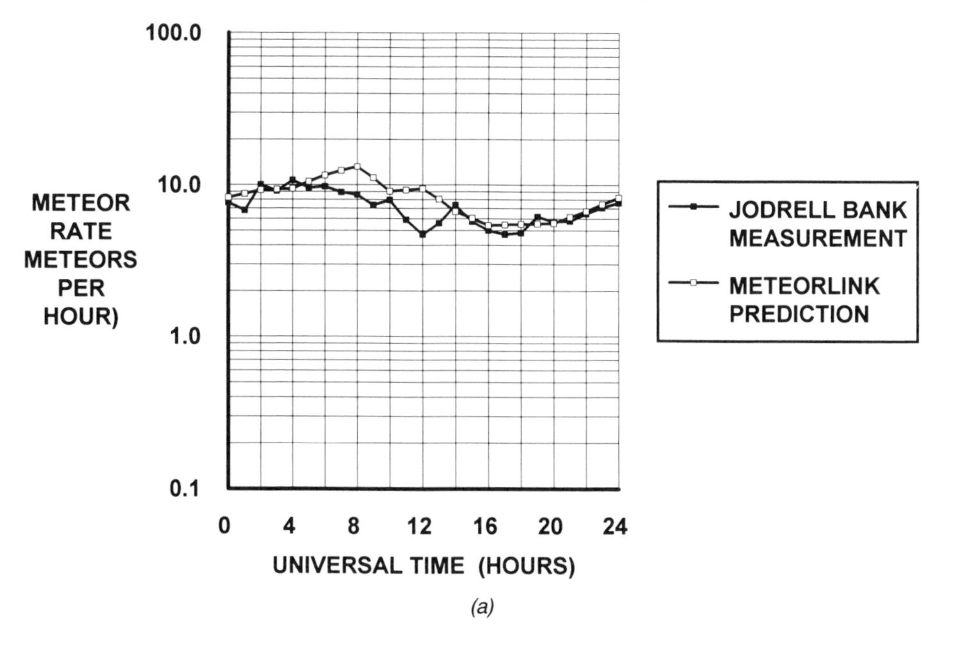

(a)

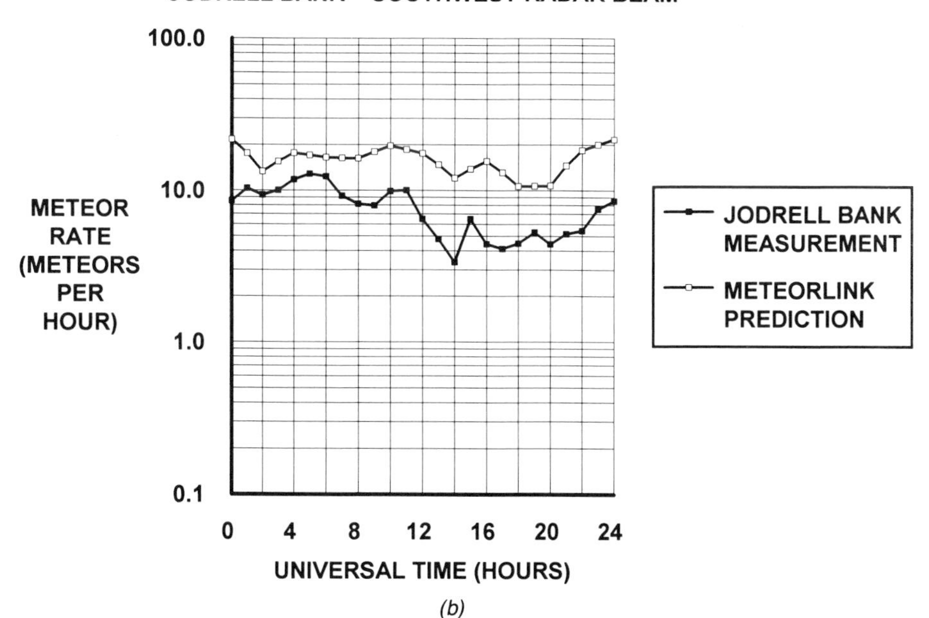

(b)

Figure 3.C.9. Jodrell Bank MR measurement-prediction comparison for September 1951 (*a*) northwest radar beam; (*b*) southwest radar beam.

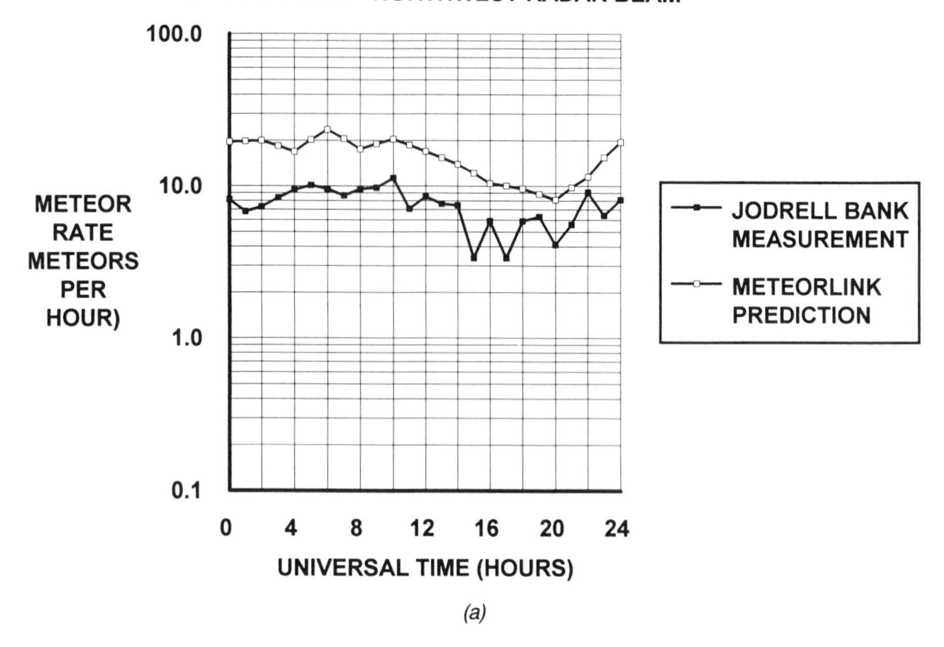

(a)

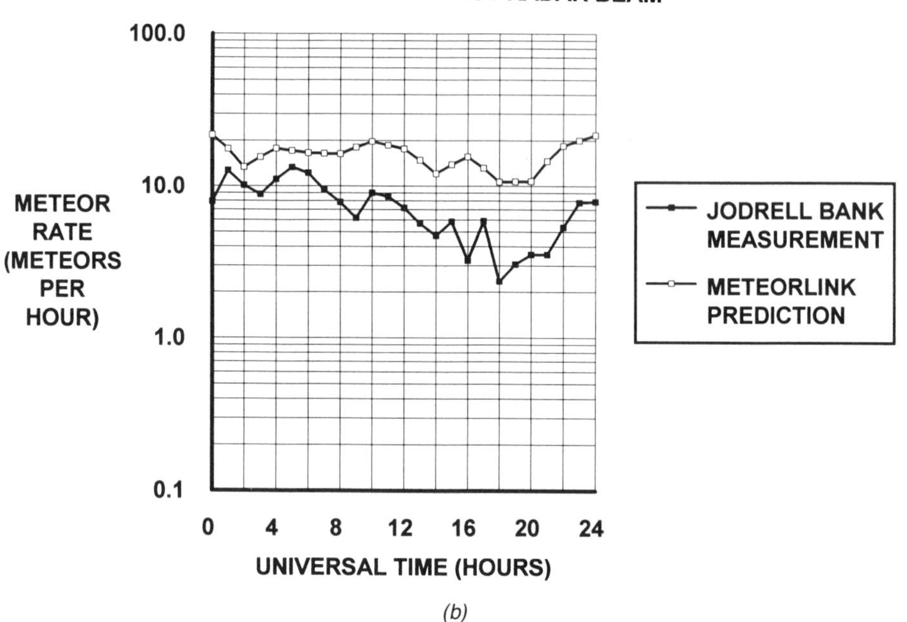

(b)

Figure 3.C.10. Jodrell Bank MR measurement-prediction comparison for October 1950 (*a*) northwest radar beam; (*b*) southwest radar beam.

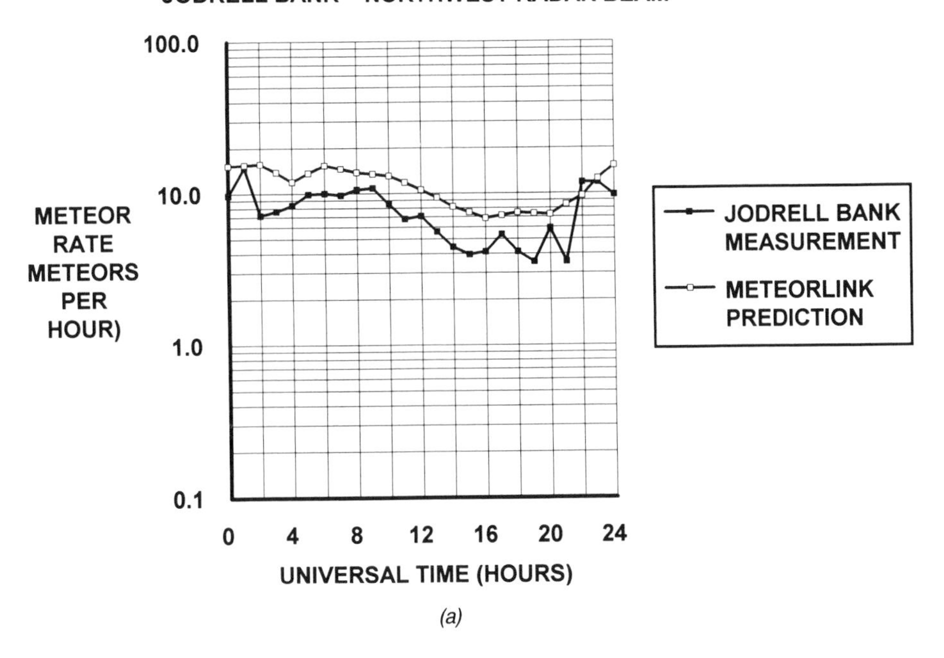

(a)

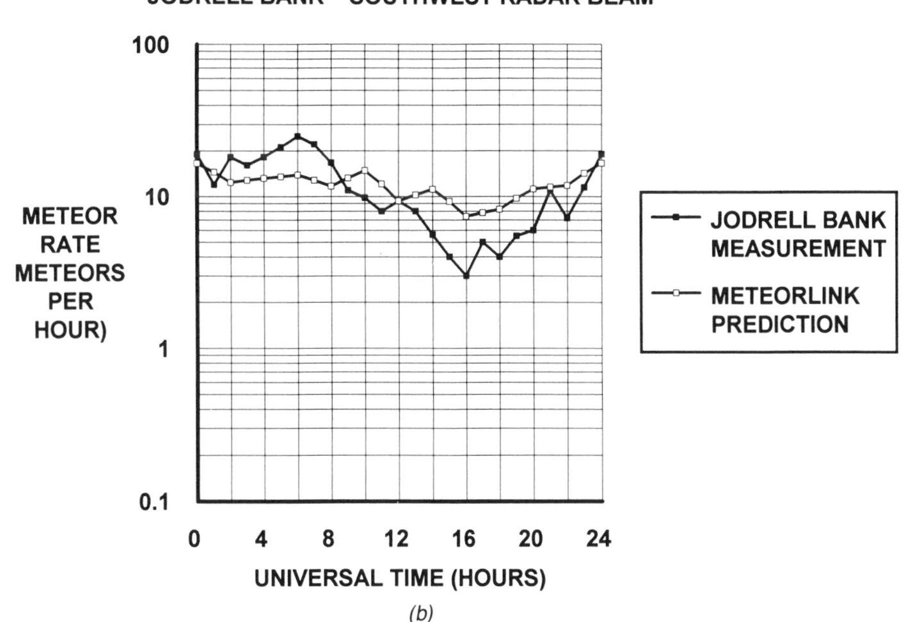

(b)

Figure 3.C.11. Jodrell Bank MR measurement-prediction comparison for November 1950 (*a*) northwest radar beam; (*b*) southwest radar beam.

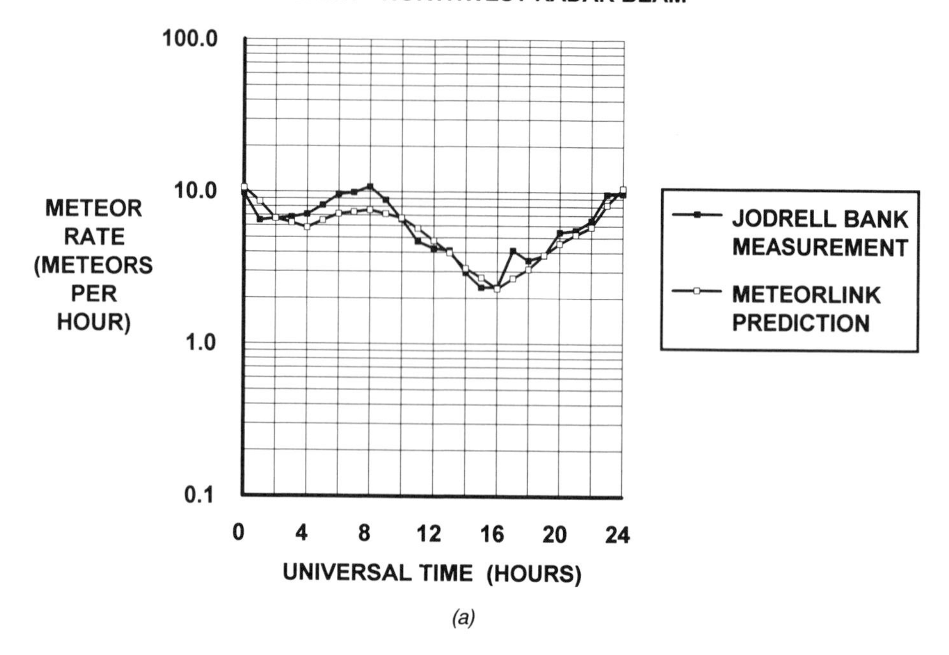

(a)

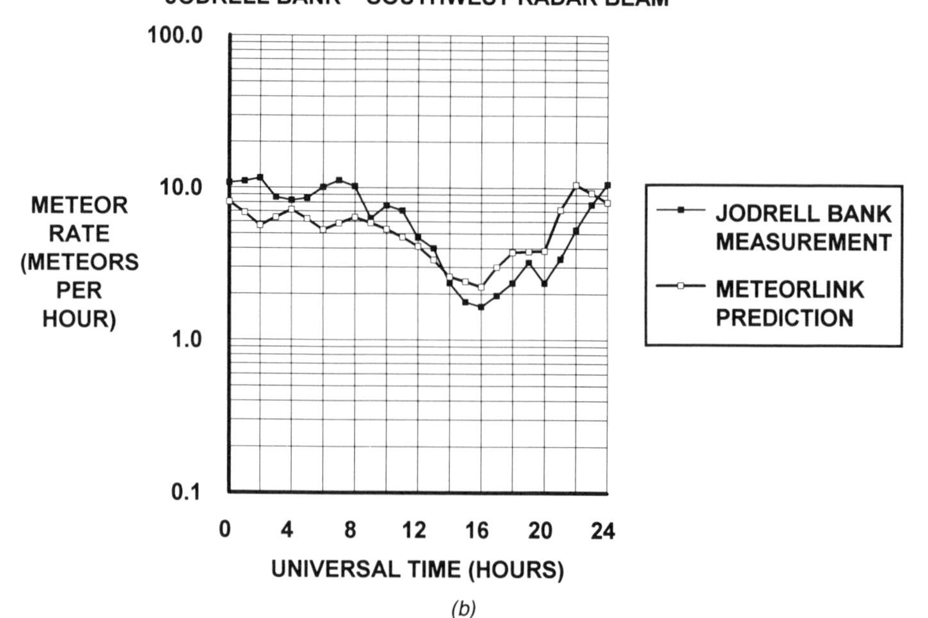

(b)

Figure 3.C.12. Jodrell Bank MR measurement-prediction comparison for December 1950 (*a*) northwest radar beam; (*b*) southwest radar beam.

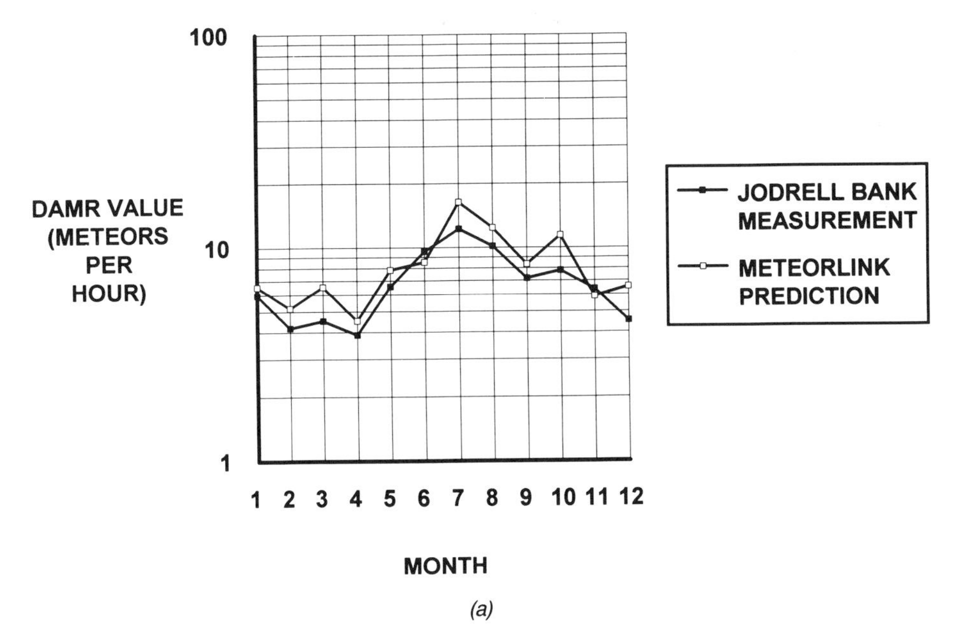

(a)

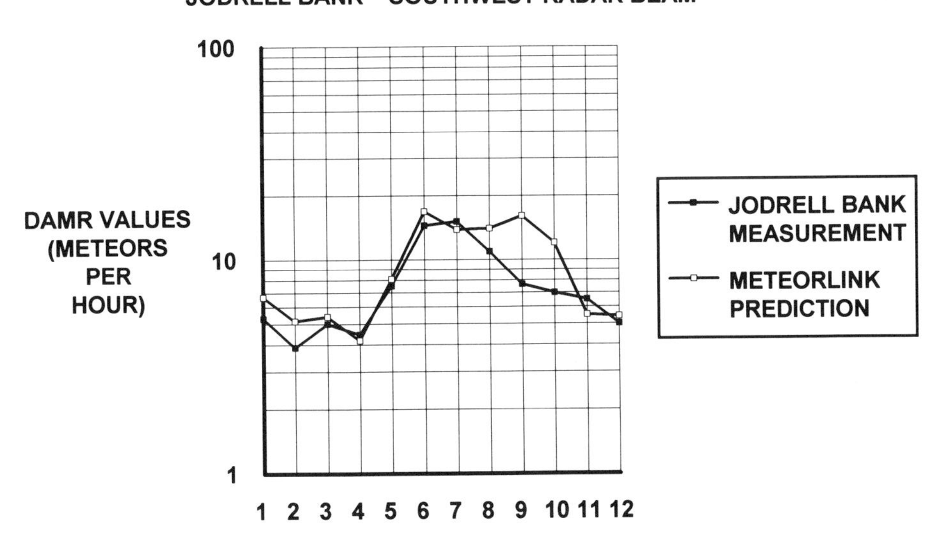

(b)

Figure 3.C.13. Jodrell Bank measurement-prediction comparison of DAMR values (*a*) northwest radar beam; (*b*) southwest radar beam.

GLOSSARY

ACQ. acquire signal in link protocol
A_j. ANTENNA coordinate system at site j, $j = 1, 2$
AMR. array meteor rate
ARQ. automatic repeat request
BER. bit error rate
BMO. Ballistic Missile Office
bps. bits per second
BW. bandwidth
CCITT. International Radio Consultative Committee
DADC. daily average duty cycle
DAMR. daily average meteor rate
DAWT. daily average waiting time
dB. decibels
dBi. decibels relative to an isotropic radiator
dBm. decibels above 1 milliWatt
dBW. decibels above 1 Watt
DC. duty cycle
DCA. Defense Communications Agency, now the Defense Informations Systems Agency (DISA)
DIF. diversity improvement factor
DTA. data block in link protocol
E. east
FDX. full-duplex
FEC. forward error correction
g. grams
GCP. Great Circle Path
G_j. GEOGRAPHIC coordinate system at site j, $j = 1, 2$
GW. Greenwich meridian
h. hour
HAM. Home Amateur Mechanic
HDX. half-duplex
HF. high frequency
HLP. horizontally polarized log periodic antenna
HLPS. dual-stacked horizontally polarized log periodic antenna
HML. hard mobile launcher
HXD. horizontally polarized crossed dipole antenna
I. inertial coordinate system
kbps. kilobits per second
kHz. kiloHertz

km. kilometer
kW. kiloWatt
L. LINK coordinate system
LO. long wavelength overdense trail model
LPA. log periodic antenna
LPD. low probability of detection
LPI. low probability of intercept
LT. local time
LU. long wavelength underdense trail model
m. meter
M. METEOR coordinate system, MB master station
max. maximum
MB. meteor burst
MCC. Meteor Communications Corporation
MHz. megaHertz
MIL-STD. military standard
min. minimum
MLCC. Mobile Launch Control Center
MR. meteor rate
MWT. message waiting time
NEC. Numerical Electromagnetics Code
NEP. north ecliptic pole
NUWC. Naval Undersea Warfare Center
O. ORBITAL coordinate system
P_j. PATTERN coordinate system at site j, $j = 1, 2$
PL. Phillips Laboratory
PRB. probe signal in link protocol
R. remote terminal
RADC. Rome Air Development Center
RO. receive-only
RSL. received signal level
RF. radio frequency
S. SCATTER coordinate system
s. second
SAFB. Sondrestrom Air Force Base
SAIC. Science Applications International Corporation
SEP. south ecliptic pole
SICBM. Small Intercontinental Ballistic Missile
SMOD. sporadic meteor orbital density
SMRD. sporadic meteor radiant density

SN. Sondrestrom-to-Narsarsuaq MB link
SNOTEL. snowpack telemetry system
SNR. signal-to-noise ratio
SO. short wavelength overdense trail model
SPC. special control frame in link protocol
ST. Sondrestrom-to-Thule
SU. short wavelength underdense trail model
T. TERRESTRIAL coordinate system
TAFB. Thule Air Force Base
TO. transmit-only
UDC. underdense duty cycle
UMR. underdense meteor rate
UT. Universal Time
V. TRAIL VECTOR coordinate system
VE. Vernal Equinox
VLP. vertically polarized log periodic antenna
VLPS. dual-stacked vertically polarized log periodic antenna
VWP. vertical whip antenna
W. Watt, west
wpm. words per minute
Y5H. horizontally polarized five-element Yagi antenna
Y12H. horizontally polarized 12-element Yagi antenna

ACKNOWLEDGMENTS

This work represents the efforts of many individuals who have contributed to the author's understanding of meteor burst communications as well as preparation of the manuscript. Joseph Wojtaszek of SAIC provided invaluable analysis and software validation expertise throughout the development of the METEORCOM computer models and edited several drafts of this paper. Steven Merrill of SAIC provided similar expertise and was primarily responsible for the development of the METEORTRAK and METEORDIV computer models from the METEORLINK model. Kelly McDonough, also of SAIC, developed the METEORWAIT computer model. The author thanks Henry Sunkenberg of SAIC for his management support of this effort and Richard Formato of SAIC, who provided invaluable antenna expertise and contributed to the initial meteor burst model development in 1983 that culminated in the METEORCOM models.

Leo Baggerly of TRW provided many valuable insights regarding link, footprint, and diversity model validation. The author thanks Dr. Baggerly for these insights and for his scientific analysis of the fundamental assumptions governing meteor scatter phenomenology and modeling. The author also thanks Dr. Baggerly for his editorial review of the draft manuscript and many qualitative suggestions.

The author thanks Jens Ostergaard of the University of Lowell for many useful discussions regarding his extensive empirical knowledge of meteor scatter and radio propagation in general. This knowledge is derived in part from high-quality measurements performed on the Greenland Meteor Burst Test Bed under the excellent guidance of John Rasmussen and Alan Bailey of the Phillips Laboratory. I also thank Mr. Bailey for his cooperation in seeking release of METEORLINK model validation results presented in Section 3.4.

The author expresses his appreciation to Michael Zugich of the Ballistic Missile Organization and Italo Fantera of the Rome Air Development Center for their leadership of model validation efforts. The author thanks Joseph Katan of the Naval Undersea Warfare Center for the contribution of valuable measured data for model validation. The author thanks Clay Miller of the Boeing Aerospace and Electronics Corporation for his careful evaluation of model accuracy. In addition, the author appreciates the efforts of Mark Rich of SRI International in verifying the description of meteor burst footprint and diversity measurements provided in Section 3.4.

The author thanks Dale Smith, Thomas Donich, Winston Rice, and James Larsen of the Meteor Communications Corporation; Robert Richmond., formerly of Hadron, Inc.; and Ronald Bauman of the Defense Advanced Research Projects Agency for many useful discussions regarding the design and operation of meteor burst terminal equipment and antennas. The author also thanks Mr. Smith and Mr. Rice for correcting details regarding the Phase 2 MB link tests.

Anatoly Goldberg and Cynthia Baird provided several useful translations of important works in meteor astronomy and trail-scatter from the former Soviet Union. Without their unselfish efforts, use of the Kazan and Mogadishu radar measurements of meteor radiant density in the link model as well as the author's understanding of velocity selective effects would not have been possible.

The author also thanks David Meisel of the American Meteor Society for his examination of the manuscript and important technical as well as editorial comments.

The author thanks Donald Schilling for the opportunity to contribute to this book. Without Dr. Schilling's interest in the documentation of meteor burst modeling and analysis techniques, this work would not have been possible.

Finally, the authors thanks his wife Betty and family Danielle, Nicole, and Amanda, for their infinite patients during the many hours of enforced quiet needed to complete this work.

DEDICATION

This work is dedicated to the memory of my father, Robert I. Desourdis, Sr., and to the faith of my mother, Margaret Gerinecz.

REFERENCES

1. D. K. Smith and R. J. Fulthorp, Transport, network, and link layer considerations in medium and large meteor burst communication networks, presented at the *Meteor Burst Communications Symposium*, Shape Technical Centre, The Hague, The Netherlands, November 1987.
2. E. J. Morgan, Meteor burst communications: an update, *Signal*, March 1988.
3. D. K. Smith and T. G. Donich, Variable data rate applications in meteor burst communications, *IEEE Electro '88*, Boston, May 10–12 1988.
4. D. W. R. McKinley, *Meteor Science and Engineering*, pp. 260–265, McGraw-Hill, New York, 1961.
5. V. R. Eshleman, On the wavelength dependence of the information capacity of meteor-burst propagation, p. 1714, *Proc. IRE*, December 1957.
6. G. W. Kronk, *Meteor Showers: A Descriptive Catalog*, Enslow Publishers, Hillside, N. J., 1988
7. A. C. B. Lovell, *Meteor Astronomy*, p. 248, University Press, Oxford, New York, 1954.
8. J. G. Davies, Radio observation of meteors, *Advances in Electronics and Electron Physics*, Vol. 9, pp. 117–123, Academic Press, New York, 1957.
9. J. A. Pupyshev, T. K. Filimonova, and T. V. Kazakova, Maps of the distribution over the entire celestial sphere of the sporadic meteor radiant density, *Radiowave Meteor Propagation*, No. 15, Kazan State U. P., Kazan, Russia, 1980, Document A82–16533, Technical Information Service, American Institute of Aeronautics and Astronautics, (in Russian).
10. O. G. Villard, Jr., V. R. Eshleman, L. A. Manning, and A. M. Peterson, The role of meteors in extended-range VHF propagation, pp. 1473–1481, *Proc. IRE*, October 1955
11. J. G. Davies, *Ibid.*, p. 118.
12. N. J. Rudie, *The Relative Distribution of Observable Meteors in Forward Scatter Meteor Communications*, pp. 77–82, Thesis, Montana State University, Bozeman, 1967.
13. D. W. R. McKinley, *Ibid.*, pp. 112–120.
14. N. T. Svetashkova, Density variations of meteor flux along the earth's orbit, *Middle Atmospheric Program: Handbook for MAP*, Vol. 25, R. G. Roper (ed.), pp. 311–320, Government Printing Office, Washington, DC, August 1987.
15. A. C. B. Lovell, *Ibid.*, pp. 91–92.
16. D. W. R. McKinley, *Ibid.*, p. 34.
17. D. W. R. McKinley, *Ibid.*
18. V. A. Bronshten, *Physics of Meteoric Phenomena*, p. 216, D. Reidel, Boston, 1983.
19. D. W. R. McKinley, *Ibid.*, pp. 190–198.
20. *Ibid.*, pp. 225–229.
21. G. R. Suger, Radio propagation by reflections from meteor trails, *Proc. IEEE*, Vol. 52, pp. 125, February 1964.
22. *Ibid.*, p. 127.

23. N. Herlofson, Plasma resonance in ionospheric irregularities, *Arkiv Fysik*, Vol. 3, pp. 247–297, 1951.
24. V. N. Lebedinetts and A. K. Sosnova, Radio reflections from meteor trails, *Physics and Dynamics of Meteors*, R. Kresak and P. M. Millman, pp. 27–44, eds., D. Reidel, Dordrecht, Holland, 1968.
25. W. Jones and J. Jones, Oblique scattering of radio waves from meteor trains: theory, *Planet. Space Sci.*, Vol. 38, No. 1, pp. 55–66, Pergamon Press, New York, 1990.
26. E. R. Billam and I. C. Browne, Characteristics of radio echoes from meteor trails, IV: polarization effects, *Proc. Phys. Soc.*, Vol. B 69, pp. 98–113, 1956.
27. V. N. Lebedinetts and A. K. Sosnova, *Ibid.*, p. 34.
28. E. M. Poulter and W. J. Baggaley, Radiowave scattering from meteoric ionization, *J. Atmos. Terr. Phys.*, Vol. 39, p. 763, Pergamon Press, 1977.
29. P. S. P. Wei, C. R. Miller and C. K. Martin, Polarization effects in radio wave scattering from meteor bursts, *IEEE MILCOM'89 Conf. Proc.*, Vol. 2, pp. 21.6.1–21.6.7, Boston, October 1989.
30. G. R. Suger, *Ibid.*, pp. 123–124.
31. H. Brysk, Electromagnetic scattering from high-density meteor trails, *IRE Trans. Ant. Prop.*, AP-7, pp. S330-S336, December 1959.
32. D. W. R. McKinley, *Ibid.*, p. 220.
33. J. A. Weitzen, W. P. Birkemeier, and M. D. Grossi, An estimate of the capacity of the meteor burst channel, *IEEE Trans. Comm.*, COM–32, pp. 972–974, August 1984.
34. V. A. Bronshten, *Ibid.*, pp. 225–233.
35. J. Z. Schanker, *Meteor Burst Communications*, p. 102, Artech House, Boston, 1990.
36. J. D. Oetting, An analysis of meteor burst communications for military applications, *IEEE Trans. Comm.*, COM–28, No. 9, p. 1594, September 1980.
37. R. I. Desourdis, Jr. and S. C. Merrill, Meteor burst signal footprint sensitivity to range, power margin and time of day, *IEEE MILCOM'89 Conf Proc.*, Vol. 2, pp. 21.4.1–21.4.5, Boston, October 1989.
38. R. I. Desourdis, Jr. and S. C. Merrill, Spatial diversity improvement in meteor burst communications, unpublished paper, November 1991.
39. R. I. Desourdis, Jr., S. C. Merrill, J. H. Wojtaszek, and K. Hernandez, Meteor burst link performance sensitivity to antenna pattern, power margin, and range, *IEEE MILCOM'88 Conf. Proc.*, Vol. 1, pp. 14.5.1–14.5.7, October 1988.
40. R. I. Desourdis, Jr. and R. Trementozzi, Optimized antenna designs for meteor burst communications, *IEEE MILCOM'88 Conf. Proc.*, pp. 40.2.1–40.2.5, Monterey, CA, October 1990.
41. J. A. Weitzen, Effects of polarization coupling loss mechanism on design of meteor scatter antennas for short and long range communications, *IEEE MILCOM'89 Conf. Proc.*, Vol. 2, pp. 21.5.1–21.5.5, Boston, October 1989.
42. J. A. Weitzen, Predicting the arrival of meteors useful for communications, *Radio Sci.*, Vol. 21, No. 6, pp. 1009–1010, December 1986.
43. F. J. Sites, Communication via meteor trails, *23rd Symposium on Aspects EM Wave Scattering on Radio Communication Conf. Proc.*, pp. 4.5.1–4.5.11, October 1977.

44. *Technical Reference Manual and User's Guide for the Meteor Burst Link Program (BLINK)*, Defense Communications Agency: Center for Command and Control, and Communication Systems, Arlington, Virginia, January 1986.
45. J. A. Weitzen, *Ibid.*, pp. 1012–1013.
46. D. W. R. McKinley, *Ibid.*, pp. 187–235.
47. D. W. Brown, A physical meteor-burst propagation model and some significant results for communication system design, *IEEE J. Sel. Areas Comm.*, SAC–3, No. 5, September 1985.
48. J. A. Weitzen, *Ibid.*, p. 1014.
49. J. D. Oetting, *Ibid.*, pp. 1598–1599.
50. J. Z. Schanker, *Ibid.*, p. 116.
51. R. I. Desourdis, Jr., V. V. Sidorov, A. V. Karpov, R. G. Huziashev, L. A. Epictetov, and D. W. Brown, A Russian meteor burst communications experiment and measurement-prediction comparison, *IEEE MILCOM'92 Conf. Proc.*, Vol. 1, pp. 1.6.1–1.6.5, San Diego, October 1992.
52. J. C. Ostergaard, J. E. Rasmussen, M. J. Sowa, J. M. Quinn, and P. A. Kossey, *The RADC High Latitude Meteor Scatter Test Bed*, RADC-TR–86-74 ADA180550, July 1986, Unlimited Distribution.
53. Defense Communications Agency: Center for Command and Control, and Communication Systems, *Ibid.*
54. J. Z. Schanker, *Ibid.*, pp. 109–111.
55. H. Alfvén and G. Arrhenius, Interplantary and transplanetary condensation, Ch. 19, *Evolution of the Solar System*, ASA SP; 345, Government Printing Office, Washington, DC, 1976
56. J. G. Davies, *Ibid.*, pp. 95–128.
57. J. A. Pupyshev, T. K. Filimonova, and T. V. Kazakova, *Ibid.*
58. P. Pecina, Sporadic flux determination from radar observations, *Bull. Astron. Inst. Czechoslovakia*, Vol. 35, No. 1, pp. 5–14, 1984.
59. L. M. Poole and D. G. Roux, Meteor radiant mapping with an all-sky radar, *Mon. Not. Roy. Astron. Soc.*, Vol. 236, pp. 645–652, 1989.
60. M. R. Owen, VHF meteor scatter-an astronomical perspective, *QST*, Vol. 65, pp. 17–24, June 1986.
61. N. T. Svetashkova, *Ibid.*
62. P. R. Escobal, *Methods of Orbit Determination*, p. 2, Wiley, New York, 1965.
63. L. Triskova, A note on the sporadic meteor radiant distribution, *Physics and Dynamics of Meteors*, R. Kresak and P. M. Millman, eds., pp. 304–307, D. Reidel, Dordrecht, Holland, 1968.
64. A. C. B. Lovell, *Ibid.*, pp. 112–122.
65. N. J. Rudie, *Ibid.*
66. J. A. Pupyshev, T. K. Filimonova, and T. V. Kazakova, *Ibid.*
67. O. I. Belkovic and J. A. Pupyshev, The variation of sporadic meteor radiant density and the mass law exponent over the Celestial Sphere, *Physics and Dynamics of Meteors*, R. Kresak and P. M. Millman, eds., pp. 373–381, D. Reidel, Dordrecht, Holland, 1968.
68. J. A. Pupyshev, Methods of statistical examination of varying sporadic meteor burst radiant density (SMRD) over the celestial sphere, *Radiowave Meteor Propagation*, Kazan State U. P., Vol. 15, Kazan, Russia, 1980, Technical Information Service, American Institute of Aeronautics and Astronautics, Document A82–16532 (in Russian).

69. N. J. Rudie, *Ibid.*, p. 77.
70. J. A. Pupyshev, *Ibid.*, p. 16.
71. J. G. Davies, Ibid., p. 102.
72. Y. I. Voloschuk and B. L. Kashcheyev, *Distribution of Meteor Bodies in the Neighborhood of the Earth's Orbit*, Nauka, Moscow, 1981 (in Russian).
73. A. C. B. Lovell, *Ibid.*, pp. 112–115.
74. T. R. Kaiser (ed.), The incident flux of meteors and the total meteoric ionization, *Meteors*, Spec. Supp. J. Atmos. Terr. Phys., Vol. II, p. 119, Pergamon Press, London, 1955.
75. P. Pecina, Meteor shower flux determination from radar observations, *Bull. Astron. Inst. Czechoslovakia*, Vol. 33, No. 1, p. 2, 1982.
76. A. C. B. Lovell, *Ibid.*, p. 141.
77. J. A. Pupyshev, T. K. Filimonova, and T. V. Kazakova, *Ibid.*, p. 27.
78. V. R. Eshleman, The theoretical length distribution of ionized meteor trails, *J. Atmos. Terr. Phys.*, Vol. 10, No. 7, Pergamon Press, London, 1957.
79. Committee on Extension to the Standard Atmosphere (COESA), *U.S. Standard Atmosphere, 1976*, p. 16, Government Printing Office, Washington, D.C., October 1976.
80. V. A. Bronshten, *Ibid.*, p. 213.
81. P. Pecina, On the determination of the mass distribution index from radar observations, *Bull. Astron. Inst. Czechoslovakia*, Vol. 35, No. 3, p. 183, 1984.
82. V. A. Bronshten, *Ibid.*, p. 217.
83. COESA, *Ibid.*, p. 15.
84. D. W. R. McKinley, *Ibid.*, p. 215.
85. *Ibid.*, p. 135.
86. V. V. Andreev and O. I. Belkovich, Models of sporadic meteor body distributions, *Middle Atmospheric Program: Handbook for MAP*, Vol. 25, R. G. Roper (ed.), pp. 298–304, Government Printing Office, Washington, D.C., August 1987.
87. P. Pecina, Sporadic flux determination from radar observations, *Ibid.*, p. 8.
88. A. C. B. Lovell, *Ibid.*, p. 93.
89. A. C. B. Lovell, *Ibid.*, pp. 221–222.
90. A. Papoulis, *Probability, Random Variables and Stochastic Processes*, p. 126, McGraw-Hill, New York.
91. J. A. Pupyshev, T. K. Filimonova, and T. V. Kazakova, *Ibid.*, p. 29–40.
92. *Ibid.*, p. 24.
93. W. J. Baggaley, The mass distribution of large meteoroids, *Mon. Not. Roy. Astron. Soc.*, Vol. 180, p. 91, 1977.
94. M. L. Meeks and J. C. James, On the influence of meteor-radiant distributions in meteor-scatter communication, *Proc. IRE*, Vol. 45, p. 1725, December 1957.
95. J. C. Ostergaard, J. E. Rasmussen, M. J. Sowa, J. M. Quinn, and P. A. Kossey, *Ibid.*
96. G. J. Burke, et al., *Numerical Electromagnetics Code (NEC)-Method of Moments*, Lawrence Livermore Lab., Rep. UCID 18834; Naval Ocean Systems Center, Rep. NOSC TD 116; reprint ed., NTIS, Springfield, VA, January 1981.
97. N. Herlofson, *Ibid.*, pp. 283–292.

98. E. R. Billam and I. C. Browne, Characteristics of radio echoes from meteor trails, IV: polarization effects, *Proc. Phys. Soc.*, B 69, p. 110., 1956.
99. D. W. R. McKinley, *Ibid.*, p. 211.
100. E. M. Poulter and W. J. Baggaley, *Ibid.*, p. 761.
101. V. N. Lebedinetts and A. K. Sosnova, *Ibid.*, p. 27–44.
102. W. Jones and J. Jones, *Ibid.*
103. H. Brysk and M. L. Buchanan, Scattering by a cylindrical Gaussian potential: exact solution, Can. J. Phys., Vol. 43, pp. 28–37, January 1965.
104. J. Jones and J. G. Collins, On the validity of certain approximations in radio meteor echo theory, *Mon. Not. Roy. Astron. Soc.*, Vol. 168, pp. 433–449, 1974.
105. A. V. Karpov and V. V. Sidorov, The calculation of radiowave meteor propagation parameters by the Monte-Carlo method for meteor links of arbitrary distance, *Radiowave Meteor Propagation*, No. 15, Kazan State U. P., Kazan, Russia, 1980 (in Russian).
106. R. G. Huziashev, A calculation of signal amplitude-phase characteristics of oblique scattering from meteor trails, *News from the Higher Schools*, Vol. 27, No. 9, pp. 1110–1115, 1984.
107. V. R. Eshleman, Meteor scatter, *The Radio Noise Spectrum*, Chap. 4, p. 62, Harvard U. P., Cambridge, MA, 1960.
108. N. J. Rudie, *Ibid.*, pp. 127–128.
109. COESA, *Ibid.*, p. 17.
110. V. A. Bronshten, *Ibid.*, p. 152.
111. *Ibid.*, p. 217.
112. COESA, *Ibid.*, Fig. 10, p. 17.
113. V. A. Bronshten, *Ibid.*, p. 225.
114. D. W. R. McKinley, *Ibid.*, p. 199.
115. J. S. Greenhow and E. L. Neufeld, The diffusion of ionized meteor trails in the upper atmosphere, *J. Atmos. Terr. Phys.*, Vol. 6, pp. 133–140, 1955.
116. E. L. Murray, Ambipolar diffusion of a meteor trail and its relation with height, *Planet. Space Sci.*, Vol. 1, pp. 125–129, 1959.
117. E. M. Poulter and W. J. Baggaley, The applications of radio-wave scattering theory to radio-meteor observations, *Plan. Space Sci.*, Vol. 26, p. 972, 1978.
118. G. R. Suger, *Ibid.*, p. 125.
119. D. W. R. McKinley, *Ibid.*, p. 211.
120. *Ibid.*, pp. 212–213.
121. H. Brysk, *Ibid.*, pp. S330–S331.
122. C. O. Hines and P. A. Forsyth, The forward scattering of radio waves from overdense meteor trails, *Can. J. Phys.*, Vol. 35, pp. 1033–1041, 1957.
123. D. W. R. McKinley, *Ibid.*, pp. 214–219.
124. G. R. Suger, *Ibid.*, pp. 123–124.
125. D. W. R. McKinley, *Ibid.*, p. 242.
126. G. R. Suger, *Ibid.*, p. 126.
127. L. A. Manning, Oblique echoes from over-dense meteor trails, *J. Atmos. Terr. Phys.*, Vol. 14, pp. 82–93, April 1959.
128. D. W. R. McKinley, *Ibid.*, pp. 242–245.
129. *Ibid.*, p. 244.

130. V. A. Bronshten, *Ibid.*
131. D. W. R. McKinley, *Ibid.*, pp. 217, 225.
132. V.A. Bronshten, *Ibid.*, p. 228.
133. *Ibid.*, pp. 225, 233.
134. D. W. R. McKinley, *Ibid.*, pp. 229
135. V. A. Bronshten, *Ibid.*, p. 233.
136. *Ibid.*, p. 231.
137. *Ibid.*, p. 232.
138. J. C. Ostergaard, J. E. Rasmussen, M. J. Sowa, J. M. Quinn, and P. A. Kossey, *Ibid.*
139. K. Simmons (ed.), *Meteor News*, No. 93, p. 4, W. L. Simmons (pub.), Callahan, FL, April 1991.
140. P. A. Forsyth, D. R. Hansen, and C. O. Hines, The principles of Janet-a meteor-burst communication system, *Proc. IRE*, p. 1646, December 1957.
141. B. A. Lindblad, Meteor radar rates and the solar cycle, *Nature*, Vol. 259, pp. 99–101, January 1976.
142. P. Prikryl, Meteor radar rates and solar activity, *Bull. Astron. Inst. Czechoslovakia*, Vol. 34, No. 1, pp. 44–50, 1983.
143. M. W. Browne, Radio system uses fiery meteor trails to transmit data, *The New York Times*, p. B8, August 22 1989.
144. R. V. Harper, Performance analysis for adaptive meteor burst communications, *Effects of the Ionosphere on C3I Systems*, J. M. Goodman (ed.), NTIS 85-600558, Alexandria, VA, May 1984.
145. R. I. Desourdis, Jr. and R. Trementozzi, *Ibid.*
146. V. R. Eshleman and R. F. Mlodnosky, Directional characteristics of meteor propagation derived from radar measurements, *Proc. IRE*, pp. 1715–1723, December 1957.
147. *Ibid.*, p. 1722.
148. J. A. Weitzen, Communicating via meteor burst at short ranges, *IEEE Trans. Comm.*, COM-35, No. 11, pp. 1217–1221, November 1987.
149. R. A. Monzingo and T. W. Miller, *Introduction to Adaptive Arrays*, p. 5, Wiley, New York, 1980.
150. V. R. Eshleman and R. F. Mlodnosky, *Ibid.*, pp. 1722–1723.
151. J. D. Larsen, R. S. Mawrey, and J. A. Weitzen, The use of antenna beam steering to improve the performance of meteor burst communication systems, *IEEE MILCOM'92 Conf. Proc.*, Vol. 1, pp. 1.3.1–1.3.6, San Diego, October 1992.
152. J. O. Oetting, *Ibid.*, p. 1594.
153. M. R. Owen, The great sporadic-E opening of June 14, 1987, *QST*, pp. 21–29, May 1988.
154. D. W. R. McKinley, *Ibid.*, p. 229.
155. L. A. Manning, Air motions and the fading, diversity and aspect sensitivity of meteoric echoes, *J. Geophys. Res.*, Vol. 64, pp. 1424, 1959.
156. R. I. Desourdis, Jr. and S. C. Merrill, Meteor burst signal footprint sensitivity to range, power margin and time of day, *Ibid.*
157. W. T. Ralston and J. A. Weitzen, Network waiting time for meteor-burst

communications, *IEEE MILCOM'92 Conf. Proc.*, Vol. 1, pp. 1.1.1–1.1.5, San Diego, October 1992.

158. M. Schwartz, W. R. Bennet, and S. Stein, *Communication Systems and Techniques*, p. 418, McGraw-Hill, New York, 1966.
159. *Ibid.*, 422, 423.
160. J. D. Oetting, *Ibid.*, p. 1598.
161. J. Bartholome and I. M. Vogt, COMET-A new meteor-burst system incorporating ARQ and diversity reception, *IEEE Trans. Comm. Tech.*, Vol. COM-16, No. 2, April 1968.
162. D. K. Smith and T. G. Donich, *Ibid.*
163. F. S. Hillier and G. J. Lieberman, *Operations Research*, pp. 635–640, Holden-Day, San Francisco, 1974.
164. A. K. McGurl, Performance analysis of meteor burst communication networks, *IEEE MILCOM'89 Conf. Proc.*, Vol. 2, pp. 21.1.1–21.1.5, Boston, October 1989.
165. J. Z. Schanker, *Ibid.*, pp. 137–148.
166. U. D. Black, *Data Communications Networks and Distributed Processing*, pp. 235–240, Reston, Englewood Cliffs, NJ, 1983.
167. T. L. Fox, *AX.25 Amateur Packet-Radio Link-Layer Protocol*, Version 2.0, American Radio Relay League, Newington, CT, October 1984.
168. S. C. Merrill, S. W. Symes, A. K. McDonough, and R. I. Desourdis, Jr., A low-cost multi-media radio system, *IEEE MILCOM'91 Conf. Proc.*, Vol. 3, pp. 41.1.1–41.1.5, McLean, Virginia, November 1991.
169. National Communications System Office of Technology and Standards, *Proposed Federal Standard 1055, Telecommunications Interoperability Requirements for Meteor Burst Communications*, Section 4.6, Washington, DC, June 1989.
170. L. Kleinrock, *Queuing Systems, Vol. 2, Computer Applications*, pp. 333–334, Wiley, New York, 1976.
171. A. Papoulis, *Ibid.*, p. 101.
172. Defense Communications Agency, Command and Control Technical Center, *Analysis and Demonstration of Airborne Meteor Burst Communications for MEECN*, p. 5–4, Defense Technical Information Center, Alexandria, VA, July 1981.
173. Science Applications International Corporation, *Meteor Burst Software Simulation for SICBM*, Stow, MA, October 1989.
174. A. Papoulis, *Ibid.*, p. 162.
175. R. I. Desourdis, Jr., V. V. Sidorov, A. V. Karpov, R. G. Huziashev, L. A. Epictetov, and D. W. Brown, *Ibid.*
176. J. D. Oetting, *Ibid.*
177. L. A. Manning and V. R. Eshleman, Meteors in the ionosphere, *Proc. IRE*, Vol. 47, p. 192, February 1959.
178. M. J. Miles, *Sample Size and Precision in Communication Performance Measurements*, NTIA Report 84–153, p. 60, U.S. Dept. of Commerce, Institute for Telecommunication Sciences, August 1984.
179. *Ibid.*, p. 19.

180. J. Z. Schanker, *Ibid.*, p. 25.
181. D. K. Smith and R. J. Fulthorp, *Ibid.*
182. R. I. Desourdis, Jr. and S. C. Merrill, *Ibid.*
183. J. C. Ostergaard, J. E. Rasmussen, M. J. Sowa, J. M. Quinn, and P. A. Kossey, *Ibid.*
184. A. C. B. Lovell, *Ibid.*, pp. 112–115.
185. A. F. Cook, A working list of meteor streams, *Smithsonian Contrib. Astrophys.*, pp. 183–191, 1973.
186. Defense Communications Agency: Center for Command and Control, and Communication Systems, *Ibid.*, pp. 1–5
187. J. C. Ostergaard, J. E. Rasmussen, M. J. Sowa, J. M. Quinn, and P. A. Kossey, *Ibid.*
188. R. I. Desourdis, Jr., J. C. Ostergaard, and A. D. Bailey, Meteor burst computer model validation using high-latitude measurements, *IEEE MILCOM'91 Conf. Proc.*, Vol. 2, pp. 22.1.1–22.1.5, McLean, Virginia, November 1991.
189. J. A. Weitzen, A data base approach to analysis of meteor burst data, *Radio Sci.*, Vol. 22, pp. 133–140, January 1987.
190. E. R. Billam and I. C. Browne, *Ibid.*, p. 110.
191. J. A. Weitzen, Communicating via meteor burst at short ranges, *Ibid.*
192. J. G. Davies, *Ibid.*, p. 120.
193. Y. I. Voloschuk and B. L. Kashcheyev, *Ibid.*
194. B. A. Lindblad, *Ibid.*
195. P. S. Cannon, Polarization rotation in meteor burst communication systems, *Radio Sci.*, Vol. 21, No. 3, pp. 501–510, May 1986.
196. Meteor Communications Corporation, *Small Intercontinental Ballistic Missile (SICBM) Meteor Burst Communication System (MBCS) Phase 2 Performance Evaluation Test Report*, Kent, Washington, January 1989.
197. Science Applications International Corporation, *Communications Engineering Laboratory, Meteor Burst Communications Simulation Software: Final Report*, Stow, MA, October 1989.
198. T. Bennett and J. E. MacCarthy, *MBC Phase 2 Data Reduction*, Transmittal IOC F523.JEM.90-010, TRW Space & Technology Group, Ballistic Missiles Division, San Bernardino, CA, October 1990.
199. L. L. Baggerly, *Antenna Noise in MBC Phase 2 Test*, F523.LLB.89–117, TRW Defense Systems Group, Ballistic Missile Division, San Bernardino, CA, November 1989.
200. Meteor Communications Corporation, *Ibid.*
201. G. W. Kronk, *Ibid.*, pp. 162–169.
202. L. L. Baggerly, *Ibid.*, p. 2.
203. A. C. B. Lovell, *Ibid.*, pp. 112–115.
204. A. Aspinall, J. A. Clegg, and G. S. Hawkins, A radio echo apparatus for the delineation of meteor radiants, *Phil. Mag.*, Vol. 7, No. 42, pp. 504–514, May 1951.
205. *Ibid.*, p. 506.

206. A. C. B. Lovell, *Ibid.*, p. 112.
207. A. Aspinall, J. A. Clegg, and G. S. Hawkins, *Ibid.*, p. 511.
208. A. C. B. Lovell, *Ibid.*, p. 114.
209. Y. I. Voloschuk and B. L. Kashcheyev, *Ibid.*, pp. 10–15.
210. A. Aspinall, J. A. Clegg, and G. S. Hawkins, *Ibid.*, p. 506.
211. J. C. Ostergaard, University of Lowell, personal communication.
212. P. M. Heilman, M. M. Murray, M. J. Rich, and B. Yetso, *Final Technical Report: Meteor Burst and Space Diversity Testing-Phase 3*, SRI Project No. 6361, SRI International, July 1989.
213. *Ibid.*, pp. 38–41.
214. *Ibid.*, pp. 40–42.
215. *Ibid.*, pp. 12–18.
216. T. Bennett and J. E. MacCarthy, *MBC Phase 3 Data Reduction*, Transmittal Letter F523.JEM.90-0014, TRW Space & Technology Group, Ballistic Missiles Division, San Bernardino, CA, October 1990.
217. *Ibid.*, pp. 26, 39.
218. R. I. Desourdis, Jr., S. C. Merrill, J. H. Wojtaszek, and K. Hernandez, *Ibid.*, p. 14.5.4.
219. *Ibid.*, p. 14.5.3.
220. Y. I. Voloschuk and B. L. Kashcheyev, *Ibid.*, p. 21.
221. J. A. Pupyshev, T. K. Filimonova, and T. V. Kazakova, *Ibid.*, pp. 23–26.
222. J. A. Pupyshev, Methods of statistical examination of varying sporadic meteor burst radiant density (SMRD) over the celestial sphere, *Ibid.*
223. P. Prikryl, *Ibid.*
224. B. A. Lindblad, *Ibid.*
225. J. C. Ostergaard, J. E. Rasmussen, M. J. Sowa, J. M. Quinn, and P. A. Kossey, *Ibid.*
226. R. I. Desourdis, Jr., V. V. Sidorov, A. V. Karpov, R. G. Huziashev, L. A. Epictetov, and D. W. Brown, *Ibid.*
227. O. I. Belkovic and J. A. Pupyshev, *Ibid.*
228. P. Pecina, *Ibid..* pp. 7–8.
229. V. V. Sidorov, personnel communication, July, 1992.
230. W. Jones and J. Jones, *Ibid.*
231. E. R. Billam and I. C. Brown, *Ibid.*, p. 110.
232. A. V. Karpov and V. V. Sidorov, *Ibid.*
233. V. A. Bronshten, *Ibid.*, p. 266.
234. COESA, *Ibid.*, p. 13.
235. J. A. Pupyshev, T. K. Filimonova, and T. V. Kazakova, *Ibid.*
236. N. T. Svetashkova, *Ibid.*
237. G. J. Burke, et al., *Ibid.*
238. J. C. Ostergaard, J. E. Rasmussen, M. J. Sowa, J. M. Quinn, and P. A. Kossey, *Ibid.*
239. A. Aspinall, J. A. Clegg, and G. S. Hawkins, *Ibid.*
240. A. C. B. Lovell, *Ibid.*, pp. 112–115.

4

EFFICIENT COMMUNICATIONS USING THE METEOR BURST CHANNEL

Sheldon Chang
Department of Electrical Engineering, State University of New York, Stony Brook, New York

Donald L. Schilling
Department of Electrical Engineering, City College of New York, New York

The research described in this chapter was supported by SCS Telecom, Inc. under Contracts with NSF and the U.S. Army, NSF ISI-866079, NSF DMC-8702465, and U.S. Army DAAB07-84-C-D039.

4.1 INTRODUCTION

As the worldwide demand for communications increases, there is an urgent need for new means of communications that would avoid the congestion that currently exists in conventional channels. This congestion is a particular problem at the lower frequencies often used in beyond-line-of-sight (BLOS) communications systems. There are also performance limitations associated with many existing BLOS systems that compound the congestion problem. HF systems, for example, are quite sensitive to solar disturbances and other galactic phenomena and are often limited by degradations due to multipath return and other ground and atmospheric conditions.

Meteor burst channels provide a relatively new and uncongested means of communications that offer a significant opportunity for overcoming many of the limitations of existing BLOS systems. The meteor burst channel operates in the relatively unused lower portion of the very high frequency

(VHF) band ranging from 30 to 100 MHz and at path lengths up to 1500 miles. It therefore avoids the degradations exhibited by HF due to noise on the low end of the HF spectrum and ionospheric phenomena at the high end. This region of the spectrum also provides a sufficiently large bandwidth for efficient data communication. A meteor burst channel is not easily destroyed and has an inherent privacy feature due to its limited footprint [1, 9].

Early uses of meteor burst channels, however, have been restricted to relatively low throughput due to the random nature of the meteor burst phenomena. Indeed, typical systems operate at bit rates of 8 kb/s and less and use modulation techniques, such as BPSK or QPSK. It should be noted that the bit rate of 8 kb/s occurs during the time that the channel is present and that the average bit rate that determines throughput is often significantly less than this rate. Rate $-1/2$ codes are typical and such code rates also limit the information transfer.

The research presented in this Chapter describes an innovative technique to enhance the performance of meteor burst communications. We call the technique the Feedback Adaptive Variable Rate (FAVR) system. Using this system approach, a feedback channel is maintained that allows the transmitted bit rate to mimic the time behavior of the received power so as to maintain a constant bit energy; that is, the bit duration varies in a reciprocal manner to the received power variation so as to maintain a constant received bit energy. This results in a constant probability of bit error in each transmitted bit.

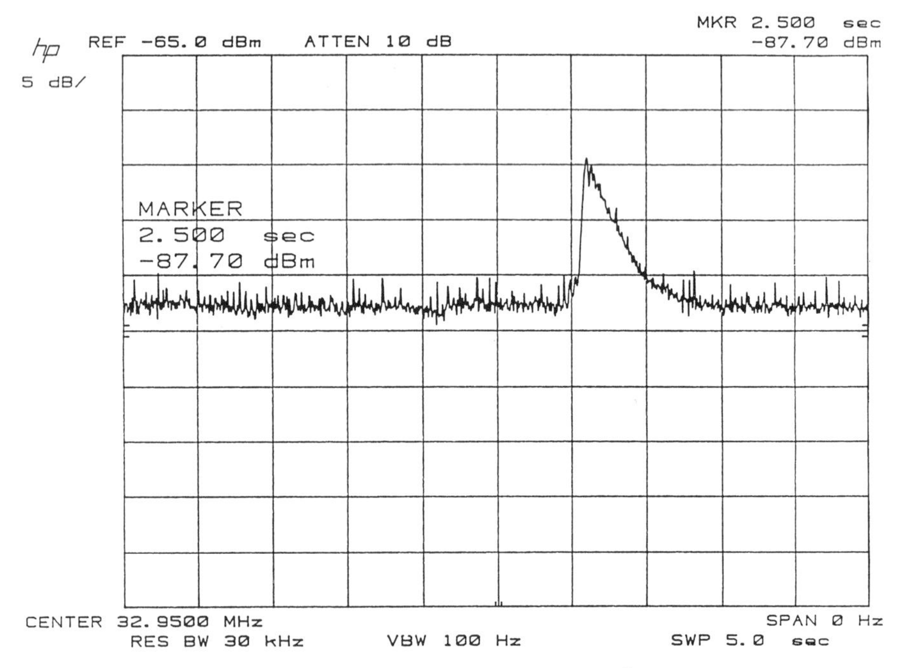

Figure 4.1. An underdense meteor trail.

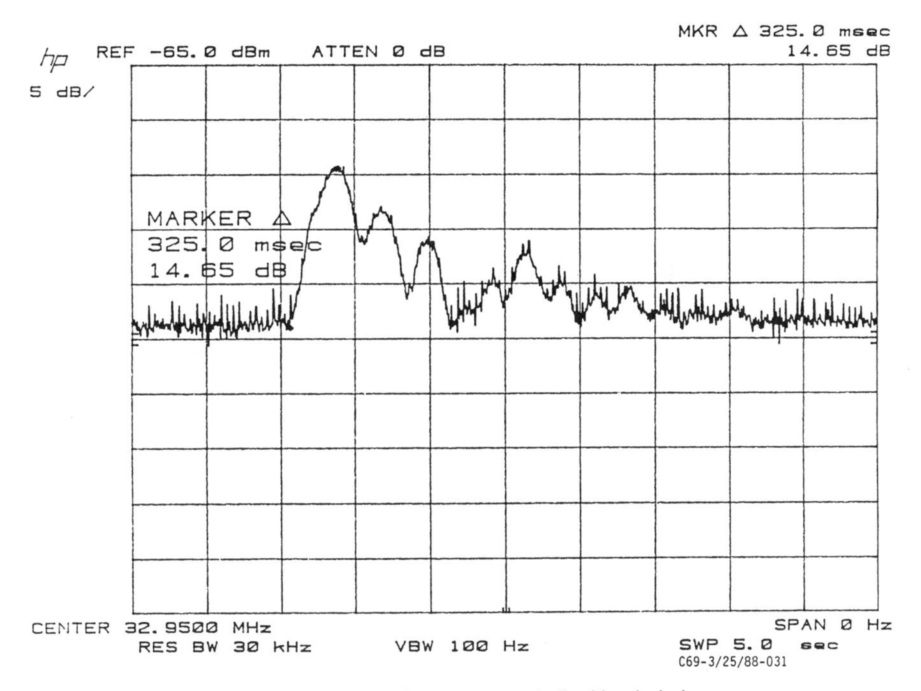

Figure 4.2. An overdense meteor trail with wind shear.

4.2 EXPERIMENTALLY DETERMINED CHANNEL CHARACTERISTICS

The meteor burst channel is a noncontinuous channel. When a channel is present the power received varies with time. Further, the duration of the channel, the peak and rms signal strengths, and the time between channels are random processes. A typical underdense waveform measured by SCS Telecom is shown in Fig. 4.1. Note that the peak SNR is about 12 dB and that the SNR decreases exponentially with time. The channel is seen to be present for about 0.4 s. Figure 4.2 shows an overdense trail with wind shear. Note that the peak SNR is about 15 dB and it last about 1 s.

These are but two of numerous waveforms that are obtained in practice [10]. Hence, to communicate efficiently using such a channel requires a robust modulation technique: FAVR.

4.3 METEOR BURST COMMUNICATION SYSTEM ALGORITHMS

4.3.1 Introduction

To demonstrate the operation of FAVR we employ the underdense meteor burst channel, since this type of channel is usually of short time duration and low received SNR. The underdense trails are modeled as

$$P_r(t) = P_o e^{-t/\tau} \tag{4.1}$$

where P_o is the power received at the start of the trail and τ is the time constant of the trail. Note that P_o and τ are random variables that differ from trail to trail. Their actual statistics are unimportant to the FAVR system.

4.3.2 The Constant Bit Rate System

For a constant bit rate system, the duration of each bit T_b is a constant. The usable time duration T_T for a meteor burst with peak power P_o and decay rate τ is determined from Eq. 4.1 as

$$P_K = P_o e^{-T_T/\tau} \tag{4.2}$$

where the minimum acceptable received power P_K is selected so that

$$P_K T_b = E_b \tag{4.3}$$

the energy required for a specified error rate. Note that if $P_o \gg P_K$, many bits are transmitted with a lower probability of error than that required.

The usable time duration is obtained from Eq. 4.2:

$$T_T = \tau \ln \frac{P_o}{P_K} \tag{4.4}$$

The number of bits N_c transmitted in the underdense trail for a constant rate system is given by

$$N_c = \frac{T_T}{T_b} = \frac{\tau}{T_b} \ln \frac{P_o}{P_K} = \frac{P_K \tau}{E_b} \ln \frac{P_o}{P_K} = \frac{P_o \tau}{E_b} \left(\frac{P_K}{P_o}\right) \ln \frac{P_o}{P_K} \tag{4.5}$$

4.3.3 The Optimum Communication System

We now define the optimum communication system for transmission, using the underdense channel characterized by Eq. 4.1, in the sense of maximizing the number of bits transmitted with a desired error rate. Such a system changes the duration of each bit transmitted during the trail in such a way as to maintain the energy of each bit, constant and equal to E_b. Then, from Eq. 4.1 the maximum energy in the meteor trail is

$$E_T = \int_0^\infty P_o e^{-t/\tau}\, dt = P_o \tau \tag{4.6}$$

Hence, the maximum number of bits having the energy E_b that can be transmitted over the trail is

$$N_o = \frac{E_T}{E_b} = \frac{P_o \tau}{E_b} \tag{4.7a}$$

and the bit durations can readily be shown to be

$$T_{bi} = -\tau \ln\left[1 - \frac{\Pi_{k=1}^{i-1} \exp(T_{bk}/\tau)}{N_o}\right] \qquad i > 1 \tag{4.7b}$$

4.3.4 ARQ System

In an ARQ system, a packet of N bits is transmitted during each meteor burst. As an approximation, we assume that the packet is successfully transmitted if the last bit is transmitted at the acceptable level P_K as defined by Eq. 4.3. Hence, the required peak power level is

$$P_R = P_K e^{NT_b/\tau} \tag{4.8a}$$

The transmitted bits in a trail are

$$N_Q = \begin{cases} N & \text{if the actual peak power level } P_o \geq P_R \\ 0 & \text{if } P_o < P_R \end{cases} \tag{4.8b}$$

If $N_Q = 0$, the entire packet of N bits is retransmitted.

Allowing transmission of multiple packets can improve the performance of the ARQ system. However, the improvement for normal sized packets is small, due to the low probability of such occurrences.

4.3.5 FAVR

The Feedback Adaptive Variable Rate (FAVR) system is a practical embodiment of the optimum system. Three important technical problems must be solved as prerequisite conditions for realizing the variable bit rate concept: (1) a variable bandwidth filter or the equivalent of it, (2) accurate and timely determination of the signal-to-noise ratio, and (3) communication of the bit rate information. The three problems are solved in the FAVR system by using a receiver with a simple analog circuit followed by digital processing of the analog signals. The variable bandwidth filter problem is solved by summing a variable number of terms of the analog signals. Signal power and noise power may be determined separately by using Kalman's optimum filtering and prediction algorithm. Changes in bit rate need not be explicitly communicated, but can be determined by the receiver digital circuit.

Cost is an important consideration in FAVR design. Once developed, digital processing is reliable and least expensive. Within the capabilities of the processors used, increased complexity of computation does not add to the cost. For this reason, the analog part of FAVR is kept at as simple a level as possible, and all the processing is done digitally. The inputs to the digital circuit are sine and cosine demodulated signal chips of duration T_1. The durations of the transmitted bits are integral multiples of T_1, or $n_b T_1$. The local oscillators for demodulation are not synchronized to the transmitter oscillators. However, their frequencies are assumed to be sufficiently close to the latter. For BPSK, an optimum phase estimate is provided digitally.

4.3.5.1 Signal Composition.

Figure 4.3 shows the received signal power. During the time interval T_i the transmitted bits will all be of the same bit rate f_{bi} having a bit duration T_{bi}. Referring to Fig. 4.4 we see that each bit is assumed to consist of a number of incremental *chips* n_{bi} of fixed duration τ_1, so that

$$T_{bi} = n_{bi}\tau_1 \tag{4.9}$$

We further assume that the bits are collected into *packet segments*. Each packet segment has a segment number appended so that if an error occurs in a segment the segment can be repeated.

We take cognizance of the fact that reciprocity exists in the channel. Thus, the signal power seen at the transmitter and receiver is to all intents and purposes the same. The above assumption is crucial to FAVR operation (just as Link Quality Analysis (LQA) is crucial to adaptive HF). The channel delay, which affects reciprocity, is only 6 ms and is assumed to be negligible. Of far greater importance is the fact that the noise level at transmitter and receiver is different and non-Gaussian external noise is often the main reason for the difference. In this paper the authors have assumed that noise cancellation, whether by null steering or other means, has been used and that the only noise present in each receiver is normal Gaussian noise.

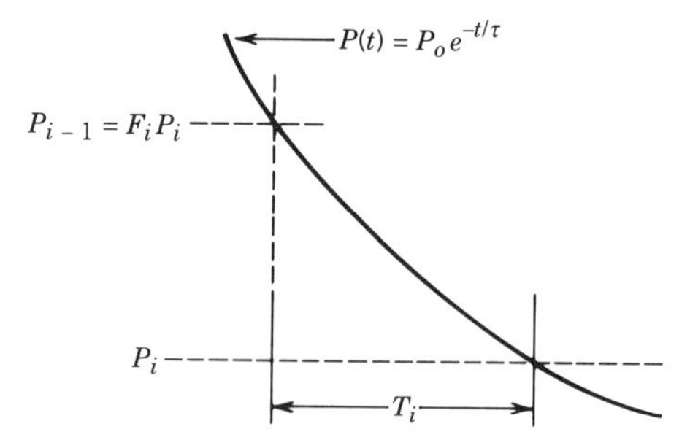

Figure 4.3. Received signal power curve showing the time interval T_i during which each bit has the same duration T_{bi}.

During the time interval T_i the transmitter and receiver each see the signal power decrease from P_{i-1} to P_i. For ease of calculation let

$$P_{i-1} = F_i P_i \tag{4.10}$$

where F_i is a factor greater the unity. For example, we see that by letting $F_i = 2$ for all i means that the transmitter decreases its bit rate whenever the signal power decreases by 3 dB.

To insure that a specified probability of error is achieved, each transmitted bit must contain an energy greater than or equal to some minimum value, say E_b. Then, in the interval T_i (see Fig. 4.3)

$$E_b = P_i T_{bi} \tag{4.11}$$

Then, since

$$P(t) = P_o e^{-t/\tau} \tag{4.12}$$

where P_o is the maximum received power from the trail, we have, using Eq. 4.10,

$$P_i = F_i P_i e^{-T_i/\tau} \tag{4.13}$$

Hence,

$$T_i = \tau \ln F_i \tag{4.14}$$

The total number of bits sent in interval T_i is then N_i, given by

$$N_i = \frac{T_i}{T_{bi}} = \frac{\tau P_i}{E_b} \ln F_i \tag{4.15}$$

Thus, the *total* number of bits sent N_F in a time interval T_K is

$$N_F = \sum_{i=1}^{K} N_i = \sum_{i=1}^{K} \frac{\tau}{E_b} P_i \ln F_i \tag{4.16}$$

where K packet segments can be practically sent. Thus, assuming a minimum acceptable received power P_K, we have from Eq. 4.10,

$$P_o = F_1 F_2 \cdots F_K P_K \tag{4.17}$$

Substituting Eq. 4.10 into Eq. 4.16 and using Eq. 4.17 yields

$$N_F = \frac{\tau P_o}{E_b}\left(\frac{\ln F_1}{F_1} + \frac{\ln F_2}{F_2 \cdot F_1} + \frac{\ln F_3}{F_3 \cdot F_2 \cdot F_1} + \cdots + \frac{\ln F_K}{F_K \cdots F_3 \cdot F_2 \cdot F_1}\right) \tag{4.18}$$

where $\tau P_o / E_b = N_o$, the maximum number of bits that can be transmitted (see Eq. 4.3). The FAVR system selects the F_i in such a manner as to allow N_F to approach N_o.

Several algorithms concerning the method to be used to select the set of factors $\{F_i\}$ immediately present themselves.

Algorithm 1 Let F_i be a constant. For example, let $F_i = 2$, so that the bit rate changes only when the power level changes by 3 dB. From Eq. 4.11 we see that in this system the bit rate will halve at each transition. This algorithm yields

$$\begin{aligned}\frac{N_{F1}}{N_o} &= \frac{\ln 2}{2}\left[1 + \frac{1}{2} + \left(\frac{1}{2}\right)^2 + \cdots + \left(\frac{1}{2}\right)^{K-1}\right] \\ &= \left[1 - \left(\frac{1}{2}\right)^K\right]\ln 2 = \left(1 - \frac{1}{P_o/P_K}\right)\ln 2\end{aligned} \tag{4.19}$$

In obtaining Eq. 4.19 we have made use of Eqs. 4.17 and 4.7 to show that $P_o/P_K = 2^K$. Thus, the maximum efficiency of this system is always less than $\ln 2 \approx 70\%$ of optimum system.

Algorithm 2 Let F_i vary so that $F_1 = 2$, $F_2 = \frac{3}{2}$, $F_3 = \frac{4}{3}$, $F_4 = \frac{5}{4}$, and so on. This algorithm is more difficult to implement but results in an increased number of bits being sent in a given trail. A particular subset of this algorithm is to select F_i to alternate between $\frac{3}{2}$ and $\frac{4}{3}$. This is the FAVR algorithm.

The efficiency of the FAVR algorithm is obtained from Eq. 4.18 and yields

$$\frac{N_{FAVR}}{N_o} = \frac{\ln \frac{3}{2}}{\frac{3}{2}} + \frac{\ln \frac{4}{3}}{\frac{4}{3} \cdot \frac{3}{2}} + \frac{\ln \frac{3}{2}}{\frac{3}{2} \cdot \frac{4}{3} \cdot \frac{3}{2}} + \cdots \tag{4.20}$$

There are K terms in the sum. If we assume, for simplicity, that K is even, we have

$$\frac{N_{FAVR}}{N_o} = 2\left(\frac{2}{3}\ln\frac{3}{2} + \frac{1}{2}\ln\frac{4}{3}\right)\left[1 - \left(\frac{1}{2}\right)^{K/2}\right] \tag{4.21}$$

In deriving Eq. 4.7*a* it was assumed that the lower limit of usable signal power is 0 and consequently that the bit duration is allowed to approach infinity. This is not possible. In practice, a lower limit P_L of usable signal power must be selected. The corresponding number of bits N_{0L} is then

$$N_{0L} = \frac{P_o - P_L}{E_b} \tag{4.22a}$$

In comparing the FAVR system with the optimum system, the same lower limit P_L is assumed. Consequently,

$$\frac{P_o}{P_L} = 2^{K/2} \quad \text{and} \quad \frac{N_{FAVR}}{N_{0L}} = 2\left(\frac{2}{3}\ln\frac{3}{2} + \frac{1}{2}\ln\frac{4}{3}\right) \tag{4.22b}$$

Equation 4.22*b* can be readily extended. Let m denote the number of terms in the cycle. The F factors are then

$$\frac{m+1}{m}, \frac{m+2}{m+1}, \ldots, \frac{2m}{2m-1}$$

Equation 4.18 gives

$$\frac{N_{FAVR}}{N_{0L}} = 2m \sum_{i=1}^{i=m} \frac{1}{m+i} \ln\left(\frac{m+i}{m+i-1}\right) \tag{4.22c}$$

For $m = 2$ and 3, the values of N_{FAVR}/N_{0L} are 0.83 and 0.88, respectively. While the ratio increases with m, the required accuracy for signal power estimation also increases with m.

Referring to Fig. 4.4 we see that each packet segment contains N_s bits, where N_s is a constant of the FAVR system. The signal bits in each packet segment are transmitted at the same bit duration T_{bi}. In general, N_s is between 40 and 200 bits and a message packet contains 10 to 20 packet segments. A strong burst may contain packet segments belonging to a number of packets, and packet segments from a few successive weak bursts may make up a single packet. Packaging the transmitted bits into packet segments facilitates the incorporation of a FAVR link into a packet switching network. It is also essential to the FAVR MODEM design. However, if the predicted signal at the end of a packet segment fails to satisfy the requirement that

$$P_R > \frac{E_b}{T_{bi}}$$

then the entire packet may be transmitted at a lower rate. Since the number of remaining bits may be anywhere between 1 and $N_s - 1$, the average loss of transmitted bits per change of bit rate is

$$N_{LS} = \frac{N_S}{2}\left(1 - \frac{1}{F_{i+1}}\right) \tag{4.23}$$

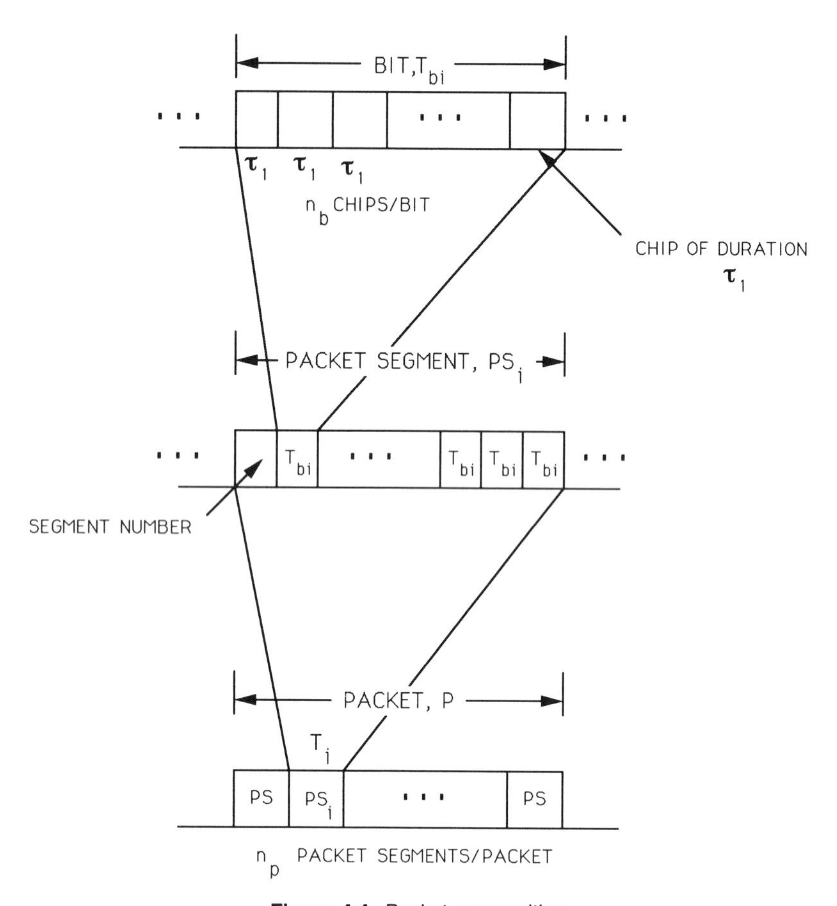

Figure 4.4. Packet composition.

4.3.5 Comparison of FAVR and the Constant Bit Rate System

The constant bit rate system transmits N_c bits during the underdense meteor trail, where from Eqs. 4.5 and 4.7

$$N_c = N_o\left(\frac{\ln P_o/P_K}{P_o/P_K}\right) \tag{4.24}$$

while the number of bits transmitted by the FAVR system is

$$N_{\text{FAVR}} = 0.83N_o\left(1 - \frac{1}{P_o/P_K}\right) - N_k\bar{N}_{\text{LS}} \tag{4.25}$$

where N_k is the number of bit rates used in the transmission. Referring to Fig. 4.5 we see that if $P_o/P_K = 20$ dB (a typical value), and if 1000 bits can

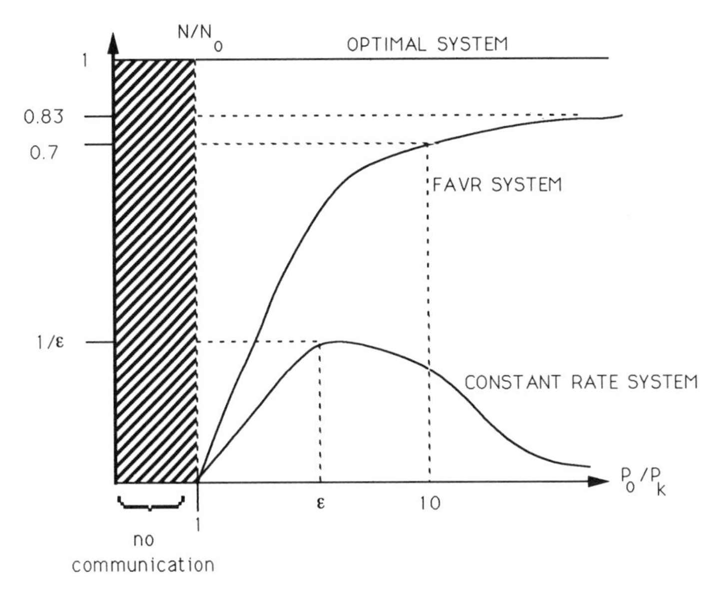

Figure 4.5. Comparison of the constant-rate and FAVR systems.

be transmitted over a trail using the traditional fixed bit rate approach, the expected transmission with FAVR is

$$N_{\text{FAVR}} = 18\,000 - 12 \times 30 \times 0.3 = 17\,883$$

4.3.6 Transmission over Arbitrary Meteor Channels

While an underdense trail is most probable, it usually disappears in several hundred milliseconds. Overdense trails are far fewer in number but each lasts for several seconds. Further, as seen from Figs. 4.1 and 4.2, neither of these trails follow the textbook equations.

The FAVR system is a piecewise approximation to each actual trail while the Constant Rate (CR) system is an on–off (binary) approximation. As a result FAVR results in a substantial improvement over the CR system for any set of trails.

4.4 STOCHASTIC PROPERTIES OF THREE TYPES OF SYSTEMS

4.4.1 Underdense Channel Model

In the following, the throughputs and waiting times of three types of MBC systems are compared: (1) the constant bit rate system, (2) the optimum communication system, and (3) the ARQ system. Since the FAVR system realizes 83% of the throughput of the optimum communication system, the results obtained below for the optimum system are similar to those of the FAVR system.

To make the comparison meaningful, the same model is used for all three types of systems. The underdense channel model is described as follows [8, 9]:

1. The received signal power during one meteor burst can be represented as in Eq. 4.1.
2. P_o and τ are random variables for the ensemble of meteor bursts. For each individual burst, P_o and τ are constants. The distributions of P_o and τ are independent:

$$f(a) = \text{average number of meteor bursts per second with } P_o \geq a$$

$$P_\tau(b) = \text{probability of } \tau \geq b$$

3. The occurrences of meteor bursts are Poisson.

The mathematical model presented above is for the purposes of comparing the three types of systems. It is not essential to the FAVR system's operation.

4.4.2 Mean Throughput

The mean throughput rate R_i can be expressed as

$$R_i = \int_0^\infty \int_{P_L}^{P_u} N_i(a, b)\, df(a)\, dP_\tau(b) \tag{4.26}$$

where $N_i(a, b)$ is the transmitted number of bits during a trail of peak power $P_o = a$, and time constant $\tau = b$. The subscript i stands for c, o, or q representing the constant bit rate system, the optimum communication system, or the ARQ system. The function $N_i(a, b)$ is given by Eqs. 4.5, 4.7b, and 4.8 for the three systems, respectively.

The limits of integration satisfy the following conditions:

$$N_i(P_L, b) = 0 \tag{4.27}$$

$$f(P_u) = 0 \tag{4.28}$$

where P_L is the minimum acceptable power level and P_u exceeds the maximum possible power level of an underdense trail. Integration by parts of Eq. 4.26 gives

$$R_i = \int_0^\infty \int_{P_L}^{P_u} f(a) \frac{\partial N_i(a, b)}{\partial a} \, dP_\tau(b) \, da \tag{4.29}$$

The following results are obtained by applying Eq. 4.24:

Constant Bit Rate System

$$\frac{\partial N_c(a, b)}{\partial a} = \frac{b}{aT_b} \tag{4.30}$$

$$R_c = \frac{\bar{\tau}}{T_b} \int_{P_L}^{P_u} \frac{1}{a} f(a) \, da \tag{4.31}$$

where $\bar{\tau}$ is the ensemble average of the decay time constant τ.

Optimum Communication System

$$\frac{\partial N_o(a, b)}{\partial a} = \frac{b}{E_b} \tag{4.32}$$

$$R_o = \frac{\bar{\tau}}{E_b} \int_{P_L}^{P_u} f(a) \, da \tag{4.33}$$

ARQ System

From Eqs. 4.26 and 4.8

$$R_q = -\int_0^\infty Nf(P_L) \, dP_\tau(b) \tag{4.34}$$

where P_L is a function of b:

$$P_L = \frac{E_b}{T_b} e^{NT_b/b} \tag{4.35}$$

The bit duration is selected to maximize the transmission. Equation 4.34 implies that maximizing R_Q is equivalent to minimizing P_L. From Eq. 4.35

$$\frac{dP_L}{dT_b} = \left[-\frac{E_b}{T_b^2} + \frac{E_b}{T_b} \left(\frac{N}{b} \right) \right] e^{NT_b/b} \tag{4.36}$$

Setting $dP_L/dT_b = 0$ gives

$$N = \frac{b}{T_b} \tag{4.37}$$

Substituting Eq. 4.37 into Eq. 4.35 gives

$$P_L = \frac{eE_b}{T_b} \tag{4.38}$$

Substituting Eqs. 4.37 and 4.38 into Eq. 4.34 gives

$$R_q = \frac{\bar{\tau} P_L}{eE_b} f(P_L) \tag{4.39}$$

The ARQ system performance can be improved by allowing multiple packet transmissions to take place during each trail. However, the average improvement is not significant since the probability of trails sufficiently long to permit multiple transmissions is small.

4.4.3 Waiting Time for Short Messages

For long messages that take many meteor burst to complete transmission, the waiting time for first usable burst to occur is not a significant part of the total transmission time. However, for short messages that are transmitted within the duration of one burst, the waiting time is the major part in the total time required for message transmission. The following simplifying assumptions are made:

1. The short message is ready to be transmitted at $t = 0$.
2. If there is a usable meteor burst in existence with received signal power $P_R(0) \geq P_L$, the waiting time is then 0; otherwise, the waiting time is the interval that elapses until a meteor burst with $P_o \geq P_L$ occurs.

Since the meteor bursts are Poisson distributed, the distribution function $F(T_w)$ for the waiting time $t_w \leq T_w$ is

$$F(T_w) = 1 - P_{nt} e^{-f(P_L) T_w} \tag{4.40}$$

where P_{nt} is the probability of *not* having a usable meteor burst in existence at $t = 0$:

$$P_{nt} = 1 - P_{ex} \tag{4.41}$$

The probability P_{ex} that a trail is present can be calculated as follows. Any trail with peak power

$$a = P_L e^{t/b} \tag{4.42}$$

occurring within the time period $-t$ to 0 would last until $t = 0$. The differential probability of such an occurrence is

$$dP_{ex} = t[-df(a)] = b \ln\left(\frac{a}{P_L}\right)[-df(a)] \tag{4.43}$$

Integrating over-all values of a and b gives

$$P_{ex} = \int_0^\infty \int_{P_L}^{P_u} b \ln\left(\frac{a}{P_L}\right) df(a)\, dP_\tau(b) \tag{4.44}$$

Integrating Eq. 4.44 by parts gives

$$P_{ex} = \bar{\tau} \int_{P_L}^{P_u} \left(\frac{1}{a}\right) f(a)\, da \tag{4.45}$$

Comparing Eq. 4.45 with Eq. 4.31 gives

$$P_{ex} = T_b R_c \tag{4.46}$$

The distribution of the waiting time $F(T_w)$ is the percentage or probability F that the waiting time is less than T_w. Solving T_w in terms of F from Eq. 4.40 gives

$$T_w = \frac{1}{f(P_L)} \ln\left(\frac{1 - P_{ex}}{1 - F}\right) \tag{4.47}$$

4.4.4 An Illustrative Example

In the above sections, the stochastic properties are derived without making any assumption concerning $f(a)$. However, from previous work much about $f(a)$ is known. The density function is of the form

$$P(a) = \begin{cases} ca^{-(n+1)} & \text{for } a \le P_u \\ 0 & \text{for } a > P_u \end{cases} \tag{4.48}$$

where the constant n is between 0.5 and 0.85 [11]. Since $P(a)$ is proportional to $-f'(a)$, Integration of Eq. 4.48 gives

$$f(a) = K_1(a^{-n} - P_u^{-n}) \qquad 0 < a \le P_u \tag{4.49}$$

where P_u is the maximum value of the peak power and $f(a) = 0$ for $a > P_u$.

Substitution of Eq. 4.49 into Eqs. 4.31, 4.33, and 4.39 gives

$$R_i = \frac{\bar{\tau} K_1}{E_b} P_u^{1-n} g_i(n, x) \tag{4.50}$$

where i stands for subscripts c, o, and 1, and x is the ratio P_L/P_u. The function $g_i(n, x)$ is the *specific throughput* of the system:

For a Constant Bit Rate System

$$g_c(n, x) = \frac{x}{n}(x^{-n} - 1 + n \ln x) \tag{4.51}$$

For an Optimum Communication System

$$g_o(n, x) = \left(\frac{n}{1-n} - \frac{x^{1-n}}{1-n} + x\right) \tag{4.52}$$

For an ARQ System

$$g_q(n, x) = e^{-1}(x^{1-n} - x) \tag{4.53}$$

A plot of the specific throughput functions $g_c(0.6, x)$, $g_o(0.6, x)$, $g_q(0.6, x)$ versus x is shown in Fig. 4.6. In this figure we note that as x decreases, both g_c and g_q attain peak values at some values of x, while $g_o(x)$ increases monotonously. The maximum values of x are obtained by differentiating Eqs. 4.51 and 4.53:

$$(1-n)(x_{oc}^{-n} - 1) + n \ln x_{oc} = 0 \tag{4.54}$$

$$(1-n)x_{oq}^{-n} - 1 = 0 \tag{4.55}$$

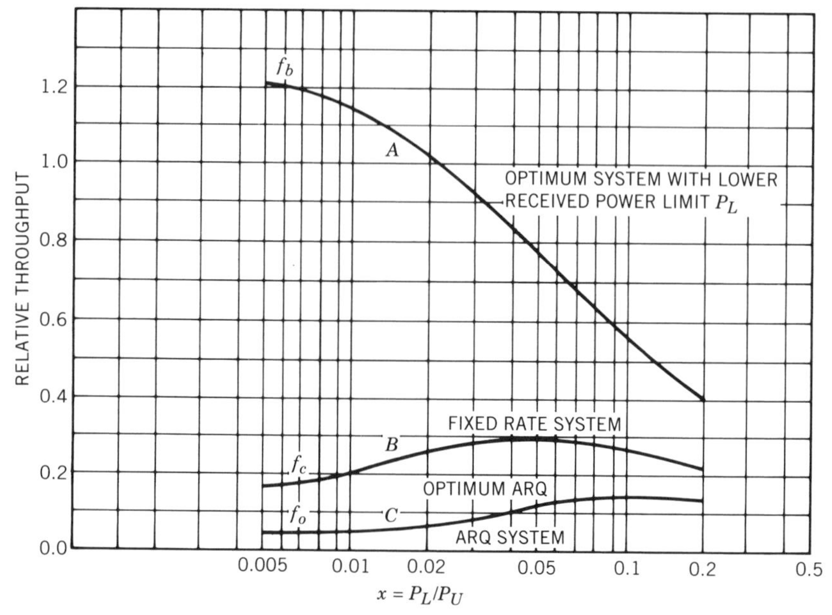

Figure 4.6. Throughputs of three types of communication systems.

TABLE 4.1 Calculated Specific Throughputs and Waiting Time Parameters

System	X	P_{ex}	$f(P_L)$	Specific Throughput	90% Waiting Time (s)
Constant bit rate	0.0673	0.0008	0.01567	0.273	146.9
Optimum	0.005	0.0133	0.0921	1.205	45.9
ARQ	0.217	0.0004	0.006	0.120	383.8

where x_{oc} and x_{oq} are optimum values of x for the constant bit rate system and ARQ system, respectively. Solving Eq. 4.54 numerically gives $x_{oc} = 0.06734$, and $g_c(0.6, x_{oc}) = 0.273$. Equation 4.55 is solved for x_{oq}:

$$x_{oq} = (1 - n)^{1/n} \tag{4.56}$$

$$g_q(n, x_{oq}) = \frac{n}{w}(1 - n)^{(1-n)/n} \tag{4.57}$$

The optimum values for $n = 0.6$ are $x_{oq} = 0.217$, and $g_q(0.6, x_{oq}) = 0.120$.

While x_{oq} and x_{oc} are very different, the optimum bit rates for the two systems are not so different. For a constant bit rate system $P_{Lc} = E_b/T_b$. For an ARQ system, P_{Lp} is the received peak power level that allows N bits to be transmitted. From Eq. 4.38,

$$\frac{T_{bc}}{T_{bq}} = \frac{P_{Lq}}{eP_{LC}} = \frac{x_{oq}}{ex_{oc}} = 1.025$$

For maximum throughput, the bit transmission time for a constant bit rate system is only slightly larger than that for an ARQ system.

Table 4.1 illustrates the calculated specific throughputs and waiting time parameters for the three types of systems for a typical meteor burst channel with $K_1 P_u^{-n} = 0.004\ \text{s}^{-1}$ and $\tau = 0.1$ s. There is a conflict between the choices of high throughput and low waiting time for the constant bit rate system and the ARQ system. To achieve maximum throughput, the bit rate must be reasonable high, which implies a high value of P_L and consequently a long waiting time. There is no such conflict for the optimum communication and the FAVR systems. For both systems, throughput increases and waiting time decreases as X decreases. The value of X in Table 4.1 is selected as 0.005 for the optimum system because of practical considerations. Its value can be decreased if waiting time is of primary concern. The FAVR system has the same waiting time as the optimum system because P_L is the same for both.

4.5 THE FAVR MODEM

Three important technical problems must be solved as prerequisite conditions for realizing the variable bit rate concept: (1) a variable bandwidth filter or the equivalent of it, (2) accurate and timely determination of the signal-to-noise ratio, and (3) communication of the bit rate information. The three problems are solved in the FAVR system by using a receiver with a simple analog circuit followed by digital processing of the analog signals. The variable bandwidth filter problem is solved by summing a variable number of terms of the analog signals. Signal power and noise power are determined separately by using Kalman's optimum filtering and prediction algorithm. Changes in bit rate are not explicitly communicated, but are determined by the receiver digital circuit.

4.5.1 Transmitter

Figure 4.7 is a block diagram of the FAVR transmitter. During the standby period, the transmitter at each end sends out signals with square wave modulation at half-period equal to E_b/P_L, which corresponds to the minimum data rate.

At the arrival of a meteor burst, the handshake processor at each receiver picks up a strong wave modulation. The FAVR system operates on the basis of reciprocity of the meteor burst channel: Assuming identical terminal equipment at the two terminals A and B, the received signal strength from A to B is close approximation of the received signal strength from B to A. From the received square wave, the handshake processor determines the required bit duration for a specified signal-to-noise ratio. Usually the required bit duration falls between two allowed values of $n_b T_1$. The larger n_b is then selected, and transmission begins.

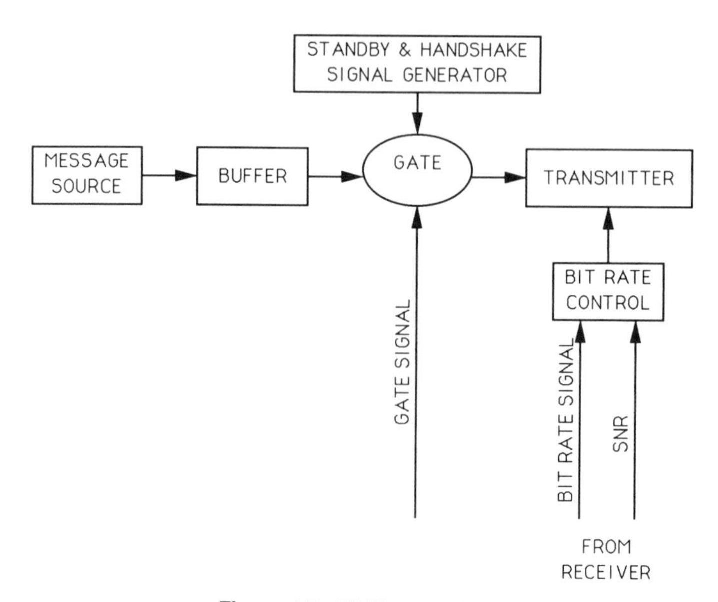

Figure 4.7. FAVR transmitter.

The channel condition is monitored through the received signal-to-noise ratio. When the latter falls below a critical level, the Bit Timing Control signals for the next lower rate of transmission or next larger n_b value. The transmitter then transmits at the new rate starting with the next packet segment without signaling the receiver, which would then follow the change of rate automatically, is described in the receiver section.

4.5.2 Receiver

A block diagram of a FSK receiver is illustrated in Fig. 4.8. It has a simple front and analog circuit which yields four numbers for each signal element: x_{1i}, y_{1i}, x_{2i}, and y_{2i}. These are the cosine and sine Fourier components in the received signal at the two alternative mark-space frequencies, ω_1 and ω_2, respectively.

Five microprocessors operate in a multiprocessing configuration to yield the detected signal and an instantaneous best estimate of the signal-to-noise ratio. Large values of any of the four numbers x_{ji} (or y_{ji}) alert the handshake processor, which computes the signal-to-noise ratio and determines the initial n_b to be used. This information is transmitted to the transmitter and also to the central track processor. The two other track processors, are then set at next higher and lower values of n_b, respectively.

The average signal-to-noise ratio over a packet segment is computed within each track. There is a track buffer within each track that stores $2n_s$ bits of information, and the demodulated bits are temporarily stored in the track buffer. The control processor compares the average signal-to-noise ratios of the three tracks and forwards only the stored signal with the highest SNR to the main buffer.

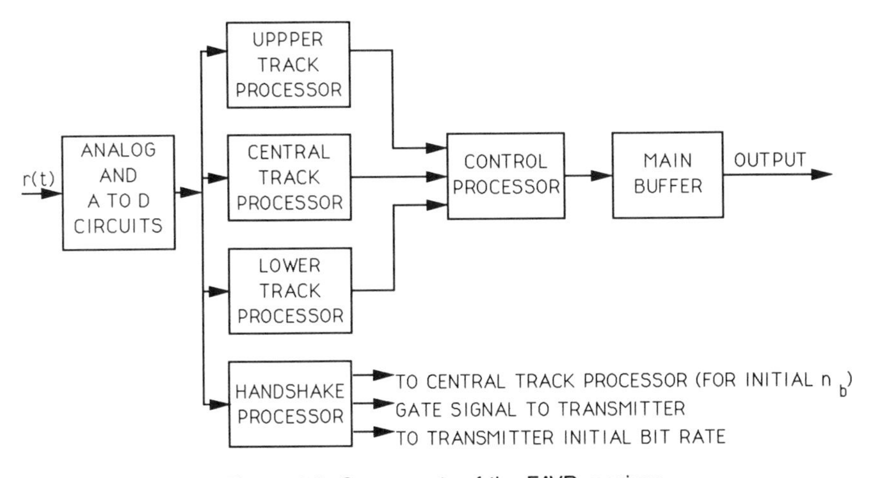

Figure 4.8. Components of the FAVR receiver.

4.5.3 Signal Processing for Variable Bit Rate BFSK

Variable bit rate BFSK is realized by using digital processing of the detected signal, as illustrated in Fig. 4.9. We assume that within the short transmission time of 1 bit the received signal phase is essentially constant.

As discussed previously, the time t_b for transmitting 1 bit can be expressed as an integer multiple of T_1:

$$t_b = n_b T_1 \tag{4.58}$$

Let $r(t)$ denote the received signal, and ω_1, ω_2 denote the alternative BFSK frequencies.

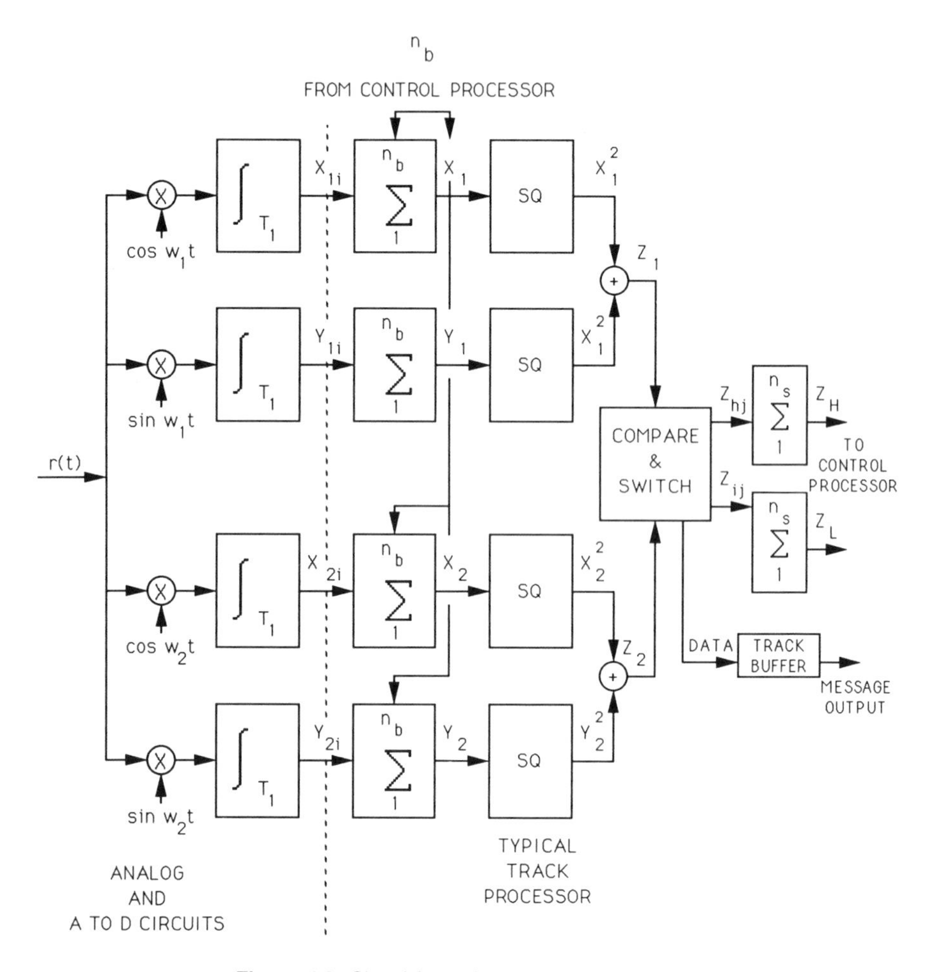

Figure 4.9. Signal formation in a track processor.

$$x_{1i} = \int_{(i-1)T_1}^{iT_1} r(t) \cos \omega_1 t \, dt \tag{4.59}$$

$$y_{1i} = \int_{(i-1)T_1}^{iT_1} r(t) \sin \omega_1 t \, dt \tag{4.60}$$

The variables x_{2i} and y_{2i} are similarly defined with ω_2 replacing ω_1. For each bit of transmitted signal, i runs from 1 to n_b:

$$x_1 = \sum_{i=1}^{i=n_b} x_{1i} \tag{4.61}$$

$$y_1 = \sum_{i=1}^{i=n_b} y_{1i} \tag{4.62}$$

The variables x_2 and y_2 are similarly defined. Let

$$z_1 = x_1^2 + y_1^2 \tag{4.63}$$

$$z_2 = x_2^2 + y_2^2 \tag{4.64}$$

If $z_1 > z_2$, then the transmitted signal is assumed to be at ω_1. The above system is readily shown to be noncoherent reception of FSK. However, the transmitter can change its bit rate by changing n_b in Eqs. 4.61 and 4.62.

Let the bits in a packet segment be numbered j, where $j = 1, 2, \ldots, n_s$. The two values z_1 and z_2 for the jth bit are designated $z_{\max.j}$ and $z_{\min.j}$, respectively, according to whichever is larger. Let

$$z_{hj} = \sqrt{z_{\max.j}} \quad \text{and} \quad z_{lj} = \sqrt{z_{\min.j}}$$

Then,

$$Z_H(n) = \sum_{j=1}^{j=n} z_{hj} \quad \text{and} \quad Z_L(n) = \sum_{j=1}^{j=n} z_{lj} \tag{4.65, 4.66}$$

The signals Z_H and Z_L are then used to determine the bit transmission rate in subsequent processing.

For *BPSK* reception, the variation in phase angle θ due to Doppler effect is significant. BPSK is possible only if θ can be estimated. A Kalman filter algorithm is used in FAVR for estimating θ. Let u denote the in-phase signal component:

$$u = x_1 \cos \theta + y_1 \sin \theta$$

The received bit is interpreted as a 0 or 1 according to the sign of u. To determine the bit transmission rate, a single signal Z is formulated, where

$$Z(n) = \sum_{j=1}^{j=n} |u| \tag{4.67}$$

4.5.4 Automatic Identification of Bit Transmission Rate

Automatic identification of the bit rate is made possible by providing three separate track processors in the receiver. The central processor has an n_b value that is the same as that of the preceding packet segment. The upper and lower track processors have n_b values one step higher and lower, respectively. For instance,

Central track processor, $n_b = 8$
Lower track processor, $n_b = 6$
Upper track processor, $n_b = 12$

The transmitted bit rate is allowed to change no more than one level in between packet segments. One and only one track is synchronous with the transmitted bit rate. Since the other two tracks are not, their signals in some bits are partly canceled. Summing over the magnitudes of the bits gives an indication of whether there are partial cancellations.

In the identification process of a BPSK system, the values of $Z(n)$ are compared. For the present explanatory example, n_b is assumed to be 48. The first comparison takes place as n reaches 48 for the lower track processor. The values of n for the central and upper tracks are then 36 and 24, respectively. If $Z(48)_{LT}$ is the largest

$$Z(48)_{LT} > Z(36)_{CT} \quad \text{and} \quad Z(48)_{LT} > Z(24)_{UT} \tag{4.68a}$$

where LT, CT, and UT stand for lower track, center track, and upper track, respectively, then n_b is determined to be 6. The decoded sequence stored in the lower track buffer is then the received signal for the packet segment. If Eq. 4.68*a* is not valid, but at a later time

$$Z(48)_{CT} > Z(64)_{LT} \quad \text{and} \quad Z(48)_{CT} > Z(32)_{UT} \tag{4.68b}$$

then n_b is determined to be 8. If Eq. 4.68*b* is not valid, but at a later time $Z(48)_{UT}$ is larger than $Z(72)_{CT}$, then n_b is determined to be 12. For a BFSK system, $Z_H(n) - Z_L(n)$ is used instead of $Z(n)$.

4.5.5 An Analysis of the Probability of an Identification Error

In a track with incorrect n_b value, the length L of one bit can be covered by two different binary digits. If the two digits are identical, no error is likely to

TABLE 4.2

Actual n_b	Candidate n_b	F
$2n+1$	$2n$	$n/(8n+4)$
$2n-1$	$2n$	$n/(8n-4)$
$2n$	$2n+1$	$(n+1)/(8n+4)$
$2n$	$2n-1$	$(n-1)/(8n-4)$

be made. If the two digits are different, an error is made for at least the length of the minority digit. Since the average length of the minority digit is $L/4$, the error fraction is given by

$$F = \frac{1}{2} \times \frac{L}{4} \times \frac{1}{L} = \frac{1}{8} \tag{4.69}$$

A detailed calculation shows that F depends on the ratio of the candidate bit length to that of the actual bit length, as shown in Table 4.2. Over a packet element, the total signal length L_T is $n_s n_b T_1$, and a length $L_e = F n_s n_b T_1$ is in error. In calculating the difference in the Z value for a BPSK system, both signal and noise are canceled in the correct length $L_T - L_e$. However, both signal and noise amplitude are doubled over the length L_e. The difference in Z, ΔZ, has a signal amplitude of $2\sqrt{P(t)}$ over a period of $F n_b n_s T_1$ and a noise power of $4P_n$. Its signal-to-noise ratio is given by

$$\left(\frac{S}{N}\right)_{\text{id}} = \frac{4[P(t)]F n_b n_s T_1}{P_n} = F n_s \frac{P(t) n_b T_1}{P_n} = \left(\frac{S}{N}\right)_{\text{bit}} (F n_s) \tag{4.70}$$

Equation 4.70 shows that the signal-to-noise ratio for identification is $F n_s$ times the signal-to-noise ratio per bit. With normally acceptable bit error rate, and an $F n_s$ value of 4 or higher, the identification error is negligible. With FSK, the identification error probability is also very small compared to the error probability of a binary digit, but larger than that for BPSK.

Table 4.2 assumes that the signs of the binary digits switch at random. If a code that has a high switching rate is used, then the value of F is improved.

4.6 CONCLUSIONS

This chapter showed experimental evidence of the random on–off nature of the meteor burst channel, and then presented an algorithm and modem design that allows communication at 83% of the maximum bit rate permitted by the channel.

Results are presented comparing the FAVR model to the optimal fixed bit rate model and to an ARQ modem. An example using typical numbers was given showing that FAVR can yield an effective data rate 17 times greater than that obtained for the optimal fixed rate system.

REFERENCES

1. G. R. Sugar, Radio propagation by reflection from meteor trails. *Proc. IEEE* 52: 116–136 (1964).
2. V. R. Eshleman and R. F. Mlodnosky, Directional characteristics of meteor propagation derived from radar measurements. *Proc. IRE* (Dec. 1957).
3. P. J. Bartholome, Results of propagation and interception experiments on the STC meteor-burst link. SHAPE Tech. Cen., The Hague, The Netherlands, *TM-173*, Dec. 1967.
4. P. A. Forsyth, E. L. Vogan, D. R. Hansen, and C. O. Hines, The principles of JANET: a meteor-burst communications system. *Proc. IRE* (Dec. 1957).
5. P. J. Bartholome and I. M. Vogt, Comet: a meteor-burst system incorporating ARQ and diversity reception. *IEEE Trans. Commun. Technol.* COM-16 (Apr. 1968).
6. V. R. Eshleman and L. A. Manning, Radio communications by scattering from meteoric ionization. *Proc. IRE* 42: 530–536 (1954).
7. A. E. Spezio, Meteor burst communication system: analysis and synthesis. *NRL Rep. 8286*, 28 Dec 1978.
8. L. B. Milstein, D. L. Schilling, R. L. Pickholtz, J. Sellman, S. Davidovici, A. Pavelchek, A. Schneider and G. Eichmann, Performance of meteor burst communication channels. *IEEE J. SAC* (Feb. 1987).
9. J. D. Oetting, An analysis of meteor burst communications for military applications. *IEEE Trans. Commun.* COM-28: 1591–1601 (1980).
10. J. C. Ostergaard et al., *AGARD Conference Proceedings No. 382.*
11. J. A. Weitzen et al., *MILCOM'88 Conference Record*, p. 0573.

5

THE FAVR METEOR BURST COMMUNICATION EXPERIMENT

D. L. Schilling
City College of New York

T. Apelewicz and G. R. Lomp
SCS Telecom, Inc., Port Washington, New York

M. Dyer and L. Lundberg
Martin Marietta Energy Systems, Oak Ridge, Tennessee

F. Rogers and J. Woodhouse
11th Air Force, Elmendorf AFB, Arkansas

5.1 INTRODUCTION

The 11th Air Force has undertaken an effort to assess the applicability and test the feasibility of state-of-the-art software and hardware technologies for upgrading the 11th Air Force's Meteor Burst Communication System (MBCS) to meet current operational requirements. Upon reviewing technological alternatives, an approach that would utilize SCS Telecom, Inc. (SCS) proprietary FAVR technology was identified as a candidate warranting further investigation in that it was believed to offer the most benefit in terms of meeting the 11th Air Force's operational requirements, while at the same time meeting the design objectives of minimizing modifications to existing hardware. As a result, a three-phase program has been undertaken to prototypically evaluate the SCS FAVR technology in the operational environment. The purpose of this program was to conduct a study and a hardware feasibility testing program to investigate the viability of extending performance of the existing 11th Air Force's MBCS to incorporate two-way voice capability while increasing data throughput.

The first phase of the program consisted of a system analysis, preliminary design, and utilization of computer simulation studies to evaluate the expected improvement in overall system performance. Phase II consisted of the development of the detailed design, fabrication and testing of a brassboard FAVR Modem Applique test unit, the design of system interface modifications, and calculation of the performance of full-duplex, single-link feasibility testing. Phase III consisted of prototype development and a field test demonstration using the Anchorage–Kotzebue MB link. This paper addresses the results obtained from the Field Test demonstration.

5.1.1 General System Description and Requirements

The primary channel between the 11th Air Force's Region Operations Control Center (ROCC) and the Long Range RADAR (LRR) sites is a commercial satellite. Figure 5.1 shows a single-link connectivity between one (of perhaps six) LRR and the ROCC. RADAR data at the LRR is collected and formatted by the RADAR processor. Each radar message contains between five and nine 13-bit words. RADAR messages are forwarded, via two synchronous 2400-bps modems, to the satellite modem for transmission. At the ROCC, the LRR data is demultiplexed and forwarded to the Hughes computer via two 2400-bps synchronous links. Two synchronous serial ports on the Hughes computer are dedicated for each LRR. In the opposite direction, control data are passed from the ROCC to the LRR via the same route.

The required radar data traffic to and from the LRR to the ROCC per 12-s intervals (each radar sweep lasts 12 s) is shown in Fig. 5.2. During any 12-s sweep interval, 26,169 radar data bits are sent from an LRR to the ROCC and 2511 control data bits are sent from the ROCC to the LRR.

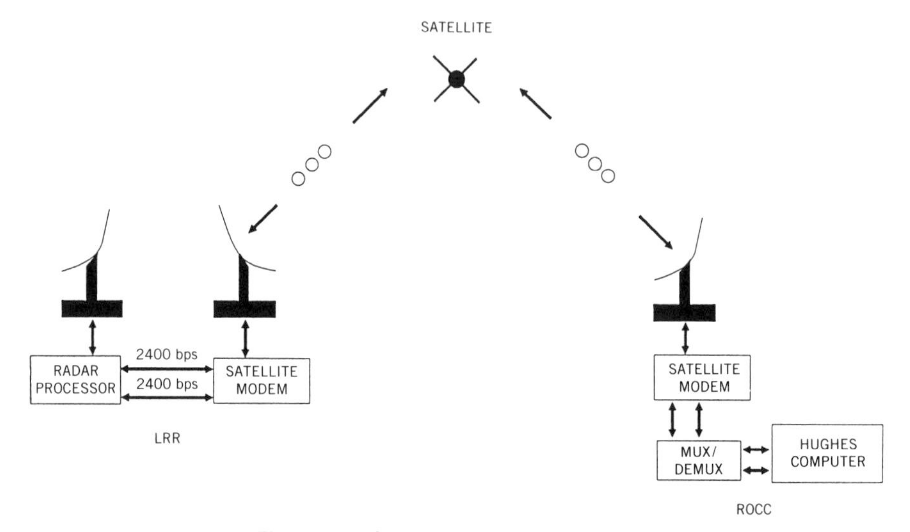

Figure 5.1. Single satellite link connectivity.

SATELLITE RADAR COMMUNICATIONS REQUIREMENTS

LRR TO ROCC

```
 65 X   3 =   195
 78 X   3 =   234
117 X   3 =   351
 65 X  24 =  1560
 78 X 250 = 19500
117 X  37 =  4329
             -----
             26169 BITS IN 12 SECONDS
                 ≈ 2181 BITS PER SECOND
```

ROCC TO LRR

```
104 X  24 = 2496
  5 X   3 =   15
            ----
            2511 BITS IN 12 SECONDS
               ≈ 210 BITS PER SECOND
```

Figure 5.2. Data traffic requirements (radar and control only).

Single link connectivity, between one LRR and the ROCC, having a primary satellite channel and a meteor burst communication (MBC) backup channel, is shown in Fig. 5.3. Whenever a MBC link is required, as requested by the ROCC, radar data will be routed away from the satellite modem to the MBC modem for transmission over the MBC channel.

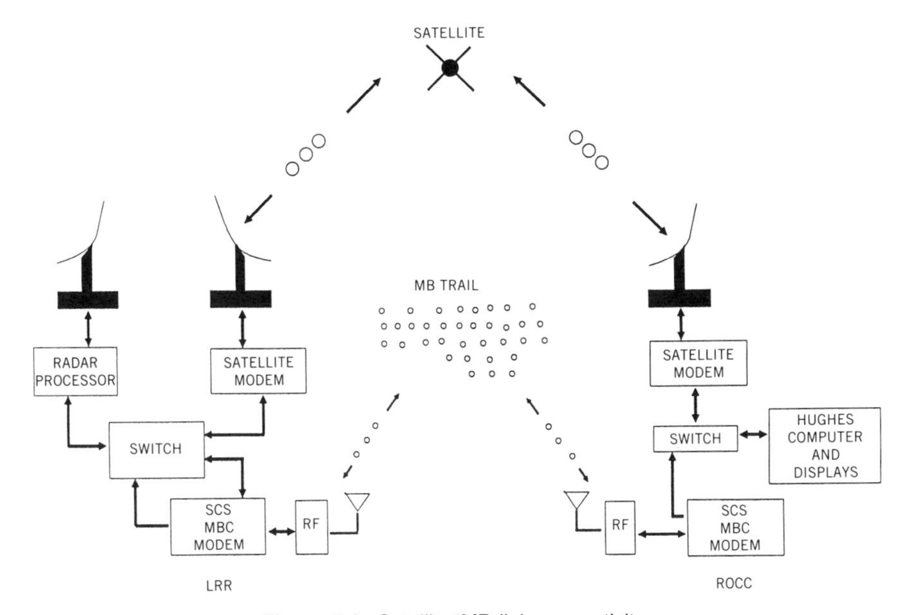

Figure 5.3. Satellite/MB link connectivity.

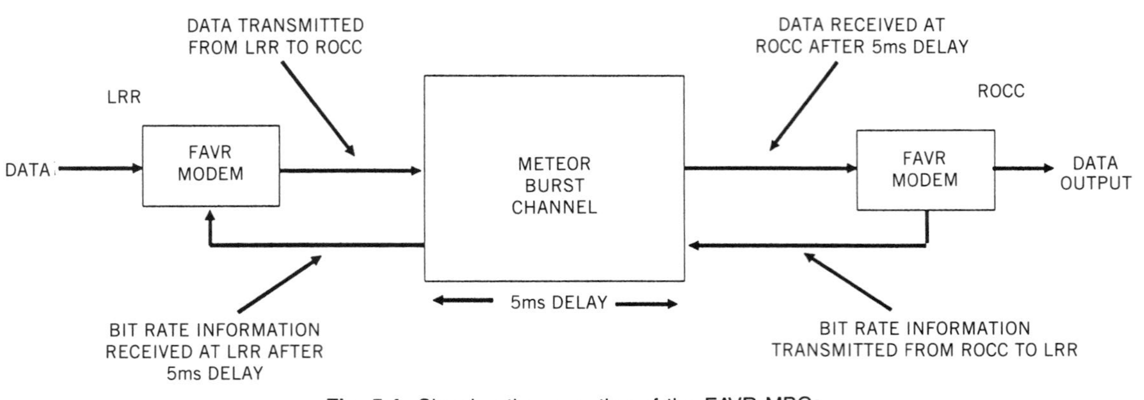

Fig. 5.4. Showing the operation of the FAVR MBCs.

Similarly, at the ROCC, radar control data will be switched away from the satellite modem to the MBC modem for transmission to the selected LRR. As shown in Fig. 5.3, it is also required that the MBC link be capable of carrying a two-way voice channel. It is anticipated that the maximum duration of the voice message will be no longer than 4 s and will be limited to no more than one message per each 12-s interval. Using linear predictive coding (LPC), a 4-s segment of voice requires 9600 bits. The overall data traffic requirements from a single MBC link in a 12-s interval is calculated to be

$$\text{ROCC to LRR:} \quad 9600 + 2511 = 12\,111 \text{ bits} \tag{5.1}$$

$$\text{LRR to ROCC:} \quad 9600 + 26169 = 35\,769 \text{ bits} \tag{5.2}$$

5.2 SIMULATION

The system to be simulated is shown in Fig. 5.3 and is detailed in Fig. 5.4. In this system, the data to be transmitted are modulated using the FAVR modem. The simulation of the model includes the trellis coded FM and the bit rate adjustment. Only five different bit rates were employed. Note that the bit rate information is received after a 5-ms delay (which is a typical delay over the meteor channel).

A typical set of underdense trails is shown in Fig. 5.5. In this figure the

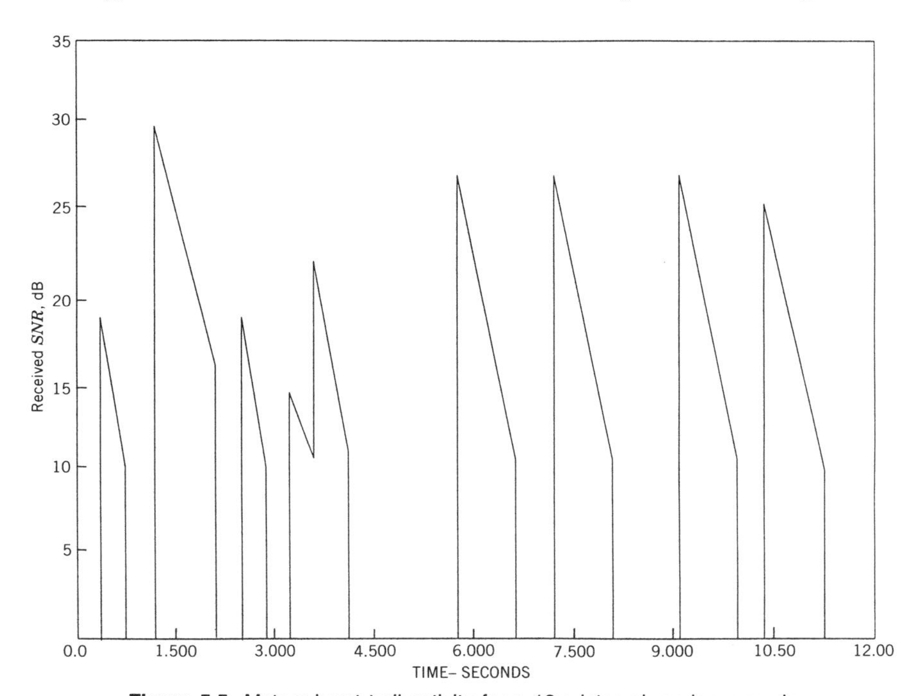

Figure 5.5. Meteor burst trail activity for a 12-s interval random sample.

received SNR (in dB) is plotted as a function of time. Since the waveforms are plotted on the dB axis, the normally seen exponential decay of an underdense trail appears as a straight diagonal line. Note that in a 12-s interval, nine trails occur on the average, each with a peak SNR between 15 and 30 dB. The FAVR system simulation used here is designed to ignore trails that result in less than 10 dB received SNR.

The simulator processes the received signal by using the FAVR modem and the error rate is determined. The FAVR model also has the capability to errors that remain after decoding. In this way the received mini-packet could be discarded.

5.2.1 Simulation Results

Figure 5.6 shows the probability of receiving N_b bits in 12 s for a FAVR system operating in a bandwidth of 15 kHz. Note that for a probability of 20%, fewer than 26,000 bits will be received successfully in 12 s. Hence, with a probability of 80%, more than 26,000 bits will be received correctly in 12 s. Figure 5.7 shows the probability of a FAVR system constrained to operate at a single fixed bit rate of 11.25 kb/s. Note that with a probability of 80%, more than 1800 bits will be received correctly in 12 s. Comparing Figs. 5.6 and 5.7, we see that FAVR can provide an improvement of approximately 15 compared to a fixed rate system.

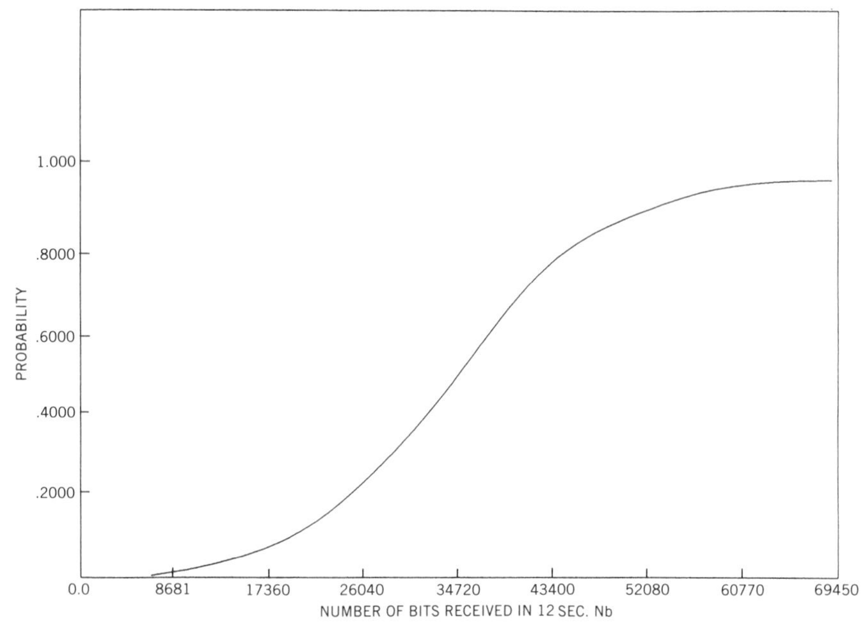

Figure 5.6. Probability of receiving N_b bits in a 12-s interval FAVR system, 15-kHz bandwidth

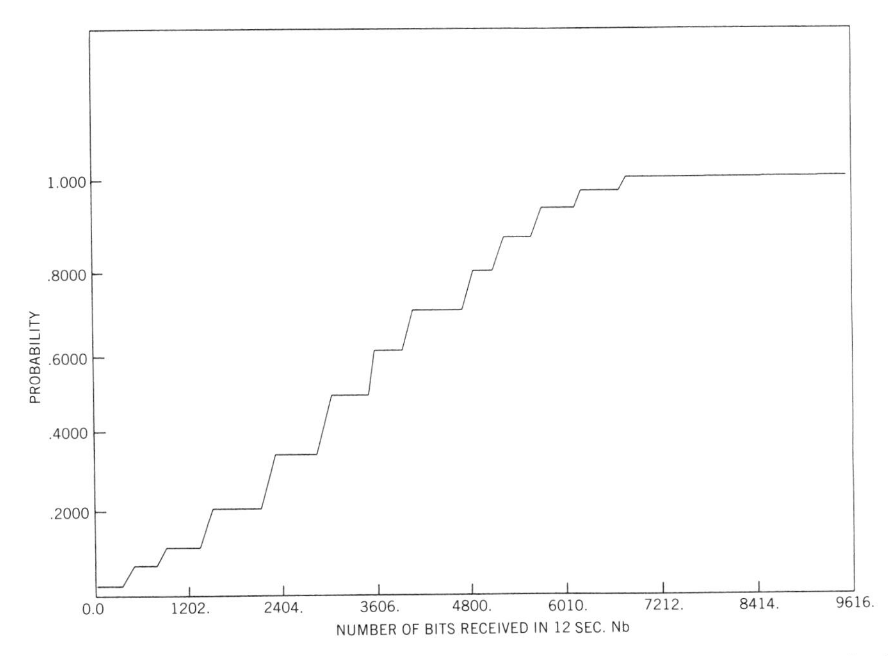

Figure 5.7. Probability of receiving N_b bits in a 12-s interval FAVR system, fixed rate = 11.25 kB/s.

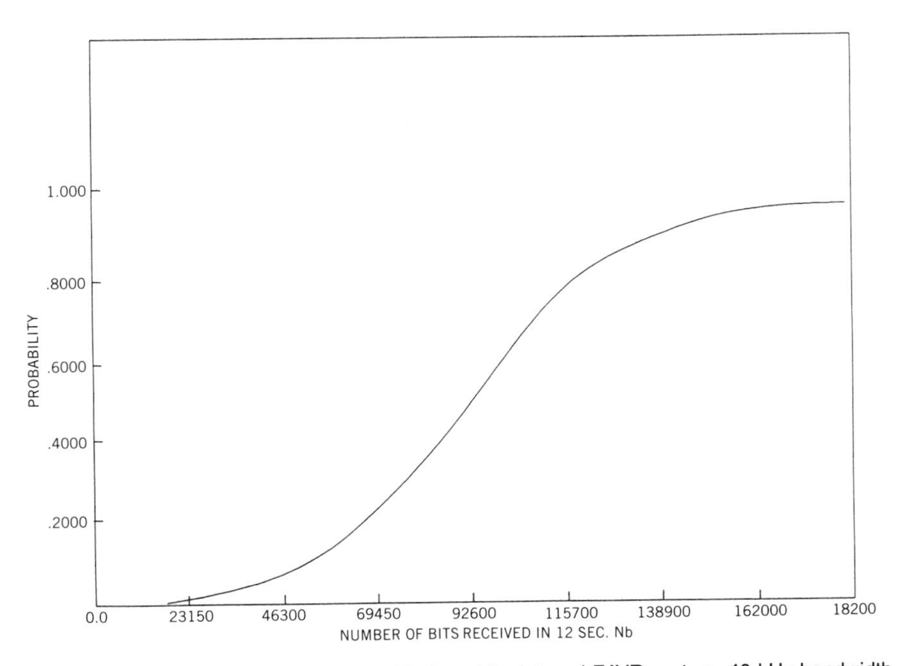

Figure 5.8. Probability of receiving N_b bits in a 12-s interval FAVR system, 40-kHz bandwidth.

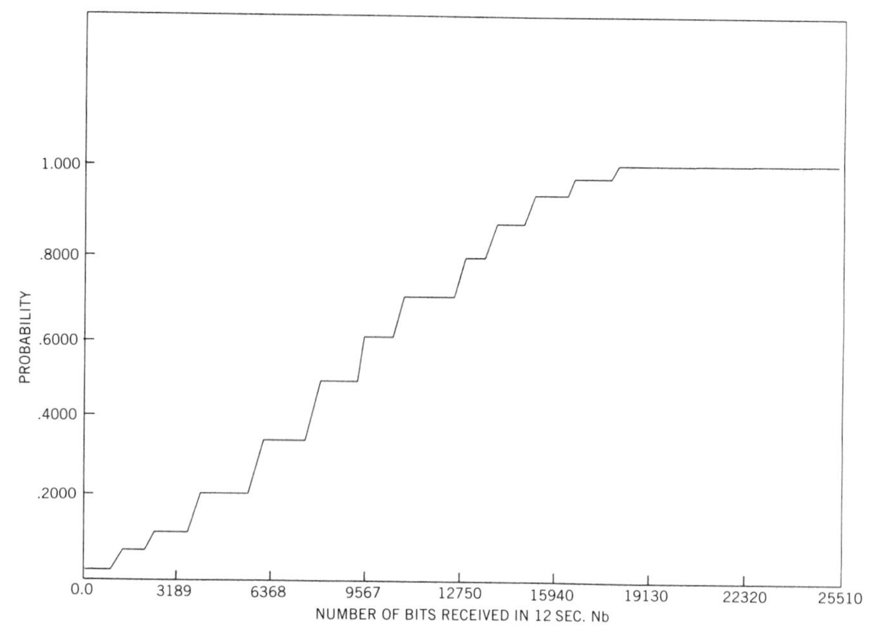

Figure 5.9. Probability of receiving N_b bits in a 12-s interval FAVR system, fixed rate–30 kB/s.

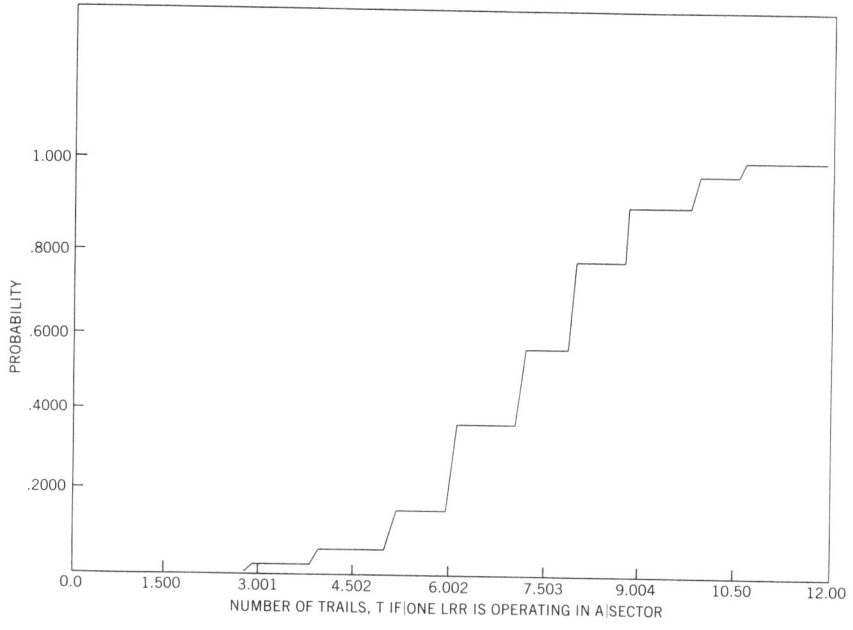

Figure 5.10. Probability of T trails occurring at ROCC 1 LRR operating in a sector.

Figure 5.8 shows the probability of receiving N_b bits in 12 s for a FAVR system operating in a bandwidth of 40 kHz. Note that with a probability of 80%, 70,000 bits will be received correctly in 12 s. Figure 5.9 shows that with a probability of 80%, more than 4500 bits are received correctly every 12 s if the FAVR system is constrained to operate at a fixed bit rate of 30 kb/s.

Figure 5.10 gives the probability of T trails occurring between the ROCC and the LRRs. Note that with a probability of 80%, 6–12 trails occur every 12 s between the ROCC and a single LRR.

5.2.2 Hardware Simulator

In addition to the software simulation, a hardware simulator was built to test the FAVR system. The simulator was developed to generate underdense trails with waiting time, peak SNR, and decay rate probabilities that could be independently programmed. An example of several simulated trails received by the FAVR receiver is shown in Fig. 5.11. Note that each of the parameters (waiting time, peak SNR, and decay rate) vary.

Two-way voice and two-way data were transmitted over the simulator. Figure 5.12 shows packets received at the ROCC receiver during a long trail. The first and last "short" packets are for signaling, while the remaining seven packets carried data at different bit rates.

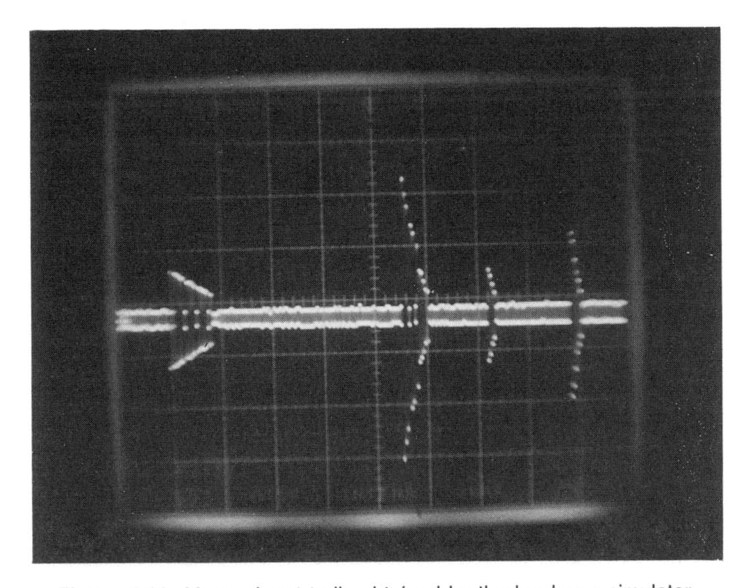

Figure 5.11. Meteor burst trails obtained by the hardware simulator.

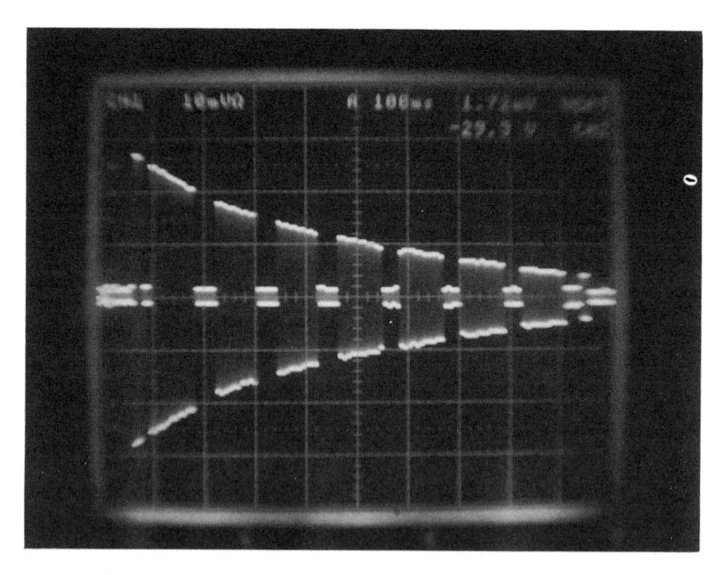

Figure 5.12. The packets received from a single MB trail.

5.3 FIELD TEST RESULTS

The SCS FAVR model was adjusted to operate using a bandwidth of 40 kHz and a data rate variable from 32 to 90 kb/s. With this configuration a minimum SNR of 12 dB, measured in the 40-kHz bandwidth, is required. Reduced waiting time (as well as reduced data rates) can be accomplished by switching between the 40-kHz bandwidth and a 12-kHz bandwidth filter. (Associated with the 12-kHz bandwidth are data rates from 8 to 32 kb/s). This permits detection of weaker trails.

Figure 5.13 shows the cumulative probability distribution of the peak SNR of the received packets. The SNR of these packets varied between 12 and 30 dB. Packets with a peak SNR of less than 12 dB were discarded and packets with a peak SNR of more than 30 dB were a rarity. Indeed, as seen from Fig. 5.13, the probability of receiving a packet of between 17 and 30 dB was 50%.

The data taken appeared as though the unconditional probability distribution was normal. To test this hypothesis, the function $\psi(x)$ was defined, where

$$\psi(x) = \frac{f'(x)}{f(x)} \tag{5.3}$$

where $f(x)$ is the probability density function (pdf).

$\psi(x)$ can be estimated using the sample data taken during the field

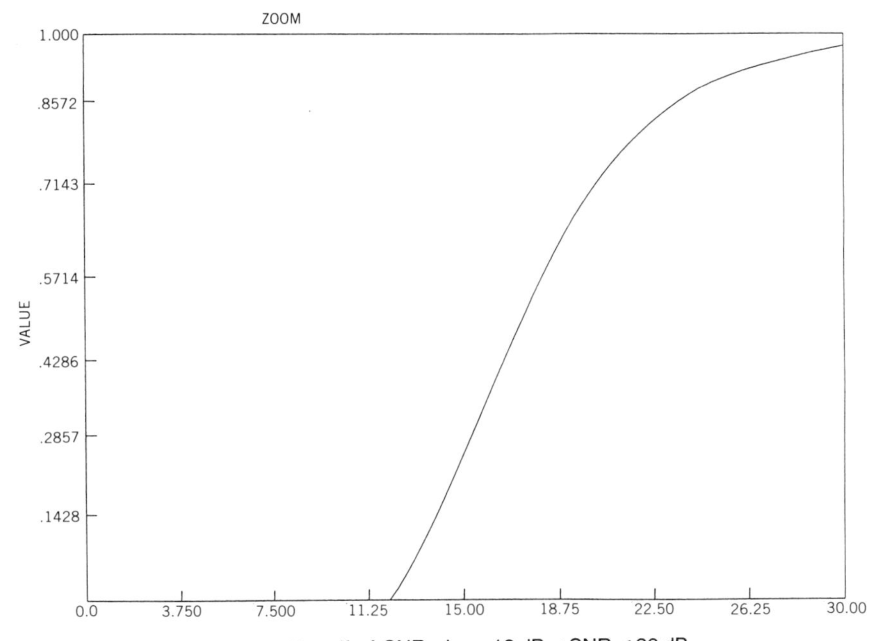

Figure 5.13. cdf of SNR given 12 dB ≤ SNR < 30 dB.

measurements. In particular, a polynomial function can be found that estimates ψ. In the normal case the function ψ is linear:

$$\psi(x) = -\frac{(x - E\{x\})}{\sigma^2} \tag{5.4}$$

Once ψ is estimated in the interval $[a, b]$ (range of SNR over which data were taken), the shape of the conditional density can be inferred. The data obtained do indeed indicate that ψ is linear in $[a, b]$. Therefore, it was concluded that a normal density is a reasonable model for determining the unconditional distribution of the SNR. The above method of extrapolation is particularly useful because the desired density was truncated (no data were recorded by the system monitor for SNR less than 12 dB).

Figure 5.14 shows the normal distribution obtained by the above procedure. Figure 5.15 shows the empirical probability distribution function overlaid on the distribution function shown in Fig. 5.13. The agreement is seen to be quite good. Note the very close comparison between the distribution obtained using the actual data and the normal distribution. Using the unconditional normal distribution it is estimated that 50% of the trails have a peak SNR of less than 13 dB. Hence, approximately 50% of the incoming trails were disregarded using a 40-kHz bandwidth. Approximately 40% more trails could be accommodated using a switched bandwidth system

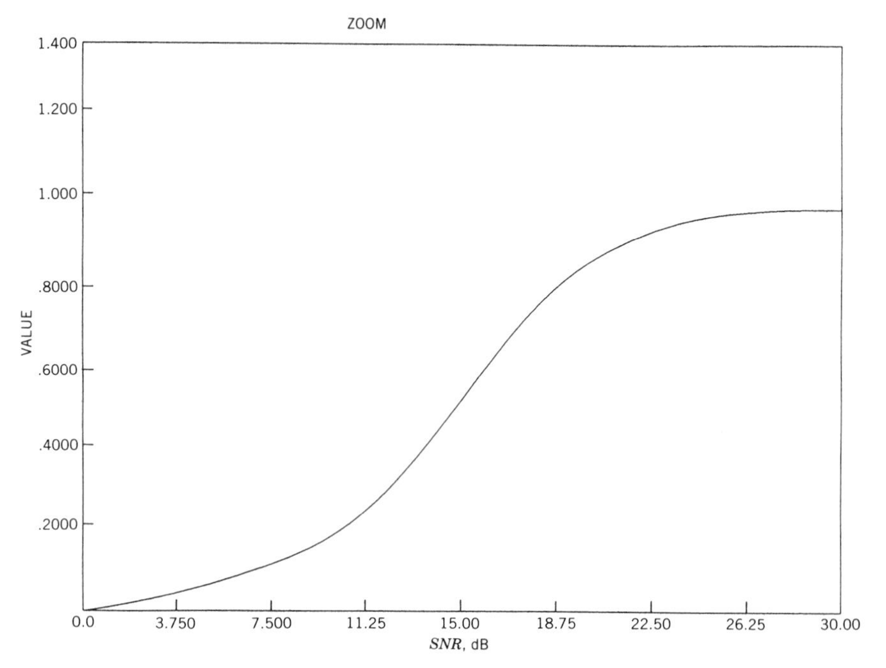

Figure 5.14. Normal distribution fit to measured df.

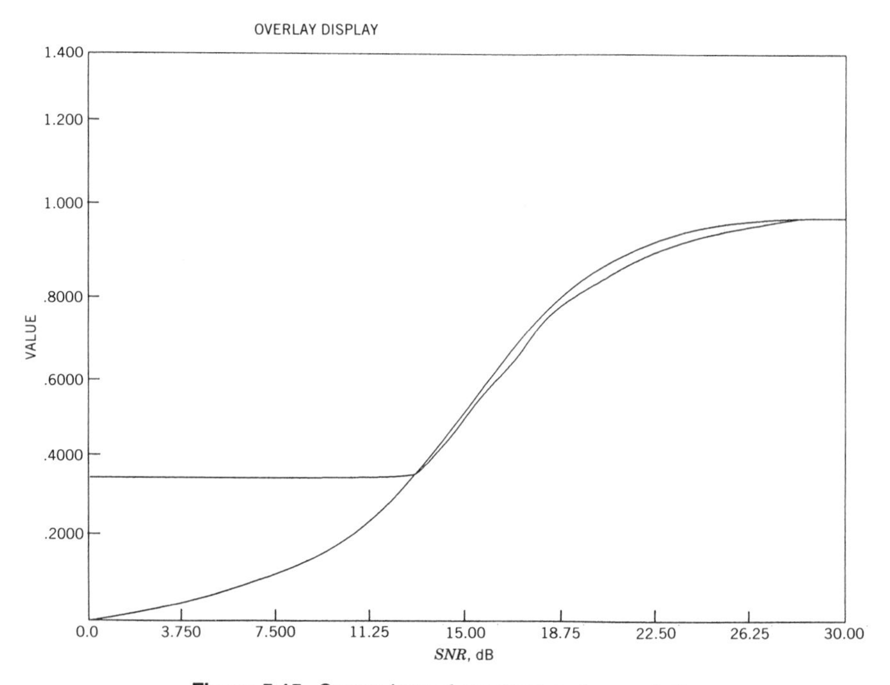

Figure 5.15. Comparison of empirical and normal df.

of 12 kHz. (The extra 10% represents the percentage of trails with peak SNR too low to detect with the 12-kHz bandwidth).

Figure 5.16 shows the empirical probability distribution function of the number of bits transmitted per trail for the FAVR system. Note that there is a 10% probability that no bits are transferred; this is due to the estimated decay of the SNR resulting in the NAK (the available signal is unsuitable for demodulation at 32 kb/s). The average number of bits transferred is 6.9 kb/trail.

Figure 5.17 shows the empirical probability distribution function of the number of packets per trail that the FAVR system processed. The inter-packet time is 100 ms. Assuming that one trail occurs every 2 s on the average, the average number of bits transferred in 12 s (the time of each radar sweep) is 41 kb. This exceeds the number of bits required to be transferred from the LRR to the ROCC.

Figure 5.18 shows the empirical probability distribution function of the number of bits transmitted by a fixed-rate system. The fixed system operates at a data rate of 8 kb/s in a 15-kHz bandwidth. Again there is a small probability that zero bits are transferred. The average number of bits transferred is 1.23 kb/trail.

Figure 5.19 is an overlay of Figs. 5.16 and 5.17, affording a comparison of throughput of the two systems on a per trail basis. While the lower (fixed)

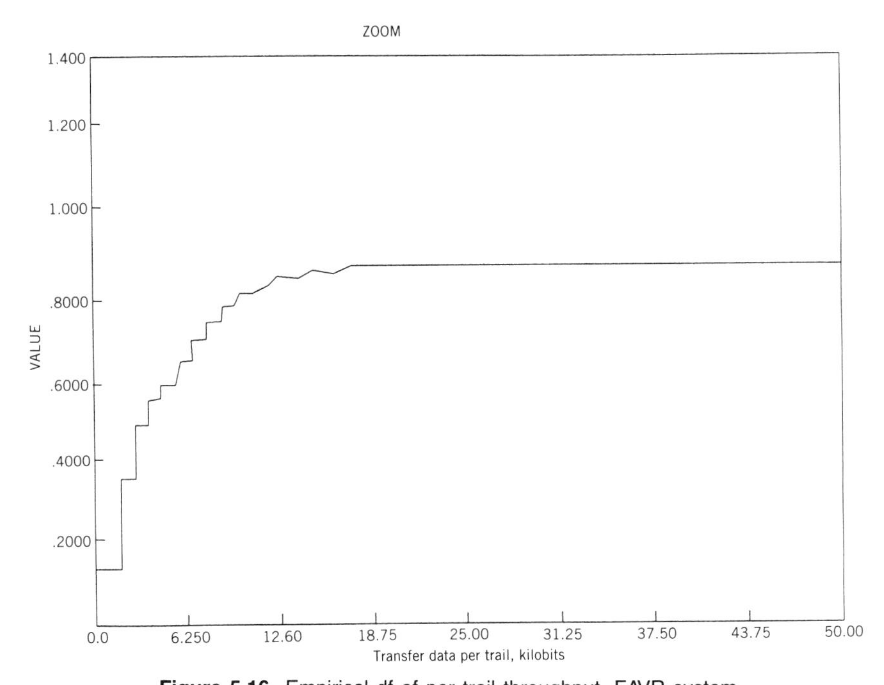

Figure 5.16. Empirical df of per trail throughput, FAVR system.

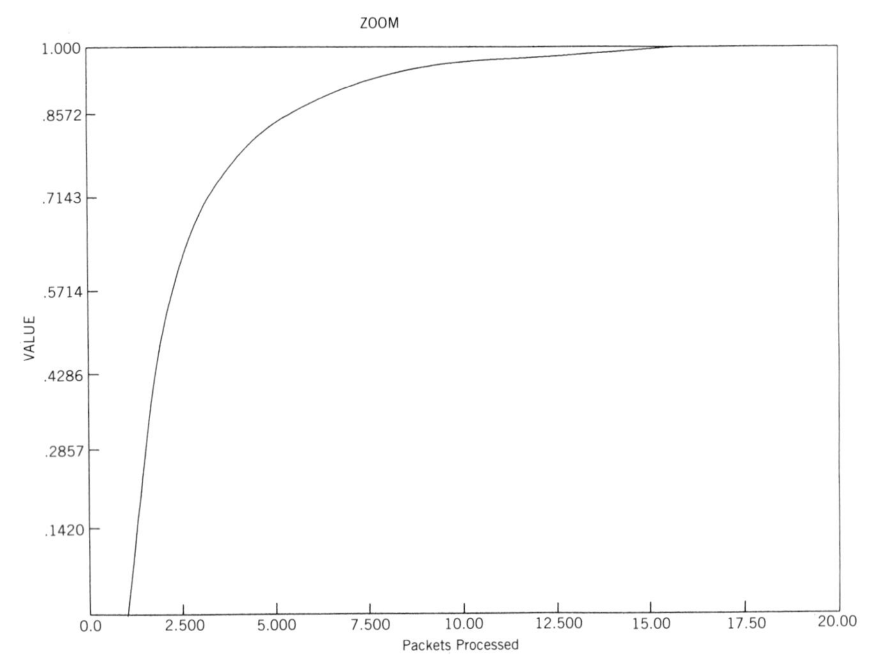

Figure 5.17. cdf of number of packets processed, FAVR system.

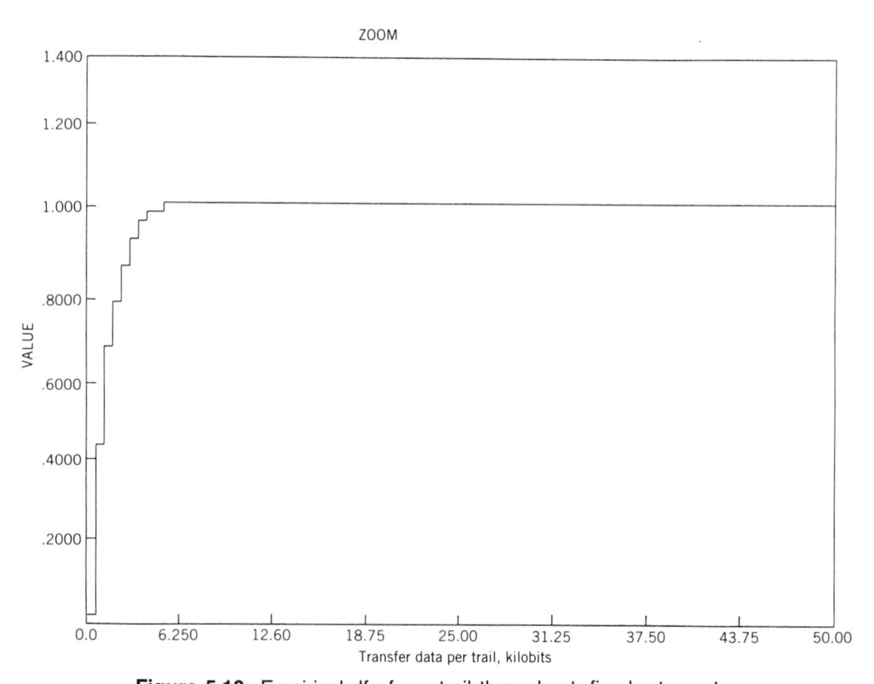

Figure 5.18. Empirical df of per trail throughput, fixed rate system.

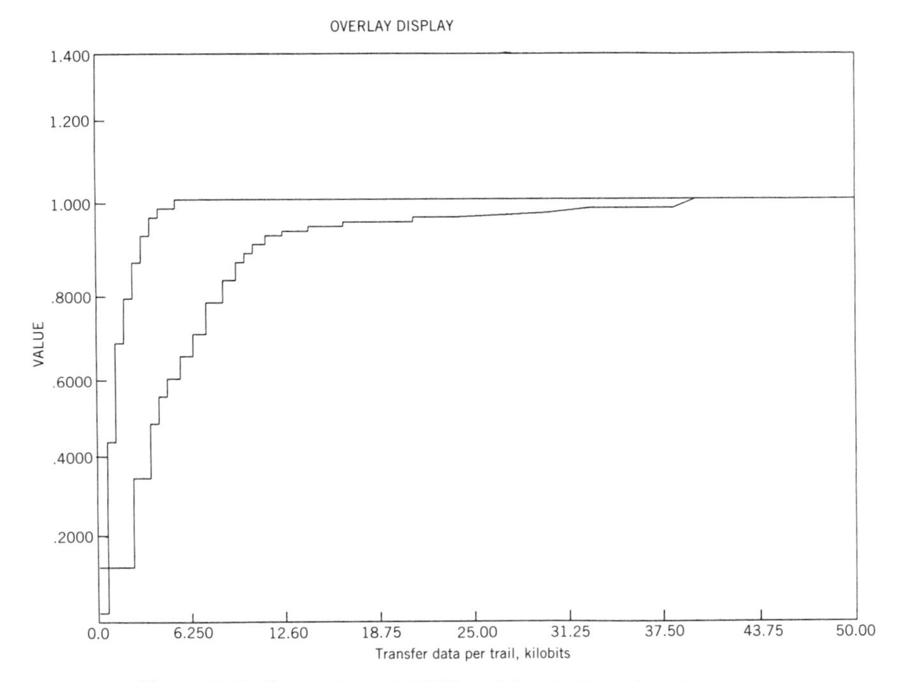

Figure 5.19. Comparison of FAVR and fixed rate system throughput.

data rate system can utilize a greater percentage of trails (about 60% more), the average throughput per trail is 3.5 times higher for the FAVR system.

5.4 CONCLUSION

It has been clearly demonstrated that SCS Telecom's FAVR modem could be used to transmit data and voice at data rates significantly higher than possible using a fixed rate modem. In addition, full-duplex operation would more than double the number of bits of data transmitted, since headers and protocols could be minimized.

6

PERFORMANCE OF METEOR BURST COMMUNICATION USING VARIABLE DATA RATES

Sorin Davidovici and Emmanuel G. Kanterakis
Rutgers University, Department of Computers and Electrical Engineering, New Brunswick, New Jersey

6.1 INTRODUCTION

The subject of meteor burst (MB) communications has received a great deal of attention since the late 1950s and early 1960s [1–3, 6–12, 16, 23]. Though interest seemed to decay for some years due to lack of adequate technology, the advance in electronics and the need for secure communications without the use of satellites has revived the interest in MB communications in recent years [4, 5, 13–15, 17–23].

As the earth orbits around the sun, a great number of meteors enter the earth's atmosphere. Upon entering the earth's atmosphere meteors are subject to ablation, producing long columns of ionized particles approximately 100 km above the earth's surface. These ionized columns (or trails) diffuse rapidly in the earth's atmosphere. Typical meteor burst trail durations are in the order of a second. When a meteor burst trail is properly oriented with respect to the transmitter and receiver antennas it can be used to reflect radio waves for the effective time duration of the trail. As implied by it's name, meteor burst communication systems transmit messages in an intermittent manner. The transmitter and the receiver must be able to detect the occurrence of a burst, use a short portion of the meteor burst trail duration for synchronization, and use the remaining effective duration of the meteor burst trail for communication. Though meteor burst trails are categorized depending on their electroline density [23], for the sake of simplicity and mathematical convenience, all meteor burst trails will be assumed to be underdensed. Section 6.4 discusses the meteor burst trail characteristics and scenario in more detail. Expressions for the system performance are developed as a function of the fixed operational parameters associated with the given scenario. Thus, optimal signal-to-noise ratios are found and plotted as a function of packet size in bits per packet for the case where the data is transmitted without the use of any forward error correction code. When error correction coding is used the optimal signal-to-noise ratio is found for the particular combination of operational parameters, as outlined in the following sections. Due to the nature of the expressions involved, numerical maximization algorithms are used to generate results. Likewise, the maximum throughput for the uncoded case is found by a numerical maximization procedure. We have attempted to improve the readability of the paper by minimizing the number of mathematical derivations in the main text. The appendix, however, contains detailed derivations of all quantities and expressions used in the main text. The results are plotted and the plots are discussed in Section 6.8. Section 6.9 summarizes the results obtained here and their significance.

6.2 PREVIOUS WORK

The most complete methodology for the design and analysis of meteor burst communication systems has been given by Oetting [18]. His approach facilitates the analysis of meteor burst communications systems as used in a number of applications. Other researchers investigated the meteor burst channel model as well as the performance of meteor burst systems under various assumptions. Because of the time-varying characteristics of the meteor burst channel, some researchers assumed worst-case bit error rate (BER) conditions when analyzing the performance of communication systems operating over this channel. This assumption circumvented the mathematical difficulties associated with an exact analysis which would take into account the time varying nature of the channel and complicate the receiver structure.

A thorough investigation of meteor burst communication systems under the assumption of worst-case BER per transmitted packet (or for the whole transmission period) is given by Hampton [20] and by Milstein et al. [19]. Hampton computed the time required to broadcast a message consisting of a set of packets to an arbitrary number of receivers over a meteor burst channel while taking into account the actual time-varying nature of the BER. He computed the broadcast time as a function of different modulation schemes and system parameters using either an ARQ or a hybrid ARQ/FEC error control method. Optimal data rates and packet sizes as functions of message length were found. To simplify his results Hampton associated with the packet a constant BER equal to the worst-case BER (which occurred at the end of the packet). The analysis given by Milstein et al. [19] quantizes the performance of a system by the probability of successfully communicating a message in a given time period. Two different protocols are used. The protocols used in this work are identical to the protocols used in Milstein et al. [19] where results are again bounds obtained under the worst-case BER assumption.

The only investigations made on systems operating under constant SNR (and thus with time varying bit rates) were performed by Abel [17] and Weitzen et al. [22]. Both investigated the maximum throughput that can be achieved using time varying bit rates. Abel derived two performance bounds, which give the maximum number of bits that can be relayed during a single burst using either an optimum constant bit rate or a continuously varying bit rate. Since Abel considered a single burst with a specific electroline density, his results do not account for the statistical nature of the channel. Weitzen et al. analyzed the capacity of the meteor burst communication channel using a time-varying bit rate approach.

The most complete investigation of the meteor burst channel and the effect of scenario-dependent factors is as performed by Spezio [21]. The major portion of the Study is devoted to the analysis of the propagation and statistical characteristics associated with the meteor ionization trails. This work uses the basic system setup and geographical parameters given in Spezio [21].

6.3 PROTOCOL DESCRIPTION

The scenario that governs the operation of the systems has been described in detail in Spezio [21]. Prior to communicating, the transmitter and the receiver must determine the presence of an available meteor burst and synchronize their systems. Successful system synchronization is followed by transmission of messages. Finally, the systems must be able to determine the disappearance of the channel, stop communicating, and restart the synchronization process if necessary. The procedure by which the above takes places is determined by the communications protocol. The performance of the communication system is determined by the message completion probabilities will be analyzed subject to the following two protocols:

1. In the first protocol the receiver continuously probes the channel. When the presence of a burst is detected by the transmitter it begins the message transmission after a delay of approximately t_0 seconds, which represents the one-way propagation delay. Transmission is maintained until the channel strength falls below a predetermined threshold.
2. In the second protocol the transmitter probes for the presence of meteor burst. When a burst is detected the receiver sends an acknow-

ledgment. After receiving an acknowledgment from the receiving station the transmitter sends a packet and restarts the synchronization cycle. Thus, only one packet is transmitted per burst.

6.4 CHANNEL CHARACTERIZATION

The mathematical modeling of the channel closely follows the models presented in Millstein et al. [19] and Spezio [21]. The scenario-related factors are based on the geographical setup, which is shown in Fig. 6.1. In addition, the following assumptions are used in the analysis:

1. All of the intercepted trails are underdense.
2. The transmitter and the receiver beam to an area halfway between them. The effective reflection area has a radius $r = 100$ km.
3. The distance R_{CT} between the transmitter and the reflection area equals the distance R_{CR} between the receiver and the reflection area.
4. The angle ϕ is taken to be worst case (the time constant is smallest and the distance largest).
5. For the channel symbol rates used the channel is assumed to have a constant characteristic in the symbol bandwidth.

For the scenario illustrated in Fig. 6.1 and subject to the assumptions given above, the power received from a meteor trail is given by [22]

$$P_{\text{rec}}(t) = A_0 q^2 e^{-t/\tau} \tag{6.1}$$

with

$$A_0 = P_t G_t G_r \lambda^3 q^2 \sigma_e \frac{\exp[-(8\pi^2 r_0^2/\lambda^2 \sec^2 \phi)]}{(4\pi)^3 2R_T^3(1 - \sin^2 \phi \cos^2 \beta)} \tag{6.1a}$$

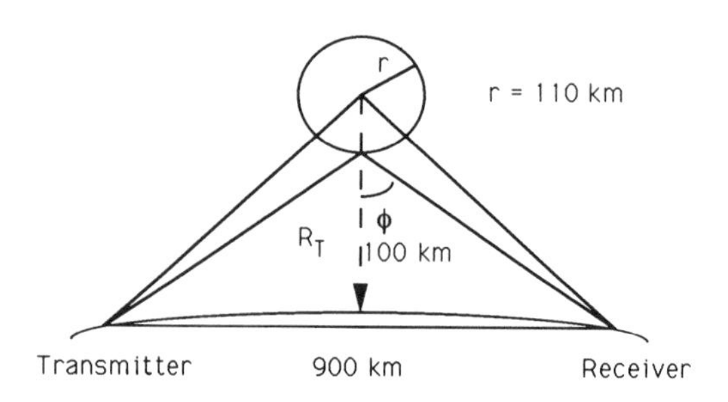

Figure 6.1. Geographical setup.

and

$$\tau = \frac{\lambda^2 \sec^2 \phi}{32\pi^2 D} \tag{6.1b}$$

where the transmitted power $P_t = 2000$ W, both transmitter and receiver antenna gains are 13 dBi, the carrier frequency corresponding to the wavelength λ is 35 MHz, q is the electroline density in (el/m), r_0 is the nominal initial radius of a trail (0.65 m), σ_e is the effective echoing area of the electron (10^{-28} m^2), and the diffusion coefficient D is (10 m^2/s). The angles ϕ and β are the angle of incidence of the transmitted planewave, and the angle between the trail and the great circle path between transmitter and receiver.

Given that the distance D_S between the transmitter and the receiver is 900 km, τ equals 0.38 s. Also, if the noise and signal bandwidth are equal, the noise power received in the signal's bandwidth can be expressed as

$$N(t) = 2kR(t) \times 9.65 \times 10^3 \text{ W} \tag{6.1c}$$

where $R(t)$ is the signaling rate and k is Boltzman's constant. The signal-to-noise ratio as a function of the data rate used, $R(t)$, then becomes

$$\text{SNR}(t) = \frac{P_{\text{rec}}(t)}{N(t)} = \frac{A_0 q^2 e^{-t/\tau}}{N(t)} = \frac{Aq^2 e^{-t/\tau}}{R(t)} \tag{6.2}$$

where, from Eqs. 6.1c and 6.2 A is given by

$$A = \frac{A_0}{2k \times 9.65 \times 10^3} = 7.10237 \times 10^{-22}$$

For the scenario assumed, Eq. 6.2 relates the signal-to-noise ratio to the actual data rate used by the system. This key relationship is used in the subsequent analysis of the performance of variable data rates meteor burst communication systems.

6.5 STATISTICS OF USEFUL CHANNEL LIFETIMES AND TRANSMISSION DATA RATES

It was shown in Eq. 6.2 that the data rate $R(t)$ and the received signal-to-noise ratio SNR(t) are related by

$$R(t) = \frac{Aq^2}{\text{SNR}(t)} e^{-t/\tau} \tag{6.3}$$

It is possible to keep the received signal-to-noise ratio constant at some arbitrary value SNR$_c$ by varying the bit rate accordingly. The bit rate $R(t)$ can then be expressed as

$$R(t) = \frac{Aq^2}{\text{SNR}_c} e^{-t/\tau} \tag{6.4}$$

where τ is a function of system and scenario-dependent parameters (which are assumed fixed even though they may exhibit small variations due to the orientation of the meteor trails) and of q, the electroline density, which is a random variable. The electroline density can be estimated by a received signal power measurement performed at the beginning of a burst (in the acquisition phase). After determining the electroline density of the meteor burst trail the value of the data rate to be used becomes a decreasing, well-defined function of time for the remainder of the burst. The burst itself will be used only for as long as the data rate $R(t)$ is above a lower limit R_{min}. Thus, the time span for which the meteor burst trail is of use is found from the minimum data rate consideration:

$$R_{\text{min}} = \frac{Aq^2}{\text{SNR}_c} e^{-\Delta t/\tau} \tag{6.5}$$

where Δt becomes

$$\Delta t = -\tau \ln\left(\frac{\text{SNR}_c \times R_{\text{min}}}{Aq^2}\right) \tag{6.6}$$

The quantity Δt is a random variable which is a function of the electroline density of the meteor burst trail. The probability density function of the electroline density function q is given in Spezio [21] as

$$f(q) = \frac{q_0}{q^2} \tag{6.7}$$

where q_0 is some minimum value of q for which the meteor trail can be considered to be useful. One way of defining an useful meteor burst channel is by its ability to support any amount of data at or above the minimum data rate R_{min} after synchronization. The electroline density of meteors that can provide a channel able to support rates greater than R_{min} must be larger than some q_0. The value of q_0 can be found from Eq. 6.5 by setting Δt equal to the synchronization time t_0. Thus,

$$q_0 = \sqrt{\frac{\text{SNR}_c \times R_{\text{min}}}{\text{A}}} e^{t_0/2\tau} \tag{6.8}$$

Note that for the second protocol the synchronization time is double that of the first protocol and thus Δt must be set equal to $2t_0$.

From elementary probability theory the probability density function of Δt is found to be

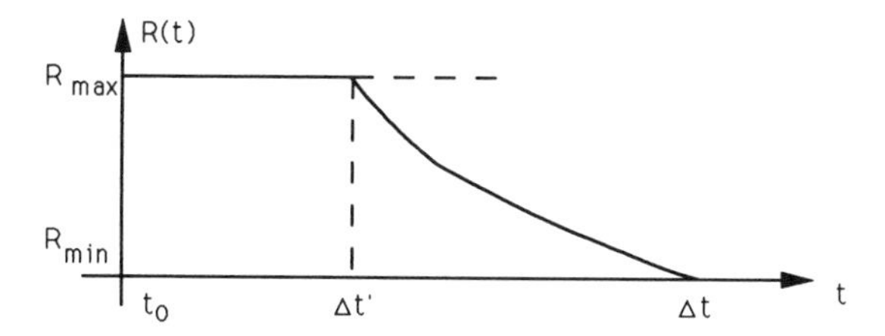

Figure 6.2. Rate (t) as a function of time when $q > q'$.

$$f(\Delta t) = \frac{1}{2\tau} e^{-(\Delta t - t_0)/2\tau} u(\Delta t - t_0) \tag{6.9}$$

where $u(t)$ is the unit step function. In a realistic system the data rate cannot vary solely as dictated by Eq. 6.4. Any practical system will have a maximum data rate at which the transmitter is able to signal. This rate is denoted here by R_{max}. Therefore, whenever Eq. 6.4 dictates that a data rate $R(t)$ greater than R_{max} should be used, the system will operate at R_{max}. Also, as stated previously, the lowest allowable signaling rate is R_{min}, after which time the link will be closed and, if necessary, the search for a new channel will restart. Throughout the following analysis R_{max} and R_{min} will be kept variable. Figure 6.2 illustrates the function $R(t)$ given some particular q such that for some time duration $\Delta t' - t_0$ the data rate $R(t)$ is greater than its maximum allowable value R_{max}.

6.6 PROBABILITY OF CORRECT PACKET RECEPTION

The preceeding sections have derived expressions for system operational variables (data rates and link statistics) as a function of meteor burst trail characteristics (the electroline density). The expected system performance in terms of the probability of correctly receiving a message can now be derived using these intermediate results. In all of the subsequent analysis it will be assumed that a message consists of exactly one packet.

When the system operates using the first protocol, it was shown in Millstein et al. [19] that the probability of correctly receiving a packet in T_D seconds is given by

$$P_{CM} = \sum_{n=1}^{\infty} \sum_{l=1}^{\infty} P\left(\frac{\bar{e}}{l}, n\right) P\left(\frac{l}{n}\right) P(n) \tag{6.10}$$

Since the protocols considered only allow an integer number of packets to be transmitted per burst, all the above probability density functions are assumed to be conditioned on the fact that l, the number of packets transmitted per meteor burst trail, is greater than or equal to one. In the above expression $\bar{e}$ represents the event (at least one packet error free), l

represents the event (l completed packets), and n represents the event (n meteor bursts in T_D seconds). The probability of receiving a correct packet out of l packets is independent of the number of trails used to transmit the l packets. Therefore,

$$P\left(\frac{\bar{e}}{l}, n\right) = P\left(\frac{\bar{e}}{l}\right) = 1 - P_p^l \tag{6.11}$$

where P_p is the probability of a packet being received in error. If the modulation used is FSK with noncoherent demodulation if there are I bits per packet, P_p is given by

$$P_p = 1 - (1 - P_b)^I \tag{6.12}$$

where P_b is the average bit error rate given as

$$P_b = \frac{1}{2} \exp\left(-\frac{1}{2} \text{SNR}\right) \tag{6.13}$$

The number of meteor bursts that occur in T_D seconds can be assumed to be Poisson distributed. The probability density function of the number of meteor bursts n that occur in T_D seconds, $P(n)$, is then completely determined by T_I, the average burst interarrival time and given by

$$P(n) = \frac{(T_D/T_I)^n e^{-T_D/T_I}}{n!} \tag{6.14}$$

The condition $l \geq 1$ can be satisfied by restricting the above expression to consider only trails whose electroline density is greater than or equal some $q_{\min}$, which is defined in the Appendix. Trails having an electroline density $q \geq q_{\min}$ can be used to transmit one or more packets. It was shown in Spezio [21] that for the geographical scenario given in this paper, the average interarrival time between bursts whose electroline density is greater than some $q_{\min}$ is given by

$$T_I = \frac{q_{\min}}{\lambda_0} \tag{6.15}$$

where λ_0 is a constant dependent on the specific geographical scenario. This is because in Spezio [21] it was shown that the average number of bursts/second whose electroline density is greater than some value of q, $q_{\min}$, is given by $\lambda_0/q_{\min}$. For this scenario λ_0 becomes $\lambda_0 = 1.056 \times 10^{12}$.

The last term to be defined in Eq. 6.10 is $P(l/n)$, the conditional probability of transmitting l packets over n meteor burst channels. The number of packets sent over n meteor burst channels is a random variable due to the randomness of the electroline density q, which in turn determines the duration and strength of the channel. The conditional density $P(l/n)$ can be obtained from $P(l/n=1)$ by use of the convolution operator. However, because of the numerical computational difficulties associated with Eq. 6.10, a different approach is used.

Since in Eq. 6.10 the occurrence of meteors is assumed to be disjoint in time (communications periods over different bursts do not overlap), an equivalent system composed of a number of subsystems disjoint in the q domain is proposed for analysis. The system is shown in Fig. 6.3 and is assumed to consist of a number of subsystems (assumed infinite for now), each of which transmits solely on a particular range of electroline density.

The total number of symbols that can be transmitted on a meteor burst trail of electroline density q is given by the integral of the symbol rate function $R(t)$ over the useful duration of the meteor burst channel (i.e., from t_0 to $t_0 + \Delta t$). The number of packets l that can be transmitted over a meteor burst trail with a specific electroline density q is then given by

$$l = \frac{1}{I} \int_{t_0}^{t_0+\Delta t} R(t)\, dt \tag{6.16}$$

where $R(t)$ is given in Eq. 6.4. Let $q^{[l]}$ be that value of q in the above equation for which a meteor burst trail can support the transmission of exactly l packets. Likewise, let $q^{[l+1]}$ be that value of q for which the meteor burst trail can support the transmission of exactly $l+1$ packets. Then the probability that a single burst can support exactly l packets must equal the probability that the value of the electroline density q lies between $q^{[l]}$ and $q^{[l+1]}$. Hence, using Eq. 6.7,

$$P(l/n=1) = q_{\min} \int_{q^{[l]}}^{q^{[l+1]}} \frac{1}{q^2}\, dq \tag{6.17}$$

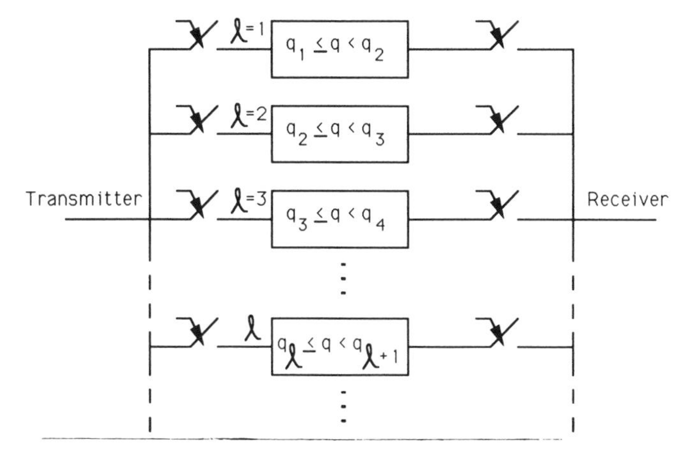

Figure 6.3. A hypothetical meteor burst communication system.

Each one of the subsystems operates in some range of electroline density Δq from $q^{[l]}$ to $q^{[l+1]}$. When a burst with an electroline density in the range between $q^{[l]}$ and $q^{[l+1]}$ occurs, l packets will be transmitted through the subsystem covering that range of electroline density q, while all other subsystems will remain idle. If the event of correctly receiving a packet using the lth subsystem is defined as P_{Cl}, then the probability of correctly receiving a packet using the overall system is the probability that a packet was correctly received through any of the subsystems. Hence,

$$\begin{aligned} P_{CM} &= 1 - P[\text{incorrectly receiving a packet in all subsystems}] \\ &= 1 - (1 - P_{C1})(1 - P_{C2}) \cdots (1 - P_{Cl}) \cdots = 1 - \prod_{l=1}^{\infty} (1 - P_{Cl}) \qquad (6.18) \end{aligned}$$

The probabilities of correct reception of a packet using the lth system P_{Cl} are

$$P_{Cl} = \sum_{n=1}^{\infty} [1 - (P_p^l)^n] P_l(n) \qquad (6.19)$$

where P_p^l is the probability of receiving all l packet in error after a single use of the lth subsystem, and $P_l(n)$ is the probability density function of n meteor bursts with an electroline density q in the range $(q^{[l]}, q^{[l+1]})$ occurring in T_D seconds.

Defining as T_{Il} the interarrival time of bursts in the $(q^{[l]}, q^{[l+1]})$ range, Eq. 6.19 can be written as

$$P_{Cl} = \sum_{n=1}^{\infty} (1 - P_p^{ln}) \frac{e^{-T_D/T_{Il}}}{n!} \left(\frac{T_D}{T_{Il}}\right)^n = 1 - \exp\left[-\left(\frac{T_D}{T_{Il}}\right)(1 - P_p^l)\right] \qquad (6.20)$$

Substituting in Eq. 6.18 we get

$$P_{CM} = 1 - \prod_{l=1}^{\infty} \exp\left[-\left(\frac{T_D}{T_{Il}}\right)(1 - P_p^l)\right] = 1 - \exp\left[-\sum_{l=1}^{\infty} (1 - P_p^l) \frac{T_D}{T_{Il}}\right] \qquad (6.21)$$

From Eqs. 6.15 and 6.17 the interarrival time of meteor burst trails that can support the transmission of exactly l packets is

$$T_{Il} = \frac{1}{\lambda_0}\left(\frac{1}{q_l} - \frac{1}{q_{l+1}}\right)^{-1} = \frac{q_{\min}}{\lambda_0}\left[q_{\min}\left(\frac{1}{q_l} - \frac{1}{q_{l+1}}\right)\right]^{-1} = \frac{q_{\min}}{\lambda_0}[P(l/n = 1)]^{-1} \qquad (6.22)$$

Hence, Eq. 6.21 becomes

$$P_{\mathrm{CM}} = 1 - \exp\left[-\sum_{l=1}^{\infty} \frac{(1-P_{\mathrm{p}}^{l})T_{\mathrm{D}}\lambda_0 P(l/n=1)}{q_{\min}}\right]$$
$$= 1 - \exp\left\{-\frac{T_{\mathrm{D}}\lambda_0}{q_{\min}}\left[1 - \sum_{l=1}^{\infty} P_{\mathrm{p}}^{l} P(l/n=1)\right]\right\} \tag{6.23}$$

Let

$$C^{-1} = \frac{\lambda_0}{q_{\min}}\left[1 - \sum_{l=1}^{\infty} P_{\mathrm{p}}^{l} P(l/n=1)\right] \tag{6.24}$$

Then

$$P_{\mathrm{CM}} = 1 - e^{-T_{\mathrm{D}}/C} \tag{6.25}$$

where now C represents the time constant for P_{CM}. The summation in Eq. 6.23 can be stopped at some large value N since $P(l/n=1)$, the probability of l packets transmitted over a meteor burst, approaches 0 for large l. The significant reduction in the computational difficulty associated with Eq. 6.23 as compared with the computational difficulty associated with Eq. 6.10 is evident. Equation 6.23 uses a single summation, and the conditional densities $P(l/n)$ for $n > 1$ are not used. The conditional density that is used, $P(l/n=1)$, is derived in the appendix. Equations 6.10 and 6.23 were numerically shown to generate identical results.

The conditional probability density function $P(l/n=1)$ is derived in the appendix and it is shown to be

$$q_{\min}\sqrt{\frac{A\tau}{\mathrm{SNR_e}}}\left[\frac{1}{\sqrt{lI + R_{\min}\tau}} - \frac{1}{\sqrt{(l+1)I + R_{\min}\tau}}\right]e^{-t_0/2\tau} \qquad \text{if } l+1 \le l_0'$$

$$q_{\min}\sqrt{\frac{A}{\mathrm{SNR_e}}}\left(\frac{1}{\sqrt{lI + R_{\min}\tau}} - \frac{1}{\sqrt{R_{\max}}}\right.$$
$$\left.\exp\left\{-\frac{1}{2}\left[\frac{(l+1)I}{R_{\max}\tau} + \frac{R_{\min}}{R_{\max}} - 1\right]\right\}\right)e^{-t_0/2\tau} \qquad \text{if } l \le l_0' \le l+1$$

$$q_{\min}\sqrt{\frac{A}{\mathrm{SNR_e} \times R_{\max}}}\left\{\exp\left(-\frac{lI}{2R_{\max}\tau}\right) - \exp\left[-\frac{(l+1)I}{2R_{\max}\tau}\right]\right\}$$
$$\exp\left[\frac{1}{3}\left(1 - \frac{R_{\min}}{R_{\max}} - \frac{t_0}{\tau}\right)\right] \qquad \text{if } l > l_0' \tag{6.26}$$

$$\text{where} \quad l_0' = \frac{\tau}{I}(R_{\max} - R_{\min}) \tag{6.27}$$

When the system operates using the second protocol (in which only one packet is transmitted per burst) the probability of receiving a message correctly depends on the probability of the electroline density that the meteor burst trail q is above the low one-packet transmission threshold $q_{\min}$, as derived in the appendix and given by Eq. 6.A.16, on the probability that all of the data bits are received correctly and on the probability of a meteor burst trail actually occurring. The probability of message completion in one meteor burst trail is given by

$$\Pr[\text{message completion}/1 \text{ burst}] = \begin{cases} 1 - P_{\mathrm{p}} & \text{if } q \geq q_{\min} \\ 0 & \text{otherwise} \end{cases} \tag{6.28}$$

The expression above is the conditional probability of successful message completion given one meteor burst trail available. The total probability of completion is found by averaging the conditional probabilities of successful message completion in any number of meteor burst trails that may occur in the observation time T_{D}. Thus,

$$P_{\mathrm{CM}} = \sum_{n=1}^{\infty} \Pr[\text{message completion}/n \text{ bursts}] \Pr[n \text{ bursts}] \tag{6.29}$$

where the probability of n meteor burst trails occurring during the observation time T_{D} is given in Eq. 6.14 and P_{CM} becomes

$$P_{\mathrm{CM}} = \sum_{n=1}^{\infty} (1 - P_{\mathrm{p}}^{n}) \left(\frac{T_{\mathrm{D}}}{T_{\mathrm{I}}} \right)^{n} \frac{e^{-T_{\mathrm{D}}/T_{\mathrm{I}}}}{n!} = 1 - e^{-(T_{\mathrm{D}}/T_{\mathrm{I}})(1-P_{\mathrm{p}})} \tag{6.30}$$

In the above expression T_{I} is again given by Eq. 6.15.

The nature of the channel is such that the errors are quite bursty in nature [19]. Therefore, the forward error control (FEC) coding schemes chosen in previous works were Reed–Solomon (RS) codes. In this work these codes are again chosen due to the nature of the channel as well as for comparison purposes.

The packets consist of I bits of information which are blocked into J blocks of mk bits per block. An (n, k, e) RS code is used to encode this data. The notation denotes a block length of n symbols, out of which k are information symbols and which can correct e errors. To receive a packet correctly all J blocks must be received correctly. The probability of this occurring is

$$1 - P_{\mathrm{p}} = (1 - P_{\mathrm{w}})^{J}$$

$$\text{where} \quad P_{\mathrm{w}} = \sum_{i=e+1}^{n} \binom{n}{i} P_{\mathrm{s}}^{i} (1 - P_{\mathrm{s}})^{n-i}$$

and P_{s} is the error probability of an RS symbol, which is given by

$$P_{\mathrm{s}} = 1 - (1 - P_{\mathrm{b}})^{m}$$

where P_b is the bit error rate defined previously. The performance of the system when FEC is used is discussed in Section 6.8.

6.7 MAXIMUM THROUGHPUT AND OPTIMUM SIGNAL-TO-NOISE RATIOS

It is useful to compute not only the probability of correct packet reception but also the system throughput when using either of the two protocols previously presented. In both protocols the average number of data bits sent per second is given by the average number of packets sent per second times the number of bits per packet. For the first protocol, the average number of packets/second transmitted will be the sum of the average number of packets per second transmitted through each subsystem. Since the throughput for the lth subsystem is given by

$$T_{H_{1l}} = l \times I \times \bar{N}_l = l \times I \times \frac{1}{T_{Il}} = \frac{lI\lambda_0}{q_{\min}} P(l/n=1) \tag{6.31}$$

where $\bar{N}_l$ = average number of bursts in the lth subsystem and T_{Il} is given in Eq. 6.22. It follows that the system throughput will then be given by

$$T_{H_1} = \frac{I\lambda_0}{q_{\min}} \sum_{l=1}^{\infty} lP(l/n=1) = \frac{I\lambda_0}{q_{\min}} E[l/n=1] \tag{6.32}$$

where $E[\cdot]$ denotes the expectation operator. When we evaluate Eq. 6.32, it will be shown that the throughput under the first protocol is maximized when the packet size is a minimum. For the second protocol, the average number of packets transmitted $\bar{l}$ is equal to the average number of bursts whose electroline density exceeds $q_{\min}$, as given in Eq. 6.A.16. Hence,

$$\bar{l} = \frac{\lambda_0}{q_{\min}} \tag{6.33}$$

as given by Eq. 6.15 and the average number of bits transmitted per second is given by

$$T_{H2} = I\bar{l} \tag{6.34}$$

When utilizing the second protocol the throughput possesses a single maximum when plotted as a function of the packet size. The maximum throughput can be found by maximizing the expression in Eq. 6.34 with respect to the packet size. Using Eq. 6.A.16, and substituting for $q_{\min}$ the expression given in Eq. 6.A.13 and dropping the terms that do not affect the maximization process yields

$$f_{H2}(I) = \frac{I}{\sqrt{R_0\tau + I}} \tag{6.35}$$

where $f_{H2}(I)$ is the function to be maximized. The above expression does not possess a finite maximum. Therefore, when the system uses the second protocol and operates without any restriction on the maximum data rate that can be transmitted, the throughput is infinite. This rather unexpected behavior is due to the mathematical model of the channel. As shown in Eq. 6.4 the data rate $R(t)$ is proportional to q^2, where the probability density function of q, $f(q)$, decays as $1/q^2$. In any numerical results the maximum rate allowed by the system R_{max} will be set such that $R_{max} \leq 10^6$. This maximum rate will not abuse the channel model considerably. The existence of a maximum data transmission rate q_{min} is given by Eq. 6.A.14 and the function to be maximized becomes

$$f_{H2}(I) = I \exp\left(-\frac{I}{2\tau R_{max}}\right) \tag{6.36}$$

The maximum occurs at $I = 2\tau R_{max}$, and the maximum throughput is given by

$$T_{H2max} = 2\tau\lambda_0 \sqrt{\frac{A \times R_{max}}{\text{SNR}_c \times e}} \exp\left[-\frac{1}{2}\left(\frac{R_{min}}{R_{max}} + \frac{t_0}{\tau}\right)\right] \tag{6.37}$$

In the limit as the synchronization is instantaneous (i.e., t_0 approaches zero) and the minimum allowed data rate R_{min} approaches zero, the throughput is seen to increase linearly with the average meteor burst trail duration and as the square root of the maximum rate allowed R_{max}. The average duration of a meteor burst trail has been calculated using Eq. 6.9 (the probability density function of the useful duration of a meteor burst) and equals $t_0 + 2\tau$.

When messages are entirely contained in one packet (as has been assumed in the analysis) it is possible to optimize the probability of successful message completion P_{CM} with respect to the constant signal-to-noise ratio selected for the system's operation SNR_c. The optimum value of the signal-to-noise ratio SNR_o is that operational signal-to-noise ratio SNR_c of Eq. 6.4 which maximizes the probability of message completion P_{CM}. The maximization of the probability of message completion P_{CM} can be done at any particular time T_D. When using the first protocol, the expression to be maximized is

$$f_1(\text{SNR}_c) = \sum_{l=1}^{\infty} (1 - P_p^l) \frac{1}{q_{min}} P(l/n = 1) \tag{6.38}$$

which is the power of the exponential in Eq. 6.23. This function does not lend itself to analytical maximization, and a numerical maximization algorithm must be used.

When using the second protocol, the function to be maximized is

$$f_2(\mathrm{SNR_c}) = \frac{T_D}{T_I}(1 - P_p) \tag{6.39}$$

which is the power of the exponential in Eq. 6.34 and T_I is given in Eq. 6.15. Since both choices of $q_{\min}$ as given in Eq. 6.A.16 are proportional to the square root of $\mathrm{SNR_c}$, they will give the same result in maximizing $f_2(\mathrm{SNR_c})$. After removing the constant terms, the function $f_2(\mathrm{SNR_c})$ becomes

$$f_2(\mathrm{SNR_c}) = \frac{(1 - P_p)}{\sqrt{\mathrm{SNR_c}}} = \frac{[1 - \frac{1}{2}\exp(-\frac{1}{2}\mathrm{SNR_c})]^I}{\sqrt{\mathrm{SNR_c}}} \tag{6.40}$$

Differentiating $f_2(\mathrm{SNR_c})$ and setting the result equal to zero, we get

$$(I \times \mathrm{SNR_o} + 1)\exp(-\tfrac{1}{2}\mathrm{SNR_o}) = 2 \tag{6.41}$$

The above equation cannot be solved analytically, and a nonlinear root-finding algorithm must be used to solve for $\mathrm{SNR_o}$ as a function of the packet size.

6.8 NUMERICAL RESULTS

The performance of the system analyzed here depends on the signal-to-noise ratio chosen for operation. The optimal signal-to-noise ratio, $\mathrm{SNR_o}$, is found from Eqs. 6.38 and 6.41. Figure 6.4 plots the optimum signal-to-noise ratio for both the first and the second protocol as a function of the packet size used in the data transmission. A threshold is seen to exist as a packet size of approximately 2×10^3 bits/packet. As the packets get smaller in size the signal to noise ratio required for optimal operation decreases. As the packet size increases, so does the optimal signal-to-noise ratio. This tendency reflects the position of the packet size I in the exponent in Eq. 6.12. As the packet size I decreases, it is possible to decrease the signal-to-noise ratio and increase the corresponding bit error rate and still get a good probability of successful reception of the data packet. As the packet size I increases, it is necessary to increase the signal-to-noise ratio and decrease the bit error rate accordingly to keep the good probability of successful packet reception. Above the threshold, an increase in the packet size from 4×10^3 bits per packet to 20×10^3 bits per packet changes the optimal signal-to-noise value by less than 1 dB. Figures 6.5 and 6.6 plot the probability of successful message completion as a function of signal-to-noise ratio at a fixed packet

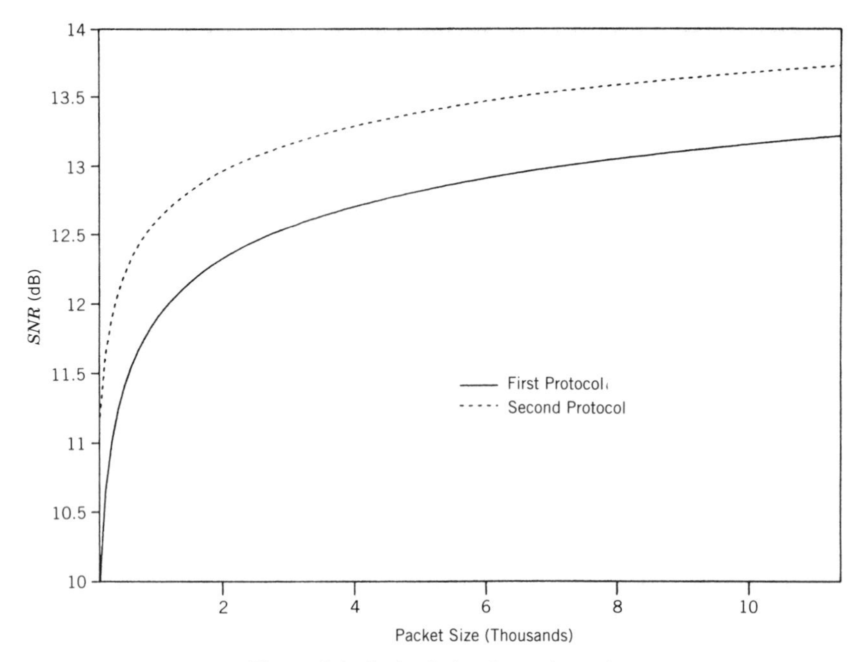

Figure 6.4. Optimal signal-to-noise ratio.

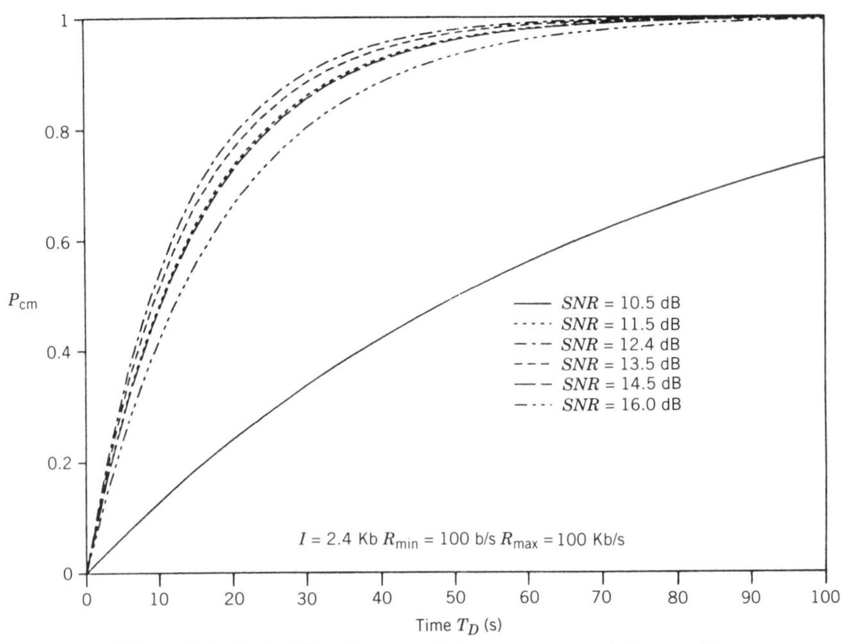

Figure 6.5. Probability of successful message completion vs. T_d.

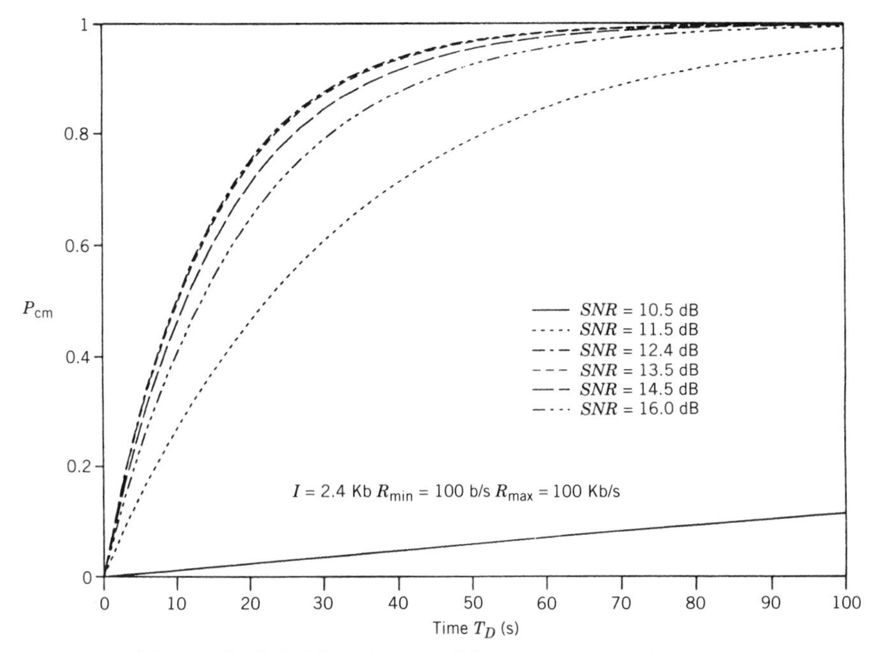

Figure 6.6. Probability of successful message completion vs. T_d.

size of 2400 bits. Provided that the operational signal-to-noise ratio is above a threshold value of approximately 12.4 dB, the system performance does not change appreciably until the operational signal-to-noise ratio increases to 16 dB. Thus, the system performance (P_{CM}) versus operational signal-to-noise ratio function does not have a sharp peak.

The system performance also depends on the minimum and maximum bit rates chosen for transmission. Figures 6.7 and 6.8 show the effect of the chosen R_{min} and R_{max} upon the probability of message completion. First, the minimum rate was chosen as 100 bps. Figures 6.7*b* and 6.8*b* show that very little improvement is to be gained in using a maximum rate in excess of 4 kbps for both protocols. Thus, the maximum rate was arbitrarily set at 100 kbps. Using this maximum data transmission rate, the probability of message completion was plotted as a function of the minimum data transmission rate used before declaring the link broken. Figures 6.7*a* and 6.8*a* show that for all practical purposes a minimum data transmission rate of 1 kbps is sufficient.

Next, the effect of forward error correction coding is considered as applied to the first and second protocol. The system again operates at an optimal signal-to-noise ratio. The optimal signal-to-noise ratio when using error correction codes is different than that found for the case where no error correction codes were used (Fig. 6.4). The optimal signal-to-noise ratio varies as a function of which ever error correction code is considered. Table 6.1 summarizes these results.

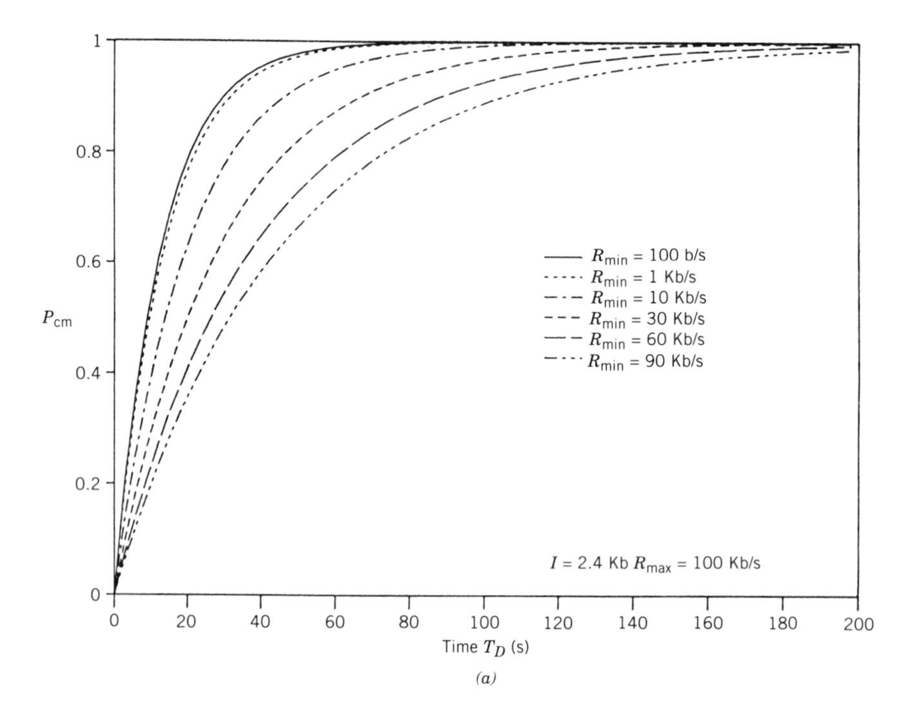

(a)

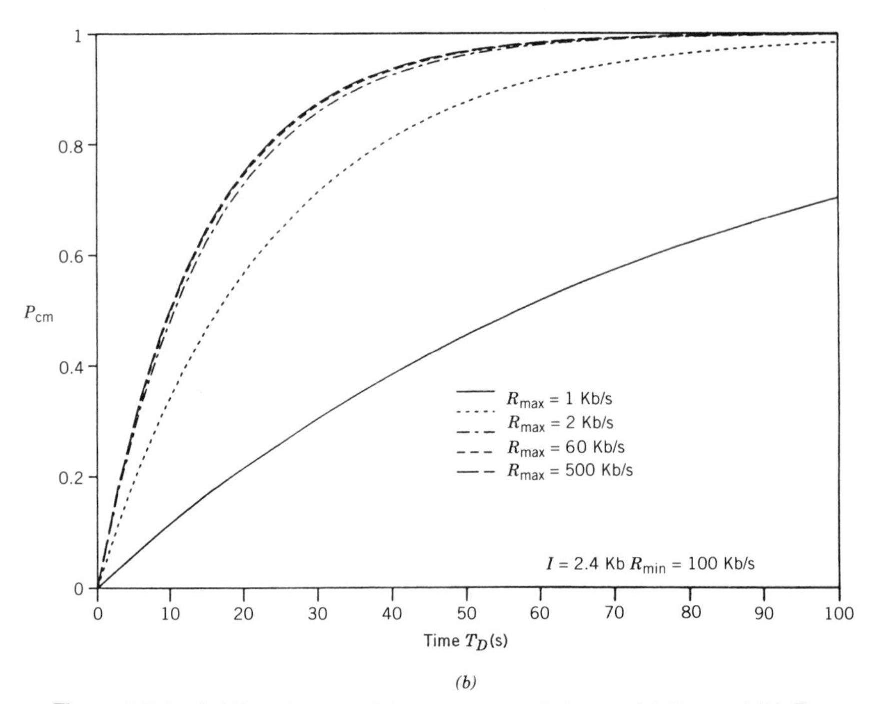

(b)

Figure 6.7. Probability of successful message completion vs. (a) R_{min} and (b) R_{max}.

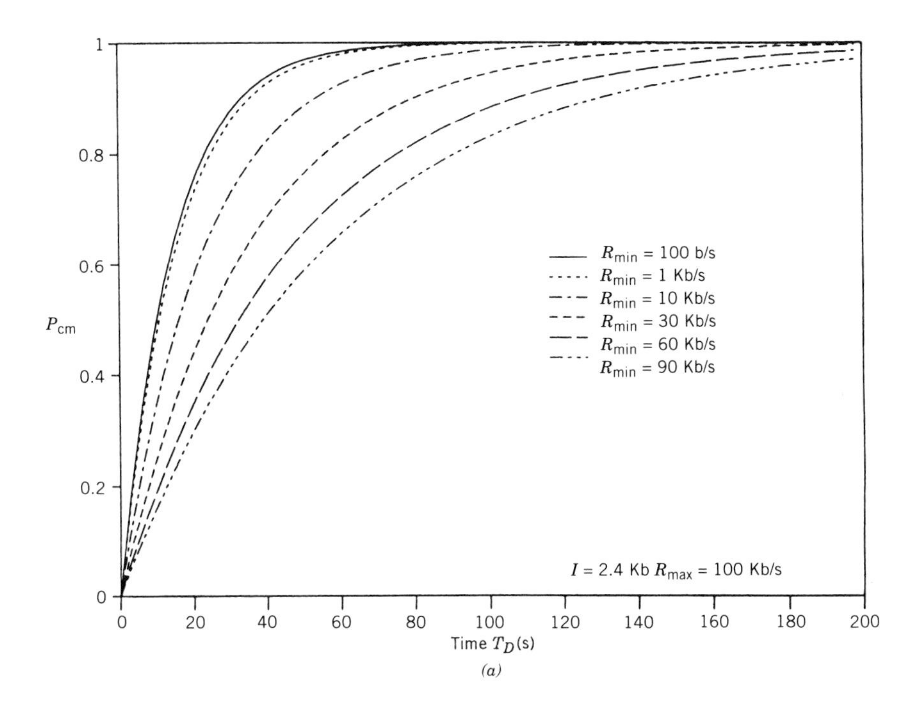

(a)

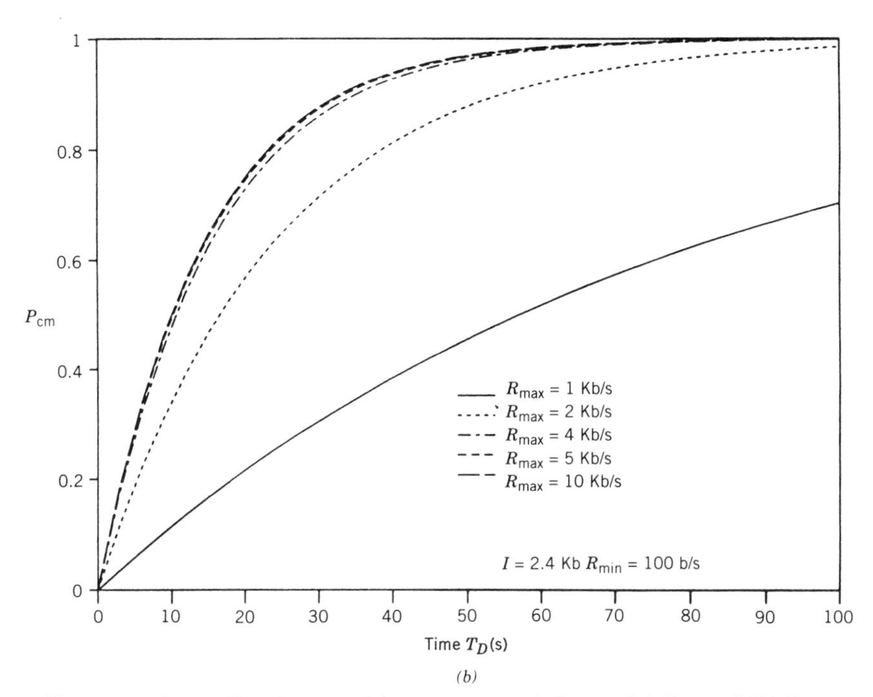

(b)

Figure 6.8. Probability of successful message completion vs. (*a*) R_{min} and (*b*) R_{max}.

TABLE 6.1 Optimal Signal-to-Noise Ratios

Code Used	First Protocol	Second Protocol
No code	12.4 dB	13.1 dB
RS(31, 11, 10)	7.6 dB	7.8 dB
RS(31, 15, 8)	8 dB	8.3 dB
RS(31, 21, 5)	8.9 dB	9.2 dB

The probability of message completion versus observation time when using different coding schemes and an optimal signal-to-noise ratio is plotted in Fig. 6.9 for the first protocol and Fig. 6.10 for the second protocol. The use of error correction coding does not appear to significantly improve the system performance. This is due to the fact that the system operates at a *fixed* signal-to-noise ratio and thus the system designer can fix this parameter a priori to its optimal value. It thus appears that the extra complexity introduced by the forward error correction scheme is not justified.

Finally, the probability of message completion of the variable data rate system is compared to that of a fixed data rate system. Figures 6.11 and 6.12 clearly show the variable data rate system to be much superior when compared to Figs. 6.5 and 6.6.

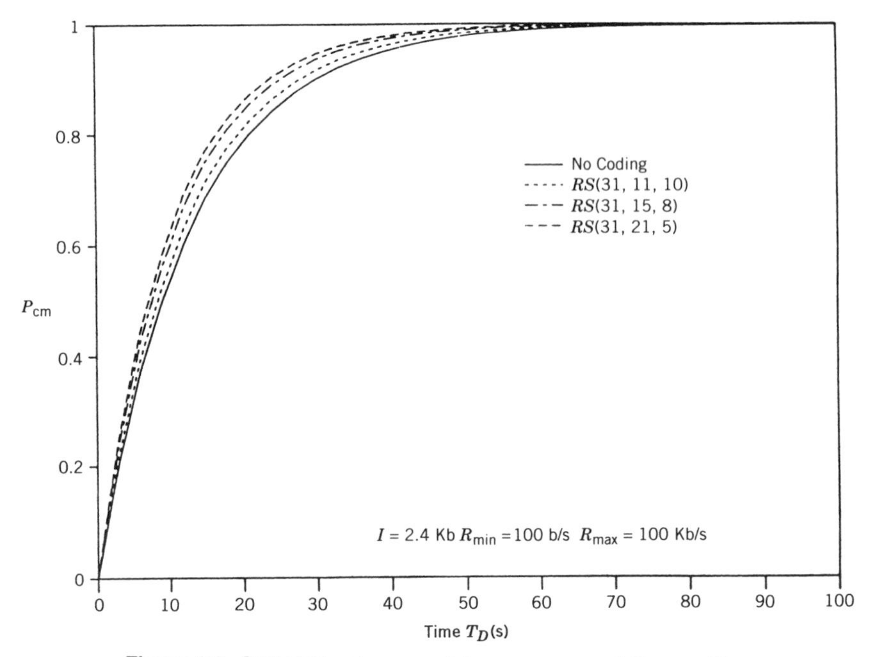

Figure 6.9. Probability of successful message completion vs. T_d.

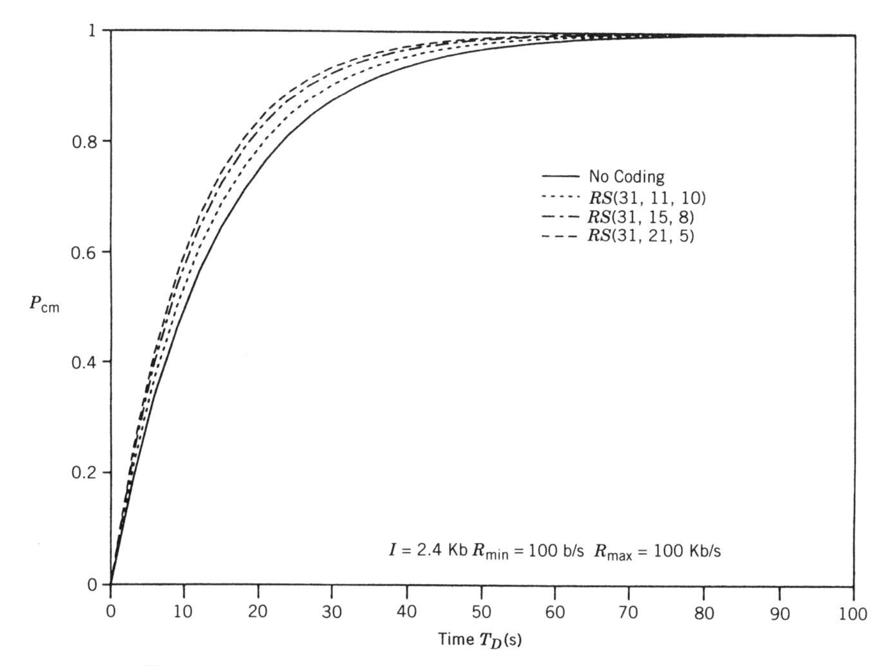

Figure 6.10. Probability of successful message completion vs. T_d.

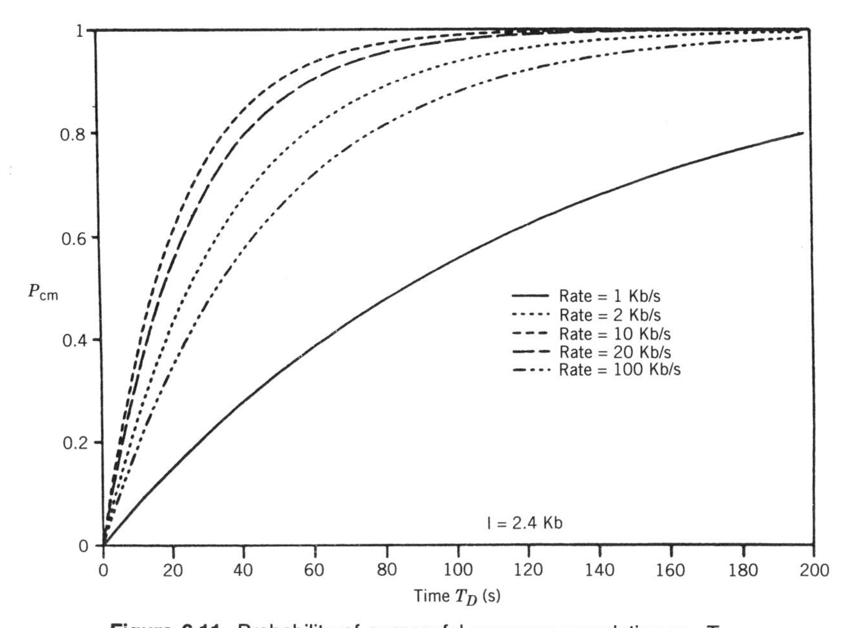

Figure 6.11. Probability of successful message completion vs. T_d.

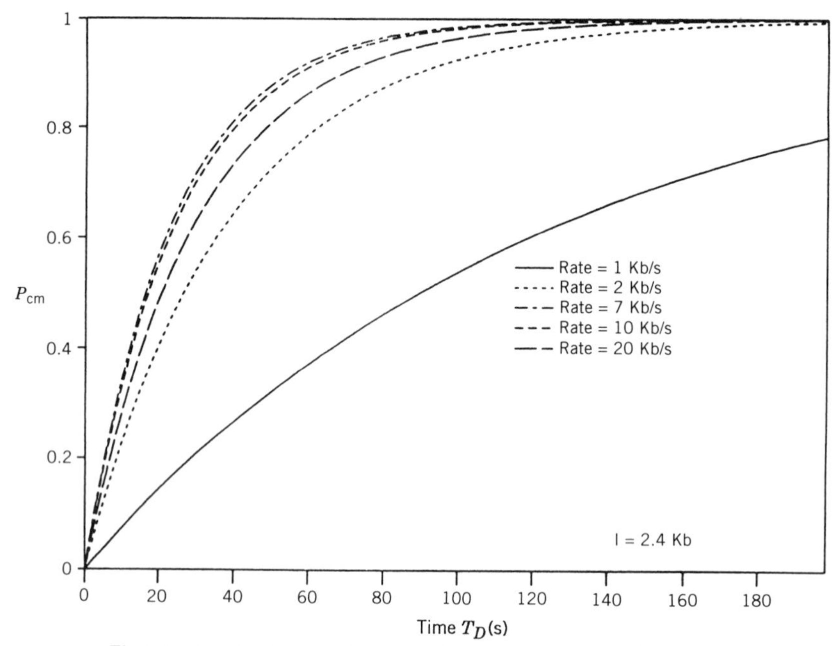

Figure 6.12. Probability of successful message completion vs. T_d.

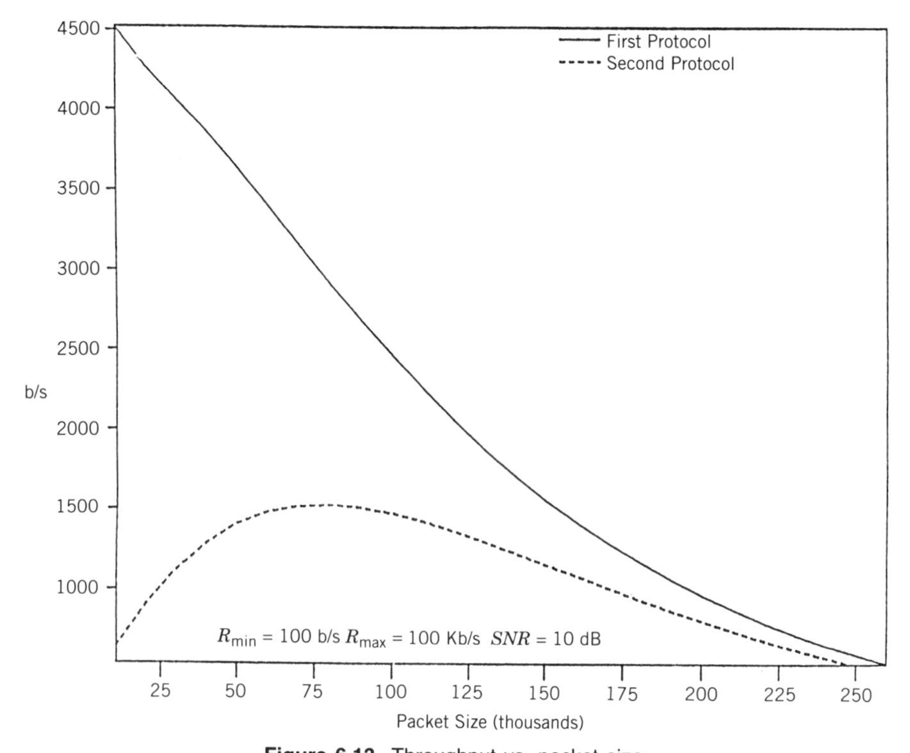

Figure 6.13. Throughput vs. packet size.

In some applications the probability of message completion may not be the desired measure of system performance. The system throughput (given in bits per second) is then a good measure of the system performance. Figure 6.13 shows the system throughput at a fixed signal-to-noise ratio (10 dB) as a function of packet size. The first protocol operates best at the minimum packet size; then it can minimize those wasted time slots at the end of every useful meteor burst trail when incomplete packets are discarded. The second protocol sends only one packet per burst; if the packet size is too small it does not efficiently use the meteor bursts. If the packet size is too large it can successfully use only very few bursts. It shows an optimum value at a packet size of 75 kb. The function is not sharply peaked, however, and, as Fig. 6.13 shows, any packet size in that range will result in reasonable throughput.

6.9 SUMMARY AND CONCLUSIONS

In this chapter, we analyzed the performance of a communication system operating with a variable transmission data rate. The channel used by the system is created by meteors that enter the atmosphere and are subject to ablation. The nature of the channel makes channel characterization possible only in probabilistic terms. It was shown that optimal functions governing the transmission rates do exist. These optimal transmission rate functions are found as a function of time, subject to a maximum and minimum transmission rate constraint, which is found to not degrade system performance. The performance of the system is quantized by the probability of successful message completion. For those applications where the throughput is of interest, the maximum throughput has been analyzed and generated as a function of the packet size in bits per packet. The probability of successful message completion was found for the cases when the system operates with and without error correction codes. The results were rather unexpected: They show that for optimal parameter settings the use of error correction codes does not significantly improve the system's performance.

APPENDIX

This appendix finds the probability of transmitting l packets upon the occurrence of a usable meteor burst, and the minimum value of q for which a packet can be transmitted using the second protocol. The maximum data rate $R_{\max}$ will be reached only during transmissions on bursts whose electroline density q exceeds some threshold value q'. Using Eq. 6.7, the value q' is shown to be

$$q' = \sqrt{\frac{\mathrm{SNR_c} \times R_{\max}}{A}}\, e^{t_0/2\tau} \tag{6.A.1}$$

Now consider the first protocol. If the system operates without any limitation on the maximum data rate $R(t)$ that can be transmitted, then the data rate to be used is given by Eq. (6.7) as

$$R(t) = \frac{Aq^2 e^{-t/\tau}}{\mathrm{SNR_c}} \tag{6.A.2}$$

However, if the system operates with the constraint that the maximum data rate $R(t)$ that can be transmitted is limited to some maximum value R_{max}, then the data rate cannot increase indefinitely. Instead, if the meteor burst electroline density q is such that the data rate as given by Eq. 6.7 is larger than the maximum data rate R_{max}, then the data rate used by the transmitter will remain at R_{max} until some time $\Delta t'$, after which the data rate will begin to decrease according to Eq. 6.7. The meteor burst trails for which the system will exhibit this data rate limiting until some time $t = \Delta t'$ seconds are those trails that have an electroline density q larger than q' as given in Eq. 6.A.1. The time $t = \Delta t'$ can be found by replacing $R(t)$ in Eq. 6.A.2 by its maximum value and then solving for $\Delta t'$. Thus,

$$R_{max} = \frac{Aq^2 e^{-\Delta t'/\tau}}{\mathrm{SNR_c}} \tag{6.A.3}$$

and

$$\Delta t' = -\tau \ln\left(\frac{\mathrm{SNR_c} \times R_{max}}{Aq^2}\right) \tag{6.A.4}$$

If a packet contains I bits and the data rate used is the maximum rate R_{max}, then the number of packets transmitted between the end of the synchronization process and that time at which the data rate falls below its maximum value (i.e., in $\Delta t' - t_0$ seconds) is given by

$$N_c = \frac{(\Delta t' - t_0)R_{max}}{I} = -\frac{\tau R_{max}}{I} \ln\left(\frac{\mathrm{SNR_c} \times R_{max}}{Aq^2}\right) - \frac{t_0 R_{max}}{I} \tag{6.A.5}$$

The data bits transmitted after time $\Delta t'$ and until time Δt (when the transmission ceases) will not be transmitted at the fixed maximum data rate R_{max}, but at a variable rate, as given by Eq. 6.A.2. The number of packets transmitted in the time span $\Delta t'$ to Δt can be found by using Eq. 6.A.2:

$$N_v = \frac{1}{I} \int_{\Delta t'}^{\Delta t} R(t)\, dt \tag{6.A.6}$$

Using Eq. 6.A.6 and the values of $\Delta t'$ and Δt obtained from Eqs. 6.A.4

and 6.9, respectively, the number of packets transmitted using a variable data rate transmission mode y given by

$$N_{\mathrm{v}} = \frac{\tau}{I}\,(R_{\max} - R_{\min}) \tag{6.A.7}$$

where, again, it is assumed that the electroline density q of the meteor burst trail is larger than the threshold value q' as given in Eq. 6.A.1.

It is interesting to note in Eq. 6.A.7 that when $q > q'$, the number of packets transmitted at a variable rate is constant. Using Eqs. 6.A.5 and 6.A.7 it can be concluded that when the electroline density q of a meteor trail is larger than the threshold q', the number of packets transmitted during the existence of the link is given by

$$N_{\mathrm{t}} = N_{\mathrm{c}} + N_{\mathrm{v}} = \frac{\tau R_{\max}}{I}\left[1 - \frac{R_{\min}}{R_{\max}} - \frac{t_0}{\tau} - \ln\left(\frac{R_{\max} \times \mathrm{SNR_c}}{Aq^2}\right)\right] \tag{6.A.8}$$

If the electroline density q is lower than q', then the data will be transmitted using a data rate always lower than $R_{\max}$. Then, using $R_{\min}$ as given by Eq. 6.8, the number of packets transmitted in this case is given by

$$N_{\mathrm{t}} = N_{\mathrm{v}} = \frac{1}{I}\int_{t_0}^{\Delta t} \frac{Aq^2}{\mathrm{SNR_c}}\, e^{-t/\tau}\, d\tau = \frac{\tau}{I}\left(\frac{Aq^2 e^{-t_0/\tau}}{\mathrm{SNR_c}} - R_{\min}\right) \tag{6.A.9}$$

where (see Eq. 6.A.2) the first term in parentheses represents some transmission data rate that is larger than $R_{\min}$ but lower than $R_{\max}$. When the value of the electroline density q of the meteor trail approaches the threshold value q' then Eq. 6.A.9 approaches Eq. 6.A.7, as expected. Equations 6.A.8 and 6.A.9 relate fixed scenario- and system-related parameters to the values of the electroline density q necessary to transmit exactly l packets using a meteor burst trail. Let $q^{[l]}$ be the electroline density of a meteor burst trail that can support the transmission of exactly l packets. Then the exact value of $q^{[l]}$ can be found from Eq. A.6.8 when $q^{[l]}$ is above the threshold q' and from Eq. 6.A.9 when $q^{[l]}$ is below the threshold q'. Therefore, using Eqs. 6.A.8 and 6.A.9, $q^{[l]}$ is given by

$$q^{[l]} = \begin{cases} \sqrt{\dfrac{\mathrm{SNR_c} \times R_{\max}}{A}}\, \exp\left[\dfrac{1}{2}\left(\dfrac{lI}{\tau R_{\max}} + \dfrac{R_{\min}}{R_{\max}} + \dfrac{t_0}{\tau} - 1\right)\right] & \text{if } q^{[l]} > q' \\ \sqrt{\left(\dfrac{lI}{\tau} + R_{\min}\right)\dfrac{\mathrm{SNR_c}}{A}}\; e^{t_0/2\tau} & \text{if } q^{[l]} < q' \end{cases} \tag{6.A.10}$$

Equation 6.A.7 finds the maximum number of packets that can be transmitted using only a variable data rate. This is the number of packets that can be transmitted on a meteor burst trail that has an electroline density $q = q'$. Thus, depending on the packet size I, Eq. 6.A.7 may generate a fractional number. Let l'_0 define this (possibly) noninteger maximum number of packets that can be transmitted using a variable data rate prior to this data rate being limited to R_{max}. Then the range of electroline densities q for which the corresponding meteor burst trails have a duration such that they can support exactly l packets is given by

$$\sqrt{\left(\frac{lI}{\tau} + R_{min}\right)\frac{SNR_c}{A}}\, e^{(t_0/2\tau)} \le q < \sqrt{\left(\frac{(l+1)I}{\tau} + R_{min}\right)\frac{SNR_c}{A}}\, e^{(t_0/2\tau)} \qquad \text{if } l'_0 - 1 \ge l$$

$$\sqrt{\left(\frac{lI}{\tau} + R_{min}\right)\frac{SNR_c}{A}}\, e^{(t_0/2\tau)} \le q < \sqrt{\frac{SNR_c \times R_{max}}{A}} \exp\left[\frac{1}{2}\left(\frac{(l+1)I}{\tau R_{max}} + \frac{R_{min}}{R_{max}} + \frac{t_0}{\tau} - 1\right)\right] \qquad \text{if } l < l'_0 \le l+1$$

$$\sqrt{\frac{SNR_c \times R_{max}}{A}} \exp\left[\frac{1}{2}\left(\frac{lI}{\tau R_{max}} + \frac{R_{min}}{R_{max}} + \frac{t_0}{\tau} - 1\right)\right] \le q < \sqrt{\frac{SNR_c \times R_{max}}{A}} \exp\left[\frac{1}{2}\left(\frac{(l+1)I}{\tau R_{max}} + \frac{R_{min}}{R_{max}} + \frac{t_0}{\tau} - 1\right)\right] \qquad \text{if } l'_0 \le l \tag{6.A.11}$$

Given the range of electroline density values for which the corresponding meteor burst trails support exactly l packets, and given the probability density function of q in Eq. 6.10, the probability of transmitting exactly l packets using a given meteor burst trail can be derived. The conditional density function of the random variable l is given below as $P(l/n = 1)$ and equals

$$q_{min}\sqrt{\frac{A\tau}{SNR_c}}\left[\frac{1}{\sqrt{lI + R_{min}\tau}} - \frac{1}{\sqrt{(l+1)I + R_{min}\tau}}\right] e^{-t_0/2\tau} \qquad \text{if } l+1 \le l'_0$$

$$q_{min}\sqrt{\frac{A}{SNR_c}}\left(\frac{1}{\sqrt{lI + R_{min}\tau}} - \frac{1}{\sqrt{R_{max}}} \times \exp\left\{-\frac{1}{2}\left[\frac{(l+1)I}{R_{max}\tau} + \frac{R_{min}}{R_{max}} - 1\right]\right\}\right) e^{-t_0/2\tau} \qquad \text{if } l \le l'_0 \le l+1$$

$$q_{min}\sqrt{\frac{A}{SNR_c \times R_{max}}}\left\{\exp\left(-\frac{lI}{2R_{max}\tau}\right) - \exp\left[-\frac{(l+1)I}{2R_{max}\tau}\right]\right\} \times \exp\left[\frac{1}{2}\left(1 - \frac{R_{min}}{R_{max}} - \frac{t_0}{\tau}\right)\right] \qquad \text{if } l > l'_0 \tag{6.A.12}$$

It can be seen that the minimum electroline density q_{min} (where $q_{min} < q'$) required of a meteor burst trail such that it can support the transmission of at least one packet is derived from Eq. 6.A.10 with l set to one as

$$q_{min} = \sqrt{\left(\frac{I}{\tau} + R_{min}\right) \frac{\text{SNR}_c}{A}}\, e^{t_0/2\tau} \qquad \text{if } q < q' \tag{6.A.13}$$

If $q_{min} > q'$ then the value of q_{min} can again be derived from Eq. 6.A.10 as

$$q_{min} = \sqrt{\frac{\text{SNR}_c \times R_{max}}{A}} \exp\left[\frac{1}{2}\left(\frac{I}{\tau R_{max}} + \frac{R_{min}}{R_{max}} + \frac{t_0}{\tau} - 1\right)\right] \tag{6.A.14}$$

Now consider the second protocol. The maximum size of a packet that can be transmitted without limiting the transmission data rate can be found from Eq. 6.A.7 with N_v set to one. Thus, the maximum packet size that can be transmitted using a meteor burst trail with an electroline density $q < q'$ is given by

$$I_{max} = \tau(R_{max} - R_{min}) \tag{6.A.15}$$

A packet larger than I_{max} above can be transmitted in its entirety only using a meteor burst trail that has an electroline density $q > q'$. Then the expressions for q_{min} can be written conditioned on the packet size I as

$$q_{min} = \begin{cases} \sqrt{\dfrac{\text{SNR}_c}{A}\left(R_{min} + \dfrac{I}{\tau}\right)}\, e^{(t_0/2\tau)} & \text{for } I \le \tau(R_{max} - R_{min}) \\ \sqrt{\dfrac{\text{SNR}_c \times R_{max}}{A}} \exp\left[\dfrac{1}{2}\left(\dfrac{I}{\tau R_{max}} + \dfrac{R_{min}}{R_{max}} + \dfrac{t_0}{\tau} - 1\right)\right] & \text{for } I > \tau(R_{max} - R_{min}) \end{cases} \tag{6.A.16}$$

REFERENCES

1. V. R. Eshleman, L. A. Manning, A. M. Peterson, and O. G. Villard, Jr., Some properties of oblique radio reflections from meteor ionization trails. *J. Geophys. Res.* 61: 233–249 (1956).
2. V. R. Eshelman and R. F. Mlodnosky, Directional characteristics of meteor propagation derived from radar measurements. *Proc. IRE* (Dec 1957).
3. P. A. Forsyth, E. L. Vogan, D. R. Hansen, and C. O. Hines, The principles of JANET: a meteor-burst communications system. *Proc. IRE* (Dec 1957).

4. K. Ain-un-Din Ahmad, M. D. Grossi, and A. Javed, Adaptive communications via meteor forward scattering. *Pakistan Int. Symp. Elec. Eng.*, 15–17 Nov 1972.
5. J. L. Heritage, Meteor burst communication application to patrol aircraft in the Mediterranean. *Naval Ocean Syst. Cent. memorandum* (Apr 1977).
6. L. L. Campell and C. O. Hines, Bandwidth considerations in a JANET system. *Proc. IRE* (Dec 1957).
7. G. W. L. Davis, S. J. Gadys, G. R. Land, L. M. Luke, and M. K. Taylor, The Canadian JANET system. *Proc. IRE* (Dec 1958).
8. P. J. Bartholome, Results of propagation and interception experiments on the STC meteor-burst link. The Hague, The Netherlands: SHAPE Tech. Cen., *TM-173* Dec 1967.
9. P. J. Bartholome and I. M. Vogt, COMET: a meteor-burst system incorporating ARQ and diversity reception. *IEEE Trans. Commun. Technol.* COM-16 (Apr 1968).
10. P. J. Bartholome, The fine structure of meteor-burst signals. The Hague, The Netherlands: SHAPE Tech. Cen., *TM-36*, Apr 1962.
11. P. J. Bartholome, The STC meteor-burst system. The Hague, The Netherlands: SHAPE Tech. Cen., *TM-156*, Feb 1967.
12. D. W. Brown, A comparison of the performance of the COMET meteor-burst communications system at 40 and 100 MHz. The Hague, The Netherlands: SHAPE Tech. Cen., *TM-66*, June 1969.
13. D. W. Brown and R. E. Schemel, An operational trial of the STC meteor-burst system between Oslo and Bodo. The Hague, The Netherlands: SHAPE Tech. Cen., *TM-290*, June 1971.
14. Western Union Gov. Sys. Div., Western Union meteor burst communications system overview. *Inform. Brochure*, Jan 1977.
15. J. E. Bickel, J. L. Heritage, and C. P. Kugel, Meteor burst communications in minimum essential emergency communications network. Naval Ocean Syst. Cen., Defence Commun. Agency (*Code 960*), Oct 1976.
16. L. A. Maynard, Meteor burst communications in the arctic. In *Proc. NATO Inst. Ionospheric Radio Commun. in Arctic*. New York: Plenum, 1967.
17. M. W. Abel, Meteor burst communications: bits per burst performance bounds. *IEEE Trans. Commun.* COM-34: 927–936 (1986).
18. J. D. Oetting, An analysis of meteor burst communications for military applications. *IEEE Trans. Commun.* COM-28: 1591–1601 (1980).
19. L. B. Milstein et al., Performance of meteor-burst communication channels. *IEEE Trans. Commun.* SAC-5: 146–154 (1987).
20. J. R. Hampton, A meteor burst model with time-varying bit error rate. Proc MILCOM'85, vol. 2, pp. 559–563.
21. A. E. Spezio, Meteor burst communication system analysis and design. *NRL Rep. 8286*, Dec 1978.
22. J. A. Weitzen, W. P. Birkemeier, and M. D. Grossi, An estimate of the capacity of the meteor burst communication channel. *IEEE Trans. Commun.* vol. COM-32, 972–974 (1984).
23. G. R. Sugar, Radio propagation by reflection from meteor trails. *Proc. IEEE*, vol. 52, 116–136 (1964).

7

FORWARD ERROR CORRECTION FOR METEOR BURST COMMUNICATIONS

Scott L. Miller
Department of Electrical Engineering, University of Florida, Gainesville, Florida

Laurence B. Milstein
Department of Electrical & Computer Engineering, University of California, San Diego, La Jolla, California

7.1 INTRODUCTION

The issue of coding for a meteor burst channel has had somewhat of a controversial aspect associated with it. On the one hand, there is the approach that says a communication system should be designed to incorporate as much error-correction coding as possible (subject to such obvious practical constraints as complexity and delay, as well as the theoretical limit of requiring the information rate to be less than the channel capacity). This approach essentially says to allow the channel error rate to be as large as necessary, but have a sufficient amount of error-correction capability built into the system so that the channel errors can be corrected.

On the other hand, it is well known that the meteor burst channel exhibits an exponential decay of signal-to-noise ratio (SNR) with time, and this suggests it might be problematic as to how extensive the benefits of coding will be. That is, in any communication system operating at a constant bandwidth, and using a fixed modulation size, the information rate must decrease when coding is added to the system because of the introduction of the redundancy inherent in the use of the code. This increase in the length of time needed to transmit a given length message (say, for example, one packet of information bits) is especially bad for a meteor burst system because of the exponential decay in SNR.

In other words, if an uncoded message takes T seconds to complete, then a coded system employing a code with rate $R < 1$ takes T/R seconds to complete. This extra $T(1 - R)/R$ seconds extends the required duration of the meteor burst, at times below the minimum acceptable SNR. Thus, one encounters an explicit problem in using coding over a meteor burst channel over and above the mere reduction of information rate. As a compromise in attempting to keep this effect small while at the same time attempting to achieve at least some of the advantages of coding, one could consider the use of high rate codes, so that the increase in the required duration of the meteor burst is, in some sense, minimized.

This uncertainty as to (1) whether coding should be used at all, and (2) what types of codes are advantageous for use over a meteor burst channel and what should their rates be has formed the core of the controversy referred to above. This chapter addresses these questions.

7.2 CAPACITY OF THE METEOR BURST (MB) CHANNEL

To obtain a true perspective on the desirability of using forward error correction in a given communication system, it is useful to have some idea of the capacity of the channel under consideration. In this section we present both upper and lower bounds on the capacity of the MB channel. Results are presented for a constant SNR model of the MB channel extended to a time-varying SNR model.

7.2.1 Bounds on Capacity: Constant SNR Model

Consider the following constant SNR model for a meteor burst (MB) channel. The channel is considered to be in one of two states (either on or off) at any given time. The channel is said to be off anytime the SNR drops below some threshold A_0, and the channel is modeled as a binary symmetric channel (BSC) with crossover probability 1/2. When the SNR is above A_0, the channel is on and is modeled as a BSC with crossover probability p, as shown in Fig. 7.1. When binary noncoherent FSK is used as the modulation format, p is given by

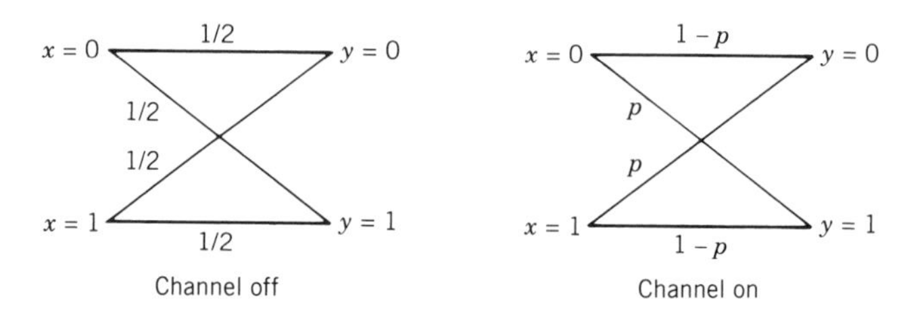

Figure 7.1. Two-state model of MB channel.

$$p = \frac{1}{2} \exp\left(-\frac{A_0}{2}\right) \tag{7.1}$$

Let the random variable S denote the state of the channel. For the two-state model described above, let S be 0 when the channel is off and 1 when the channel is on. A meteor burst channel will be in the off state most of the time, with periodic bursts into the on state. Even with this crude model, finding an exact expression for the capacity requires an enormous amount of computation; however, both upper and lower bounds on capacity that are reasonably tight can be found without too much computational effort.

To bound the capacity, consider grouping transmitted bits into blocks of length m. Then, for each group of m bits, assign a channel state (either on or off) for the entire group. The block of m bits is said to be on if the SNR is greater than A_0 for *all* bits in the group, and is said to be off otherwise. Thus, each block of m bits has associated with it a value of the random variable S. For the MB channel under consideration, the value of S for a given block is, in general, dependent on the value of S for the previous block. In other words, the channel has memory, and it is the presence of this memory that makes determining the capacity of the channel difficult. Bounds on capacity can be obtained by making the following assumptions which destroy the memory:

Assumption 1: The channel state assigned to any given block is independent of the states of any other blocks, and the receiver gets no information about which state is active at any given time.

Assumption 2: The state assigned to each block is known to the receiver.

Neither of these two assumptions are, in general, true for the MB channel, but consider computing the capacity separately under each of these assumptions In Miller and Milstein [8], the following result is proven:

THEOREM. Let $C^{(1)}(m)$ be the capacity of the channel if Assumption 1 is true, $C^{(2)}(m)$ be the capacity if Assumption 2 is true, and C be the capacity of the actual channel. Then

$$\max_m C^{(1)}(m) \leq C \leq C^{(2)}(1) \tag{7.2}$$

Under the two assumptions above, the model for the MB channel is just that of the block interference (BI) channels considered by McEliece and Stark [5], and defined as follows. When a sequence of binary digits is to be sent over the channel, each block of m consecutive digits is, in fact, sent over one of the component channels (such as those shown in Fig. 7.1). The random variable S determines which component channel is used to transmit a given m-block. In McEliece and Stark [5], two channels were considered: one in which the component channel that was used for each m-block was known to the receiver (called the BI channel with side information), and one in which the receiver gets no information about which component channel is active for any block (called the BI channel without side information). $C^{(2)}(m)$ is then the capacity of the BI channel with side information, and $C^{(1)}(m)$ is the capacity without side information.

The capacity of the channel with side information is just the weighted average of the capacities of each of the individual channels, with the weighting given by the random variable S. Thus, if C_s is the capacity of the component channel when $S = s$, then $C^{(2)}(m)$ is shown in McEliece and Stark [5] to be given by

$$C^{(2)}(m) = E[C_s] \tag{7.3}$$

where the expectation is with respect to S. For the channel model being considered here,

$$C^{(2)}(m) = \Pr(S = 0)C_0 + \Pr(S = 1)C_1 \tag{7.4}$$

where C_0 is the capacity of the channel in the off state, and C_1 is the capacity of the channel in the on state. But the capacity of the channel in the off state is 0, so that

$$C^{(2)}(m) = \Pr(S = 1)[1 - \mathrm{H}(p)] \tag{7.5}$$

where

$$\mathrm{H}(p) = -p \log(p) - (1 - p) \log(1 - p) \tag{7.6}$$

To complete the derivation, we need the distribution of the random variable S. If it is assumed that the SNR threshold A_0 is chosen to be large enough so that the probability of overlapping bursts is negligible, then S will be 1 if the entire block falls within a single burst. The SNR of a single burst is of the form [11]

$$\mathrm{SNR}(t) = Ae^{(t/\tau)} \tag{7.7}$$

Thus, if the SNR at the beginning of a block is A, then at the end of the block it will have decayed to $Ae^{-mT/\tau}$, where T is the duration of a bit, and hence

$$\Pr(S = 1) = \Pr(Ae^{-mT/\tau} \geq A_0) = \Pr(A \geq A_0 e^{mT/\tau}) \tag{7.8}$$

Now let $\lambda(A_0)$ be the arrival rate of bursts with SNR $\geq A_0$. Using the model of [6], the average duration of bursts with SNR $\geq A_0$ is independent of A_0, and $\lambda(A_0)$ is given by

$$\lambda(A_0) = \frac{K}{\sqrt{A_0}} \tag{7.9}$$

where K is a constant. The probability that the SNR at any given time is greater than some threshold η is given by

$$\Pr(A \geq \eta) = \frac{\text{Average burst length}}{\text{Average time between bursts}} \tag{7.10}$$

Equation 7.10 is based on the assumption that the average burst length is much less than the average time between bursts, and can be determined by using the results of queueing theory. The MB channel can be thought of as an $M|M|\infty$ queue, with the presence of a meteor trail being analogous to a customer in an infinite server queueing system. Gross and Harris [3] show that the steady-state distribution of the number of customers in an $M|M|\infty$ queue is given by

$$p_n = \frac{1}{n!}\left(\frac{\lambda}{\mu}\right)^n e^{-\lambda/\mu} \tag{7.11}$$

where p_n is the probability of n customers being in the system, λ is the arrival rate of customers, and μ is the service rate. Under the assumption that $\lambda \ll \mu$, Eq. 7.10 can be thought of as the probability that one meteor trail is present.

The average burst length is equal to 2τ, and the average time between bursts is $1/\lambda(A_0)$, so that Eq. 7.8 becomes

$$\Pr(S=1) = 2\tau\lambda(A_0 e^{mT/\tau}) = \frac{2\tau K e^{-mT/2\tau}}{\sqrt{A_0}} \tag{7.12}$$

Thus, by combining Eqs. 7.1, 7.5, and 7.12, the capacity with side information is given by

$$C^{(2)}(m) = \left[1 - \mathrm{H}\left(\frac{1}{2}\, e^{-A_0/2}\right)\right] \frac{2\tau K e^{-mT/2\tau}}{\sqrt{A_0}} \tag{7.13}$$

For the channel without side information, the capacity is derived in [8]. However, using these results, bounds on capacity are shown as a function of A_0 in Fig. 7.2. According to this figure, the capacity is maximized by choosing the SNR threshold A_0 to be about 8 dB. As shown in [6], the constant SNR model predicted the best performance for uncoded systems to be at about 13 dB. According to this model, the capacity of the channel with A_0 at 8 dB is about 50% higher than at 13 dB. This would seem to indicate that there is room for significant performance improvement if a code can be found such that the system can operate well at a threshold of 8 dB.

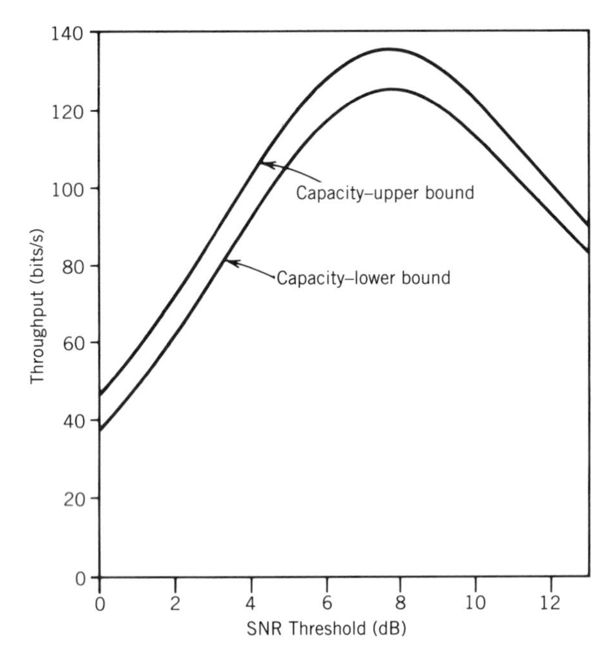

Figure 7.2. Capacity bounds for constant SNR model.

7.2.2 Bounds on Capacity: Time-Varying SNR Model

The results for the constant SNR model can easily be extended to a time-varying SNR model. The models are similar, but the time-varying SNR model has N on states instead of just one. The channel is said to be in state $i(i = 1, 2, \ldots, N)$ if the SNR is such that $A_i \leq \text{SNR} < A_{i+1}$, in which case the channel is modeled as a BSC with crossover probability p_i given by

$$p_i = \tfrac{1}{2} e^{-A_{i-1}/2} \tag{7.14}$$

A_{N+1} can be taken to be infinite, so that the channel is in state N if $\text{SNR} \geq A_N$. If the SNR is less than A_1, then the channel is in the off state and again is modeled as a BSC with crossover probability $p = 1/2$. This provides a reasonably accurate model for the meteor burst channel if the A_i are spaced close together and N is chosen to be large enough so that p_N is very small.

As before, the transmitted bits are grouped into blocks of m bits and each block is assigned a channel state. Each block is said to be in state i if i is the largest integer such that $\text{SNR} \geq A_i$ for the entire block. The capacity of the channel is then bounded by using the two assumptions made above. If the random variable S is again used to indicate the state of a given m block, then the results of [8] can be used to show

$$C^{(2)}(m) = \sum_{i=1}^{N} \Pr(S = i)[1 - \mathrm{H}(p_1)] \tag{7.15}$$

It remains to determine the distribution of the random variable S. If the SNR is $\geq A_i$ for the entire block, then $S \geq i$. Thus, using the previous results,

$$\Pr(S = i) = \Pr(S \geq i) - \Pr(S > i + 1) = 2\pi K e^{-mT/2\tau}\left(\frac{1}{\sqrt{A_i}} - \frac{1}{\sqrt{A_{i+1}}}\right) \tag{7.16}$$

At this point, how to choose the A_i is totally arbitrary. For mathematical convenience, the A_i are taken to have the relationship

$$A_{i+1} = z^2 A_i \qquad i = 1, 2, \ldots, N - 1 \tag{7.17}$$

where $z > 1$, so that Eq. 7.16 becomes

$$\Pr(S = 1) = \frac{2\tau K e^{-mT/2\tau}(z - 1)}{z^i \sqrt{A_1}} \qquad i = 1, 2, \ldots, N - 1 \tag{7.18}$$

and

$$\Pr(S = N) = \frac{2\tau K e^{-mT/2\tau}}{2^{N-1}\sqrt{A_1}} \tag{7.19}$$

We refer the reader to [8] for the derivation of capacity for the channel without side information. However, using these results, the capacity of the MB channel using the time-varying SNR model is bounded as in Eq. 7.2. The accuracy of this model increases at z decreases and the number of states increases. These bounds are shown in Fig. 7.3 and are obtained for a system with the following parameters: average burst length = 0.2 s, packet length = 2400 bits, packet duration = 0.2 s. Comparing this figure with Fig. 7.2 shows that the capacity of the channel is not maximized at any SNR threshold, as opposed to the results predicted by the constant SNR model, but is instead a monotonic function of A_0. The results of [6] showed that the performance of the uncoded system leveled off as the threshold was lowered below about 9 dB, whereas Fig. 7.3 shows that the capacity is still increasing as the threshold is lowered past 9 dB. Also note that the best protocol in [6] achieved only 60 bits/s. Even at 9 dB, the capacity given by Figure 7.3 is at least 125 bits/s and is up to 200 bits/s at lower thresholds. Thus, there is clearly room for much improvement.

At this point, a few words should be said about the meaning of the capacity as derived here so that the results are not misunderstood. In general, the capacity of the channel represents an upper bound on the

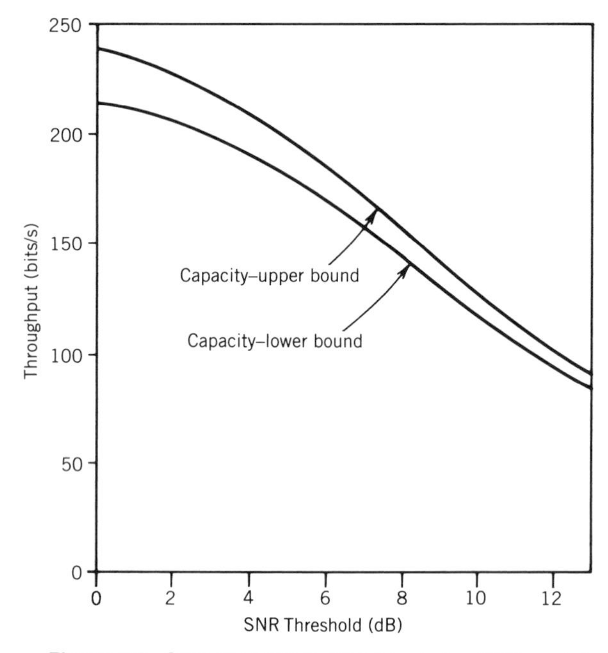

Figure 7.3. Capacity bounds for time-varying SNR model.

throughput achievable on that channel. In deriving the capacity of the MB channel, several assumptions were made. First, a particular modulation scheme was assumed (i.e., binary noncoherent FSK). Second, a specific arrival rate of meteor bursts was used. This was done to give a direct comparison with the results of [6]. If these restrictions are relaxed, the capacity of the channel will change. Thus, the numerical results on capacity derived here should not be considered as an upper bound for any MB system, but only for those that use noncoherent binary FSK with a fixed transmission rate of 10 kbits/s and the assumed arrival rate of bursts.

7.3 PERFORMANCE OF FIXED-RATE CODING

The capacity results of the previous section indicate that it is possible to gain significant performance improvement through the use of error correction coding. This has been verified by several researchers who have analyzed the performance of MB communication systems employing coding. This section focuses on those results that have been obtained using fixed-rate coding schemes. Fixed rate means that the code rate (or the code itself) does not change with time.

The results obtained vary according to the assumed channel model, the assumed transmission protocol, the performance criteria used, and the actual codes used. This discussion is organized according to the channel model assumed in each of the various research efforts. Starting with the simplest, the results of Milstein et al. [9] illustrate that coding has only a minimal effect if a two-state (on/off) channel model is used. Using the commonly assumed underdense channel model, Miller and Milstein [7, 8] and Pursley and Sandburg [10] show that coding does give a significant improvement. Finally, empirical results derived from data obtained from an actual MB link [1] verify some of the conclusions obtained analytically.

7.3.1 Two-State Channel Model

Milstein et al. [9] studied the performance of a MB communication system using Reed–Solomon (RS) codes. The channel was assumed to be in one of two states (either on or off). Any time the received SNR was below some threshold, it was assumed that no communication could take place. If the received SNR was above the threshold, the channel was modeled as an additive Gaussian noise channel with a worst-case SNR (i.e., equal to the threshold). The arrival of suitable meteor trails was modeled as a Poisson process with T_I as the average time between trails for which the instantaneous SNR exceeded the SNR threshold at the start of the burst. The value of T_I depends on the SNR threshold used and the assumed variation is given in Table 7.1. This relationship is described by

$$T_I = K\sqrt{\text{SNR}} \tag{7.20}$$

where the constant of proportionality K is about 12.4 s. The duration of each burst was taken to have an exponential distribution with mean T_L.

Two transmission protocols were considered in this study, protocol 1 and protocol 2. Most of the numerical results apply to protocol 1 and so only that protocol is described here. The terminal that is to receive data is assumed to be continually broadcasting so as to probe for a channel opening. The terminal sending the message data begins transmitting as soon as it hears the probing signal. Once the channel closes, the probing signal disappears and data transmission ceases; the search for another channel opening now begins.

TABLE 7.1 Average Interval Between Bursts Versus SNR

SNR (dB)	T_I (s)
7	28
7.5	29
8	31
10	39
12	50

Using this protocol, the transmitted station has knowledge of both the beginning and end of each usable burst. The message to be sent is contained in a single packet which consists of several codewords from a RS code. This packet is repeated as long as the received SNR exceeds the threshold. At the beginning of each trail, a certain amount of time t_0 is needed for acquisition purposes. The performance criteria used is the probability of successfully completing the message (on at least one of the successive repeats of the packet) in a specified time interval T_D. The duration of a packet T_P depends on the transmission rate, the message length, and the specific code used.

The probability of completing the message P_{CM} is given by [9]

$$P_{CM} = 1 - \exp\left(-\alpha(1 - P_P)\exp\left(-\frac{T_P + t_0}{T_L}\right)\right) \tag{7.21}$$

where $\alpha = T_D/T_I$ and P_P is the probability of making an error in a packet. Assuming that the packet consists of J codewords, then

$$1 - P_P = (1 - P_W)^J \tag{7.22}$$

where P_W is the probability of a codeword being decoded incorrectly. Assuming that each codeword has a block length of N symbols and corrects up to E errors, then

$$P_W = \sum_{i=E+1}^{N} \binom{N}{i} P_s^i (1 - P_s)^{N-i} \tag{7.23}$$

where P_s is the probability of symbol error in the RS symbol. Each symbol consists of m bits, where $N = 2^m - 1$ (for RS codes), so that

$$P_s = 1 - (1 - P_b)^m \tag{7.24}$$

The quantity P_b is the bit error rate and depends on what kind of modulation is used. This work assumed FSK modulation with noncoherent detection so that

$$P_b = \frac{1}{2}\exp\left(-\frac{\text{SNR}}{2}\right) \tag{7.25}$$

where SNR is the received SNR threshold assumed.

By using Eqs. 7.21–7.25, the probability of completing a message P_{CM} was determined for a message consisting of 2400 information bits which was divided into Reed–Solomon codewords of block length 31 (each RS symbol is thus 5 bits). Figure 7.4 shows P_{cm} as a function of time for several different SNR thresholds when the message is encoded into 32 codewords from a (31, 15, 8) RS code. Note that an optimum SNR threshold occurs at about 8 dB and P_{cm} drops off rapidly as the SNR threshold is decreased below 8 dB. To obtain these results, the following parameter values were used: $t_0 = 0.03$ s, $T_L = 0.2$ s, $T_p = 0.2$ s (instantaneous data rate of 24.8 kb/s).

By using a lower rate (31, 11, 10) code, more errors can be tolerated, but a longer packet length is needed ($T_p = 0.3$ s). Figure 7.5 shows P_{cm} when this code is used for several SNR thresholds. Again the optimum threshold occurs at 8 dB, but in this case the performance is slightly worse than that of the (31, 15, 8) coded system. Figure 7.6 shows the performance when a higher rate (31, 21, 5) code is used. In this case, the packet length is reduced to 0.15 s and the optimum SNR threshold occurs at about 10 dB. Note that the performance of this code using the optimum threshold is about the same as that of the (31, 15, 8) coded system. Finally, it is pointed out in Milstein et al. [9] that an uncoded system operating at a SNR threshold of 13 dB will give about the same performance as either the (31, 15, 8) system operating at 8 dB or the (31, 21, 5) system operating at 10 dB. Thus, for the two-state channel model, the use of a fixed-rate code provides little, if any, performance improvement.

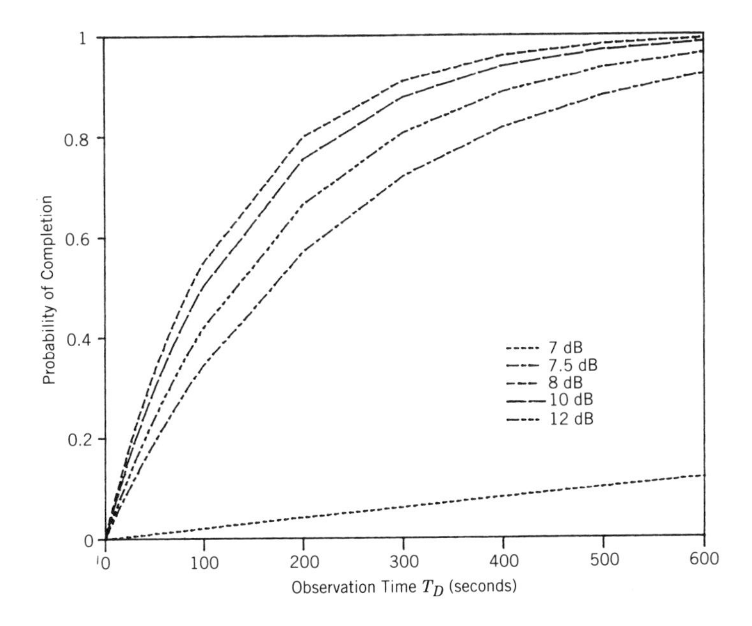

Observation Time T_D (seconds)

Figure 7.4. Probability of completion versus observation time: protocol 2 with RS(31, 15, 8) code.

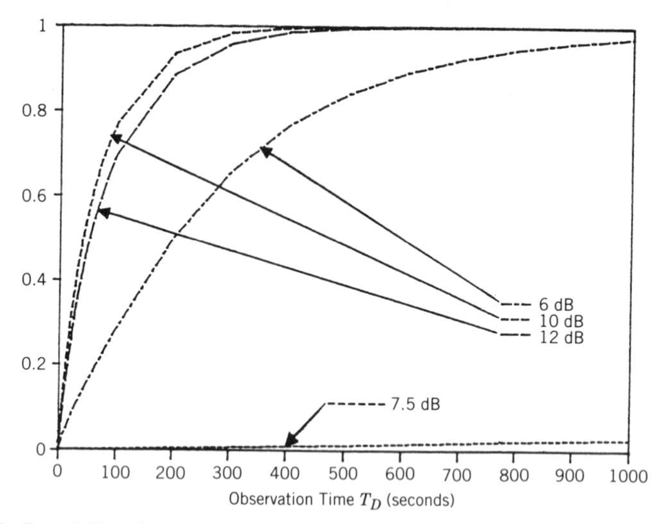

Figure 7.5. Probability of completion versus observation time: protocol 1 with RS(31, 21, 5) code.

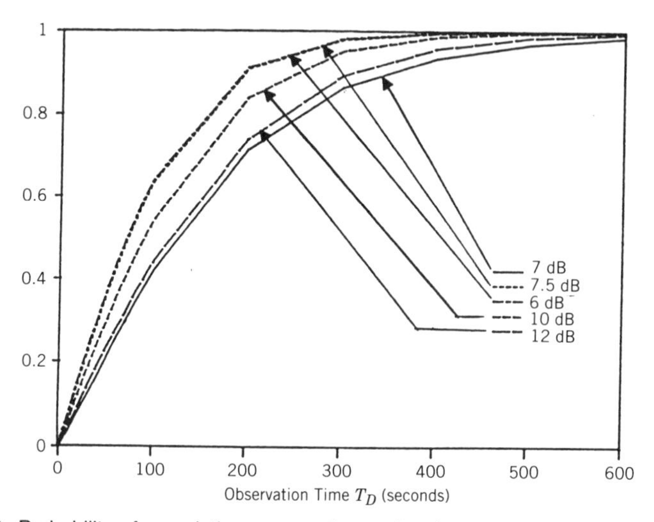

Figure 7.6. Probability of completion versus observation time: protocol 1 with RS(31, 11, 10) code.

7.3.2 Underdense Channel Model

The two-state channel model of Milstein et al. [9] makes the MB channel appear to have more errors than it really does and thus puts a larger error-correcting requirement than is needed on any coding scheme that is to be employed. By using a time-varying SNR model that more closely represents the actual MB channel, a better understanding of the merits of coding can be obtained. Of course, using a channel model that is time-varying makes the analysis much more difficult. Several researchers have undertaken this problem. The results of Miller and Milstein [7, 8], Pursley and Sandberg [10], and Jacobsmeyer [4] are briefly described in the following sections.

7.3.2.1 BCH Codes with Stop-and-Wait ARQ. Miller and Milstein [7, 8] have considered the performance of a MB communication system that encodes the message to be sent into several codewords from a binary BCH code. By using an underdense channel model, it is shown that there is indeed some benefit to using forward error correction techniques.

In this work, as with the work in Milstein et al. [9], the arrival of usable meteor trails is modeled as a Poisson process with a mean time between trails as given by Eq. 7.20. The SNR for each trail is taken to decay exponentially according to

$$\mathrm{SNR}(t) = A_0 \exp\left(\frac{2}{T_{\mathrm{L}}}(t_{\mathrm{e}} - t)\right) \tag{7.26}$$

The trail is assumed to start at $t = 0$ and end at $t = t_{\mathrm{e}}$. In Eq. 7.26 A_0 represents the SNR threshold and the channel is assumed to be off any time the SNR decays below A_0. Also, t_{e} represents the random length of the MB trail and is taken to have an exponential distribution with a mean value of T_{L}. This model for the decaying SNR follows from the classical underdense MB trail model [11].

In the work of [7] and [8], protocol B of [6], which is briefly described as follows, was assumed. When the transmitter has a message to send, it probes for a channel by sending a continuous tone. When the receiver hears the tone, it knows that a channel is available and sends a message instructing the transmitter to begin sending the data. Upon receipt of the transmit command, the transmitter sends one packet and then waits for an acknowledgment (ACK) from the receiver. When the ACK is heard, the transmitter sends the next packet and continues the process until the message is completed. If, at any time, the transmitter receives a no acknowledge (NACK) message from the receiver, it resends the previous packet until the ACK message is received for that packet. If, after sending a packet, the transmitter does not hear a response, it assumes that the channel has disappeared and that the last packet transmitted was not received correctly.

The transmitter then goes back to sending a tone until the channel reopens again. In summary, this protocol uses a stop-and-wait ARQ scheme together with channel probing to efficiently use the MB channel. The message to be sent is assumed to be divided into multiple packets, where each packet consists of a number of codewords from a binary BCH code.

By using the results of [6], the average waiting time $E[w]$ and the throughput θ are given by

$$E[w] = \frac{T_{\mathrm{I}}}{1 - Q_0}\left(1 + \sum_{k=1}^{N-1}\sum_{j=k}^{N-1} a_{jk}\frac{T_{\mathrm{I}}}{(1 - Q_0)^k}\right) \tag{7.27}$$

and

$$\theta = \frac{p}{T_{\mathrm{I}}}\sum_{i=1}^{\infty} iQ_i \tag{7.28}$$

In these equations, N is the number of packets in the message, Q_k is the probability that k packets are successfully received during a single meteor trail, p is the number of information bits per packet, and the $\{a_{j,k}\}$ are given recursively by

$$a_{j,k} = Q_j \qquad k = 1 \tag{7.29a}$$

$$a_{j,k} = \sum_{n=1}^{j-k+1} Q_n a_{j-n,k-1} \qquad k = 2, 3, \ldots, j \tag{7.29b}$$

Finding expressions for Q_k based on a time-varying SNR model is quite tedious and the derivations are rather uninteresting, so the reader is referred to [7] for details on how to evaluate these quantities.

In [7], the expression for throughput in Eq. 7.28 was evaluated for several different codes; these results are shown in Fig. 7.7. The assumed parameter values are as follows:

BCH code block length	$n = 2047$ bits
Waiting time for stop-and-wait ARQ	$t_{\mathrm{a}} = 50$ ms
Time from formation of trail to start of first packet	$t_0 = 30$ ms
Average burst length	$T_{\mathrm{L}} = 0.5$ s
Instantaneous transmission rate	$R = 10$ kb/s

The main conclusions are that an improvement of about 25% over the uncoded system can be obtained and that an optimum code rate exists in the range 0.85–0.90. Also note that the performance does not degrade as the SNR threshold is lowered. This is a result of using the more accurate channel model.

When compared with the capacity of the MB channel as given in Section

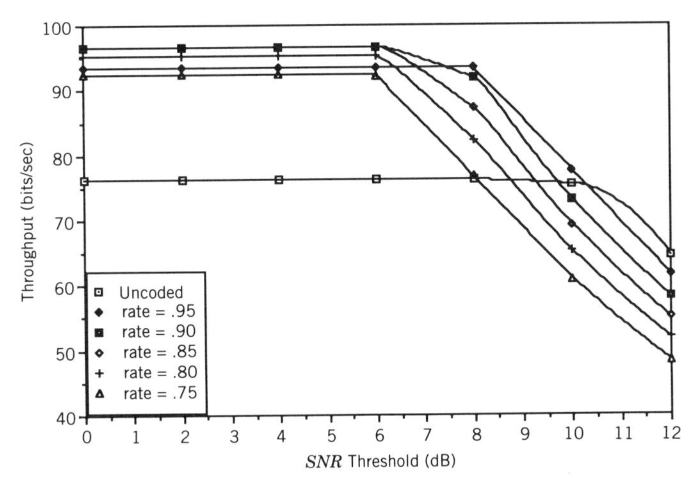

Figure 7.7. Performance of various BCH codes using protocol B of (1).

7.2, it is seen that there is still a lot of room for improvement. It is shown in [7] that by using a selective repeat ARQ scheme and shorter packet lengths, it is possible to approach the capacity of the channel for SNR thresholds above about 9 dB. By lowering the SNR threshold, the capacity increases further, while the throughput of the coded MB systems does not increase nearly as much. It is the feeling of the authors that, to approach the capacity of the MB channel at lower SNR thresholds, an adaptive-rate coding scheme is needed.

While the throughput of the channel is a convenient performance measure to compare with the channel capacity, the average time to completion of a message is often more interesting to practical users of the MB channel. The average waiting time as given by Eq. 7.27 was determined for several different systems in [8]. Figure 7.8 shows the performance when three different code rates are used. The specifics of the codes used are shown in Table 7.2. In Table 7.2 r is the code rate, n is the block length, k is the number of information bits per codeword, C is the number of codewords per packet, and L is the message length in bits. Other system parameters are the same as those used to generate Fig. 7.7, with the exception of the BCH code block length. From Fig. 7.8, it is seen that there is no significant difference in the average waiting time for any of the three systems. The slight

TABLE 7.2 Code Parameters Used in Figure 7.8

r	n	k	C	L
0.53	255	135	8	1080
0.69	255	175	6	1050
0.84	255	215	5	1075

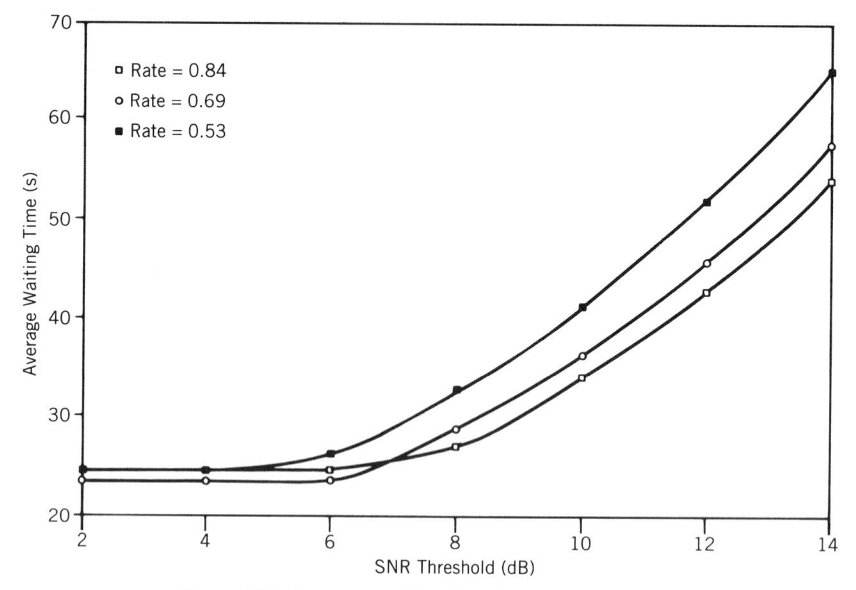

Figure 7.8. Average waiting time for 3 coded systems.

difference in performance can be accounted for by the slightly different message lengths. In Fig. 7.9, the rate 0.84 coded system is compared with an uncoded system (message length = 1075 bits) and it is seen that the coded system gives about a 21% reduction in average waiting time. It was also

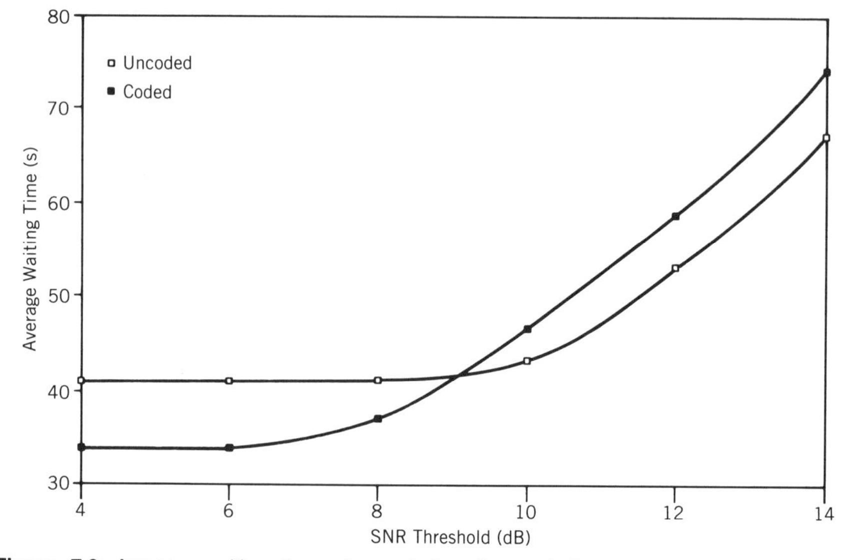

Figure 7.9. Average waiting times for coded and uncoded systems; message length = 2150 bits.

determined in [8] that varying the message length did not significantly affect the relative performance of the coded and uncoded systems, but that varying the average burst length did have a significant effect. When the average burst length was reduced to 0.1 s, the uncoded system performed just as well as the coded system.

7.3.2.2 Reed–Solomon Codes with MFSK. Pursley and Sandberg [10] have studied the problem of finding the optimum code rate to be used in conjunction with a coded meteor burst communication system. Each trail was assumed to have the underdense form whose SNR decays exponentially in time as in Eq. 7.26. Noncoherent n-ary FSK was used together with a block length n singly extended Reed–Solomon code. The code rate was chosen by attempting to optimize the probability of successfully receiving one packet of information over a single meteor trail. This is similar to the study performed by Milstein et al. [9], with the exception that the time-varying SNR model is now being considered.

A packet of information consisted of C codewords from an (n, k) Reed–Solomon code for a total of $L = Ck \log_2(n)$ information bits per packet. The received energy in the jth symbol in the packet was taken to have the form [10]

$$E_j = E_1 \exp\left[-\frac{2(j-1)T}{T_L}\right] \tag{7.30}$$

where E_1 is the energy in the first symbol, T is the time duration of a symbol and, as before, T_L is the average length of a trail (or the decay constant of the received signal amplitude). The probability of symbol error for the jth n-ary noncoherent FSK symbol is given by [10]

$$P_j = \frac{1}{n} \sum_{i=1}^{n-1} (-1)^{i+1} \binom{n}{i+1} \exp\left[-\frac{iE_j}{(i+1)N_0}\right] \tag{7.31}$$

As with the work in [8], the error rate is different for different symbols within the same codeword. The expressions for the probability of decoding a given received word correctly are rather involved and the reader is referred to Pursley and Sandberg [10] for the details.

Several different code rates were compared on the basis of equal number of information bits per packet. Thus, a rate 1/2 code would need half the number of codewords as would a rate 1/4 code. The following parameter values assumed:

Number of information bits per packet	700
Block length	32
Transmission rate (symbols/second)	3200
Decay constant, T_L (s)	0.5

TABLE 7.3 Code Parameters Used in Figure 7.10

r	n	k	C	L
0.88	32	28	5	700
0.63	32	20	7	700
0.44	32	14	10	700
0.13	32	4	35	700

The performance of the coding schemes listed in Table 7.3 is shown in Fig. 7.10. The probability of successfully receiving the entire packet is shown as a function of the initial E_b/N_0 (given by $E_1/N_0 \log_2(n)$). It is seen that of those codes considered, the rate 0.44 (32, 14) code gives the best performance. The same sort of curve for $T_L = 0.1$ s is also given in [10] and it is found that in this case the highest rate code gives the best performance. Thus, the optimum coding rate is sensitive to the actual value of the decay constant, which changes from trail to trail.

7.3.3 Empirical Results

Although the underdense model of the meteor burst channel gives more fidelity than a simple on–off model, in practice the actual form of the received signal power may be quite different from the theoretical model. One could argue that the underdense model does not always apply and thus the results derived using that model are questionable. Brayer and Natarajan [1] and Crane [2] have applied the use of error correction techniques to a set of data obtained from a meteor burst link operating between Thule Air Base and Sondrestrom Air Base in Greenland. Data were collected in 30-min intervals every 2 h for a period of several months. A subset of this data

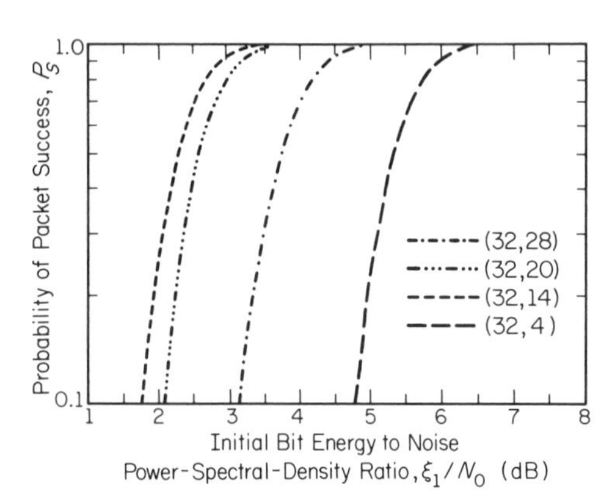

Figure 7.10. Probability of success versus E_b/N_0 for various fixed-rate codes (decay constant $\tau = 0.5$).

consisting of about 10^9 bits was used for the study. The data consisted of the actual error patterns that were encountered.

Once the data had been recorded, the performance of several schemes was determined, based on the given error patterns. The assumed protocol grouped data into k bit blocks and encoded using a $(63, k, t)$ BCH code. An entire message consisted of C codewords ($L = kC$ information bits per message). If the error pattern encountered caused the message to be received incorrectly, the same message was repeated until correct reception occurred. Figure 7.11 shows a quantity that is proportional to the probability of successfully completing a message as a function of transmission delay for several different rate codes. The assumed message length was $L = 400$ bits and the codes considered were BCH codes with parameters (63, 57, 1), (63, 51, 2), (63, 45, 3), (63, 49, 4), and (63, 36, 5). These results show that the optimum code rate was 0.81. This is consistent with the results of [7], which indicate the need for a high rate code.

This study cited a raw bit error rate on the order of 0.01 near the end of the useful portion of the trail. For the DPSK modulation used, that would imply a SNR threshold on the order of 6 dB. This is consistent with the results of Section 7.2, which indicate that at a SNR threshold of 6 dB, the optimum code rate should be 0.84.

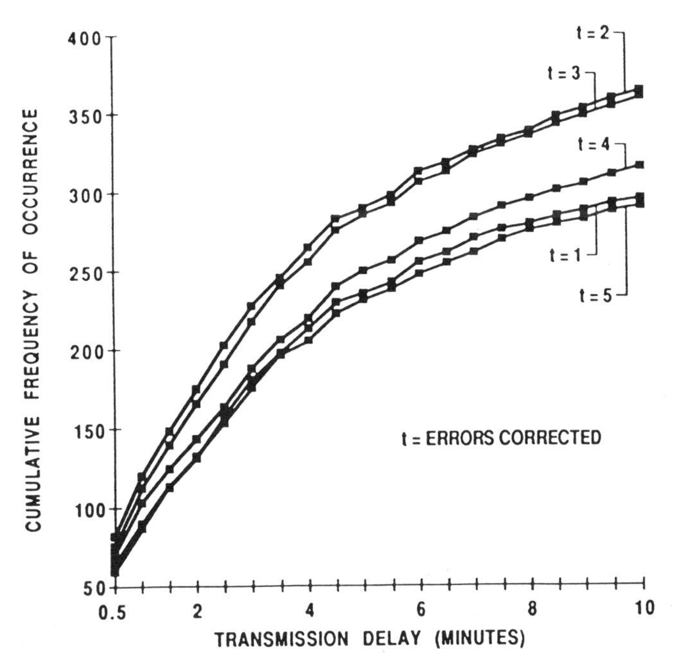

Figure 7.11. Message delivery versus code correction capability for set 2, 400-bit message length.

7.4 VARIABLE-RATE CODING SCHEMES

The results of the previous two sections indicate that the best code rate to use is a function of the SNR on the channel. Indeed, this is true of any channel. In the case of the meteor burst channel, the SNR is time-varying and so the code rate must be chosen so that the system works efficiently over a range of possible SNRs. If the code rate is very high, the system will work efficiently when the SNR is large, but will not utilize the extra channel capacity available during the frequent times when the SNR is low. On the other hand, if the code rate is low, the system will work efficiently when the SNR is low, but will send too much redundancy (and thus waste valuable channel time) when the SNR is high. One way to try to obtain the best of both worlds is to implement a system whereby the code rate changes to match the prevailing SNR. This section discusses some methods that have been proposed to achieve variable-rate coding.

7.4.1 Multiple Codes per Packet

Pursley and Sandberg [10] have analyzed a scheme whereby information is grouped into packets of L codewords. Each codeword within a packet is allowed to have a different rate. In particular, each codeword is taken from a singly extended Reed–Solomon code of block length n. The codewords at the beginning of the packet (where the SNR is higher) use a larger value of k, the number of information symbols per packet, while the codewords toward the end of the packet (where the SNR is lower) use a smaller k (and thus have a smaller code rate).

One of the key contributions of Pursley and Sandberg [10] is an algorithm that can determine the optimum packet configuration for a specific set of conditions. In particular, given that the packet must contain K information symbols, and all codewords must be of blocklength n, the algorithm will determine the optimum number of codewords L^* and the optimum number of information symbols in each codeword, $k_1^*, k_2^*, \ldots, k_L^*$, as a function of the SNR at the beginning of the burst and the decay constant T_L (the standard underdense channel model is used in this work). The packet configuration is optimum in the sense that it gives the highest probability of success P_s.

The algorithm is briefly described as follows: At the beginning of the first stage, the initial assignments $L = K/2$ (K is assumed to be even) and $k_i = 2$, $i = 1, 2, \ldots, K/2$ are made. Then, for a given step of the first stage, call it step j, two codewords, ρ and λ, are selected and k_ρ is increased by two and k_λ is decreased by two, while the other k_i remain unchanged. This new packet configuration is used in step $j+1$. The codewords λ and ρ are selected to obtain a packet configuration for step $j+1$ that has the largest value of P_s possible with a change of this type to the configuration at step j. If at some step $k_i = 0$ for some i, then L is decremented by 1. The first stage

TABLE 7.4 Packet Configurations for Optimal Variable-Rate Coding ($T_L = 0.5$ s)

SNR_i	P_s	L^*	k_i^*, $i = 1, \ldots, L^*$
1.0	0.0018	12	18, 16, 16, 14, 14, 12, 12, 10, 8, 8, 6, 6
1.5	0.1002	12	18, 18, 16, 14, 14, 12, 12, 10, 8, 8, 6, 4
2.0	0.5649	13	18, 18, 16, 14, 14, 12, 12, 10, 8, 6, 6, 4, 2
3.0	0.9934	13	18, 18, 16, 14, 14, 12, 12, 10, 8, 6, 6, 4, 2

is completed when the packet configuration cannot be improved by going to the next step. The second stage checks for configurations possibly overlooked in the first stage. At each step, the packet is altered by deleting the last codeword and distributing its information symbols among the remaining codewords in such a way that P_s is maximized. If this configuration offers improvement, then it is saved. The second stage ends after checking the configuration with the minimum possible number of codewords [10].

Table 7.4 shows the optimum packet configuration as given by Pursley and Sandberg [10] for a trail with a decay constant of $T_L = 0.5$ s. Results show that any packet configuration that is optimum at a particular SNR will be nearly optimum over a fairly large range of SNRs. On the other hand, the packet configuration is fairly sensitive to the value of the decay constant T_L. To obtain a feeling for how much improvement is achieved by using variable-rate coding instead of fixed-rate coding, the performance of the optimum variable-rate scheme is given in Fig. 7.12 along with the performance of two fixed-rate schemes. As can be seen, for this case, the difference between variable rate and fixed-rate is about 0.25 dB. In general, the improvement given by variable-rate coding is greater for trails with smaller values of T_L.

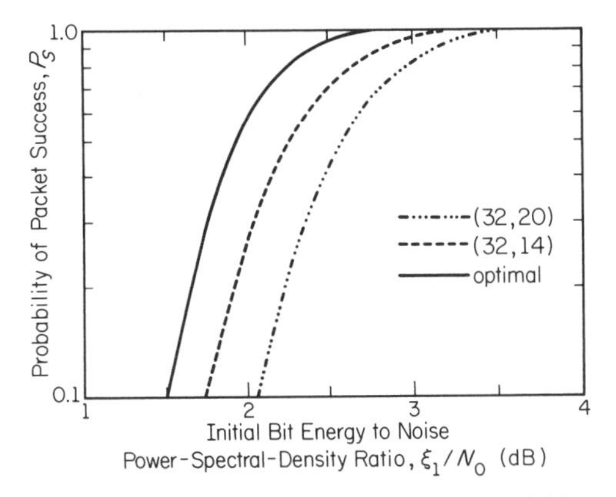

Figure 7.12. Probability of success versus E_b/N_0 for optimal variable-rate coding and for two fixed-rate codes (decay constant $\tau = 0.5$).

TABLE 7.5 Throughput Improvement of Adaptive TCM

Modulation	Improvement	Incremental Improvement
Uncoded 2-PSK	1.0	—
Coded 4-PSK	1.80	80%
Coded 4-8-PSK	2.82	57%
Coded 4-8-16-PSK	3.40	21%

7.4.2 Adaptive Trellis-Coded Modulation

Another way to vary the transmission rate to match the channel conditions is to change the size of the signal constellation that is used. Jacobsmeyer [4] has used this idea in his work which employs trellis-coded modulation with an M-ary PSK signal constellation, where the value of M, and thus the rate of the trellis code, are changed to match the channel conditions. As with most adaptive schemes, the channel requires side information which gives the current E_s/N_0 of the channel. When E_s/N_0 is above 14.4 dB, a rate 3/4 trellis code is used with 16-PSK modulation; for E_s/N_0 between 9.4 and 14.4 dB, a rate 2/3 trellis code is used with 8-PSK; and if E_s/N_0 is between 4.5 and 9.4 dB, a rate 1/2 code with QPSK is used.

The improvement in throughput over a system using uncoded PSK is given in Table 7.5. In an ideal situation (including perfect side information), this scheme promises a factor of 3.4 improvement in throughput. To the best of our knowledge, there is no work detailing how this performance improvement might diminish under realistic conditions.

7.5 CONCLUSION

In this chapter, we have presented an overview of the effect that forward error-correction coding has on the performance of a meteor burst channel. As was indicated in the Introduction, there is a basic trade-off with respect to the use of coding on a meteor burst channel because, for a fixed bandwidth, the packet duration must increase as coding is introduced, and that increased packet duration results in a greater probability that the meteor burst channel will decay below the minimum signal-to-noise ratio threshold. It is seen in this chapter, however, that the benefit of using coding more than outweighs the concern expressed above. Indeed, it is immediately seen just from the capacity arguments that coding could provide a noticeable improvement in performance. The types of coding schemes suggested in the chapter, both fixed-rate and adaptive, then illustrated more precisely just how much improvement could realistically be obtained.

REFERENCES

1. K. Brayer and S. Natarajan, An investigation of ARQ and hybrid FEC-ARQ on an experimental high latitude meteor burst channel. *IEEE Trans. Commun.* 37: (Nov 1989).
2. P. C. Crane, An empirical analysis of the application of forward error correction to meteor burst communication. *Proceedings of MILCOM 1988*, Sec. 14.3, 24 Oct 1988.
3. D. Gross and C. Harris, *Fundamentals of Queueing Theory*. New York: Wiley, 1985.
4. J. M. Jacobsmeyer, Adaptive trellis coded modulation for bandlimited meteor burst channels. *Proceedings of MILCOM 1989*, Sec. 21.3, 17 Oct 1989.
5. R. J. McEliece and W. E. Stark, Channels with block interference. *IEEE Trans. Inform. Theory* IT-30: 44–53 (1984).
6. S. L. Miller and L. B. Milstein, A comparison of protocols for a meteor-burst channel based on a time-varying channel model. *IEEE Trans. Commun.* 37: 18–30 (1989).
7. S. L. Miller and L. B. Milstein, Performance of a coded meteor burst system. *Proceedings of MILCOM 1989*, Sec. 21.1, 17 Oct 1989.
8. S. L. Miller and L. B. Milstein, Error correction coding for a meteor burst channel. *IEEE Trans. Commun.* 38: 1520–1529 (1990).
9. L. B. Milstein, D. L. Schilling, R. L. Pickholtz, J. Sellman, S. Davidovici, A. Pavelcheck, A. Schneider, and G. Eichmann, Performance of meteor-burst communication channels. *IEEE J. Select Areas Commun.* SAC-5: 146–154 (1987).
10. M. B. Pursley and S. D. Sandberg, Variable rate coding for meteor burst communications. *IEEE Trans. Commun.* 37: 1105–1112 (1989).
11. G. R. Sugar, Radio propagation by reflection from meteor trails. *Proc. IEEE* 52: 116–136 (1964).

8

PROBABILITY OF ERROR, ENERGY-TO-NOISE RATIO, AND CHANNEL CAPACITY IN *M*-ARY METEOR BURST COMMUNICATION SYSTEMS

Goran M. Djuknic
Department of Electrical Engineering and Computer Science, Stevens Institute of Technology, Hoboken, New Jersey

Donald L. Schilling
Department of Electrical Engineering, City College, The City University of New York, New York

8.1 INTRODUCTION

Meteor burst is a useful technique for low data rate digital communications at ranges from 800 to 2000 km, and if the antenna patterns are designed properly, the communication is possible even as the range decreases below 400 km [1]. What makes the communication feasible is the capability of trails of ionized particles, produced by meteors burned upon entering the Earth's atmosphere, to reflect radio waves in a specular manner. The channel is intermittent, and its duration and time of appearance are random. An additional difficulty for the analytical treatment is that the received signal power decays with time, as a consequence of ionized particles starting to diffuse into the surrounding air immediately upon the trail formation.

To conform to the nature of the channel, the communication is performed in a bursty manner, where packets of data are buffered on the transmitting side and then sent to the receiver every time the channel opens, what is determined by monitoring the probe signal from the receiver, or the transmitter itself, depending on the protocol. Consequently, a lot of effort in recent years has been invested in determining the probability of transmitting a data packet correctly [2, 10], since it is of basic importance in analyzing system throughput, waiting time to receive a packet correctly, and probability of receiving the packet within certain time period.

In this work we examine other meteor burst channel properties of interest to the system designer – symbol error probability, energy-to-noise ratio, and channel capacity – for M-ary phase- and differential phase-shift keying (MPSK and DPSK), as well as M-ary frequency-shift keying (MFSK). With respect to the previous work, appearing in [3–7], we use more accurate channel models and our analysis leads to closed-form expressions for the mentioned properties.

Meteor burst channel model is summarized in Section 8.2. Expressions for MFSK, MPSK, and DPSK symbol error probabilities are derived in Section 8.3. Energy-to-noise ratio and channel capacity are analyzed in Sections 8.4 and 8.5. Numerical results for a sample meteor burst communication system are presented in Section 8.6, and analysis is summarized in Section 8.7.

8.2 METEOR BURST CHANNEL MODEL

It is common to divide meteor trails into two classes, underdense and overdense, according to their initial electron line density, with a threshold set at 10^{14} electrons per meter. Underdense trails are much more frequent than the overdense but are of short duration, on the order of a fraction of a second, while overdense trails may last tens of seconds. Nevertheless, because of their frequency of appearance, underdense trails are considered to be the "carriers" of meteor burst communication, and in the channel model used in this work it is assumed that all the trails encountered in transmission are underdense. Furthermore, when overdense trails are used in transmission, signal-to-noise ratio remains large during their entire duration, due to much higher electron line density. This results in virtually error-free transmission and, consequently, there is little point in analyzing the error statistics. The additional assumption for underdense trails is that the scattered signal power is a function of only one random variable, the electron line density at the instant of a trail formation.

8.2.1 Received Power

The signal power received from a meteor trail is [8]

$$P_s(q, t) = A_0 q^2 e^{-t/\tau} \tag{8.1}$$

where q is the electron line density of a trail, and t is the time measured from the instant of a trail formation. The proportionality constant is given by [9]

$$A_0 = \frac{P_T G_T G_R \lambda^3 \sigma_e}{16\pi^2 R_T R_R (R_T + R_R)(1 - \sin^2 \phi \cos^2 \beta)} \exp\left(-\frac{8\pi^2 r_0^2}{\lambda^2 \sec^2 \phi}\right) \tag{8.2}$$

and the power decay constant is equal to

$$\tau = \frac{\lambda^2 \sec^2 \phi}{32\pi^2 D}$$

P_T is the transmitter power, G_T and G_R are the transmitter and receiver antenna gains, λ is the carrier wavelength, σ_e is the effective echoing area of the electron, R_T and R_R are distances of the transmitter and receiver from the point on a trail at which the reflection requirement is satisfied, ϕ is one-half the included angle between R_T and R_R, β is the angle between the trail and the great circle path between transmitter and receiver, r_0 is the nominal initial radius of a trail, and D is the diffusion coefficient of the air.

8.2.2 Electron Line Density

Initial electron line density of a trail q' is a random variable with probability density function [10]

$$f(q') = \frac{Q}{q'^2} = \frac{q_t q_u}{q_u - q_t} \frac{1}{q'^2} \qquad q_t \le q' \le q_u = 10^{14}$$

where q_t is the threshold electron line density, determined by the receiver sensitivity or the minimum received power requirement. Very small particles do not burn up upon entering the atmosphere. The minimum value for q_t is 10^8 e/m [11]. q_u is the maximum electron line density for underdense trails.

The physical duration of a burst, measured from the instant of a trail formation to the point where its electron line density reaches the threshold value q_t is a random variable,

$$t'_B = 2\tau \ln\left(\frac{q'}{q_t}\right)$$

but the duration of a burst time used for information transmission is less than t'_B for the amount t_0, the sum of the one-way propagation time between the receiver and the transmitter and the time required for resynchronization before transmitting data over a new trail. Thus, the useful duration of a burst is

$$t_B = 2\tau \ln\left(\frac{q}{q_t}\right) \tag{8.3}$$

and we use Eq. 8.3 to define a new random variable q as

$$q = q' U(q - q_0) \tag{8.4}$$

$U(\cdot)$ is the unit step function, and

$$q_0 = q_t e^{t_0/2\tau}$$

The probability density function of the new random variable is then

$$f(q) = \frac{Q}{q^2} + Q_1 \delta(q - q_0) \qquad q_0 \leq q \leq q_u \tag{8.5}$$

where

$$Q_1 = \frac{q_u(q_0 - q_t)}{q_0(q_u - q_t)}$$

8.2.3 Received Noise

In the frequency spectrum of interest to meteor burst communication, the wideband noise is either of galactic origin, picked by receiver antenna, or is due to thermal agitations in the receiver itself. Its power spectral density is of the form [10]

$$N_0 = kT_0\left[\frac{104}{L_R}\left(\frac{\lambda}{15}\right)^{2.3} + F\right] \tag{8.6}$$

where $k = 1.38 \times 10^{-23}$ J/K is the Boltzmann's constant, $T_0 = 290$ K is the room temperature, L_R is the power loss between the antenna and the receiver, and F is the receiver noise figure.

8.3 PROBABILITY OF ERROR

8.3.1 Positions of Signaling Intervals Within the Burst

If the transmitter is provided with the information about the beginning and the end of a burst, that is, when it continuously monitors the probe signal from the receiver, the entire burst duration may be used for communication. But since the received signal power varies from one signaling interval to the other, so will the resulting symbol error probability. Those variations take part within a particular burst, and symbol error probabilities will also differ among different bursts, so their instantaneous values are of little significance. What is of interest is the mean-statistical symbol error probability, and to be able to evaluate it we take that symbol intervals T_s have random starting times z uniformly distributed within burst duration. The conditional probability density function for z is thus

$$f(z|q) = \frac{1}{2\tau \ln(q/q_t)} \qquad 0 \leq z \leq 2\tau \ln(q/q_t)$$

where q is the random electron line density from Eq. 8.5.

8.3.2 *M*-ary Orthogonal Frequency-Shift Keying

M-ary orthogonal FSK signal scattered from a meteor trail is of the form

$$s_i(t) = \sqrt{\frac{2}{T_s}}\, E_s(q, z) \cos(2\pi f_i t + \phi) \qquad z \leq t \leq z + T_s \qquad i = 1, \ldots, M$$

where the separation of frequencies f_i is $1/T_s$ for noncoherent and $1/(2T_s)$ for coherent reception. From Eq. 8.1, the energy per symbol is

$$E_s(q, z) = \int_z^{z+T_s} P_s(q, t)\, dt = A_0 q^2 \tau e^{-z/\tau}(1 - e^{-T_s/\tau})$$

a function of random variables q and z. An upper bound on the symbol error probability for coherent as well as noncoherent reception of orthogonal signals is [12]

$$P_{\mathrm{E\,MFSK}} < \frac{M-1}{2} \exp\left(-\frac{E_s}{2N_0}\right)$$

Consequently, the symbol error probability in our case, conditioned on random variables q and z, is

$$P_{\mathrm{E\,MFSK}}(q, z) = P\{\epsilon|q, z\} < \frac{M-1}{2} \exp(-Aq^2 e^{-z/\tau}) \tag{8.7}$$

$$\text{where} \quad A = \frac{A_0 \tau}{2N_0}(1 - e^{-T_s/\tau}) \cong \frac{A_0 T_s}{2N_0},$$

since $\tau \gg T_s$. To find the mean-statistical symbol error $\bar{P}_{\mathrm{E\,MFSK}}$, we average Eq. 8.7 first with respect to z,

$$P_{\mathrm{E\,MFSK}}(q) < \int_z P\{\epsilon|q, z) f(z|q)\, dz$$

$$< \frac{M-1}{4\tau \ln(q/q_t)} \int_0^{2\tau \ln(q/q_t)} \exp(-Aq^2 e^{-z/\tau})\, dz$$

With a substitution, $x = e^{-z/\tau}$, we have

$$P_{\mathrm{E\,MFSK}}(q) < \frac{M-1}{4\ln(q/q_t)} \int_{(q_t/q)^2}^{1} \frac{1}{x} e^{-Aq^2x}\, dx$$
$$< \frac{M-1}{4\ln(q/q_t)} [\mathrm{Ei}(-Aq^2) - \mathrm{Ei}(-Aq_t^2)]$$

where Ei(·) is the exponential-integral function.* Averaging further over q, we obtain†

$$\bar{P}_{\mathrm{E\,MFSK}} < \int_q P\{\epsilon|q\} f(a)\, dq$$
$$< \frac{Q(M-1)}{4} \int_{q_0}^{q_u} \frac{\mathrm{Ei}(-Aq^2)}{q^2 \ln(q/q_t)}\, dq$$
$$- \frac{Q(M-1)\mathrm{Ei}(-Aq_t^2)}{4q_t} \left[\mathrm{Ei}\left(-\ln \frac{q_u}{q_t}\right) - \mathrm{Ei}\left(-\ln \frac{q_0}{q_t}\right)\right]$$
$$+ \frac{Q_1(M-1)}{8\ln(q_0/q_t)} [\mathrm{Ei}(-Aq_0^2) - \mathrm{Ei}(-Aq_t^2)] \tag{8.8}$$

Functions constituting the integrand of the remaining integral in Eq. 8.8, $1/[q^2 \ln(q/q_t)]$ and $\mathrm{Ei}(-Aq^2)$, are separately integrable over $[q_0, q_u]$ and since the product of two integrable functions is also an integrable function, the integral converges [15]. Still, it cannot be solved in a closed form, but since the exponential-integral function from the integrand, $\mathrm{Ei}(-Aq^2)$, is nonnegative and monotonic decreasing in $[q_0, q_u]$, and the remaining part of the integrand is bounded in the same interval,

$$0 \le \frac{1}{q^2 \ln(q/q_t)} \le 1$$

we can use Steffensen's inequality‡ to estimate the value of the integral [16].

*Defined as the principal value of the following integrals [13]:

$$\mathrm{Ei}(x) = -\int_{-x}^{\infty} \frac{e^{-t}}{t}\, dt = \int_{-\infty}^{x} \frac{e^{t}}{t}\, dt \qquad [x > 0]$$

†From Korn [24] we have that for $x_0 = x_1$

$$\int_{x_1}^{x_2} \phi(x)\delta(x - x_0)\, dx = \tfrac{1}{2}\phi(x_0^+)$$

‡If $f(x)$ is nonnegative and monotonic decreasing in $[a, b]$ and $g(x)$ is such that $0 \le g(x) \le 1$ in $[a, b]$, then

$$\int_{b-r}^{b} f(x)\, dx \le \int_a^b f(x)g(x)\, dx \le \int_a^{a+r} f(x)\, dx$$

where

$$r = \int_a^b g(x)\, dx$$

The constant r in the integration limits of the inequality is

$$r = \int_{q_0}^{q_u} \frac{1}{q^2 \ln(q/q_t)} = \frac{1}{q_t}\left[\mathrm{Ei}\left(-\ln \frac{q_u}{q_t}\right) - \mathrm{Ei}\left(-\ln \frac{q_0}{q_t}\right)\right]$$

For all the practical values of the parameters involved, r is at least ten orders of magnitude less than either of the constants in the integral limits, $a = q_0$ or $b = q_u$, and the value of the integral from Eq. 8.8 tends to zero. The resulting bound on the mean-statistical symbol error is then

$$\bar{P}_{\mathrm{E\ MFSK}} < \frac{Q_1(M-1)}{8\ln(q_0/q_t)}\left[\mathrm{Ei}(-Aq_0^2) - \mathrm{Ei}(-Aq_t^2)\right] + \frac{Q(M-1)\mathrm{Ei}(-Aq_t^2)}{4q_t}\left[\mathrm{Ei}\left(-\ln\frac{q_0}{q_t}\right) - \mathrm{Ei}\left(-\ln\frac{q_u}{q_t}\right)\right]$$

By substituting the appropriate constants, and having in mind that in all practical cases of interest $q_u \gg q_t$, which means that both $\mathrm{Ei}[-\ln(q_u/q_t)]$ and q_t/q_u tend to zero, the expression for the mean-statistical symbol error can be simplified to

$$\bar{P}_{\mathrm{E\ MFSK}} < \frac{(M-1)\tau}{4t_0}\left(1 - e^{-t_0/(2\tau)}\right)\left[\mathrm{Ei}\left(-\frac{A_0 q_t^2}{2N_0R_s}e^{t_0/\tau}\right) - \mathrm{Ei}\left(-\frac{A_0 q_t^2}{2N_0R_s}\right)\right] + \frac{M-1}{4}\mathrm{Ei}\left(-\frac{t_0}{2\tau}\right)\mathrm{Ei}\left(-\frac{A_0 q_t^2}{2N_0R_s}\right)$$

The inequality $q_u \gg q_t$ certainly holds, because a system designer will always try to hold the received signal threshold as low as practically possible, for it is a way to increase the available communication time. This inequality is also the expression of the often used assumption that all the trails are underdense [e.g., 17]. If we define a threshold energy-to-noise ratio as

$$\left(\frac{E_b}{N_0}\right)_{\min} = \frac{Q_0 q_t^2}{kN_0R_s}$$

where $k = \log_2 M$, we have that

$$\bar{P}_{\mathrm{E\ MFSK}} < \frac{(M-1)\tau}{4t_0}\left(1 - e^{-t_0/(2\tau)}\right) \times \left\{\mathrm{Ei}\left[-\frac{k}{2}e^{t_0/\tau}\left(\frac{E_b}{N_0}\right)_{\min}\right] - \mathrm{Ei}\left[-\frac{k}{2}\left(\frac{E_b}{N_0}\right)_{\min}\right]\right\} + \frac{M-1}{4}\mathrm{Ei}\left(-\frac{t_0}{2\tau}\right)\mathrm{Ei}\left[-\frac{k}{2}\left(\frac{E_b}{N_0}\right)_{\min}\right] \tag{8.9}$$

This closed-form expression can be extremely useful for a meteor burst communication designer, since the MFSK mean-statistical symbol error is expressed solely in terms of the threshold energy-to-noise ratio, a parameter that is simple to control and measure.

8.3.3 *M*-ary Phase-Shift Keying and Differential Phase-Shift Keying

For large energy-to-noise ratios, the symbol error for equally likely coherently detected M-ary PSK signaling is [14]

$$P_{\mathrm{E\,MPSK}} \cong 2Q\left(\sqrt{\frac{2E_s}{N_0}} \sin \frac{\pi}{M}\right)$$

where $E_s = E_b \log_2 M$ is the energy per symbol, and $M = 2^k$ is the size of the symbol set. The error function* $Q(x)$ can be upper-bounded as [18]

$$Q(x) < \frac{1}{\sqrt{2\pi}} \frac{1}{x} e^{-x^2/2} \qquad x > 0 \tag{8.10}$$

and, consequently, an upper bound for the instantaneous symbol error probability is

$$P_{\mathrm{E\,MPSK}}(q, z) < \sqrt{\frac{2}{\pi}} \frac{1}{Bq} e^{z/(2\tau)} \exp\left(-\frac{1}{2} B^2 q^2 e^{-z/\tau}\right) \tag{8.11}$$

$$\text{where} \quad B = \sqrt{\frac{2A_0}{N_0 R_s}} \sin \frac{\pi}{M}$$

To obtain the mean-statistical symbol error probability $\bar{P}_{\mathrm{E\,MPSK}}$ we average Eq. 8.11 first over z to obtain

$$\begin{aligned} P_{\mathrm{E\,MPSK}}(q) &< \sqrt{\frac{2}{\pi}} \frac{1}{2Bq\tau \ln(q/q_t)} \int_0^{2\tau \ln(q/q_t)} e^{z/(2\tau)} \exp\left(-\frac{1}{2} B^2 q^2 e^{-z/\tau}\right) dz \\ &< \sqrt{\frac{2}{\pi}} \frac{1}{B \ln(q/q_t)} \left(\frac{1}{q_t} e^{-B^2 q_t^2/2} - \sqrt{\frac{\pi}{2}} B \operatorname{erfc}\sqrt{\frac{B^2 q_t^2}{2}} - \frac{1}{q} e^{-B^2 q^2/2} \right. \\ &\quad \left. + \sqrt{\frac{\pi}{2}} B \operatorname{erfc}\sqrt{\frac{B^2 q^2}{2}}\right) \end{aligned}$$

and then over q, which results in

*In the notation used in this work, $\operatorname{erfc}(x) = 2Q(x\sqrt{2})$.

$$
\begin{aligned}
\bar{P}_{\mathrm{E\,MPSK}} < {} & K_1 \operatorname{erfc}\left(\sqrt{k\left(\frac{E_b}{N_0}\right)_{\min}} \sin\frac{\pi}{M} \right) \\
& - \frac{1}{\sqrt{\pi}} \frac{K_1}{\sqrt{k(E_b/N_0)_{\min}} \sin(\pi/M)} \exp\left[-k\left(\frac{E_b}{N_0}\right)_{\min} \sin^2\frac{\pi}{M} \right] \\
& + K_2 \operatorname{erfc}\left(\sqrt{k\left(\frac{E_b}{N_0}\right)_{\min}} e^{t_0/(2\tau)} \sin\frac{\pi}{M} \right) \\
& - \frac{1}{\sqrt{\pi}} \frac{K_3}{\sqrt{k(E_b/N_0)_{\min}} \sin(\pi/M)} \exp\left[-k\left(\frac{E_b}{N_0}\right)_{\min} e^{t_0/\tau} \sin^2\frac{\pi}{M} \right]
\end{aligned}
\tag{8.12}
$$

with the constants

$$
K_1 = \mathrm{Ei}\left(-\frac{t_0}{2\tau}\right) - \frac{\tau}{t_0}\left(1 - e^{-t_0/(2\tau)}\right)
$$

$$
K_2 = \frac{\tau}{t_0}\left(1 - e^{-t_0/(2\tau)}\right)
$$

$$
K_3 = K_2 e^{-t_0/(2\tau)}
$$

As before, in deriving the expression 8.12 we have assumed that $q_u \gg q_t$, that is, $q_t/q_u \to 0$ and Steffensen's inequality was used to prove that integrals

$$
\int_{q_0}^{q_u} \frac{1}{q^3 \ln(q/q_t)} e^{-B^2q^2/2}\, dq \quad \text{and} \quad \int_{q_0}^{q_u} \frac{1}{q^2 \ln(q/q_t)} \operatorname{erfc}\left(\frac{Bq}{\sqrt{2}}\right) dq
$$

tend to zero. And, as in the MFSK case, the symbol error for MPSK signaling is expressed solely in terms of the threshold energy-to-noise ratio and constant parameters, namely τ, the decay constant, and t_0, the sum of the propagation and synchronization time. Bound 8.12 is also valid for M-ary DPSK, when the term $\sin(\pi/M)$ is replaced with $\sin(\pi/\sqrt{2}M)$, and the energy-to-noise ratio is large.

8.4 ENERGY-TO-NOISE RATIO

The received energy-to-noise ratio is constantly changing value with time inside a particular burst, as well as over the ensemble of all possible bursts, so its instantaneous value has little meaning in the design process. Thus, in addition to error probability, the other parameter of interest is the long-term average energy-to-noise ratio. If the signal and noise bandwidths are equal, the instantaneous energy-to-noise ratio is

$$
\frac{E_b(q, z)}{N_0} = \frac{A_0}{N_0 R_b} q^2 e^{-z/\tau}
\tag{8.13}
$$

where $E_b = E_s/k$ is the energy per bit, and R_b is the bit rate. Averaging Eq. 8.13 over z, we obtain

$$\frac{E_b(q)}{N_0} = \frac{1}{N_0}\int_z E_b(q, z) f(z|q)\, dz$$
$$= \frac{A_0}{2N_0R_b}\,\frac{q^2 - q_t^2}{\ln(q/q_t)}$$

Taking further the expectation with respect to q, the average energy-to-noise ratio becomes

$$\frac{\bar{E}_b}{N_0} = \frac{1}{N_0}\int_q E_b(q) f(q)\, \mathrm{d}q$$
$$= \frac{A_0 Q q_t}{2N_0R_b}\left[\mathrm{Ei}\left(\ln\frac{q_u}{q_t}\right) - \mathrm{Ei}\left(-\ln\frac{q_u}{q_t}\right) - \mathrm{Ei}\left(\ln\frac{q_0}{q_t}\right) + \mathrm{Ei}\left(-\ln\frac{q_0}{q_t}\right)\right]$$
$$+ \frac{A_0 Q_1 (q_0^2 - q_t^2)}{4N_0R_b \ln(q_0/q_t)} \tag{8.14}$$

As before, both $\mathrm{Ei}[-\ln(q_u/q_t)]$ and q_t/q_u tend to zero, and by rearranging Eq. 8.14 we obtain the average energy-to-noise ratio as

$$\frac{\bar{E}_b}{N_0} = \frac{A_0 q_t^2}{2N_0R_b}\left[\mathrm{Ei}\left(\ln\frac{q_u}{q_t}\right) - \mathrm{Ei}\left(\frac{t_0}{2\tau}\right) + \mathrm{Ei}\left(-\frac{t_0}{2\tau}\right)\right.$$
$$\left. + \frac{\tau}{t_0}\left(e^{t_0/(2\tau)} - 1\right)^2\left(e^{t_0/(2\tau)} + 1\right)\right]$$

Trailing terms in the brackets can always be neglected as compared to the first term and we have that

$$\frac{\bar{E}_b}{N_0} = \frac{1}{2}\left(\frac{E_b}{N_0}\right)_{\min} \mathrm{Ei}\left(\ln\frac{q_u}{q_t}\right) \tag{8.15}$$

The threshold energy-to-noise ratio $(E_b/N_0)_{\min}$ and the bit rate R_b are not independent variables, and to relate them we require that the maximum bit error rate

$$P_{b\,\max} = \frac{1}{2}\exp\left[-\frac{1}{2}\left(\frac{E_b}{N_0}\right)_{\min}\right]$$

does not exceed certain value. (The approach similar to the one in Hibshoosh [10]). The threshold electron line density is then

$$q_t = \sqrt{\frac{2N_0R_b}{A_0}\ln\frac{1}{2P_{b\,\max}}}$$

which can be substituted in Eq. 8.15 to obtain the final expression for the average energy-to-noise ratio

$$\frac{\bar{E}_b}{N_0} = \ln \frac{1}{2P_{b\,max}} \, \mathrm{Ei}\left[-\frac{1}{2} \ln\left(\frac{2N_0 R_b}{A_0 q_u^2} \ln \frac{1}{2P_{b\,max}}\right)\right] \tag{8.16}$$

In describing fading channels it is customary to express the mean-statistical error probability in terms of the average energy-to-noise ratio, but we do not follow that course here, although it is possible to combine expressions 8.16 and 8.12, or 8.16 and 8.9 in that sense. The meteor burst channel is a fading channel, but the received power is monotonically decaying from the moment of a trail formation, in contrast to channels with Rayleigh distribution of the received signal amplitude. Thus, it is more meaningful to express the error probabilities in terms of the threshold energy-to-signal ratio, the value determined by the communication system designer, who can vary the system parameters to obtain the desired threshold. The choice is limited, of course, by the channel's physical parameters.

8.5 METEOR BURST CHANNEL CAPACITY

8.5.1 Binary Frequency-Shift Keying

For binary FSK, the bit error probability is exactly

$$P_{b\,BFSK} = \frac{1}{2} \exp\left(-\frac{E_b}{2N_0}\right)$$

and the corresponding channel capacity is

$$C_{BFSK} = 1 + P_{b\,BFSK} \log_2 P_{b\,BFSK} + (1 - P_{b\,BFSK}) \log_2(1 - P_{b\,BFSK}) \; \frac{\text{bits}}{\text{symbol}}$$

C_{BFSK} is a concave function and, according to the theory of convex functions,* the lower bound on the mean-statistical capacity value is

$$\bar{C}_{BFSK} \geq 1 + \bar{P}_{b\,BFSK} \log_2 \bar{P}_{b\,BFSK} + (1 - \bar{P}_{b\,BFSK}) \log_2(1 - \bar{P}_{b\,BFSK}) \; \frac{\text{bits}}{\text{symbol}} \tag{8.17}$$

where the mean-statistical bit error probability,

*If a function $f(x)$ is real-valued and continuous, and if $f''(x) \geq 0$, $x > 0$, then $f(x)$ is said to be *concave* over $[0, \infty]$. If x is a random variable with finite expectation, for concave functions $E[f(x)] \geq f(E[x])$ [19].

$$\bar{P}_{\text{b BFSK}} = \frac{\tau}{4t_0}\,(1 - e^{-t_0/(2\tau)})\left\{\text{Ei}\left[-\frac{1}{2}\,e^{t_0/\tau}\left(\frac{E_\text{b}}{N_0}\right)_{\min}\right] - \text{Ei}\left[-\frac{1}{2}\left(\frac{E_\text{b}}{N_0}\right)_{\min}\right]\right\}$$
$$+ \frac{1}{4}\,\text{Ei}\left(-\frac{t_0}{2\tau}\right)\text{Ei}\left[-\frac{1}{2}\left(\frac{E_\text{b}}{N_0}\right)_{\min}\right]$$

is obtained by setting $k = 1$ in Eq. 8.9. This and the following bound are also valid for automatic-repeat-request transmission, since the channel capacity is not increased by introduction of feedback [20].

8.5.2 Binary Phase-Shift Keying

Bit error probability in this case is

$$P_{\text{b BPSK}} = Q\left(\sqrt{\frac{2E_\text{b}}{N_0}}\right)$$

and for large values of the energy-to-noise ratio the bound from Eq. 8.10 may serve as an approximation to $Q(x)$, that is, when $x > 3$ [21]. The lower bound for the channel capacity is then found as

$$\bar{C}_{\text{BPSK}} \geq 1 + \bar{P}_{\text{b BPSK}} \log_2 \bar{P}_{\text{b BPSK}} + (1 - \bar{P}_{\text{b BPSK}}) \log_2(1 - \bar{P}_{\text{b BPSK}})\ \frac{\text{bits}}{\text{symbol}} \tag{8.18}$$

where the bit error probability is derived from Eq. 8.12 by setting k to 1 and dividing the whole expression with 2, namely

$$\bar{P}_{\text{b BPSK}} \cong \frac{K_1}{2}\,\text{erfc}\sqrt{\left(\frac{E_\text{b}}{N_0}\right)_{\min}} - \frac{1}{2\sqrt{\pi}}\,\frac{K_1}{\sqrt{(E_\text{b}/N_0)_{\min}}}\,\exp\left[-\left(\frac{E_\text{b}}{N_0}\right)_{\min}\right]$$
$$+ \frac{K_2}{2}\,\text{erfc}\left[e^{t_0/\tau}\sqrt{\left(\frac{E_\text{b}}{N_0}\right)_{\min}}\right] - \frac{1}{2\sqrt{\pi}}\,\frac{K_3}{\sqrt{(E_\text{b}/N_0)_{\min}}}$$
$$\times \exp\left[-\left(\frac{E_\text{b}}{N_0}\right)_{\min} e^{t_0/\tau}\right]$$

8.6 NUMERICAL RESULTS

Numerical results presented here are for the sample meteor burst communication system used in Davidovici and Kanterakis [22] and Spezio [23]. Parameter values are $P_\text{T} = 2000$ W, $G_\text{T} = G_\text{R} = 13$ dBi, carrier frequency $f = 35$ MHz, $\sigma_\text{e} = 10^{-28}$ m^2, $R_\text{T} = R_\text{R} = 514$ km, $\phi = 76.56°$, $\overline{\cos \beta} = 2/\pi$, $r_0 = 0.65$ m, $L_\text{R} = 1.3$, $F = 2.5$, $\tau = 0.4$ s, and $t_0 = 0.1$ s.

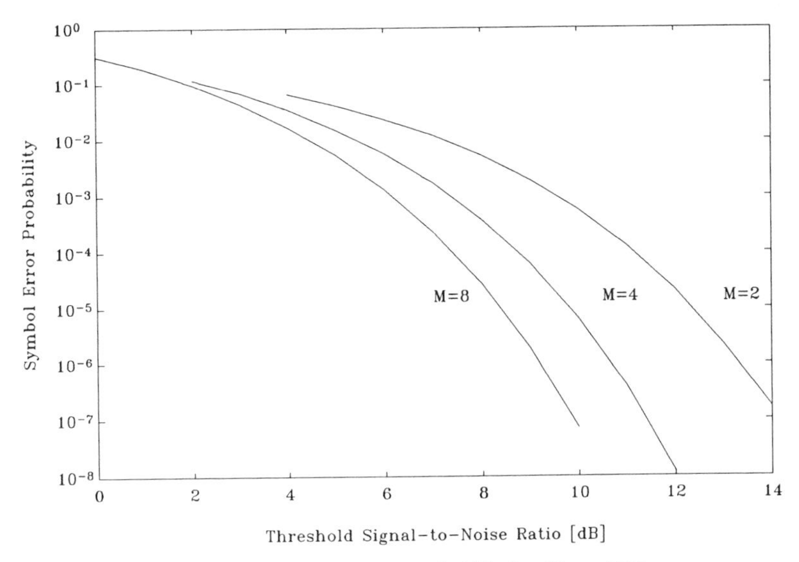

Figure 8.1. Symbol error probability for M-ary FSK.

MFSK symbol error probability as a function of the threshold energy-to-noise ratio, for 2-, 4-, and 8-level signaling, is shown in Fig. 8.1. Symbol error probability for M-ary PSK and DPSK versus the threshold energy-to-noise ratio are shown in Figs. 8.2 and 8.3 for 2-, 4-, 8-, and 16-level signaling (values for $M = 4$ were almost indistinguishable from values for $M = 2$ in case of MPSK).

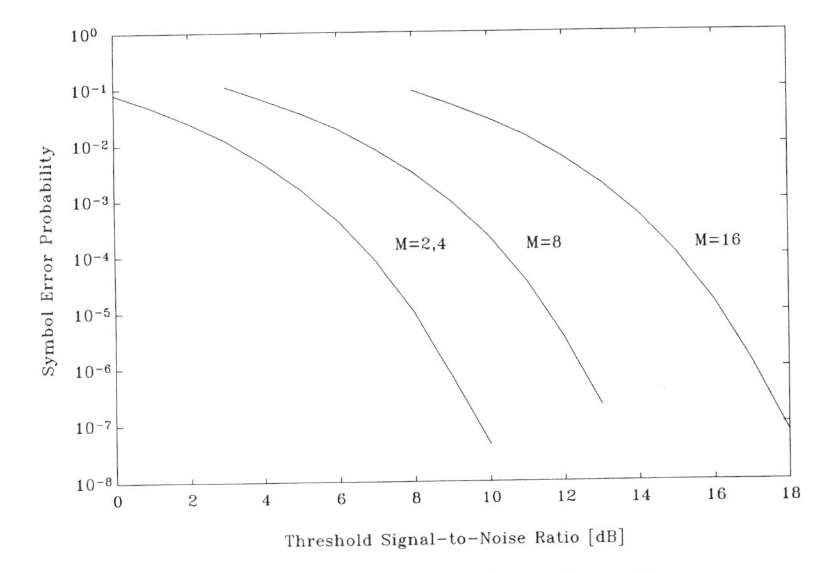

Figure 8.2. Symbol error probability for M-ary PSK.

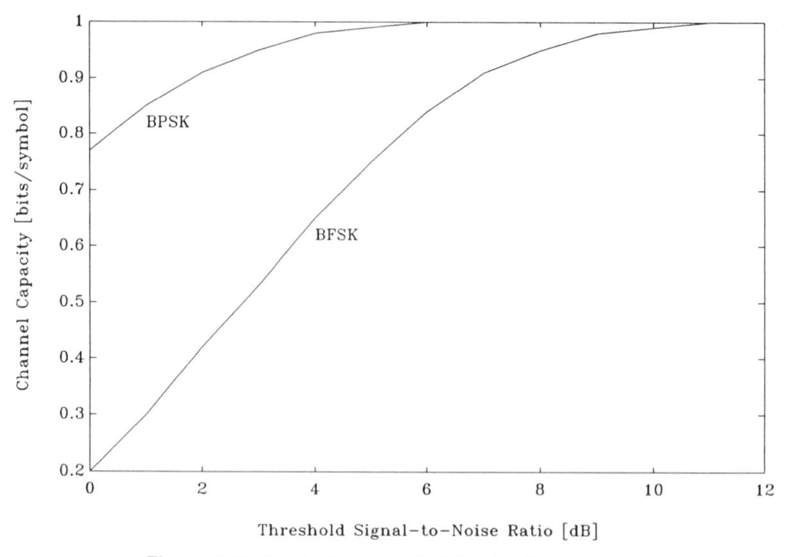

Figure 8.3. Symbol error probability for *M*-ary DPSK.

Capacity values, when meteor burst channel is considered to be a binary symmetric FSK or PSK channel, are shown in Fig. 8.4 as functions of the threshold energy-to-noise levels. From these curves it can be concluded that to expect the operation at channel capacity, one has to maintain the threshold of at least 6 dB for PSK signaling, and at least 10 dB for FSK signaling.

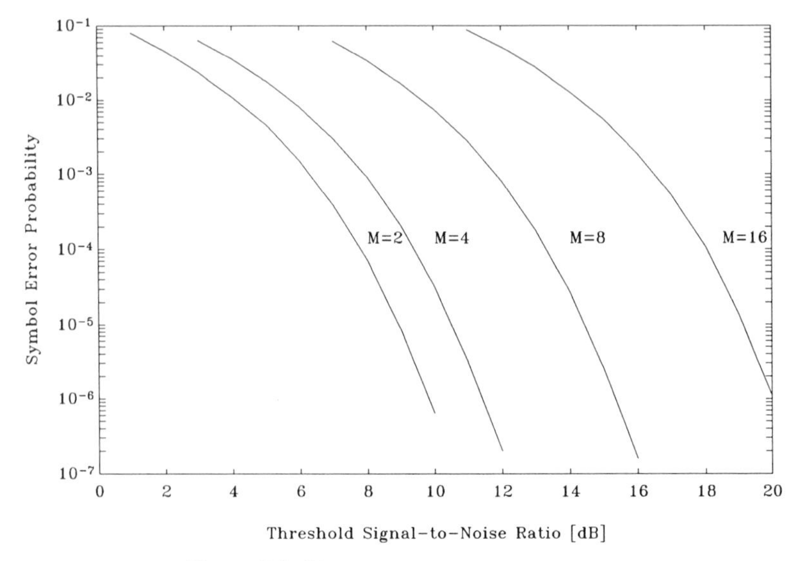

Figure 8.4. Binary-symmetric channel capacity.

8.7 SUMMARY AND CONCLUSIONS

In the meteor burst channel model considered in this work, signal-to-noise ratio within the trail (i.e., when the communications channel is open) exponentially decays with time elapsed from the trail formation. It is assumed that the initial value of the received signal power is a function of one random variable, the electron line density of a trail, and that all the trails suitable for communication are of underdense type.

We have derived closed form expressions for the mean-statistical MFSK, MPSK, and DPSK symbol error probabilities, as functions of threshold signal-to-noise ratios. The resulting graphs have the familiar waterfall-like shape, and relative positions of curves for different numbers of signaling levels are similar to those in time non-varying channels. When characterizing communications channels with fading, it is common to represent error probabilities in terms of average signal-to-noise ratios. We chose here a different approach since the received signal power is monotonically decaying function of time, and threshold signal-to-noise ratio is a more meaningful parameter. Threshold SNR is either determined by the receiver equipment sensitivity, or is enforced by the system designer. Once set, it determines the beginning and end points of communication intervals, that is, the duty cycle and burst occurrence rate, two most important parameters to characterize a meteor burst communication system. The expressions for BFSK and BPSK capacity, derived for the case when meteor burst channel is considered to be binary symmetric, help determine the minimum required threshold SNR value for signaling at capacity.

REFERENCES

1. R. D. Sinnot et al., Meteor burst communications with a buried antenna, *Proc. MILCOM '85*, paper 32.J. (Oct. 1985).
2. L. B. Milstein et al., Performance of meteor-burst communication channels, *IEEE J. Sel. Areas Commun.*, SAC-5, No. 2, 146–154 (Feb. 1987).
3. V. Ye. Dyrda, Error probability and transmission rate in a meteor communication channel using minimal multiposition codes, *Telecomm. and Radio Eng.*, No. 8, 29–33 (Aug. 1967).
4. N. P. Khvorostenko, Noise immunity of a method of receiving fading signals with differential phase keying, *Telecomm. and Radio Eng.*, No. 9, 68–72, (Sep. 1962).
5. N. P. Khvorostenko, Noise immunity of multiplex phase telegraphy, *Telecomm. and Radio Eng.*, No. 12, 76–82 (Dec. 1964).
6. Yu. F. Korobov, Noise immunity of an RPSK telegraph meteor channel, *Telecomm.*, 22, No. 11, 30–36 (1968).
7. A. L. Peisikhman, Noise immunity and relative efficiency of multiposition frequency-shift telegraphy in interrupted channels, *Telecomm. and Radio Eng.*, No. 12, 17–23 (Dec. 1966).

8. G. R. Sugar, Radio propagation by reflection from meteor trails, *Proc. IEEE*, 116–136 (Feb. 1964).
9. J. Weitzen, An estimate of the capacity of the meteor burst channel, *IEEE Trans. Comm.* COM-32, No. 8, 972–974 (Aug. 1984).
10. E. Hibshoosh, A study of a meteor burst communication system, *Ph. D. Dissertation*, City College of the City University of New York, New York, June 1987.
11. L. A. Manning and U. R. Eshleman, Meteors in the ionosphere, *Proc. IRE*, 47: 186–199 (Feb. 1959).
12. A. J. Viterbi, *Principles of Coherent Communications*, McGraw-Hill Book Company, New York, 1966.
13. Handbook of mathematical functions with formulas, graphs, and mathematical tables, M. Abramowitz and I. A. Stegun, Dover Publications, Inc., New York, 1970.
14. I. Korn, *Digital Communications*, Van Nostrand Reinhold Company, Inc., New York, 1985.
15. E. T. Whittaker and G. N. Watson, A *Course of Modern Analysis*, 4th ed., Cambridge University Press, London, 1952.
16. I. S. Gradshteyn and I. M. Ryzhik, *Table of Integrals, Series, and Products*, Academic Press, Inc., Orlando, Florida, 1980.
17. S. L. Miller and L. B. Milstein, A comparison of protocols for a meteor-burst channel based on a time-varying channel model, *IEEE Trans. Comm.*, 37, No. 1, 18–30 (Jan. 1989).
18. P. O. Börjesson and C.-E. W. Sundberg, Simple approximations of the error function $Q(x)$ for communications applications, *IEEE Trans. Comm.* COM-27, No. 3, 639–643 (Mar. 1979).
19. J. M. Wozencraft and I. M. Jacobs, *Principles of Communication Engineering*, John Wiley, New York, 1965.
20. C. E. Shannon, Two-way communication channels, Proc. 4*th Berkeley Symp. Prob. and Stat.*, 1, 611–644, University of California Press, Berkeley, 1961.
21. B. Sklar, *Digital Communications – Fundamentals and Applications*, Prentice Hall, Englewood Cliffs, 1988.
22. S. Davidovici and E. G. Kanterakis, Performance of a meteor-burst communication system using packet messages with variable data rates, *IEEE Trans. Comm.* 37: 6–17 (Jan. 1989).
23. A. E. Spezio, Meteor burst communication system analysis and design, *NRL Rep. 8286* (Dec. 1978).
24. G. A. Korn and T. M. Korn, *Mathematical Handbook for Scientists and Engineers*, McGraw-Hill Book Company, New York, 1961.

INDEX